TENSOR-VALUED RANDOM FIELDS FOR CONTINUUM PHYSICS

Many areas of continuum physics pose a challenge to physicists; for instance, what are the most general, admissible statistically homogeneous and isotropic tensor-valued random fields (TRFs)? Previously, only TRFs of rank 0 were completely described. This book assembles a complete description of all such fields in terms of one- and two-point correlation functions for tensors of rank 1 through 4. Working from the standpoint of invariance of physical laws with respect to the choice of coordinate system, both spatial domain representations and their wavenumber domain counterparts are rigorously given in full detail. The book also provides an introduction to a range of continuum theories requiring TRFs, an introduction to the mathematical theories necessary for the description of homogeneous and isotropic TRFs and a range of applications, including a strategy for the simulation of TRFs, ergodic TRFs, scaling laws of stochastic constitutive responses and applications to stochastic partial differential equations. It is invaluable for mathematicians looking to solve stochastic problems of continuum physics and for physicists aiming to enrich their knowledge of the relevant mathematical tools.

ANATOLIY MALYARENKO is Professor of Mathematics at Mälardalen University in Sweden. Previously he was a researcher at the Statistical Research Centre at Taras Shevchenko National University, and at the International Mathematical Centre of the National Academy of Science of Ukraine. His research interests include random functions of several variables, actuarial and financial mathematics and computer simulation of real-world systems.

MARTIN OSTOJA-STARZEWSKI is a professor in the Department of Mechanical Science and Engineering at the University of Illinois and also at the Institute for Condensed Matter Theory and Beckman Institute. He serves as Editor of *Acta Mechanica*, Editor-in-Chief of the *Journal of Thermal Stresses* and Chair Managing Editor of *Mathematics and Mechanics of Complex Systems*. His research interests include random and fractal media, continuum theories and aerospace, bio- and geo-physical applications.

CAMBRIDGE MONOGRAPHS ON MATHEMATICAL PHYSICS

S. J. Aarseth *Gravitational N-Body Simulations: Tools and Algorithms*†
J. Ambjørn, B. Durhuus and T. Jonsson *Quantum Geometry: A Statistical Field Theory Approach*†
A. M. Anile *Relativistic Fluids and Magneto-fluids: With Applications in Astrophysics and Plasma Physics*
J. A. de Azcárraga and J. M. Izquierdo *Lie Groups, Lie Algebras, Cohomology and Some Applications in Physics*†
O. Babelon, D. Bernard and M. Talon *Introduction to Classical Integrable Systems*†
F. Bastianelli and P. van Nieuwenhuizen *Path Integrals and Anomalies in Curved Space*†
D. Baumann and L. McAllister *Inflation and String Theory*
V. Belinski and M. Henneaux *The Cosmological Singularity*†
V. Belinski and E. Verdaguer *Gravitational Solitons*†
J. Bernstein *Kinetic Theory in the Expanding Universe*†
G. F. Bertsch and R. A. Broglia *Oscillations in Finite Quantum Systems*†
N. D. Birrell and P. C. W. Davies *Quantum Fields in Curved Space*†
K. Bolejko, A. Krasiński, C. Hellaby and M-N. Célérier *Structures in the Universe by Exact Methods: Formation, Evolution, Interactions*
D. M. Brink *Semi-Classical Methods for Nucleus-Nucleus Scattering*†
M. Burgess *Classical Covariant Fields*†
E. A. Calzetta and B.-L. B. Hu *Nonequilibrium Quantum Field Theory*
S. Carlip *Quantum Gravity in 2+1 Dimensions*†
P. Cartier and C. DeWitt-Morette *Functional Integration: Action and Symmetries*†
J. C. Collins *Renormalization: An Introduction to Renormalization, the Renormalization Group and the Operator-Product Expansion*†
P. D. B. Collins *An Introduction to Regge Theory and High Energy Physics*†
M. Creutz *Quarks, Gluons and Lattices*†
P. D. D'Eath *Supersymmetric Quantum Cosmology*†
J. Dereziński and C. Gérard *Mathematics of Quantization and Quantum Fields*
F. de Felice and D. Bini *Classical Measurements in Curved Space-Times*
F. de Felice and C. J. S Clarke *Relativity on Curved Manifolds*†
B. DeWitt *Supermanifolds, 2nd edition*†
P. G. O. Freund *Introduction to Supersymmetry*†
F. G. Friedlander *The Wave Equation on a Curved Space-Time*†
J. L. Friedman and N. Stergioulas *Rotating Relativistic Stars*
Y. Frishman and J. Sonnenschein *Non-Perturbative Field Theory: From Two Dimensional Conformal Field Theory to QCD in Four Dimensions*
J. A. Fuchs *Affine Lie Algebras and Quantum Groups: An Introduction, with Applications in Conformal Field Theory*†
J. Fuchs and C. Schweigert *Symmetries, Lie Algebras and Representations: A Graduate Course for Physicists*†
Y. Fujii and K. Maeda *The Scalar-Tensor Theory of Gravitation*†
J. A. H. Futterman, F. A. Handler, R. A. Matzner *Scattering from Black Holes*†
A. S. Galperin, E. A. Ivanov, V. I. Ogievetsky and E. S. Sokatchev *Harmonic Superspace*†
R. Gambini and J. Pullin *Loops, Knots, Gauge Theories and Quantum Gravity*†
T. Gannon *Moonshine beyond the Monster: The Bridge Connecting Algebra, Modular Forms and Physics*†
A. García-Díaz *Exact Solutions in Three-Dimensional Gravity*
M. Göckeler and T. Schücker *Differential Geometry, Gauge Theories, and Gravity*†
C. Gómez, M. Ruiz-Altaba and G. Sierra *Quantum Groups in Two-Dimensional Physics*†
M. B. Green, J. H. Schwarz and E. Witten *Superstring Theory Volume 1: Introduction*
M. B. Green, J. H. Schwarz and E. Witten *Superstring Theory Volume 2: Loop Amplitudes, Anomalies and Phenomenology*
V. N. Gribov *The Theory of Complex Angular Momenta: Gribov Lectures on Theoretical Physics*†
J. B. Griffiths and J. Podolský *Exact Space-Times in Einstein's General Relativity*†
T. Harko and F. Lobo *Extensions of f(R) Gravity: Curvature-Matter Couplings and Hybrid Metric-Palatini Gravity*
S. W. Hawking and G. F. R. Ellis *The Large Scale Structure of Space-Time*†
F. Iachello and A. Arima *The Interacting Boson Model*†
F. Iachello and P. van Isacker *The Interacting Boson-Fermion Model*†
C. Itzykson and J. M. Drouffe *Statistical Field Theory Volume 1: From Brownian Motion to Renormalization and Lattice Gauge Theory*†
C. Itzykson and J. M. Drouffe *Statistical Field Theory Volume 2: Strong Coupling, Monte Carlo Methods, Conformal Field Theory and Random Systems*†

G. Jaroszkiewicz *Principles of Discrete Time Mechanics*
G. Jaroszkiewicz *Quantized Detector Networks*
C. V. Johnson *D-Branes*†
P. S. Joshi *Gravitational Collapse and Spacetime Singularities*†
J. I. Kapusta and C. Gale *Finite-Temperature Field Theory: Principles and Applications, 2nd edition*†
V. E. Korepin, N. M. Bogoliubov and A. G. Izergin *Quantum Inverse Scattering Method and Correlation Functions*†
J. Kroon *Conformal Methods in General Relativity*
M. Le Bellac *Thermal Field Theory*†
Y. Makeenko *Methods of Contemporary Gauge Theory*†
S. Mallik and S. Sarkar *Hadrons at Finite Temperature*
A. Malyarenko and M. Ostoja-Starzewski *Tensor-Valued Random Fields for Continuum Physics*
N. Manton and P. Sutcliffe *Topological Solitons*†
N. H. March *Liquid Metals: Concepts and Theory*†
I. Montvay and G. Münster *Quantum Fields on a Lattice*†
P. Nath *Supersymmetry, Supergravity, and Unification*
L. O'Raifeartaigh *Group Structure of Gauge Theories*†
T. Ortín *Gravity and Strings, 2nd edition*
A. M. Ozorio de Almeida *Hamiltonian Systems: Chaos and Quantization*†
M. Paranjape *The Theory and Applications of Instanton Calculations*
L. Parker and D. Toms *Quantum Field Theory in Curved Spacetime: Quantized Fields and Gravity*
R. Penrose and W. Rindler *Spinors and Space-Time Volume 1: Two-Spinor Calculus and Relativistic Fields*†
R. Penrose and W. Rindler *Spinors and Space-Time Volume 2: Spinor and Twistor Methods in Space-Time Geometry*†
S. Pokorski *Gauge Field Theories, 2nd edition*†
J. Polchinski *String Theory Volume 1: An Introduction to the Bosonic String*†
J. Polchinski *String Theory Volume 2: Superstring Theory and Beyond*†
J. C. Polkinghorne *Models of High Energy Processes*†
V. N. Popov *Functional Integrals and Collective Excitations*†
L. V. Prokhorov and S. V. Shabanov *Hamiltonian Mechanics of Gauge Systems*
S. Raychaudhuri and K. Sridhar *Particle Physics of Brane Worlds and Extra Dimensions*
A. Recknagel and V. Schiomerus *Boundary Conformal Field Theory and the Worldsheet Approach to D-Branes*
M. Reuter and F. Saueressig *Quantum Gravity and the Functional Renormalization Group*
R. J. Rivers *Path Integral Methods in Quantum Field Theory*†
R. G. Roberts *The Structure of the Proton: Deep Inelastic Scattering*†
C. Rovelli *Quantum Gravity*†
W. C. Saslaw *Gravitational Physics of Stellar and Galactic Systems*†
R. N. Sen *Causality, Measurement Theory and the Differentiable Structure of Space-Time*
M. Shifman and A. Yung *Supersymmetric Solitons*
Y. M. Shnir *Topological and Non-Topological Solitons in Scalar Field Theories*
H. Stephani, D. Kramer, M. MacCallum, C. Hoenselaers and E. Herlt *Exact Solutions of Einstein's Field Equations, 2nd edition*†
J. Stewart *Advanced General Relativity*†
J. C. Taylor *Gauge Theories of Weak Interactions*†
T. Thiemann *Modern Canonical Quantum General Relativity*†
D. J. Toms *The Schwinger Action Principle and Effective Action*†
A. Vilenkin and E. P. S. Shellard *Cosmic Strings and Other Topological Defects*†
R. S. Ward and R. O. Wells, Jr *Twistor Geometry and Field Theory*†
E. J. Weinberg *Classical Solutions in Quantum Field Theory: Solitons and Instantons in High Energy Physics*
J. R. Wilson and G. J. Mathews *Relativistic Numerical Hydrodynamics*†

† Available in paperback

Tensor-Valued Random Fields for Continuum Physics

ANATOLIY MALYARENKO
Mälardalen University Sweden

MARTIN OSTOJA-STARZEWSKI
University of Illinois

CAMBRIDGE
UNIVERSITY PRESS

CAMBRIDGE
UNIVERSITY PRESS

University Printing House, Cambridge CB2 8BS, United Kingdom

One Liberty Plaza, 20th Floor, New York, NY 10006, USA

477 Williamstown Road, Port Melbourne, VIC 3207, Australia

314–321, 3rd Floor, Plot 3, Splendor Forum, Jasola District Centre, New Delhi – 110025, India

79 Anson Road, #06–04/06, Singapore 079906

Cambridge University Press is part of the University of Cambridge.

It furthers the University's mission by disseminating knowledge in the pursuit of education, learning, and research at the highest international levels of excellence.

www.cambridge.org
Information on this title: www.cambridge.org/9781108429856
DOI: 10.1017/9781108555401

First published 2019

Printed in the United Kingdom by TJ International Ltd. Padstow Cornwall

A catalogue record for this publication is available from the British Library.

ISBN 978-1-108-42985-6 Hardback

To all our collaborators and students

Contents

Introduction

Why Tensor-Valued Random Fields in Continuum Physics?

In this book, we use the term *continuum physics* to refer to continuum mechanics and other classical (non-quantum, non-relativistic) field-theoretic models such as continuum thermomechanics (e.g. thermal conductivity, thermoelasticity, thermodiffusion), electromagnetism and electromagnetic interactions in deformable media (e.g. piezoelectricity). Most tensor-valued (or, in what follows, just 'tensor') fields appearing in these models fall into one of two categories: fields of dependent quantities (displacement, velocity, deformation, rotation, stress...) or fields of constitutive responses (conductivity, stiffness, permeability...). All of these fields take values in linear spaces of tensors of first or higher rank over the space $\mathbb{R}^d$, $d = 2, 3$ and, generally, of random nature (i.e. displaying spatially inhomogeneous, random character), indicating that the well-developed theory of scalar random fields has to be generalised to tensor random fields (TRFs).

In deterministic theories of continuum physics we typically have an equation of the form

$$\mathcal{L}\boldsymbol{u} = \boldsymbol{f},$$

defined on some subset $\mathcal{D}$ of the d-dimensional affine Euclidean space E^d, where $\mathcal{L}$ is a differential operator, $\boldsymbol{f}$ is a *source or forcing function*, and $\boldsymbol{u}$ is a solution field. This needs to be accompanied by appropriate boundary and/or initial conditions. (We use the symbolic ($\boldsymbol{u}$) or, equivalently, the subscript ($u_{i\dots}$) notations for tensors, as the need arises; also, an overdot will mean the derivative with respect to time, $\mathrm{d}/\mathrm{d}t$.)

A field theory is stochastic if either the operator $\mathcal{L}$ is random, or there appears an apparent randomness of $\boldsymbol{u}$ due to an inherent non-linearity of $\mathcal{L}$, or the forcing and/or boundary/initial conditions are random. While various combinations of these basic cases are possible, in this book we focus on the first and second cases.

The first case is typically due to the presence of a spatially random material microstructure; see Ostoja-Starzewski (2008). For example, the coefficients of $\mathcal{L}(\omega)$, such as the elastic moduli $\boldsymbol{C}$, form a tensor-valued random field, and the stochastic equation

$$\mathcal{L}(\omega)\,\boldsymbol{u} = \boldsymbol{f}$$

governs the response of a *random medium* $\mathcal{B}$, that is, the set of possible states of a deterministic medium.

The second case is exemplified by solutions of the Navier–Stokes equation, which becomes so irregular as to be treated in a stochastic way (Batchelor 1951; Monin & Yaglom 2007*a*; Monin & Yaglom 2007*b*; Frisch 1995). In both cases, $\mathcal{B}$ is taken as a set of all the realisations $B(\omega)$ parameterised by elementary events ω of the Ω space

$$\mathcal{B} = \{ B(\omega) : \omega \in \Omega \} . \tag{0.1}$$

In principle, each of the realisations follows deterministic laws of classical mechanics; probability is introduced to deal with the set (0.1). The ensemble picture is termed *stochastic continuum physics*. Formally speaking, we have a triple $(\Omega, \mathfrak{F}, \mathsf{P})$, where Ω is the set of elementary events, $\mathfrak{F}$ is its σ-field and P is the probability measure defined on it.

Besides turbulence, another early field of research where *stochastic continuum physics* replaced the deterministic picture has been *stochastic wave propagation*: elastic, acoustic and electromagnetic. A paradigm of wave propagation in random media is offered by the *wave equation* for a scalar field u in a domain $\mathcal{D}$:

$$\nabla^2 \varphi = \frac{1}{c^2(\omega, \boldsymbol{x})} \frac{\partial^2 \varphi}{\partial t^2}, \qquad \omega \in \Omega, \quad \boldsymbol{x} \in \mathcal{D}.$$

Here c is the wave speed in a linear elastic, isotropic medium, so that, effectively, $\mathcal{B}$ is described by a random field $\{ c(\omega, \boldsymbol{x}) \colon \omega \in \Omega, \boldsymbol{x} \in \mathcal{D} \}$. Given that we simply have a Laplacian on the left-hand side, this model accounts for spatial randomness in mass density ρ only.

In order also to account for randomness in the elastic modulus E, we should consider this partial differential equation:

$$\boldsymbol{\nabla} \cdot [E(\omega, \boldsymbol{x}) \nabla u] = \rho(\omega, \boldsymbol{x}) \frac{\partial^2 u}{\partial t^2}, \qquad \omega \in \Omega, \quad \boldsymbol{x} \in \mathcal{D}. \tag{0.2}$$

Clearly, we are now dealing with two scalar random fields: E and ρ. This model's drawback, however, is the assumption of an inhomogeneous but locally isotropic second-rank stiffness (or elasticity) tensor field $\mathsf{E} = E\mathsf{I}$ instead of E $(= E_{ij}\boldsymbol{e}_i \otimes \boldsymbol{e}_j)$ with full anisotropy. In fact, extensive studies on upscaling of various mechanical and physical phenomena have shown (Ostoja-Starzewski et al. 2016) that the local anisotropy goes hand in hand with randomness: as the smoothing scale (i.e. scale on which the continuum is set up) increases, the anisotropy and random fluctuations in material properties jointly go to zero. Thus, Equation (0.2) should be replaced by

$$\boldsymbol{\nabla} \cdot [\boldsymbol{E}(\omega, \boldsymbol{x}) \cdot \boldsymbol{\nabla} u] = \rho(\omega, \boldsymbol{x}) \frac{\partial^2 u}{\partial t^2}, \qquad \omega \in \Omega, \quad \boldsymbol{x} \in \mathcal{D}.$$

The same arguments apply to a diffusion equation of, say, heat conduction

$$\boldsymbol{\nabla} \cdot [\boldsymbol{K}(\omega, \boldsymbol{x}) \cdot \boldsymbol{\nabla} T] = c(\omega, \boldsymbol{x}) \rho(\omega, \boldsymbol{x}) \frac{\partial T}{\partial t}, \qquad \omega \in \Omega, \quad \boldsymbol{x} \in \mathcal{D},$$

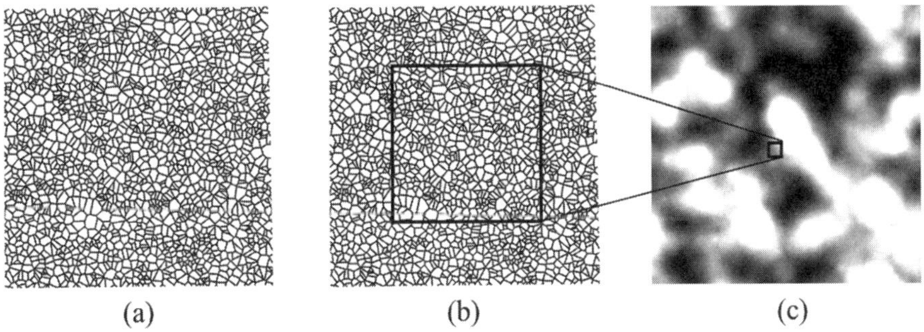

Figure 0.1 (a) A realisation of a Voronoi tesselation (or mosaic); (b) placing a mesoscale window leads, via upscaling, to a mesoscale random continuum approximation in (c). Reproduced from Malyarenko & Ostoja-Starzewski (2017*b*).

in which $\boldsymbol{K}$ is the thermal conductivity tensor (again with anisotropy present), while the specific heat c and mass density ρ jointly premultiply the first derivative of temperature T on the right-hand side.

This line of reasoning also applies to elliptic problems: consider Figure 0.1, showing a planar Voronoi tessellation of E^2 which serves as a planar geometric model of a polycrystal (although the same arguments apply in E^3). Each cell may be occupied by a differently oriented crystal, with all the crystals belonging to any specific crystal class. The latter include:

- transverse isotropy modelling, say, sedimentary rocks at long wavelengths;
- tetragonal modelling, say, wulfenite ($PbMoO_4$);
- trigonal modelling, say, dolomite ($CaMg(CO_3)_2$);
- orthotropic modelling, say, wood;
- triclinic modelling, say, microcline feldspar.

Thus, we need to be able to model fourth-rank tensor random fields, point-wise taking values in any crystal class. While the crystal orientations from grain to grain are random, they are not spatially independent of each other – the assignment of crystal properties over the tessellation is not white noise. This is precisely where the two-point characterisation of the random field of elasticity tensor is needed. While the simplest correlation structure to admit would be white noise, a (much) more realistic model would account for any mathematically admissible correlation structures as dictated by the statistically wide-sense homogeneous and isotropic assumption. A specific correlation can then be fitted to physical measurements.

Note that it may also be of interest to work with a mesoscale random continuum approximation defined by placing a mesoscale window at any spatial position, as shown in Figure 0.1(b). Clearly, the larger the mesoscale window, the weaker the random fluctuations in the mesoscale elasticity tensor: this is the trend to homogenise the material when upscaling from a statistical volume element (SVE) to a representative volume element (RVE). A simple paradigm of this upscaling, albeit only in terms of a scalar random field, is the opacity of a sheet of paper held against light: the further away the sheet is from our eyes, the more homogeneous it appears. Similarly, in the case of upscaling of elastic properties, on any finite scale there is almost certainly anisotropy, and this anisotropy, with mesoscale increasing, tends to zero hand-in-hand with the fluctuations, and it is in the infinite mesoscale limit (i.e. RVE) that material isotropy is obtained as a consequence of the statistical isotropy.

Another motivation for the development of TRF models is to have a realistic input of elasticity random fields into stochastic field equations such as stochastic partial differential equations (SPDE) and stochastic finite elements (SFE). The classical paradigm of SPDE can be written in terms of the anti-plane elastostatics (with $u \equiv u_3$):

$$\boldsymbol{\nabla} \cdot (C(\boldsymbol{x}, \omega) \nabla u) = 0, \quad \boldsymbol{x} \in \mathbb{E}^2, \quad \omega \in \Omega, \tag{0.3}$$

with $C(\cdot, \omega)$ being spatial realisations of a scalar RF. In view of the foregoing discussion, Equation (0.3) is well justified for a piecewise-constant description of realisations of a random medium such as a multiphase composite made of locally isotropic grains. However, in the case of a boundary value problem set up on coarser (i.e. mesoscale) scales, having continuous realisations of properties, a second-rank tensor random field (TRF) of material properties would be much more appropriate: see Figure 0.1(b). The field equation should then read

$$\boldsymbol{\nabla} \cdot (\boldsymbol{C}(\boldsymbol{x}, \omega) \cdot \boldsymbol{\nabla} u) = 0, \qquad \boldsymbol{x} \in E^2, \quad \omega \in \Omega,$$

where $\boldsymbol{C}$ is the second-rank tensor random field.

Moving to the in-plane or 3D elasticity, the starting point is the *Navier equation* of motion (written in symbolic and tensor notations):

$$\mu \nabla^2 \boldsymbol{u} + (\lambda + \mu) \boldsymbol{\nabla} (\boldsymbol{\nabla} \cdot \boldsymbol{u}) = \rho \ddot{\boldsymbol{u}} \quad \text{or} \quad \mu u_{i,jj} + (\lambda + \mu) u_{j,ji} = \rho \ddot{u}_i. \tag{0.4}$$

Here $\boldsymbol{u}$ is the displacement field, λ and μ are two Lamé constants and ρ is the mass density. This equation is often (e.g. in stochastic wave propagation) used as an Ansatz, typically with the pair (λ, μ) taken *ad hoc* as a 'vector' random field with some simple correlation structure for both components. However, in order to properly introduce the smooth randomness in λ and μ, one has to go one step back in derivation of (0.4) and write

$$\mu \nabla^2 \boldsymbol{u} + (\lambda + \mu) \boldsymbol{\nabla} (\boldsymbol{\nabla} \cdot \boldsymbol{u}) + \boldsymbol{\nabla} \mu \left(\nabla \boldsymbol{u} + (\boldsymbol{\nabla} \boldsymbol{u})^\top \right) + \boldsymbol{\nabla} \lambda \boldsymbol{\nabla} \cdot \boldsymbol{u} = \rho \ddot{\boldsymbol{u}},$$

or

$$\mu u_{i,jj} + (\lambda + \mu) u_{j,ji} + \mu_{,j} (u_{j,i} + u_{i,j}) + \lambda_{,i} u_{j,j} = \rho \ddot{u}_i. \tag{0.5}$$

While two extra terms are now correctly present on the left-hand side, this equation still suffers from the drawback (just as did Equation (0.3)) of local isotropy so that, again by micromechanics upscaling arguments, should be replaced by

$$\nabla\cdot(\boldsymbol{C}\nabla\cdot\boldsymbol{u})^\top=\rho\ddot{\boldsymbol{u}}\quad\text{or}\quad\left(C_{ijkl}u_{(k,l)}\right)_{,j}=\rho\ddot{u}_i.\tag{0.6}$$

Here $\boldsymbol{C}$ ($= C_{ijkl}\boldsymbol{e}_i\otimes\boldsymbol{e}_j\otimes\boldsymbol{e}_k\otimes\boldsymbol{e}_l$), which, at any scale finitely larger than the microstructural scale, is almost surely (a.s.) anisotropic. Clearly, instead of Equation (0.5) one should work with this SPDE (0.6) for $\boldsymbol{u}$.

The foregoing arguments motivate the main goal of this book: to obtain explicit representations of correlation functions of TRFs of ranks 1 through 4, so as to enable their simulation and the construction of models of various field phenomena, subject to the restrictions imposed by the field equations dictated by physics. Briefly, in the case of dependent TRFs, we have, say, the linear momentum equation restricting the Cauchy stress or the angular momentum equation restricting the Cauchy and couple stresses. In the case of material property fields (elasticity, diffusion, permeability...) there are conditions of positiveness of either the energy density or the entropy production, as the case may be. In turn, any such conditions lead to restrictions on the respective correlation functions. An introduction to a wide range of continuum physics theories where tensor random fields are needed is given in Chapter 1.

What Mathematical Background is Required?

Random functions of more than one real variable, or *random fields*, appeared for the very first time in applied physical papers. We would like to mention papers by Friedmann & Keller (1924), von Kármán (1937), von Kármán & Howarth (1938), Kampé de Fériet (1939), Obukhov (1941*a*), Obukhov (1941*b*), Robertson (1940), Yaglom (1948), Yaglom (1957), Lomakin (1964) and Lomakin (1965). The physical models introduced in the above papers follow the same scheme, which we explain below. The mathematical tools we use are described in detail in Chapter 2; see also Olive & Auffray (2013) and Auffray, Kolev & Petitot (2014).

Let $(E,\mathbb{R}^d,+)$ be the d-dimensional affine space. The underlying linear space $V=\mathbb{R}^d$ consists of vectors $\boldsymbol{x}=(x_1,\dots,x_d)^\top$. We are mainly interested in the case of $d=2$, which corresponds to *plane problems of continuum physics* as well as in the case of $d=3$ that corresponds to *space problems*. Let $(\cdot,\cdot)$ be the standard inner product in $\mathbb{R}^d$:

$$(\boldsymbol{x},\boldsymbol{y})=\sum_{i=1}^d x_iy_i.$$

Let r be a non-negative integer. The above inner product induces inner products in the space $V^{\otimes r}$ as follows: $(\alpha,\beta)=\alpha\beta$ when $r=0$ and α, $\beta\in V^{\otimes 0}=\mathbb{R}^1$ and

$$(\boldsymbol{S},\boldsymbol{T})=\sum_{j_1=1}^d\cdots\sum_{j_r=1}^d \boldsymbol{S}_{j_1\cdots j_r}\boldsymbol{T}_{j_1\cdots j_r}.$$

The linear transformations of the space $\mathbb{R}^d$ that preserve the above inner product, constitute the *orthogonal group* $\mathrm{O}(d)$. The pair $(g^{\otimes r},V^{\otimes r})$ is an orthogonal representation of the group $\mathrm{O}(d)$ (trivial when $r=0$). Let V_0 be an invariant subspace of the above representation of positive dimension. Let ρ be the restriction of the representation $g\mapsto g^{\otimes r}$ of the group $\mathrm{O}(d)$ to the subspace V_0. Consider the representation (ρ,V_0) as a group action. There are finitely many, say N, orbit types for this action. Let $[G_0]$, ..., $[G_{N-1}]$ be the corresponding conjugacy classes of the closed subgroups of the group $\mathrm{O}(d)$. Physicists call them *symmetry classes*. The representatives of conjugacy classes are *point groups*.

Let G_i be a representative of the conjugacy class $[G_i]$. Let V be the subspace of V_0 where the isotypic component of the representation ρ_0 that corresponds to the trivial representation of the group G_i acts. Let $N(G_i)$ be the normaliser of the group G_i in $\mathrm{O}(d)$. Let G be a subgroup of $N(G_i)$ such that G_i is a subgroup of G. Call G the *symmetry group* of a physical material, or the group of material symmetries. The space V is an invariant subspace for the representation $(g^{\otimes r},(\mathbb{R}^d)^{\otimes r})$ of the group G. Let ρ be the restriction of the above representation to V.

Let $\mathcal{B}$ be a material body that occupies a subset $D\subset E^d$. Consider a physical property of $\mathcal{B}$ that is described by a mapping $\boldsymbol{T}\colon D\to\mathsf{V}$. Examples are given in Subsection 3.1 and include the temperature, where $\mathsf{V}=\mathbb{R}^1$, the velocity of a turbulent fluid, where $\mathsf{V}=\mathbb{R}^d$, the strain tensor of a deformable body, where $\mathsf{V}=\mathsf{S}^2(\mathbb{R}^d)$, the space of symmetric matrices, and the elasticity (or stiffness) tensor, where $\mathsf{V}=\mathsf{S}^2(\mathsf{S}^2(\mathbb{R}^d))$.

To randomise this model, consider a random field $\boldsymbol{T}\colon E\to\mathsf{V}$. Assume that $\mathsf{E}[\|\boldsymbol{T}(A)\|^2]<\infty$, $A\in E$. Assume also that the field $\boldsymbol{T}(A)$ is *mean-square continuous*, that is,

$$\lim_{\|B-A\|\to 0}\mathsf{E}[\|\boldsymbol{T}(B)-\boldsymbol{T}(A)\|^2]=0$$

for all $A\in E$. Under the translation, the *one-point correlation tensor*

$$\langle\boldsymbol{T}(A)\rangle=\mathsf{E}[\boldsymbol{T}(A)]$$

and the *two-point correlation tensor*

$$\langle\boldsymbol{T}(A),\boldsymbol{T}(B)\rangle=\mathsf{E}[(\boldsymbol{T}(A)-\langle\boldsymbol{T}(A)\rangle)\otimes(\boldsymbol{T}(B)-\langle\boldsymbol{T}(B)\rangle)]$$

do not change. Such a field is called *wide-sense homogeneous*.

Fix a place $O\in D$. Under the rotation of the body about O by a material symmetry $g\in G$, an arbitrary place $A\in D$ becomes the place $O+g(A-O)$. Evidently, the tensor $\boldsymbol{T}(A)$ becomes the tensor $\rho(g)\boldsymbol{T}(A)$. The one-point correlation tensor of the transformed field must be equal to that of the original field:

$$\langle\boldsymbol{T}(O+g(A-O))\rangle=\rho(g)\langle\boldsymbol{T}(A)\rangle.$$

The two-point correlation tensors of both fields must be equal as well:

$$\langle\boldsymbol{T}(O+g(A-O)),\boldsymbol{T}(O+g(B-O))\rangle=(\rho\otimes\rho)(g)\langle\boldsymbol{T}(A),\boldsymbol{T}(B)\rangle.$$

A random field that satisfies the two last conditions is called *wide-sense isotropic*. In what follows we omit the words 'wide-sense'.

The main mathematical problem that is solved in this book is as follows. We would like to *find the general form of the one-point and two-point correlation tensors of a homogeneous and isotropic tensor-valued random field* $\mathsf{T}(A)$ *as well as the spectral expansion of the above field.*

To explain what we mean, consider the simplest example. Let ρ be the trivial representation of the symmetry group $G = \mathrm{O}(d)$, and let $\tau(A)$ be the corresponding homogeneous and isotropic random field. Schoenberg (1938) proved that the equation

$$\langle \tau(A), \tau(B) \rangle = 2^{(d-2)/2}\Gamma(d/2) \int_0^\infty \frac{J_{(d-2)/2}(\lambda\|B-A\|)}{(\lambda\|B-A\|)^{(d-2)/2}}\,\mathrm{d}\Phi(\lambda)$$

establishes a one-to-one correspondence between the set of two-point correlation functions of homogeneous and isotropic random fields on the space E and the set of finite Borel measures Φ on $[0,\infty)$. Here Γ denotes the gamma function and J denotes the Bessel function of the first kind.

The paper by Schoenberg (1938) was not mentioned before. The reason is that this paper does not treat random fields at all. Instead, the problem of description of all continuous positive-definite functions $B(\|\boldsymbol{y}-\boldsymbol{x}\|)$ with $\boldsymbol{x}, \boldsymbol{y} \in \mathbb{R}^d$ is considered. Thus, there exists a link between the theory of random fields and the theory of positive-definite functions.

The result by Schoenberg (1938) does not help to perform a computer simulation of sample paths of a homogeneous and isotropic random field. The following result is useful for the above purposes. Yaglom (1961) and M. Ĭ. Yadrenko, in his unpublished PhD thesis, proved that a homogeneous and isotropic random field has the following spectral expansion:

$$\begin{aligned}\tau(A-O) = \langle \tau(A) \rangle + \sqrt{2^{d-1}\Gamma(d/2)\pi^{d/2}} \sum_{\ell=0}^{\infty} \sum_{m=1}^{h(d,\ell)} S_\ell^m(\theta_1, \dots, \theta_{d-2}, \varphi)\\ \times \int_0^\infty \frac{J_{\ell+(d-2)/2}(\lambda\rho)}{(\lambda\rho)^{(d-2)/2}}\,\mathrm{d}Z_\ell^m(\lambda),\end{aligned}$$

where $(\rho, \theta_1, \dots, \theta_{d-2}, \varphi)$ are the spherical coordinates of the vector $A-O$, S_ℓ^m are real-valued spherical harmonics and Z_ℓ^m is a sequence of uncorrelated real-valued orthogonal stochastic measures on $[0,\infty)$ with the measure Φ as their common control measure. To simulate the field, we truncate the integrals and use an arbitrary quadrature formula in combination with Monte Carlo simulation.

As the reader can see, the spectral expansion of the field includes an arbitrary choice of the place $O \in E$. There is nothing strange here, because the affine space E does not contain any distinguished places. More explanation is given in Section 2.9. To avoid frequent repetitions of the same words, we vectorise the affine space E by a choice of the origin $O \in E$ once and forever, and denote the vector space E_O by $\mathbb{R}^d$.

The next interesting case is when $\rho(g) = g$. Robertson (1940) proved that in this case the two-point correlation matrix of the field has the form

$$\langle \boldsymbol{v}(\boldsymbol{x}), \boldsymbol{v}(\boldsymbol{y})\rangle_{ij} = A(\|\boldsymbol{z}\|)z_i z_j + B(\|\boldsymbol{z}\|)\delta_{ij},$$

where $\boldsymbol{z} = \boldsymbol{y} - \boldsymbol{x}$. Note that δ_{ij} is the only covariant tensor of degree 0 and of order 2 of the group $\mathrm{O}(3)$, while $z_i z_j$ is its only covariant tensor of degree 2 and of order 2. Thus, another link has been established, this time between the theory of random fields and the classical invariant theory. A review of the invariant theory is given in Section 2.7.

In Section 3.1 we continue to describe the results obtained by our predecessors. As the reader will see, the list of results is impressively short. The complete solution to the problem formulated above requires a combination of tools from different areas of mathematics. No book that describes all necessary tools in a short form is known to the authors. Therefore, in Chapter 2 we collected all of them together. The choice of material was dictated by the solution strategy, and we describe it below.

The main idea is quite simple; see Malyarenko (2013). We *describe the set of homogeneous random fields and reject those that are not isotropic.* Trying this way, we immediately meet the first obstacle: there exist no complete description of two-point correlation tensors of homogeneous random fields taking values in a *real* finite-dimensional linear space. The only known result is as follows. Let $\tilde{\mathsf{V}}$ be a *complex* finite-dimensional linear space. The equation

$$\langle \boldsymbol{T}(\boldsymbol{x}), \boldsymbol{T}(\boldsymbol{y})\rangle = \int_{\hat{V}} \mathrm{e}^{\mathrm{i}(\boldsymbol{p},\boldsymbol{y}-\boldsymbol{x})}\,\mathrm{d}F(\boldsymbol{p})$$

establishes a one-to-one correspondence between the set of $\tilde{\mathsf{V}}$-valued homogeneous random fields on the *space domain* V and the set of measures F on the Borel σ-field $\mathfrak{B}(\hat{V})$ taking values in the set of Hermitian non-negative-definite operators on $\tilde{\mathsf{V}}$. Here $\hat{V}$ denote the *wavenumber domain.*

Now we have to define a *real* subspace V of the complex space $\tilde{\mathsf{V}}$. The easiest way to do that is to introduce coordinates in $\tilde{\mathsf{V}}$. We do not want to proceed this way, however, for the following reason. The formulae that describe the solution are basis-dependent. Therefore, the choice of the most convenient basis is a part of the proof. The idea is to make the above choice at the latest possible stage of proof: that is, to write as many formulae as possible in a coordinate-free form.

To start with, we introduce a real structure J in the space $\tilde{\mathsf{V}}$. The eigenvectors of J that correspond to the eigenvalue 1, form a *real* linear space V. The linear space of all Hermitian operators on $\tilde{\mathsf{V}}$ is isomorphic to $\mathsf{V}\otimes\mathsf{V} = \mathsf{S}^2(\mathsf{V})\oplus\Lambda^2(\mathsf{V})$. Let $\top$ be the linear operator in $\mathsf{V}\otimes\mathsf{V}$ for which $\mathsf{S}^2(\mathsf{V})$ is the set of eigenvectors with eigenvalue 1, and $\Lambda^2(\mathsf{V})$ is the set of eigenvectors with eigenvalue -1 (this is just the coordinate-free definition of the transposed matrix). We have the following *necessary condition*: if a homogeneous random field takes values in V, then the measure F satisfies the *reality condition*:

$$F(-A) = F(A)^\top, \qquad A \in \mathfrak{B}(\hat{V}),$$

where $-A = \{\,-\boldsymbol{T} : \boldsymbol{T} \in A\,\}$. If one rejects away all Radon measures F that do not satisfy the above condition, no V-valued homogeneous random fields are lost (but some $\tilde{\mathsf{V}}$-valued fields may still remain).

The above method dictates the content of Section 2.1, where we explain many results of linear and tensor algebra in both coordinate and coordinate-free form.

Next, we prove that the one-point correlation tensor of an isotropic random field is a tensor lying in the isotypic subspace of the representation ρ that corresponds to its trivial component, while the measure F must satisfy the condition

$$F(gA) = (\rho \otimes \rho)(g)F(A), \qquad A \in \mathfrak{B}(\hat{V}).$$

The next idea is as follows. We find a group $\tilde{G}$ and its orthogonal representation $(\tilde{\rho}, \tilde{\mathsf{V}})$ in a real finite-dimensional space $\tilde{\mathsf{V}}$ such that the above condition *and* the reality condition together are equivalent to the condition

$$F(A) \in \tilde{\mathsf{V}}, \qquad F(\tilde{g}A) = \tilde{\rho}(\tilde{g})F(A).$$

Lemma 1 solves this problem. Proof of Lemma 1 requires both general knowledge of group representations and specific knowledge of orthogonal representations, that are given in Section 2.5.

The next step is to introduce the measure $\mu(A) = \operatorname{tr} F(A)$, $A \in \mathfrak{B}(\hat{V})$, noting that F is absolutely continuous with respect to μ, and to write the two-point correlation tensor of the field as

$$\langle \boldsymbol{T}(\boldsymbol{x}), \boldsymbol{T}(\boldsymbol{y}) \rangle = \int_{\hat{V}} \mathrm{e}^{\mathrm{i}(\boldsymbol{p}, \boldsymbol{y}-\boldsymbol{x})} f(\boldsymbol{p}) \, \mathrm{d}\mu(\boldsymbol{p}),$$

where the density $f(\boldsymbol{p})$ is a measurable function on $\hat{V}$ taking values in the *convex compact* set of all Hermitian non-negative-definite operators on $\tilde{\mathsf{V}}$ with unit trace. The measure μ and the density $f(\boldsymbol{p})$ must satisfy the following conditions:

$$\mu(\tilde{g}A) = \mu(A), \qquad f(\tilde{g}\boldsymbol{p}) = \tilde{\rho}(\tilde{g})f(\boldsymbol{p}).$$

The description of all possible measures μ is well known. It includes a detailed description of the *stratification* of the space $\hat{V}$ induced by the group action of the group G by the matrix-vector multiplication. In particular, the measure μ is uniquely determined by a Radon measure Φ on the Borel σ-field of the orbit space $\hat{V}/\tilde{G}$. All necessary tools from topology are presented in Section 2.2.

To find all measurable functions $f \colon \hat{V} \to \tilde{\mathsf{V}}$ satisfying the second condition, we proceed as follows. Let $[\tilde{G}_0]$, $\dots$, $[\tilde{G}_{M-1}]$ be the symmetry classes of the representation $(g, \hat{V})$ of the group $\tilde{G}$, where $[\tilde{G}_0]$ is the minimal symmetry class, and $[\tilde{G}_{M-1}]$ is the principal symmetry class. Let $(\hat{V}/\tilde{G})_0$, $\dots$, $(\hat{V}/\tilde{G})_{M-1}$ be the corresponding stratification of the orbit space $\hat{V}/\tilde{G}$. For simplicity of notation, assume that there is a chart $\boldsymbol{\lambda}_m$ of the manifold $(\hat{V}/\tilde{G})_m$ that covers a dense subset of the above manifold, and there is a chart $\boldsymbol{\varphi}_m$ of the orbit $\tilde{G}/H_m$ that covers a dense subset of the orbit. Let $(\boldsymbol{\lambda}_m, \boldsymbol{\varphi}^0_m)$ be the coordinates of the intersection of the orbit $\tilde{G}\cdot\boldsymbol{\lambda}_m$ with the set $(\hat{V}/\tilde{G})_m$. We have $\tilde{g}(\boldsymbol{\lambda}_m, \boldsymbol{\varphi}^0_m) = (\boldsymbol{\lambda}_m, \boldsymbol{\varphi}^0_m)$ for all $\tilde{g} \in H_m$. It follows that

$$f(\boldsymbol{\lambda}_m, \boldsymbol{\varphi}^0_m) = \tilde{\rho}(\tilde{g}) f(\boldsymbol{\lambda}_0, \boldsymbol{\varphi}^0_m), \qquad \tilde{g} \in H_m.$$

In other words, the tensor $f(\boldsymbol{\lambda}_m, \boldsymbol{\varphi}^0_m)$ lies in the isotypic subspace W_m of the trivial component of the representation $\tilde{\rho}$ of the group H_m. The intersection of the space W_m and the convex compact set of Hermitian non-negative-definite operators in $\tilde{\mathsf{V}}$ is a convex compact set, say $\mathcal{C}_m$. No other restriction exists; that is, the restriction of f to $(\hat{V}/\tilde{G})_m$ is an arbitrary measurable function taking values in $\mathcal{C}_m$.

Now we introduce coordinates. The space V consists of tensors of rank r. Let $\boldsymbol{T}^1_{i_1\cdots i_r}, \dots, \boldsymbol{T}^{\dim \mathsf{V}}_{i_1\cdots i_r}$ be an orthonormal basis in V. The space $\mathsf{V}\otimes\mathsf{V}$ can be represented as the direct sum of the subspace of symmetric tensors and the subspace of skew-symmetric tensors over V:

$$\mathsf{V}\otimes\mathsf{V} = \mathsf{S}^2(\mathsf{V}) \oplus \Lambda^2(\mathsf{V}).$$

Put $\tau(\boldsymbol{T}^1 \oplus \boldsymbol{T}^2) = \boldsymbol{T}^1 \oplus \mathrm{i}\boldsymbol{T}^2$, where $\boldsymbol{T}^1 \in \mathsf{S}^2(\mathsf{V})$, $\boldsymbol{T}^2 \in \Lambda^2(\mathsf{V})$. The map τ is an isomorphism between $\mathsf{V}\otimes\mathsf{V}$ and the real linear space H of Hermitian operators on $\tilde{\mathsf{V}}$. The *coupled basis* of the space H is formed by the tensors

$$\tau(\boldsymbol{T}^i_{i_1\cdots i_r}) \otimes \tau(\boldsymbol{T}^j_{j_1\cdots j_r}), \qquad 1 \leq i, j \leq \dim \mathsf{V},$$

while the mth *uncoupled basis* of the above space consists of the rank $2r$ tensors

$$\boldsymbol{T}^{0k}_{i_1\cdots j_r}, \qquad 1 \leq k \leq (\dim \mathsf{V})^2,$$

where the first $\dim \mathsf{W}_m$ tensors constitute an orthonormal basis in W_m.

Let $(\boldsymbol{\lambda}_m, \boldsymbol{\varphi}^0_m)$ be the coordinates of the intersection of the orbit $\tilde{G}\cdot\boldsymbol{\lambda}_m$ with the set $(\hat{V}/\tilde{G})_m$. Let $f^k_{i_1\cdots j_r}(\boldsymbol{\lambda}_m, \boldsymbol{\varphi}^0_m)$ be the value of the linear form $f(\boldsymbol{\lambda}_0, \boldsymbol{\varphi}^0_m)$ on the basis tensor $\boldsymbol{T}^{0k}_{i_1\cdots j_r}$. Then we have $f^k_{i_1\cdots j_r}(\boldsymbol{\lambda}_0, \boldsymbol{\varphi}^0_m) = 0$ when $k > \dim \mathsf{W}_0$. The value of the linear form $f(\boldsymbol{\lambda}_m, \boldsymbol{\varphi}_m)$ on the above basis tensor is then

$$f^k_{i_1\cdots j_r}(\boldsymbol{\lambda}_m, \boldsymbol{\varphi}_m) = \sum_{l=1}^{\dim \mathsf{W}_0} \tilde{\rho}^0_{kl}(\boldsymbol{\varphi}_m) f^l_{i_1\cdots j_r}(\boldsymbol{\lambda}_m, \boldsymbol{\varphi}_m),$$

where $\tilde{\rho}^0_{kl}(\boldsymbol{\varphi}_m) = (\tilde{\rho}(\tilde{g})\boldsymbol{T}^{0k}_{i_1\cdots j_r}, \boldsymbol{T}^{0l}_{i_1\cdots j_r})$ are the matrix entries of the operator $\tilde{\rho}(\tilde{g})$ in the zeroth uncoupled basis, with $\tilde{g}$ being an arbitrary element of $\tilde{G}$ that transforms the point $\boldsymbol{\lambda}_m \in (\hat{V}/\tilde{G})_m$ to the point $(\boldsymbol{\lambda}_m, \boldsymbol{\varphi}_m) \in \hat{V}$.

The tensors of the coupled basis are linear combinations of the tensors of the zeroth uncoupled basis:

$$\tau(\boldsymbol{T}^i_{i_1\cdots i_r}) \otimes \tau(\boldsymbol{T}^j_{j_1\cdots j_r}) = \sum_{k=1}^{(\dim \mathsf{W})^2} c^{mk}_{ij} \boldsymbol{T}^{0k}_{i_1\cdots j_r},$$

and the matrix entries of the Hermitian non-negative-definite operator $f(\boldsymbol{\lambda}_m, \boldsymbol{\varphi}_m)$ in the coupled basis take the form

$$f_{i_1\cdots i_r, j_1\cdots j_r}(\boldsymbol{\lambda}_m, \boldsymbol{\varphi}_m) = \sum_{k=1}^{(\dim \mathsf{W})^2} c_{ij}^{mk} \sum_{l=1}^{\dim \mathsf{W}_0} \tilde{\rho}_{kl}^0(\boldsymbol{\varphi}_m) f_{i_1\cdots j_r}^l(\boldsymbol{\lambda}_m).$$

As we will see in Chapter 3, previous ideas lead to the expansion of the two-point correlation tensor of the field into a sum of finitely many double integrals. The inner integrals can be calculated in a closed form. The quickest way to do that is to expand the *plane wave* $\mathrm{e}^{\mathrm{i}(\boldsymbol{p},\boldsymbol{y}-\boldsymbol{x})}$ into a Fourier series (a Fourier sum if the group $\tilde{G}$ is finite) with respect to the matrix entries of some special irreducible orthogonal representations of the group $\tilde{G}$, and use a fundamental theorem of the representation theory called the *Fine Structure Theorem*. Many classical formulae of the theory of special functions, reviewed in Section 2.10, are particular cases of the above expansion. For example, if $\tilde{G}$ contains two elements, we obtain the de Moivre formula, if $\tilde{G} = \mathrm{O}(2)$, we obtain the Jacobi–Anger expansion, and so on.

Many properties of the fields' spectral expansions are encoded in the geometry of convex compacta $\mathcal{C}_m$, $0 \leq m \leq M-1$. In particular, the number of integrals in the expansion of the two-point correlation tensor of the field is equal to the number of connected components of the set of extreme points of $\mathcal{C}_0$. If a component is not a one-point set, then the corresponding integrand contains an arbitrary measurable function taking values in the closed convex hull of the component. The integrals are calculated with respect to measures Φ that may satisfy additional constraints. The eventual constraints are encoded in the configuration of the sets $\mathcal{C}_m$, $1 \leq m \leq M-1$ inside $\mathcal{C}_0$, and so on. Section 2.8 introduces necessary tools from geometry of finite-dimensional convex compacta.

The coefficients c_{ij}^{mk} are expressed in terms of the so-called *Clebsch–Gordan coefficients* of the *real* orthogonal irreducible representations of the group $\tilde{G}$. Note that the classical Clebsch–Gordan coefficients familiar from quantum mechanics are different. They are connected with *complex* unitary irreducible representations of the group SU(2) of 2×2 unitary matrices with unit determinant. Both the coupled and uncoupled bases must be chosen in such a way that the calculation of the coefficients c_{ij}^{nk} becomes as easy as possible.

Finally, the expansion of the field in term of stochastic integrals is obtained as follows. Write the plane wave in the form

$$\mathrm{e}^{\mathrm{i}(\boldsymbol{p},\boldsymbol{y}-\boldsymbol{x})} = \mathrm{e}^{\mathrm{i}(\boldsymbol{p},\boldsymbol{y})}\mathrm{e}^{-\mathrm{i}(\boldsymbol{p},\boldsymbol{x})}$$

and apply the Fourier expansion to each term *separately*. As a result, the two-point correlation tensor of the field takes the form

$$\langle \boldsymbol{T}(\boldsymbol{x}), \boldsymbol{T}(\boldsymbol{y})\rangle = \int_{\Lambda} h(\boldsymbol{x},\lambda)\overline{h(\boldsymbol{y},\lambda)}\,\mathrm{d}\nu(\lambda),$$

where ν is a measure on a σ-field $\mathfrak{L}$ of subsets of a set Λ taking values in the set of Hermitian non-negative-definite linear operators on $\tilde{\mathsf{V}}$. Moreover, the set of functions $\{\, h(\boldsymbol{x}, \lambda) \colon \boldsymbol{x} \in \mathbb{R}^d \,\}$ is *total* in the Hilbert space $L^2(\Lambda, \nu)$, that is, its linear span is dense in the above space. Using the fundamental *Karhunen's theorem*, we write down the spectral expansion of the random field $\boldsymbol{T}(\boldsymbol{x})$ in the form

$$\boldsymbol{T}(\boldsymbol{x}) = \mathsf{E}[\boldsymbol{T}(\boldsymbol{x})] + \int_{\Lambda} h(\boldsymbol{x}, \lambda)\, \mathrm{d}\boldsymbol{Z}(\lambda),$$

where $\boldsymbol{Z}$ is a centred $\tilde{\mathsf{V}}$-valued random measure with control measure ν, that is,

$$\mathsf{E}[\boldsymbol{Z}(A) \otimes \boldsymbol{Z}(B)] = \nu(A \cap B), \qquad A, B \in \mathfrak{L}.$$

As we noticed before, the class of random fields we obtained still may contain the fields taking values outside the space V. Such fields should be thrown away separately in each case.

The mathematical tools introduced in Chapter 2 give us an opportunity to solve the problem formulated above. In Section 3.1 we give the exact formulation of the problem. Before going into technicalities, we consider a non-trivial example which illustrates our methods, in Section 3.2. The general part of the proof is presented in Section 3.3. In the following five sections we prove both existing and new results using the above-described methods. Sections 3.4 to 3.8 describe homogeneous and isotropic random fields of rank r, for values of r from 0 to 4 respectively.

In Chapter 4 we apply the mathematical theory to TRFs of dependent quantities and constitutive responses. The topics include: a strategy for simulation of TRFs of ranks 1–4, ergodicity of TRFs, consequences of field equations (in 2D and 3D) for classical and micropolar continua, constitutive elastic-type responses, applications to stochastic partial differential equations, stochastic damage phenomena, isotropic but inhomogeneous TRFs modelling planetary rings and a very short selection of future research directions.

Note that an alternative approach was elaborated by Guilleminot & Soize (2011), Guilleminot, Noshadravan, Soize & Ghanem (2011), Guilleminot & Soize (2012), Guilleminot & Soize (2013*a*), Guilleminot & Soize (2013*b*), Noshadravan, Ghanem, Guilleminot, Atodaria & Peralta (2013), Guilleminot, Le & Soize (2013), Guilleminot & Soize (2014) and Staber & Guilleminot (2018). Using a stochastic model alternative to our model, they constructed a generator for random fields that have prescribed symmetry properties, take values in the set of symmetric non-negative-definite tensors, depend on a few real parameters and may be easily simulated and calibrated.

1

Introduction to Continuum Theories

This chapter provides a simple introduction to a range of continuum theories where spatial randomness is very likely to go hand in hand with local anisotropy. The introduction is simple because (i) a complete picture of continuum physics would be on the order of a thousand pages and (ii) we can quickly make the case for stochastic continuum physics by going from classical (Cauchy-type) to micropolar (Cosserat-type) and then non-local theories. At the same time, we introduce several single- and multi-field models: elastic, dissipative (parabolic or hyperbolic), thermoelastic, viscothermoelastic, piezoelectric and plasticity/damage phenomena. In order to see all these directions from one common perspective, we employ thermomechanics with internal variables (TIV), an approach which hinges on the free energy and dissipation functions. Treating the free energy as stochastic corresponds to modelling a randomly inhomogeneous hyperelastic material. The dissipation function, while very well known in deterministic continuum mechanics, can also be made stochastic – a useful property for modelling macroscopic, say geological, media on macro scales. The stochastic dissipation function also provides a link to nanoscale violations of the Second Law of Thermodynamics, as studied in contemporary statistical physics.

1.1 Deterministic versus Stochastic Models

1.1.1 Elementary Considerations

Why random fields?

In this book we understand *continuum physics* to mean continuum mechanics and other classical (non-quantum, non-relativistic) field theoretic models such as continuum thermomechanics (e.g. thermal conductivity, thermoelasticity) and multi-field interactions (e.g. piezoelectricity, thermodiffusion). In deterministic theories of continuum physics we typically have an equation of the form

$$\mathcal{L}\boldsymbol{u} = \boldsymbol{f}, \tag{1.1}$$

defined on some subset $\mathcal{D}$ of the d-dimensional affine Euclidean space $E = E^d$, where $\mathcal{L}$ is a differential operator, $\boldsymbol{f}$ is a source or forcing function and $\boldsymbol{u}$ is a solution field. This needs to be accompanied by appropriate boundary and/or initial conditions. (Hereinafter we interchangeably, whichever is more convenient, use the symbolic ($\boldsymbol{f}$) and subscript ($f_{i\ldots}$) notations for tensors; an overdot means the material time derivative.)

A field theory is *stochastic* if either the operator $\mathcal{L}$ is random, the forcing function $\boldsymbol{f}$ is random or the boundary/initial conditions are random. In the vein of the theory of random processes, this is indicated by the dependence on an elementary event ω (an element of the sample space Ω) and, thus, we have:

- randomness of the operator:

$$\mathcal{L}(\omega)\,\boldsymbol{u} = \boldsymbol{f}; \tag{1.2}$$

- randomness of the forcing function:

$$\mathcal{L}\boldsymbol{u} = \boldsymbol{f}(\omega);$$

- randomness of the boundary and/or initial conditions.

There is also a possibility of inherent non-linearity of $\mathcal{L}$, such as in fluid turbulence, leading to

- apparent randomness of $\boldsymbol{u}$ expressed by

$$\mathcal{L}\boldsymbol{u}(\omega) = \boldsymbol{f}. \tag{1.3}$$

While various combinations of these four basic cases are further possible, this book is focused on $\boldsymbol{u}$ and $\boldsymbol{f}$ being tensor-valued random fields (TRFs) of statistically homogeneous (i.e. wide-sense stationary) and isotropic type.

The first case is typically due to the presence of a spatially random material microstructure, e.g. Ostoja-Starzewski (2008). For example, the coefficients of $\mathcal{L}(\omega)$, such as the elastic moduli $\boldsymbol{C}$, form a tensor-valued random field (TRF), and the stochastic equation (1.2) governs the response of a *random medium* $\mathcal{B}$. Note that a formal solution to (1.3) is

$$\left\langle \mathcal{L}^{-1} \right\rangle^{-1} \langle \boldsymbol{u} \rangle = \boldsymbol{f}. \tag{1.4}$$

The operator $\left\langle \mathcal{L}^{-1} \right\rangle^{-1}$ is generally unwieldy to obtain in an explicit form, which explains the statistical averaging involved in writing a deterministic field problem:

$$\langle \mathcal{L} \rangle \langle \boldsymbol{u} \rangle = \boldsymbol{f}. \tag{1.5}$$

Basically, this form directly averages the spatial randomness and is at the basis of classical models of continuous media. It needs to be noted that the straightforward averaging of, say, elastic coefficients does not lead to correct elasticity field equations of average responses.

The second case may be viewed as a continuum generalisation of a random vibration problem, itself a special case of a random dynamical system. Thus, $\boldsymbol{f}(\omega)$ may represent a realisation of a random force (or source) field applied to a continuous body.

The third case is not very interesting as it can usually be handled through a solution of the corresponding deterministic problem.

The fourth case (1.3) is exemplified by solutions of the Navier–Stokes equation, which become so irregular as to be treated in a stochastic way (Batchelor 1951, Monin & Yaglom 2007*a*, Monin & Yaglom 2007*b*, Frisch 1995). In both cases $\mathcal{B}$ is taken as the set of all realisations $B(\omega)$ parametrised by elementary events ω of the space Ω:

$$\mathcal{B} = \{B(\omega); \omega \in \Omega\}. \tag{1.6}$$

In principle, each of the realisations follows deterministic laws of classical mechanics; probability is introduced to deal with the set (1.6).

In all the cases, the ensemble picture is termed *stochastic continuum physics.*

Why tensor-valued random fields?

Besides turbulence, another early (and continuing) field of research where *stochastic continuum physics* was found to be necessary has been stochastic wave propagation, both acoustic and electromagnetic. The starting point in classical analyses of wave propagation in random media is offered by the wave equation for a scalar field u in a spatial domain $\mathcal{D}$:

$$\nabla^2 u = \frac{1}{c^2(\omega, \boldsymbol{x})} \frac{\partial^2 u}{\partial t^2}, \qquad \omega \in \Omega, \qquad \boldsymbol{x} \in \mathcal{D}. \tag{1.7}$$

To be more specific, and without loss of generality in the discussion, we can consider (1.7) to model the anti-plane elastic wave propagation in the (x_1, x_2)-plane, so $\mathcal{D} \subset \mathbb{R}^2$, while $u \equiv u_3$ component of the *displacement vector* $\boldsymbol{u} = u_i \boldsymbol{e}_i$ in three dimensions (3D). Here the *phase velocity* c is a random field, i.e. an ensemble $\{c(\omega, \boldsymbol{x}), \omega \in \Omega, \boldsymbol{x} \in \mathcal{D}\}$. Formally speaking, we have a triple $(\Omega, \mathfrak{F}, \mathsf{P})$, where Ω is the space of elementary events, $\mathfrak{F}$ is the σ-field of possible events and P is the probability measure defined on it; ω is written explicitly to indicate the random nature of any given quantity.

It is observed that Equation 1.7 stems from

$$\nabla^2 u = \frac{\rho}{E} \frac{\partial^2 u}{\partial t^2}, \qquad \omega \in \Omega, \qquad \boldsymbol{x} \in \mathcal{D},$$

so, in principle, it can account for spatial randomness in mass density ρ only. In order also to account for randomness in the elastic modulus μ, one has to go to an earlier step in the derivation of a (linear) hyperbolic equation for a generally inhomogeneous medium. Its form, again written for one realisation, is

$$\boldsymbol{\nabla} \cdot [\mu(\omega, \boldsymbol{x}) \boldsymbol{\nabla} u] = \rho(\omega, \boldsymbol{x}) \frac{\partial^2 u}{\partial t^2}, \qquad \omega \in \Omega, \qquad \boldsymbol{x} \in \mathcal{D}. \tag{1.8}$$

Clearly, we deal with two scalar random fields: μ and ρ. While the mass density ρ is naturally a scalar, this model has a drawback in that the shear elastic modulus μ is only the isotropic characteristic (per $C_{ij} = \mu\delta_{ij}$; $i, j = 1, 2$) of a generally anisotropic, second-rank anti-plane *stiffness tensor* C $(= C_{ij}\boldsymbol{e}_i \otimes \boldsymbol{e}_j)$:

$$C_{ij} = \mathsf{C}_{3i3j}.$$

Here C_{3i3j} stems from the general fourth-rank stiffness tensor C of 3D elasticity, defined below. The tensor C is defined as a linear mapping of the space V such that:

$$C : V \to V, \quad C \in \mathsf{S}^2(V). \tag{1.9}$$

In the physical picture, this is a mapping of the anti-plane shear *strain* $u_{,i}$ into the anti-plane shear *stress* σ_j $(= \sigma_{3j})$.

With reference to Figure 0.1 of the Introduction or Figure 1.1 here, the tensor-valued random field C, adopted with whatever resolution is preferred for a problem at hand, reflects the presence of spatial inhomogeneities – either at a single crystal/grain level or at a polycrystal level. These inhomogeneities are random and locally anisotropic, with randomness vanishing in the large mesoscale (many grains) limit, assuming a homogeneity of spatial statistics, and approaching the isotropy of an effective stiffness tensor providing the isotropy of spatial statistics holds. This is the scaling trend of properties of the statistical volume element (SVE) to the deterministic property of the representative volume element (RVE),

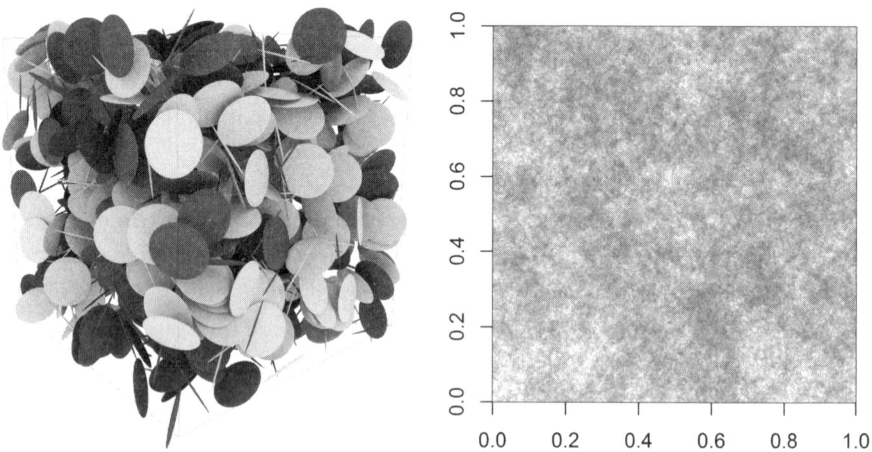

Figure 1.1 Upscaling from a finite domain (SVE) of a random two-phase composite (a) to a TRF on macroscale (b). With the SVE mesoscale increasing, the fluctuations tend to vanish and a homogeneous deterministic continuum is obtained. Figure (a) A multicomponent nanocomposite with prolate and oblate ellipsoids (Kale et al. 2018). Figure (b) was generated by the R package (gencauchy 2D function with $\alpha = 1.8$ and $\beta = 0.2$).

e.g. Ostoja-Starzewski (2008); Ostoja-Starzewski et al. (2016). Basically, the TRF hinges on the SVE concept, just as the deterministic tensor field hinges on the RVE concept; see Section 4.7.

The same arguments apply to other continuum physics problems in 2D (two dimensions, i.e. $\mathbb{R}^2$) or 3D, be they of rank 2 as well as higher tensor ranks. For example, the *piezoelectricity tensor* $\boldsymbol{D}$ $(= \boldsymbol{D}_{ijk}\boldsymbol{e}_i \otimes \boldsymbol{e}_j \otimes \boldsymbol{e}_k)$ is defined as a linear mapping of the electric field vector into the Cauchy stress:

$$\boldsymbol{D}\colon \mathsf{S}^2(V) \to V, \qquad \boldsymbol{D} \in \mathsf{S}^2(V) \otimes V. \tag{1.10}$$

Next, the *stiffness tensor* $\boldsymbol{C}$ $(= \boldsymbol{C}_{ijkl}\boldsymbol{e}_i \otimes \boldsymbol{e}_j \otimes \boldsymbol{e}_k \otimes \boldsymbol{e}_l)$, is defined as a symmetric linear mapping of the strain into the Cauchy stress:

$$\boldsymbol{C}\colon \mathsf{S}^2(V) \to \mathsf{S}^2(V), \qquad \boldsymbol{C} \in \mathsf{S}^2\left(\mathsf{S}^2(V)\right). \tag{1.11}$$

It being now agreed that various continuum physics models of constitutive properties are rank 2, 3 and 4 tensor-valued random fields, we face this challenge:

What specific models to assume?

In classical theories of wave propagation in random media, one assumes harmonic time dependence ($e^{i\gamma t}$), which, when applied to (1.7), readily leads to the *stochastic Helmholtz equation*:

$$\nabla^2 u + k(\omega, \boldsymbol{x})u = 0, \qquad \omega \in \Omega, \qquad \boldsymbol{x} \in \mathcal{D}. \tag{1.12}$$

Next, setting $k(\omega, \boldsymbol{x}) = k_0^2 n^2(\omega, \boldsymbol{x})$, one introduces a random *wave number* to deal with the spatial randomness of the medium. Thus, $k_0 = \gamma/c_0$ is the wave number of a *reference homogeneous medium* where c_0 is its *phase velocity*, and $n(\boldsymbol{x}, \omega)$ is a random *index of refraction*. Hereinafter, we employ γ for the frequency, rather than the conventional ω, which has already been reserved to denote an outcome ω (i.e. a random medium's realisation) from the sample space Ω.

Equation (1.12) is a valid Ansatz whenever the time variation in the refractive properties of the medium is much slower than the wave propagation itself; thus, for example, swirling as rapid as the wave motion violates the monochromaticity assumption. The *random field* $n(\omega, \boldsymbol{x})$, $\omega \in \Omega$, $\boldsymbol{x} \in \mathcal{D}$ is determined from experimental measurements. The conventional model has the form (with $\langle\cdot\rangle$ indicating the ensemble/statistical average):

$$n^2(\omega, \boldsymbol{x}) = 1 + \varepsilon\mu(\omega, \boldsymbol{x}), \tag{1.13a}$$

$$\langle\mu(\omega, \boldsymbol{x})\rangle = 0, \tag{1.13b}$$

$$\mu = O(1), \tag{1.13c}$$

so that all the randomness is present in the zero-mean random field μ. Note that Rytov, Kravtsov & Tatarskiĭ (1987) take $n(\omega, \boldsymbol{x}) = 1 + \epsilon\mu(\omega, \boldsymbol{x})$. This classical notation employing μ is not to be confused with the shear modulus above.

The key rôle in setting up a random field is played by a correlation function of μ which, given (1.13b), is

$$C_\mu(\boldsymbol{x}, \boldsymbol{x}') = \langle \mu(\boldsymbol{x}) \mu(\boldsymbol{x}') \rangle.$$

Usually, μ is taken as a *wide-sense homogeneous* random field

$$C_\mu(\boldsymbol{x}, \boldsymbol{x}') = C_\mu(\boldsymbol{x} - \boldsymbol{x}') < \infty, \quad \forall \boldsymbol{x}, \boldsymbol{x}',$$

possessing an ergodic property almost surely in Ω and $\mathcal{X}$

$$\langle \mu(\boldsymbol{x}) \rangle = \overline{\mu(\omega)}.$$

Rytov et al. (1987) also discuss more general random field models such as, say, those with stationary increments.

A special class of *statistically isotropic random fields* occurs when $\rho(\boldsymbol{x})$ depends only on the magnitude, but not direction, of the vector $\boldsymbol{x}$:

$$C_\mu(\boldsymbol{x}) = C_\mu(r), \quad r = \|\boldsymbol{x}\| = \sqrt{x_i x_i}.$$

A very common model for the correlation coefficient is the Gaussian form

$$C_\mu(r) = \langle \mu^2 \rangle \exp[-r^2/a^2],$$

where a is the so-called *correlation radius*. Another common model fit is

$$C_\mu(r) = \langle \mu^2 \rangle \exp[-r/a],$$

although one must bear in mind that it corresponds to random fields with discontinuous, rather than continuous, realisations (e.g. a granular/cellular structure of a polycrystal).

Considering that a wide range of natural phenomena – including geological formations, atmospheric turbulence or seismic motions, to name a few – possess fractal and Hurst characteristics (Helland & Atta 1978; Mandelbrot 1982), two powerful random field models have been developed over the past two decades: Cauchy and Dagum (Gneiting & Schlather 2004; Mateu et al. 2007; Porcu et al. 2007; Porcu & Stein 2012). In general, the fractal dimension can be described as a roughness measure that ranges from 2 to 3 in value for 2D systems. Typically, the larger the fractal dimension, the rougher the field's realisations. The Hurst parameter, or Hurst exponent, is the long-range persistence of a system. In general, if $H \in (0, 0.5)$ the system is said to be anti-persistent (e.g. an increase in value is typically followed by a decrease, or vice versa). If $H \in (0.5, 1)$, the system is said to be persistent (e.g. an increase is typically followed by an increase, or vice versa). For $H = 0.5$ the system is said to reflect a true random walk without any long-range persistence.

The Cauchy RF covariance function is

$$C_C(r) = (1 + r^\alpha)^{-\frac{\beta}{\alpha}}, \quad r \geq 0,\ \alpha \in (0, 2],\ \beta > 0.$$

The Dagum RF covariance function is

$$C_D(r) = 1 - (1 + r^{-\beta})^{-\frac{\alpha}{\beta}}, \quad r \geq 0,\ \alpha \in (0,1),\ \beta \in (0,1].$$

These two random fields are unique in that they capture and decouple spatial features with a fractal dimension (D) and Hurst parameter (H). The relationships linking D, H with d, α and β for both Cauchy and Dagum RFs (Mateu et al. 2007) are given by

$$D = d + 1 - \frac{\alpha}{2}, \quad \text{and} \quad H = 1 - \frac{\beta}{2},$$

where d, for two dimensions, is equal to 2. Unfortunately, due to the restrictions on α and β, Dagum RFs cannot capture as many pairs of D and H as compared to Cauchy RFs.

These considerations indicate another challenge: find the most realistic models of correlation functions of tensor-valued random fields.

1.1.2 Equations of Continuum Mechanics

In this book we focus on two types of tensor fields where randomness and local anisotropy occur point-wise with probability one: (i) dependent fields, subject to *balance (governing) equations* and (ii) fields of material properties, appearing in *constitutive models* (e.g. conductivity or stiffness).

Balance equations

The formulation of balance equations of continuum mechanics first involves the writing of a conservation law for a finite body of a continuum (integral form) and then applying a localisation procedure to obtain an equation on infinitesimal level (local form). Proceeding in the spatial (Eulerian) description, we first give the *Reynolds transport theorem*:

$$\frac{d}{dt}\int_{\mathcal{B}} P dV = \int_{\mathcal{B}} \left[\frac{\partial P}{\partial t} + (v_k P)_{,k} \right] dV. \tag{1.14}$$

The conservation of mass

$$\frac{d}{dt}\int_{\mathcal{B}} \rho dV = 0, \tag{1.15}$$

with v_k being the velocity field, leads to the local form

$$\frac{\partial \rho}{\partial t} + (v_k \rho)_{,k} = \frac{d\rho}{dt} + \rho v_{k,k} = 0. \tag{1.16}$$

The conservation of linear momentum

$$\frac{d}{dt}\int_{\mathcal{B}} \rho v_i dV = F_i^{\mathcal{B}} + F_i^{S} \quad \text{with} \quad F_i^{\mathcal{B}} = \int_{\partial\mathcal{B}} b_i dV, \quad F_i^{S} = \int_{\partial\mathcal{B}} \sigma_{ki} n_k dS, \tag{1.17}$$

where the last equality relies on the Cauchy theorem, invoking the concept of the Cauchy stress σ_{ki}. This leads to the local form

$$\rho \frac{dv_i}{dt} = b_i + \sigma_{ki,k} . \tag{1.18}$$

The conservation of angular momentum

$$\frac{d}{dt} \int_{\mathcal{B}} \rho \boldsymbol{\varepsilon}_{ijk} x_j v_k dV = \int_{\mathcal{B}} \boldsymbol{\varepsilon}_{ijk} x_j b_k dV + \int_{\partial \mathcal{B}} \boldsymbol{\varepsilon}_{ijk} x_j \sigma_{lk} n_l dS, \tag{1.19}$$

where $\boldsymbol{\varepsilon}_{ijk}$ is the Levi-Civita permutation tensor, on account of (1.14) and (1.18), leads to the symmetry of the Cauchy stress:

$$\sigma_{jk} = \sigma_{kj}. \tag{1.20}$$

While the above employs the standard approach to deriving (1.18) and (1.20), an independent way of arriving at the same equations is through the invariance of energy with respect to arbitrary rigid body translations and rotations.

The conservation of energy

$$\frac{d\mathcal{K}}{dt} + \frac{d\mathcal{U}}{dt} = \frac{1}{2} \int_{\mathcal{B}} \rho \frac{d(v_i v_i)}{dt} dV + \int_{\mathcal{B}} \rho \frac{du}{dt} dV, \tag{1.21}$$

where the kinetic and internal energies are

$$\mathcal{K} = \frac{1}{2} \int_{\mathcal{B}} \rho v_i v_i dV, \quad \mathcal{U} = \int_{\mathcal{B}} \rho u dV,$$

while introducing the heat flux q_i $(= \boldsymbol{q})$ and (for simplicity) neglecting any heat sources/sinks, leads to the local form

$$\rho \frac{du}{dt} = \sigma_{lk} \left(v_{k,l} \right) - q_{i,i} . \tag{1.22}$$

The second law of thermodynamics written for a finite body $\mathcal{B}$, involving the entropy density s,

$$\frac{d}{dt} \int_{\mathcal{B}} \rho s dV \geq - \int_{\mathcal{B}} \frac{q_k}{T} n_k dS, \tag{1.23}$$

leads to the local form in terms of the reversible ($\dot{s}^{(r)}$) and irreversible ($\dot{s}^{(i)}$) entropy production rates

$$\begin{gathered} \dot{s} = \dot{s}^{(r)} + \dot{s}^{(i)} \quad \text{with} \\ \dot{s}^{(r)} = - \left(q_i / T \right)_{,i} \quad \text{and} \quad \dot{s}^{(i)} := \frac{1}{\rho T} \left(\sigma_{ij}^{(d)} d_{ij} + \beta_{lk}^{(d)} \dot{\alpha}_{lk} - \frac{T_{,i}}{T} q_i \right) \geq 0. \end{gathered} \tag{1.24}$$

This form, typical of the so-called *thermomechanics with internal variables* (TIV) (see Section 1.4), relies on the definition of the free energy density (ψ) at the local level

$$\psi := u - Ts, \tag{1.25}$$

which serves as the potential for the quasi-conservative Cauchy stress, the entropy and the quasi-conservative internal stress:

$$\sigma_{ij}^{(q)} := \rho \frac{\partial \psi}{\partial \varepsilon_{ij}}, \qquad s := - \frac{\partial \psi}{\partial T}, \qquad \beta_{ij}^{(q)} := \rho \frac{\partial \psi}{\partial \varepsilon_{ij}}.$$

The inequality (1.24) also involves the dissipative parts of the total Cauchy stress (σ_{ij}) and the total internal stress (β_{ij})

$$\sigma_{ij}^{(d)} := \sigma_{ij} - \sigma_{ij}^{(q)}, \qquad \beta_{ij}^{(d)} := -\beta_{ij}^{(q)}.$$

The kinematic quantities of these, conjugate to $\sigma_{ij}^{(d)}$ $(= \boldsymbol{\sigma}^{(d)})$ and $\beta_{ij}^{(d)}$ $(= \boldsymbol{\beta}^{(d)})$, respectively, are the deformation rate $d_{ij} := v_{(i,j)}$ $(= \boldsymbol{d})$ and the rate $\dot{\alpha}_{ij}$ $(= \dot{\boldsymbol{\alpha}})$ of internal strain tensor α_{ij} $(= \boldsymbol{\alpha})$.

The basic continuum equations (1.16), (1.18), (1.20) and (1.22) are to be satisfied by the dependent fields, with (1.24) serving as the restriction on the constitutive behaviours.

The above scheme, while commonly accepted in classical continuum mechanics, hides several basic assumptions:

- Neglect of the inner structure of a continuum point. Various more realistic models are possible within the deterministic continuum physics; see Maugin (2017) for a comprehensive review. For example, working with a micropolar continuum, one treats the continuum point as analogous to a rigid body and has to admit its microinertia, body moment, couple stress tensor, curvature-torsion tensor, etc.
- The separation of scales

$$d \ll L \ll L_{\text{macro}}, \tag{1.26}$$

  where d is the microscale, L the scale of a continuum point, and L_{macro} the macroscale (macroscopic dimension of the body). The left-side inequality (sometimes simply replaced by $<$) allows one to postulate the existence of a Representative Volume Element (RVE) of continuum mechanics, so that $L = \sqrt[3]{dV}$, the size of infinitesimal volume. The second inequality covers the range of length scales where conventional continuum mechanics applies – this is the domain of spatial dependence of stress, strain and displacement fields one is interested in when solving an initial-boundary value problem. Also, this is where (1.1), with $\mathcal{L}$ being a deterministic operator, applies; the classical Navier equation of elasticity is an example.
- Neglect of any possibility of spontaneous violations of the Second Law of thermodynamics on nanoscales, where continuum physics models may still be applied.

Constitutive relations

The specification of these relations involves the physics of the medium. Henceforth, the review of constitutive models is being done from one common standpoint based on the free (or internal) energy and the dissipation function, this pair being typical of TIV, which itself is based on the *thermomechanics of irreversible processes* (de Groot & Mazur 1984). This strategy allows the treatment of a very wide range of continuum-type phenomena within the Second Law of thermodynamics *as well as* beyond that law, as dictated by the developments in statistical mechanics beginning with Evans, Cohen & Morriss (1993): on very small scales where spontaneous violations of that law occur (Sections 4.2 and 4.3). Adopting the TIV approach also allows (i) scale-dependent homogenisation in the vein of Hill–Mandel condition, and (ii) accounting for transmission of signals at finite wave speeds such as in hyperbolic thermoelasticity. Proceeding with two basic TIV formulations – Ziegler's thermodynamic orthogonality and Edelen's primitive thermomechanics – we give the basic governing equations of continua.

The following sections give a review of basic equations of several continuum physics theories, with a goal of providing a reference on tensor fields with locally anisotropic realisations, which are candidates for stochastic continuum generalisations. There is no attempt at giving a comprehensive background for all of continuum physics. The outlined models provide a reference for determining, in Chapter 4, the restrictions on TRFs dictated by the continuum physics, while the most general representations of TRFs will first be obtained in Chapter 3, based on a necessary mathematical background outlined in Chapter 2.

1.2 Conductivity

Consider the thermal conduction in a rigid (undeformable) material occupying a domain $\mathcal{D} \subset E^d$ $(d = 1, 2, 3)$ with boundary $\partial\mathcal{D}$. The energy balance equation (1.22) reduces to

$$q_{i,i} = \rho c \dot{T} \tag{1.27}$$

with the Fourier law

$$q_i = -k_{ij} T_{,j} \tag{1.28}$$

where k_{ij} is a generally anisotropic conductivity, T is the absolute temperature and $T_{,j}$ its gradient.

These two equations lead to a *diffusion equation* for heat conduction in terms of the temperature T:

$$-\left(k_{ij} T_{,j}\right)_{,i} = \rho c \dot{T}. \tag{1.29}$$

Dropping the time dependence and assuming an isotropic response analogous to what was implied in (1.8)

$$k_{ij} = k\delta_{ij}, \tag{1.30}$$

leads to the classical elliptic equation $\left(kT_{,i}\right)_{,i} = f$. However, if we keep the local anisotropy present, a boundary value problem with Dirichlet boundary conditions is (Lord et al. 2014):

$$\begin{aligned} -\left(k_{ij} T_{,j}\right)_{,i} &= f(\boldsymbol{x}) \quad && \forall \boldsymbol{x} \in \mathcal{D}, \\ T(\boldsymbol{x}) &= g(\boldsymbol{x}) && \forall \boldsymbol{x} \in \partial\mathcal{D}. \end{aligned} \tag{1.31}$$

With $r \in C^1\left(\overline{\mathcal{D}}\right)$, the classical solution is of the class

$$T \in C^2(\mathcal{D}) \cap C^1\left(\overline{\mathcal{D}}\right).$$

Table 1.1 *Diverse physical problems involving rank 2 constitutive tensors and governed by the elliptic equation or corresponding generalisations to diffusion or wave equations*

Physical phenomenon	T	∇T	$\boldsymbol{k}$	$\boldsymbol{q}$
Thermal conductivity	temperature	thermal gradient	thermal conductivity	heat flux
Anti-plane elasticity	temperature	strain	elastic moduli	stress
Torsion	stress function	strain	shear moduli	stress
Electrical conduction	potential	intensity	electrical conductivity	current density
Electrostatics	potential	intensity	permittivity	electric induction
Magnetostatics	potential	intensity	magnetic permeability	magnetic induction
Diffusion	concentration	gradient	diffusivity	flux

In practical applications, oftentimes, $r \in L^2\left(\overline{\mathcal{D}}\right)$, so that the derivatives in (1.31) have to be interpreted in the weak sense (i.e. belonging to the Sobolev space). Then, $T \in H^2\left(\mathcal{D}\right) \cap H_0^1\left(\overline{\mathcal{D}}\right)$ and $g \in H^{1/2}\left(\partial\mathcal{D}\right)$.

In stochastic problems, our functions are defined on a Cartesian product of the domain $\mathcal{D} \subset E^d$ with the probability space $(\Omega, \mathfrak{F}, \mathsf{P})$. Then, $\boldsymbol{k} = k_{ij}\boldsymbol{e}_i \otimes \boldsymbol{e}_i$ is the second-rank tensor field and also a second-order random field taking values in $\mathsf{S}^2(V)$ (recall Equation (1.9)), such that

$$\boldsymbol{k}\colon \mathcal{D} \times \Omega \to \mathsf{V}^2 \tag{1.32}$$

are functions belonging to $L^2\left(\Omega, \mathfrak{F}, L^2\left(\mathcal{D}\right)\right)$ that satisfy

$$\|\boldsymbol{k}\|^2_{L^2(\Omega,\mathfrak{F},L^2(\mathcal{D}))} = \int_\Omega \int_\mathcal{D} \boldsymbol{k}\left(\boldsymbol{x},\omega\right) : \boldsymbol{k}\left(\boldsymbol{x},\omega\right) d\boldsymbol{x} d\mathsf{P} = \mathsf{E}\left[\|\boldsymbol{k}\|^2_{L^2(\mathcal{D})}\right] < \infty.$$

Next, the source/sink function is a real-valued second-order random field

$$r\colon \mathcal{D} \times \Omega \to \mathbb{R}$$

belonging to $L^2\left(\Omega, \mathfrak{F}, L^2\left(\mathcal{D}\right)\right)$ and having the norm

$$\|r\|^2_{L^2(\Omega,\mathfrak{F},L^2(\mathcal{D}))} = \int_\Omega \int_\mathcal{D} r^2\left(\boldsymbol{x},\omega\right) d\boldsymbol{x} d\mathsf{P} = \mathsf{E}\left[\|r\|^2_{L^2(\mathcal{D})}\right] < \infty.$$

Then the stochastic boundary value problem is to find $T\colon \mathcal{D} \times \Omega \to \mathbb{R}$ such that P-a.s. (1.31) is satisfied. We return to stochastic partial differential equations in Section 4.8.

Several analogues of the stochastic conductivity problems described by elliptic equations of this type are given in Table 1.1. All of these theories are of local type, while various, more complex models are continued in Sections 1.3–1.5. Basic aspects of generalised continuum theories are introduced in Section 1.7.

1.3 Elasticity

1.3.1 General 3D Case

The linear elastic body occupies the domain $\mathcal{D}$ with the boundary $\partial\mathcal{D}$. It is defined by a generally anisotropic *Hooke law*

$$\sigma_{ij} = \mathsf{C}_{ijkl}\varepsilon_{kl} \quad \text{or} \quad \sigma = \mathsf{C} : \varepsilon \tag{1.33}$$

where the strain is subject to the strain-displacement equation

$$\varepsilon_{ij} = u_{(i,j)} \, . \tag{1.34}$$

The stiffness tensor satisfies two minor and one major symmetry relations

$$\mathsf{C}_{ijkl} = \mathsf{C}_{jikl} = \mathsf{C}_{ijlk} = \mathsf{C}_{klij}, \tag{1.35}$$

where the first and second equalities are dictated by symmetries of the strain and stress tensor, respectively, while the third is the consequence of the hyperelasticity postulate. C_{ijkl} satisfies the positive-definiteness condition

$$\varepsilon_{ij}\mathsf{C}_{ijkl}\varepsilon_{kl} > 0. \tag{1.36}$$

In the picture dual to (1.33), we have

$$\varepsilon_{ij} = \mathsf{S}_{ijkl}\sigma_{kl} \quad \text{or} \quad \varepsilon = \mathsf{S} : \boldsymbol{\sigma} \tag{1.37}$$

where $\mathsf{S} \colon \mathcal{D} \times \Omega \to V^{\otimes 4}$ is the compliance tensor. With

$$\mathsf{C}_{ijkl}\mathsf{S}_{ijkl} = 6 \quad \text{or} \quad \mathsf{CS} = \mathsf{I}, \tag{1.38}$$

S satisfies the same symmetries as those in (1.35).

The linear momentum equation (1.18) reduces to a vector equation for the displacement field u_i:

$$\left(\mathsf{C}_{ijkl}u_{(k,l)}\right)_{,j} + f_i = \rho\ddot{u}_i. \tag{1.39}$$

Assuming an isotropic elastic response

$$\mathsf{C}_{ijkl} = \lambda\delta_{ij}\delta_{kl} + \mu\left(\delta_{ik}\delta_{jl} + \delta_{il}\delta_{jk}\right)$$

where λ and μ are the *Lamé constants*, simplifies (1.33)

$$\sigma_{ij} = \lambda\varepsilon_{kk}\delta_{ij} + 2\mu\varepsilon_{ij}, \tag{1.40}$$

and also simplifies (1.39) to (recall Equation (0.5)):

$$\mu u_{i,jj} + (\lambda + \mu)\, u_{j,ji} + \mu_{,j}\left(u_{j,i} + u_{i,j}\right) + \lambda_{,i}\, u_{j,j} + F_i = \rho\ddot{u}_i. \tag{1.41}$$

While this equation still allows a smooth spatial inhomogeneity of λ and μ, assuming them to be constant leads to the classical *Navier equation of motion*,

$$\mu u_{i,jj} + (\lambda + \mu)\, u_{j,ji} + f_i = \rho\ddot{u}_i. \tag{1.42}$$

It is generally required that

$$u_i(\boldsymbol{x},t) \in C^2(\mathcal{D},T) \cap C^1(\overline{\mathcal{D}},T)$$
$$f_i(\boldsymbol{x},t) \in C^1(\overline{\mathcal{D}},T).$$

While the above formulation of elastodynamics in terms of displacements (and its various, e.g. anisotropic generalisations) harks back to the memoir read to the French Academy of Science on 14 May, 1821 by C. L. M. H. Navier and published in Navier (1827), a pure stress-based formulation was developed in the 1950s and 1960s, e.g. Ignaczak (1963). Thus, elimination of u_i and ε_{ij} from Equations (1.34), (1.18) and (1.37) leads to the *Ignaczak equation of elastodynamics* (Ostoja-Starzewski 2018):

$$\left(\rho^{-1}\sigma_{(ik,k}\right)_{,j)} - \boldsymbol{S}_{ijkl}\ddot{\sigma}_{kl} + \left(\rho^{-1}F_{(i}\right)_{,j)} = 0, \tag{1.43}$$

where

$$F_{ij} = F_{i,j} + F_{j,i} + \frac{\lambda}{\lambda+2\mu}F_{k,k}\,\delta_{ij}.$$

In the case of pointwise-isotropy, (1.43) reduces to

$$\left(\rho^{-1}\sigma_{(ik,k}\right)_{,j)} - \frac{1}{2\mu}\left(\ddot{\sigma}_{ij} - \frac{\lambda}{3\lambda+2\mu}\ddot{\sigma}_{kk}\delta_{ij}\right) + \left(\rho^{-1}F_{(i}\right)_{,j)} = 0.$$

Note that, in contradistinction to the displacement formulation (1.39) and (1.41), this formulation avoids gradients of compliance but introduces gradients of mass density.

1.3.2 2D Cases

In-plane elasticity

Assuming the body forces and all the dependent fields to be independent of one coordinate, say, x_3, the linear momentum equation (1.18) and the Hooke law – either (1.33) or (1.40) – hold providing $i, j = 1, 2$. However, care has to be exercised in distinguishing between the *plane strain* and *plane stress*. In the first case, assuming isotropy, these equations as well as (1.42) hold. In the second case, also assuming isotropy, the Hooke law is typically written in the form

$$\sigma_{ij} = \frac{2\mu\lambda}{\lambda+3\mu}\varepsilon_{kk}\delta_{ij} + 2\mu\varepsilon_{ij}.$$

Note that both cases can be written jointly as special cases of the planar elasticity involving the strains $(\varepsilon_{11}, \varepsilon_{22}, \varepsilon_{12})$, the stresses $(\sigma_{11}, \sigma_{22}, \sigma_{12})$, and one compatibility equation:

$$\varepsilon_{22,11} + \varepsilon_{11,22} = 2\varepsilon_{12,12}. \tag{1.44}$$

The isotropic Hooke law can be written as

$$\sigma_{ij} = \lambda_{2D}\varepsilon_{kk}\delta_{ij} + 2\mu_{2D}\varepsilon_{ij}, \qquad i,j,k = 1,2,$$

where λ_{2D} and μ_{2D} are the *planar Lamé constants*; see Ostoja-Starzewski (2008, chapter 5) for a more complete discussion of planar (in-plane) versus 3D elasticity theories. Henceforth, one can proceed with simpler models than those in the 3D case.

Anti-plane elasticity

As discussed earlier, there is only one degree of freedom per continuum point: the anti-plane displacement $u \equiv u_3(x_1, x_2)$, and hence there is only one non-zero component of the body force: $f \equiv f_3$. The balance of linear momentum (1.18) reduces to $\sigma_{j,j} + f = \rho\ddot{u}$ $(j = 1, 2)$ leading to the governing equation

$$[(C_{ij}u_{,j})_{,i} + f = \rho\ddot{u}, \qquad j = 1, 2, \tag{1.45}$$

where C_{ij} is the second-rank second-order tensor random field (1.9), i.e. the anti-plane part $\boldsymbol{C}_{i3j3}$ of $\boldsymbol{C}_{ikjl}$. In the static case, this model is essentially the same as (1.29) for the in-plane conductivity. For an isotropic material $(C_{ij} = \mu\delta_{ij})$,

$$(\mu u_{,i})_{,i} + f = \rho\ddot{u},$$

which, in the special case of spatial homogeneity of μ (i.e. no randomness), reduces to a scalar wave equation with body force field: $\mu u_{,ii} + \rho b = \rho\ddot{u}$. This is the dynamic version of the elasticity of the second row in Table 1.1. Assuming $\ddot{u} = 0$ gives the anti-plane elastostatics. Assuming $f = 0$, the linear wave equation $c^2 u_{,ii} = \ddot{u}$ (with the wave speed $c = \sqrt{\mu/\rho}$) is obtained, which is the basis of the stochastic Helmholtz equation (1.12) above.

1.4 Thermomechanics with Internal Variables (TIV)

1.4.1 Dissipation Function in Thermomechanics within Second Law

This type of continuum theory builds on Onsager's (1931) irreversible thermodynamics, which is based on the assumption that the entropy and internal energy in the equilibrium state serve as a reference for the non-equilibrium state. The approach originated by Ziegler (1983) has led to thermodynamics of irreversible processes (TIP) (de Groot & Mazur 1984) and then to the thermomechanics with internal variables (TIV) relying on two functionals

$$\text{free energy} \quad \psi(\boldsymbol{\varepsilon}, T) \quad \text{dissipation function} \quad \phi(\boldsymbol{d}, \dot{\boldsymbol{q}}, \dot{\boldsymbol{\alpha}}). \tag{1.46}$$

It is postulated that these two functionals describe the entire constitutive behaviour of the material: $\psi(\boldsymbol{\varepsilon}, T)$ the conservative and $\phi(\boldsymbol{d}, \dot{\boldsymbol{q}}, \dot{\boldsymbol{\alpha}})$ the dissipative.

Note: Relations (1.46) display a generic choice of arguments – depending on the phenomenon studied, there may be other/more variables involved.

The *Second Law of thermodynamics* is written in terms of the reversible $(\dot{s}^{(r)})$ and irreversible $(\dot{s}^{(i)})$ parts of entropy production rate $(\dot{s})$ (Ziegler 1983; Ziegler & Wehrli 1987; Maugin 1999):

$$\dot{s} = \dot{s}^{(r)} + \dot{s}^{(i)} \quad \text{with} \tag{1.47a}$$

$$\dot{s}^{(r)} = -\frac{\partial\,(q_i/T)}{\partial x_i} \quad \text{and} \tag{1.47b}$$

$$\dot{s}^{(i)} \geq 0, \tag{1.47c}$$

where T is the absolute temperature while q_i ($\boldsymbol{q}$) is the heat flux. This form of writing the Second Law naturally involves the dissipation function $\phi = \dot{s}^{(i)}$, which, in principle, provides a constitutive law for the irreversible part of the response.

Setting $\boldsymbol{Y} = \left(\boldsymbol{\sigma}^{(d)}, \boldsymbol{\nabla} T, \boldsymbol{\beta}^{(d)}\right)$ as the vector of *dissipative forces* (with $\boldsymbol{\sigma}^{(d)}$ being the *dissipative stress* and $\boldsymbol{\beta}^{(d)}$ the *dissipative internal stress*) dual to the vector of velocities $\boldsymbol{V} = (\boldsymbol{d}, \dot{\boldsymbol{q}}, \dot{\boldsymbol{\alpha}})$, the *dissipation function* ϕ is defined by

$$\phi(\boldsymbol{V}) = \boldsymbol{Y} \cdot \boldsymbol{V} = \dot{s}^{(i)} \geq 0. \tag{1.48}$$

$\boldsymbol{Y}$ is given by the gradient in the velocity space:

$$\boldsymbol{Y} = \lambda \nabla_{\boldsymbol{v}} \phi \quad \text{i.e. } Y_i = \lambda \frac{\partial \phi\,(\boldsymbol{V})}{\partial V_i} \quad \text{where} \quad \lambda = \left(V_i \frac{\partial \phi\,(\boldsymbol{V})}{\partial V_i} \right)^{-1} \phi. \tag{1.49}$$

Here, λ is the Lagrangean multiplier in an extremum principle underlying this approach, while the classical *Onsager reciprocity relations* hold:

$$\frac{\partial Y_i(\boldsymbol{V})}{\partial V_j} = \frac{\partial Y_j(\boldsymbol{V})}{\partial V_i}.$$

Clearly, the linear relation $\boldsymbol{Y} = \boldsymbol{L} \cdot \boldsymbol{V}$ of Onsager's thermodynamics is obtained as a special case.

Typically, $\phi = \boldsymbol{Y} \cdot \boldsymbol{V}$ is taken as a functional of the velocity $\boldsymbol{V}$ so as to obtain the dissipative force $\boldsymbol{Y}$:

$$Y_i = \lambda \frac{\partial \phi\,(\boldsymbol{V})}{\partial V_i} \quad \text{where} \quad \lambda = \left(V_i \frac{\partial \phi\,(\boldsymbol{V})}{\partial V_i} \right)^{-1} \phi. \tag{1.50}$$

Effectively, this means that, provided the dissipative force $\boldsymbol{Y}$ is prescribed, the actual velocity maximises the *dissipation rate* $\dot{s}^{(i)} = \boldsymbol{Y} \cdot \boldsymbol{V}$ subject to the side condition (1.48).

This is not the only extremum-type interpretation of (1.49), others being e.g. the principle of least velocity, the principle of least dissipative force. The approach hinges on the so-called *thermodynamic orthogonality* postulating a maximisation of entropy production in non-linear processes, just as it is maximised in linear processes.

While (1.49) applies to a very wide range of linear and non-linear material behaviours, it does not cover all of them. This is accomplished by the *primitive thermodynamics* of (Edelen 1973; Edelen 1974), in which the most general solution of the inequality (1.48c) is based on a decomposition theorem: assuming $\boldsymbol{Y} = \boldsymbol{Y}\,(\boldsymbol{V})$, there always exist functions $\varphi\,(\boldsymbol{V})$ and $\boldsymbol{U}\,(\boldsymbol{V})$ such that

$$Y_i = \frac{\partial \varphi (\boldsymbol{V})}{\partial V_i} + U_i \tag{1.51}$$

with

$$\varphi (\boldsymbol{V}) \equiv \int_0^1 V_i Y_i (\tau \boldsymbol{V}) \, d\tau$$

and

$$\boldsymbol{Y} \cdot \boldsymbol{U} = 0, \quad U_i (\boldsymbol{V}) \equiv \int_0^1 \tau V_j \left[\frac{\partial Y_j (\tau \boldsymbol{V})}{\partial (\tau V_i)} - \frac{\partial Y_i (\tau \boldsymbol{V})}{\partial (\tau V_j)} \right] d\tau. \tag{1.52}$$

Given (1.52), $\boldsymbol{U}(\boldsymbol{V})$ is called the *non-dissipative* (or *powerless*) vector; also $\boldsymbol{U}(\boldsymbol{0}) = \boldsymbol{0}$. The Maxwell–Cattaneo heat conduction is an example of a process derivable by this approach.

By analogy to the rôle played by the free energy ψ for quasi-conservative processes (such as hyperelasticity), ϕ or φ plays the rôle of a potential for all the dissipative processes, which are then called *hyperdissipative* (Goddard 2014). Effectively, the functional $\phi(\boldsymbol{V})$ or $\phi(\boldsymbol{Y})$ is employed to derive the constitutive laws of continua. In general, there is no perfectly-established (and uniformly agreed-upon) rule to decide whether $\boldsymbol{V}$ or $\boldsymbol{Y}$ should be the argument of the functional φ or ϕ.

Note that ϕ, but not φ, is directly equal to the dissipation $s^{*(i)}$. Clearly, in the linear response regime, $\varphi = \phi/2$ providing $\boldsymbol{U} = \boldsymbol{0}$.

1.4.2 Dissipation Function in Statistical Physics Beyond Second Law

The inequality in (1.48) is assumed to hold always (i.e. for $\forall t$) in conventional continuum physics, whereas in contemporary statistical physics (e.g. Evans & Searles 2002) the Second Law is replaced by the *fluctuation theorem*. It gives the ratio of probabilities of observing processes that have, respectively, positive (A) and negative ($-A$) total dissipations in non-equilibrium systems:

$$\frac{\mathsf{P}(\phi_t = A)}{\mathsf{P}(\phi_t = -A)} = e^{At}. \tag{1.53}$$

The theorem is illustrated with the help of Figure 1.2(e) and (f) and explained a little further below. In (1.53) ϕ_t is the total dissipation for a trajectory $\boldsymbol{\Gamma} \equiv \{q_1, p_1, \ldots, q_N, p_N\}$ of N particles originating at $\boldsymbol{\Gamma}(0)$ and evolving for a time t:

$$\phi_t (\boldsymbol{\Gamma}(0)) = \int_0^t \phi (\boldsymbol{\Gamma}(s)) \, ds. \tag{1.54}$$

The integral in (1.54) involves an *instantaneous dissipation function*:

$$\phi (\boldsymbol{\Gamma}(t)) = \frac{d\phi_t (\boldsymbol{\Gamma}(0))}{dt}. \tag{1.55}$$

The Second Law of thermodynamics is recovered upon ensemble averaging, time averaging or upscaling.

To better explain the mechanics behind Figure 1.2, consider a molecular dynamics simulation using LAMMPS software (Plimpton 1995) involving 50 particles, with Lennard–Jones (LJ) (6–12) potential interactions, in a Couette flow, Figure 1.2(a). In a channel of width L, the bottom plate is stationary while the top plate is moving at a constant speed ($v_1 = d_{12}L$) to the right (x_1), with periodic boundary conditions assumed on the left and right vertical boundaries. The fluid is thermostatted at a target temperature of $T_o = 1$ in LJ units. Figure 1.2(c) gives the time history of the resulting shear stress σ_{21} (directly obtained from LAMMPS), clearly showing its spontaneous, random-like fluctuations exhibiting negative(!) excursions. In a continuum approximation, the dissipation (i.e. entropy production) rate is computed, in general, from the trace of deformation rate $\boldsymbol{d}$ (d_{ij}) with the Cauchy stress tensor $\boldsymbol{\sigma}$ (σ_{ij}), i.e. $\dot{s} = \boldsymbol{d} : \boldsymbol{\sigma}$, which given the simple geometry here takes the simple form

$$\dot{s} = \boldsymbol{d} : \boldsymbol{\sigma} = d_{12}\sigma_{12}.$$

Since σ_{12} sometimes becomes negative, the probability of negative entropy production is non-zero, as reflected by the shear stress history in the plot of Figure 1.2(e).

Next, if the same type of experiment is conducted in a channel containing a larger number of particles – Figure 1.2(b) – the spontaneous, random-like fluctuations decrease – Figure 1.2(d) – and the negative excursions and negative entropy production are statistically less likely – Figure 1.2(f). As the number of particles increases further, the fluctuations become ever smaller and the shear stress history tends to a constant – the violations of Second Law tend to be improbable: the deterministic Stokesian continuum is recovered.

In view of (1.49) above, the dissipation function is a stochastic (not deterministic) quantity which possibly and spontaneously takes negative values, so that the positive-definiteness does not absolutely hold. Therefore, we change (1.48) to

$$\phi\left(\boldsymbol{V},\omega\right) = \boldsymbol{Y}\left(\omega\right)\cdot\boldsymbol{V} = s^{*(i)}, \qquad \omega \in \Omega, \tag{1.56}$$

where $\boldsymbol{Y}\left(\omega\right)$ are the dissipative forces conjugate to $\boldsymbol{V}$, while Ω is the set of all possible outcomes. Thus, the argument ω in (1.56) indicates that $\phi(\boldsymbol{V},\omega)$ is a stochastic functional, while $\boldsymbol{Y}\left(\omega\right)$ is a random quantity for a non-random (prescribed) velocity $\boldsymbol{V}$. An analogous picture holds for $\boldsymbol{Y}$ being prescribed and $\boldsymbol{V}$ being the random outcome. It is tacitly assumed that Ω is equipped with a σ-field of observable events $\mathfrak{F}$ and a probability measure P defined on $\mathfrak{F}$.

The fluctuation theorem as expressed by (1.53) states that (i) positive dissipation is exponentially more likely to be observed than negative dissipation, and (ii) ensemble averaging of ϕ_t leads to

$$\langle \Delta\phi_t \mid \mathfrak{F}_t \rangle \geq 0.$$

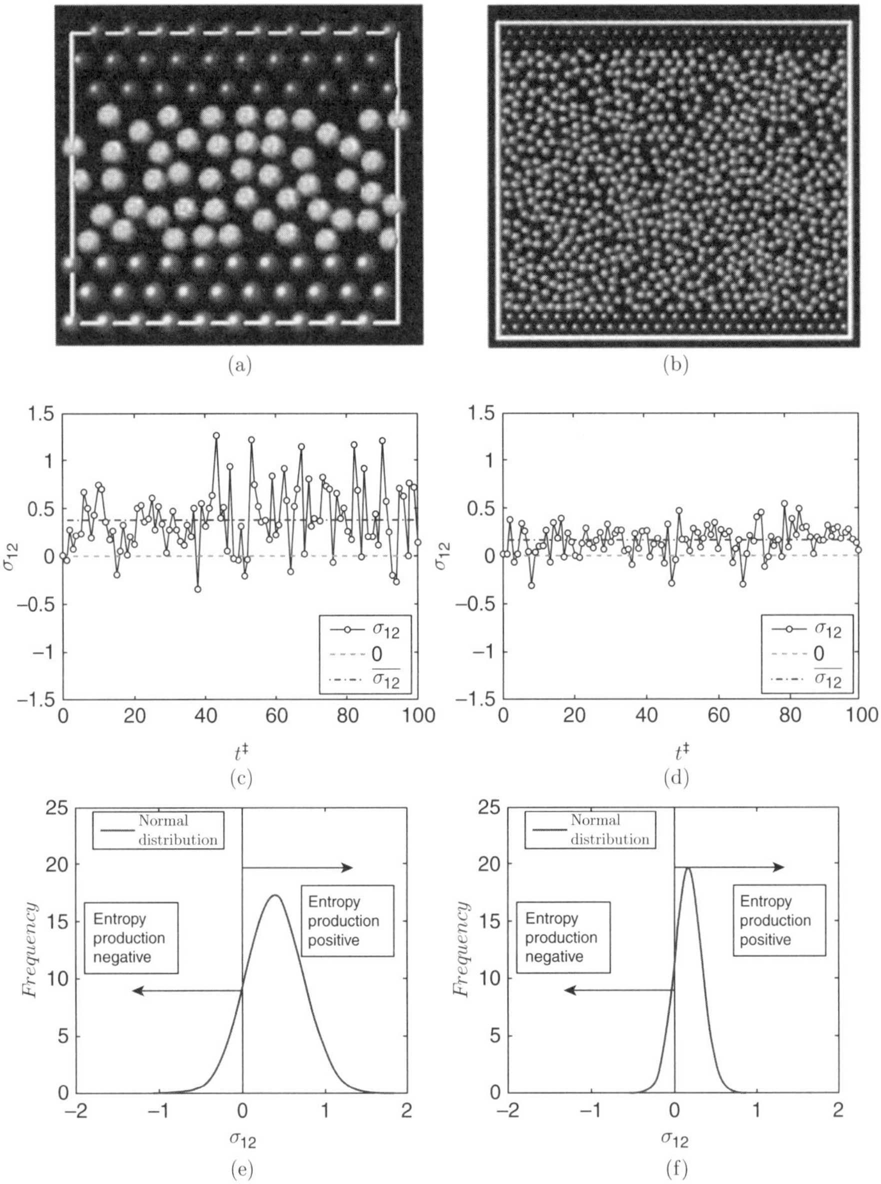

Figure 1.2 Molecular dynamics simulations using (Plimpton 1995) of Couette flows in channels with (a) 50 and (b) 750 particles. (c)–(d) Respective sample shear stress histories with evident negative excursions. (e)–(f) Histograms of shear stresses, showing the trend, with system size increasing, to deterministic fluid mechanics without violations of the Second Law of thermodynamics.

Here $\mid \mathfrak{F}_t$ indicates the conditioning on the past history and is discussed below, while $\langle f \rangle := \int f \, d\mathsf{P}$. Thus, the entropy production rate is non-negative on average. In view of the random fluctuations, ϕ_t is a stochastic process with a specific type of memory effect: a *submartingale* (Doob 1990, Ostoja-Starzewski & Malyarenko 2014). Treating time as a continuous parameter and noting that ϕ_t is deterministic for a forward evolution, we have

$$\langle \phi_{t+dt} \mid \mathfrak{F}_t \rangle \geq \phi_t.$$

Clearly, this is a weaker statement than a purely deterministic one

$$\phi_{t+dt} \mid \mathfrak{F}_t \geq \phi_t \quad \text{or, equivalently,} \quad \frac{\phi_{t+dt} - \phi_t}{\Delta t} \geq 0.$$

The latter inequality is the same as writing (1.47c) in a finite difference form.

Next, we recall the *Doob Meyer decomposition* to write ϕ_t as a sum of a martingale (M) and a 'drift' process (G):

$$\phi_t(\boldsymbol{V}, \omega) = M + G \quad \text{and} \quad \phi(\boldsymbol{V}, \omega) = \dot{M} + \dot{G}.$$

Thus, $M \neq 0$ reflects the fluctuations of entropy production about the zero level $\left\langle s^{(i)} \right\rangle = 0$. The four different cases, depending on whether $M = 0$ or $M \neq 0$ and $G = 0$ or $G > 0$, have been discussed in Ostoja-Starzewski & Malyarenko (2014). Overall, the deterministic continuum mechanics is smoothly recovered as the time and/or spatial scale increases (so that $M \to 0$) or via ensemble averaging: a result that is consistent with intuition. A corresponding re-examination of axioms of continuum mechanics has been given in Ostoja-Starzewski (2016).

Observe that one might also work with a discrete time formulation, making the mathematical analysis of martingales simpler. It is fitting here to note that mathematical physics – also in the classical, i.e. non-quantum, regime – may be formulated from the standpoint of discrete, rather than continuous, time (Jaroszkiewicz 2014).

There are three types of phenomena in classical physics where the fluctuation theorem is applicable: viscous, thermal and electrical (Evans & Searles 2002; Searles & Evans, 2001). If we concern ourselves with the first two, a contact with continuum thermomechanics is made by writing the scalar product $\boldsymbol{Y} \cdot \boldsymbol{V}$ as one involving the intrinsic mechanical dissipation (viscous effects) and thermal dissipation in spatial (Eulerian) description:

$$\phi(\boldsymbol{V}, \omega) = \phi_{\text{th}}(\boldsymbol{V}_1, \omega) + \phi_{\text{mech}}(\boldsymbol{V}_2, \omega), \quad \boldsymbol{V} \equiv (\boldsymbol{V}_1, \boldsymbol{V}_2) = \left(\frac{-\boldsymbol{\nabla} T}{T}, \boldsymbol{d} \right). \tag{1.57}$$

Thus, the generalised velocity vector $\boldsymbol{V}$ is made up of two parts: the negative temperature gradient divided by the temperature (i.e. $-\boldsymbol{\nabla} T/T$) and the deformation rate $\boldsymbol{d}$. The reason we take the former as the argument of ϕ_{th} is that the fluctuation theorem for heat flow was derived for controllable temperature differences, with the heat flux being the stochastic outcome. Analogously, the fluctuation theorem for Couette and Poiseuille flows was derived for controllable velocities, with the Cauchy stress being the stochastic outcome. Thus, the dissipative force corresponding to $\boldsymbol{V}$ is made up of the heat flux and the dissipative stress:

$$\boldsymbol{Y} \equiv (\boldsymbol{Y}_1, \boldsymbol{Y}_2) = \left(\boldsymbol{q}, \boldsymbol{\sigma}^{(d)}\right).$$

It is largely a matter of convenience whether $-\boldsymbol{\nabla}T/T$ or $\boldsymbol{q}$ should be taken as a velocity or a dissipative force. In the section on thermoviscous fluids we work with the setup outlined above, while in the section on inviscid thermoelastic solids we invert the roles of $-\boldsymbol{\nabla}T/T$ and $\boldsymbol{q}$.

There are two basic possibilities here:

- both processes in (1.57) may independently exhibit spontaneous random violations of the Second Law;
- both processes in (1.57) are coupled, implying that the Clausius–Duhem inequality holds for thermal and viscous violations jointly; the relevant statistical physics has not yet been studied.

In Ostoja-Starzewski (2017), we considered the first possibility, focusing on: (i) thermoviscous fluids with parabolic or hyperbolic type heat conduction, (ii) thermoelasticity with parabolic or hyperbolic type heat conduction and (iii) poromechanics with dissipation within the skeleton, the fluid and the temperature field. The reason we considered parabolic or hyperbolic cases is that the statistical physics has established the spontaneous violations of the Fourier-type law, but a hyperbolic heat conduction in fluids and solids can still be modelled in continuum mechanics provided that two relaxation times – one in the mechanical and another in the entropy constitutive law – are introduced (recall the 'thermoelasticity with two relaxation times'). The theoretical developments hinge on the fact that the balance laws apply irrespective of the conventional Second Law being obeyed or not. At the same time, we are interested in formulating models which are hyperelastic and hyperdissipative in ensemble average sense (or for long time averages), thereby extending such classes beyond the deterministic media fully obeying the Second Law.

1.4.3 Stochastic Dissipation Function

Basics

In view of the preceding section, the constitutive relation linking $\boldsymbol{Y}$ with $\boldsymbol{V}$ should be stochastic. Therefore, we replace the deterministic picture by a stochastic one so the internal energy density u and the entropy s are real-valued random fields over the material ($\mathcal{D}$) and time ($\mathcal{T}$) domains. For example, in the case of heat conduction in a rigid (undeformable) conductor,

$$u : \mathcal{D} \times T \times \Omega \to \mathbb{R}, \qquad s : \mathcal{D} \times T \times \Omega \to \mathbb{R}. \tag{1.58}$$

The randomness disappears as the time and/or spatial scales become large: the field quantities simplify to deterministic functions of a homogeneous continuum. Focusing on the thermal dissipation in (1.57), we have

$$\phi_{\mathrm{th}}\left(\frac{-\boldsymbol{\nabla}T}{T},\omega\right)=-q_k\frac{T_{,k}}{T}\equiv-\boldsymbol{q}\cdot\frac{\boldsymbol{\nabla}T}{T}. \tag{1.59}$$

Given the stochastic violations of the Second Law, $\phi_{\mathrm{th}}(\boldsymbol{q},\omega)$ takes the form:

$$\phi_{\mathrm{th}}(\boldsymbol{q},\omega)=\dot{G}(\boldsymbol{q})+\dot{M}(\boldsymbol{q},\omega).$$

For the linear Fourier-type conductivity, it becomes more explicit with

$$\dot{G}(\boldsymbol{q})=\frac{1}{T}q_i\kappa_{ij}q_j,\quad \dot{M}(\boldsymbol{q},\omega)=\frac{1}{T}q_i\mathcal{M}_{ij}(\omega)\,q_j.$$

Here $\dot{G}(\boldsymbol{q})$ involves the thermal resistivity λ_{ij}, which is positive definite, and $\dot{M}(\boldsymbol{q},\omega)=dM(\boldsymbol{q},\omega)/dt$, with M being the martingale modelling the random fluctuations according to (1.53). Clearly, the randomness residing in $M(\boldsymbol{d},\omega)$ allows the total thermal conductivity $\kappa_{ij}+\mathcal{M}_{ij}$ to become negative since $\mathcal{M}_{ij}$ is not required to be positive definite, thus signifying the violations of the Second Law. More specifically, $\mathcal{M}_{ij}\colon V\to V$ (where V is a real vector space) is a second-order rank 2 tensor random field (Malyarenko & Ostoja-Starzewski 2014; Malyarenko & Ostoja-Starzewski 2016*b*)

$$\mathcal{M}_{ij}:\mathcal{D}\times\Omega\to V^{\otimes 2}. \tag{1.60}$$

In the linear regime the fluctuations are Gaussian, so that $\mathcal{M}_{ij}$ is a Gaussian TRF. Above the critical deformation rate (equivalently called 'strain rate'), an ordering transition from an amorphous phase to a 'string-phase' in which particles align with the direction of the flow. Beyond that critical deformation rate, correlations develop, thus indicating the development of internal structure and anisotropy within the fluid, along with a departure from Gaussianity (Raghavan et al., 2018). Using the principle of maximum entropy, the shear stress σ_{12} has been found to follow a variance-gamma probability density function:

$$\begin{aligned}f\left(\sigma_{12};a,b,\beta,\langle\sigma_{12}\rangle\right)=&\frac{2a^b}{\Gamma(b)\,\sigma^2\sqrt{2\pi}\left(\beta^2+2a\sigma^2\right)^{b-1/2}}\\&\times\left|\langle\sigma_{12}\rangle-\sigma_{12}\right|^{b-1/2}\\&\times K_{b-1/2}\left(\frac{\sqrt{\beta^2+2a\sigma^2}}{\sigma^2}\left|\langle\sigma_{12}\rangle-\sigma_{12}\right|\right)\\&\times\exp\left(\beta\left|\langle\sigma_{12}\rangle-\sigma_{12}\right|/\sigma^2\right).\end{aligned}$$

Here, $K_{b-1/2}$ is the modified Bessel function of the second kind, a and b are adjustable constants, while $\sigma=\sqrt{\langle\sigma_{11}\rangle\langle\sigma_{22}\rangle-\langle\sigma_{12}\rangle^2}$.

Atomic Fluid in Couette Flow

In Raghavan & Ostoja-Starzewski (2017), based on a kinetic theory and Non-Equilibrium Molecular Dynamics we have derived a shear-thinning model equation of state for an atomic fluid with interactions of Lennard–Jones type. This, in turn, leads to a dissipation function $\varphi(\boldsymbol{Y})$, in which the dissipative force $\boldsymbol{Y}$ is the symmetric Cauchy stress tensor $\sigma_{ij}^{(d)}$ and $\boldsymbol{V}$ is the deformation rate d_{ij}. For a uniform shear flow, the dissipative force is just the shear stress $\sigma_{12}^{(d)}$ corresponding to the applied deformation rate $\dot{\gamma}$, so that

$$\sigma_{ij}^{(d)}(d_{ij})\, d_{ij} = \phi(\boldsymbol{d}) = \frac{\eta_0 \dot{\gamma}^2}{1 + \frac{2}{3}(\dot{\gamma}/\nu)^2},$$

where $\nu = p/\eta_0$ with η_0 being the Newtonian viscosity and p being the pressure. Adopting the thermodynamic orthogonality (1.49), on account of the side conditions ($d_{(1)} = 0$, $d_{(2)} = \dot{\gamma}^2$, and $d_{(3)} = 0$) and several intermediate steps, we obtain a quasi-linear fluid model

$$\sigma_{ij}^{(d)} = \lambda \frac{\partial \phi}{\partial d_{(2)}} d_{ij} = 2\eta\left(d_{(2)}\right) d_{ij}, \tag{1.61}$$

with the constant of proportionality in (1.61) being

$$\lambda = \frac{1 + \frac{2}{3}(\dot{\gamma}/\nu)^2}{2}.$$

The fluid viscosity η needs to at least depend on $d_{(2)} = \dot{\gamma}^2$.

Owing to the aforementioned formation of the string-phase, the viscosity should more realistically be treated as a locally anisotropic non-Gaussian random field.

1.5 Multi-Field Theories

1.5.1 Thermoelasticity

Classical thermoelasticity

The free energy and dissipation functions are:

$$\text{free energy:}\quad \psi(\varepsilon_{ij}, T) = \frac{1}{2}\varepsilon_{ij}\mathsf{C}_{ijkl}\varepsilon_{kl} + M_{ij}\varepsilon_{ij}\vartheta - \frac{C_E}{2T_0}\vartheta^2$$

$$\text{dissipation function:}\quad \phi(q_i) = \frac{1}{T}\lambda_{ij} q_i q_j.$$

In the linear thermoelastic body the generally anisotropic Hooke law (1.33) is replaced by

$$\sigma_{ij} = \mathsf{C}_{ijkl}\varepsilon_{kl} + M_{ij}\vartheta,$$

whereby the strain-displacement equation (1.34) holds as before. Here M_{ij} denotes the *stress–temperature tensor*, while $\vartheta = T - T_0$, with T_0 being the *reference temperature*. There are two more constitutive equations: one for entropy density,

$$s = -M_{ij}\varepsilon_{ij} + \frac{C_E}{T_0}\vartheta,$$

where C_E denotes the *specific heat at zero strain*, and the second for thermal conductivity already given as (1.28). In addition to the symmetries (1.35), there hold these inequalities:

$$M_{ij} = M_{ji}, \qquad C_E > 0. \tag{1.62}$$

Any thermoelastic process involves a mechanical field and a thermal field, which translates into four possibilities of describing the entire process: in terms of either a pair (u_i, T), or (σ_{ij}, q_i), or (u_i, q_i), or (σ_{ij}, T). For example, in the first case,

$$\begin{aligned} (\mathsf{C}_{ijkl}u_{k,l})_{,j} - \rho\ddot{u}_i + (M_{ij}\vartheta)_{,j} &= -b_i, \\ (k_{ij}\vartheta_{,j})_{,i} - C_E\dot{\vartheta} + T_0 M_{ij}\dot{u}_{i,j} &= -r, \end{aligned} \tag{1.63}$$

where r is the external heat source field.

Assuming locally isotropic material responses [(1.40), (1.30) and $M_{ij} = M\delta_{ij}$], one obtains a generalisation of the classical Navier equation of motion (1.42) coupled with a Duhamel equation:

$$\begin{aligned} (\lambda u_{j,j})_{,i} + (\mu u_{(i,j})_{,j)} - \rho\ddot{u}_i + (M\vartheta)_{,i} &= -b_i, \\ (\kappa\vartheta_{,i})_{,i} - C_E\dot{\vartheta} + T_0 M\dot{u}_{i,i} &= -r, \end{aligned}$$

where $M = (3\lambda + 2\mu)\alpha$, with α being the thermal expansion coefficient.

When $M_{ij} = 0$ these equations decouple into (1.39) and (1.29). Another situation when the mechanical field decouples from the thermal fields, resulting in this procedure: the temperature field is determined first and then used as, effectively, a body force field driving the elasticity problem; the temperature field is unaffected by the stress and strain fields. See Hetnarski (2013) for very extensive information on this and related topics in continuum mechanics with thermal stresses.

Hyperbolic thermoelasticity with one relaxation time

The free energy and dissipation functions are:

$$\begin{aligned} \text{free energy:} \quad \psi(\varepsilon_{ij}, q_i, T) &= \frac{1}{2}\varepsilon_{ij}\mathsf{C}_{ijkl}\varepsilon_{kl} + M_{ij}\varepsilon_{ij}\vartheta - \frac{C_E}{2T_0}\vartheta^2 \\ &\quad + \frac{t_0}{2T_0}\lambda_{ij}q_i q_j, \\ \text{dissipation function:} \quad \varphi(q_i) &= \frac{1}{2T_0}\lambda_{ij}q_i q_j, \end{aligned} \tag{1.64}$$

where $\vartheta = T - T_0$. This theory has its roots in a modification of the Fourier heat conduction (1.29), first proposed by Maxwell (1867) in the context of theory of gases, and later by Cattaneo (1949) in the settings of rigid bodies:

$$Lq_i = -\kappa_{ij}T_{,j},$$

where L is an operator defined by $L = 1 + t_0 \partial/\partial t$, with $t_0 > 0$ being the so-called relaxation time. Employing the primitive thermodynamics (1.51), leads to the constitutive equations for stress and entropy:

$$T_0 \dot{s} = -q_{i,i} + r, \tag{1.65a}$$

$$\sigma_{ij} = \mathbf{C}_{ijkl}\varepsilon_{kl} + M_{ij}\vartheta, \tag{1.65b}$$

$$T_0 s = -T_0 M_{ij}\varepsilon_{ij} + C_E\vartheta. \tag{1.65c}$$

Note: This thermoelasticity model was originally formulated via a free energy function of the form (Lord & Shulman 1967):

$$\psi(\varepsilon_{kl}, q_i, T) = \frac{1}{2}\varepsilon_{ij}\mathbf{C}_{ijkl}\varepsilon_{kl} + M_{ij}\varepsilon_{ij}\vartheta - \frac{C_E}{2T_0}\vartheta^2 + \frac{t_0}{2T_0}\lambda_{ij}q_i q_j,$$

but, since the heat conduction is a dissipative process, a TIV approach explicitly involving dissipation in the (1.64) is more natural.

By analogy to (1.63) of the classical thermoelasticity, one can formulate the displacement–temperature field equations:

$$\begin{aligned}(\mathbf{C}_{ijkl}u_{k,l})_{,j} - \rho\ddot{u}_i + (M_{ij}\vartheta)_{,j} &= -b_i,\\ (k_{ij}\vartheta_{,j})_{,i} - C_E\dot{\hat{\vartheta}} + T_0 M_{ij}\dot{\hat{u}}_{i,j} &= -\hat{r},\end{aligned} \tag{1.66}$$

where a hat denotes action of the operator L on any function f on $\mathcal{D} \times (0, \infty)$: $\hat{f} = Lf$.

Alternatively, the field equations can be written in terms of the stress–heat flux pair (σ_{ij}, q_i):

$$\begin{aligned}(\rho^{-1}\sigma_{(ik,k})_{,j)} - \mathbf{S}_{ijkl}\ddot{\sigma}_{kl} + C_S^{-1}A_{ij}\dot{q}_{k,k} &= -(\rho^{-1}b_{(i)})_{,j)} + C_S^{-1}A_{ij}\dot{r},\\ (C_S^{-1}q_{k,k})_{,i} - \lambda_{ij}\dot{\hat{q}}_j + T_0(C_S^{-1}A_{pq}\dot{\sigma}_{pq})_{,i} &= (C_S^{-1}r)_{,i},\end{aligned}$$

which generalises the Ignaczak equation of classical elasticity (1.43). Here C_S is the specific heat at zero stress.

Thermoelasticity with two relaxation times

The Ansatz is now based on:

$$\begin{aligned}\text{free energy:}\quad \psi = \psi(\varepsilon_{ij}, T, \dot{T}) &= \frac{1}{2}\varepsilon_{ij}\mathbf{C}_{ijkl}\varepsilon_{kl} + M_{ij}\varepsilon_{ij}\vartheta - \frac{C_E}{2T_0}\vartheta^2\\ &\quad - \frac{C_E}{T_0}t_0\vartheta\dot{\vartheta},\\ \text{dissipation function:}\quad \varphi(\dot{\varepsilon}_{ij}, q_i) &= t_1 M_{ij}\dot{\varepsilon}_{ij}\dot{\vartheta} + \frac{\lambda_{ij}}{T}q_i q_j,\end{aligned} \tag{1.67}$$

which involves two relaxation times, t_0 and t_1. Employing the thermodynamic orthogonality leads to the constitutive equations for stress and entropy:

$$\begin{aligned}\sigma_{ij} &= \mathbf{C}_{ijkl}\varepsilon_{kl} + M_{ij}(\vartheta + t_1\dot{\vartheta}),\\ T_0 s &= -T_0 M_{ij}\varepsilon_{ij} + C_E(\vartheta + t_0\dot{\vartheta}),\end{aligned} \qquad t_1 \geq t_0 > 0,$$

with (1.65a) unchanged and the Fourier heat conduction (1.28). Effectively, all the disturbances propagate at finite wave speeds.

Note: This thermoelasticity model was originally formulated via a free energy function of the form (Green & Lindsay 1972):

$$\psi(\varepsilon_{ij}, T, \dot{T}, T_{,i}) = \frac{1}{2}\varepsilon_{ij}\mathbf{C}_{ijkl}\varepsilon_{kl} + M_{ij}\varepsilon_{ij}(\vartheta + t_1\dot{\vartheta}) \\ - \frac{C_E}{2T_0}\vartheta^2 - \frac{C_E}{T_0}t_1\vartheta\dot{\vartheta} - \frac{C_E}{2T_0}t_0 t_1\dot{\vartheta}^2 + \frac{t_1}{2T_0}k_{ij}\vartheta_{,i}\vartheta_{,j},$$

but, since the heat conduction is a dissipative process, a TIV approach explicitly involving dissipation in the (1.67) is more natural.

By analogy to the preceding two theories, one can formulate coupled field equations in terms of the pair (u_i, ϑ):

$$(\mathbf{C}_{ijkl}u_{k,l})_{,j} - \rho\ddot{u}_i + [M_{ij}(\vartheta + t_1\dot{\vartheta})]_{,j} = -b_i,$$
$$(k_{ij}\vartheta_{,j})_{,i} - C_E(\dot{\vartheta} + t_0\ddot{\vartheta}) + T_0 M_{ij}\dot{u}_{i,j} = -r,$$

or in terms of the pair (σ_{ij}, q_i):

$$(\rho^{-1}\sigma_{(ik,k})_{,j)} - \mathbf{S}'_{ijkl}\ddot{\sigma}_{kl} + C_S^{-1}A_{ij}\dot{q}_{k,k} = -(\rho^{-1}b_{(i})_{,j)} + C_S^{-1}A_{ij}\dot{r},$$
$$(C_S^{-1}q_{k,k})_{,i} - \lambda_{ij}\hat{\dot{q}}_j + T_0(C_S^{-1}A_{pq}\dot{\sigma}_{pq})_{,i} = (C_S^{-1}r)_{,i}.$$

An extensive study of these hyperbolic theories of thermoelasticity is in Ignaczak & Ostoja-Starzewski (2010).

1.5.2 Viscothermoelasticity

Viscoelasticity via Boltzmann superposition principle

This classical approach is based on the *Boltzmann superposition principle* (Staverman & Schwarzl 1956; Christensen 2003), where the past causes in the history of loading at a material point are reflected as a summation (i.e. superposition) of all the effects up to the present. Mathematically, if $\mathcal{F}$ is a linear tensor-valued functional which transforms each strain history $\{\varepsilon_{kl}(t), -\infty \leq t \leq \infty\}$ into a corresponding stress history $\{\sigma_{ij}(t), -\infty \leq t \leq \infty\}$, on account of the Riesz representation theorem, we have

$$\sigma_{ij}(t) = \int_0^\infty \mathbf{G}_{ijkl}(t-\tau)\frac{d\varepsilon_{kl}}{d\tau}d\tau, \tag{1.68}$$

where $\mathbf{G}_{ijkl}$ is a *relaxation modulus*. Alternatively (and equivalently), (1.68) may be expressed as

$$\varepsilon_{ij}(t) = \int_0^\infty \mathbf{J}_{ijkl}(t-\tau)\frac{d\sigma_{kl}}{d\tau}d\tau, \tag{1.69}$$

where $\mathbf{J}_{ijkl}$ is the *creep compliance*. A number of relations for $\mathbf{G}_{ijkl}$ and $\mathbf{J}_{ijkl}$, as well as between them are dictated by the properties of Laplace transformation such as (with s being the Laplace transform variable)

$$\mathcal{L}(\mathbf{J}_{ijkl}) = \left[s^2\mathcal{L}(\mathbf{G}_{ijkl})\right]^{-1},$$

which shows that the intuitive extension analogous to (1.38) does not hold. However, the initial and final value theorems of the Laplace transformation imply that

$$\lim_{t\to 0} \boldsymbol{J}_{ijkl} = \lim_{t\to 0} \boldsymbol{G}^{-1}_{ijkl},$$

and for solids

$$\lim_{t\to \infty} \boldsymbol{J}_{ijkl} = \lim_{t\to \infty} \boldsymbol{G}^{-1}_{ijkl}.$$

Anisotropy can also be handled when a Fourier, rather than Laplace, transformation (γ being the frequency) is introduced:

$$\widehat{f}\,(\gamma) = \int_{-\infty}^{\infty} f\,(t)\, e^{-i\gamma t} dt, \qquad f\,(t) = \frac{1}{2\pi} \int_{-\infty}^{\infty} \widehat{f}\,(t)\, e^{i\gamma t} dt,$$

we can write the stress–strain relations (1.68) and (1.69) as

$$\widehat{\sigma}_{ij}(t) = \boldsymbol{G}^{*}_{ijkl}\,(i\gamma)\,\widehat{\varepsilon}_{kl}, \qquad \widehat{\varepsilon}_{ij}(t) = \boldsymbol{J}^{*}_{ijkl}\,(i\gamma)\,\widehat{\sigma}_{kl}.$$

Here $\boldsymbol{G}^{*}_{ijkl}$ is the *complex modulus* and $\boldsymbol{J}^{*}_{ijkl}$ is the *complex compliance.*

Parabolic formulation of viscothermoelasticity

It is well known that, as an alternative to the integral formulation above, the viscoelastic laws may also be written in differential forms. In general, if an isotropic material response is assumed and n internal parameters are introduced, then one arrives at (i) a scalar differential equation for the first basic invariants (denoted by $_{(1)}$) of the stress and strain tensors:

$$\sigma_{(1)} + p^{(1)}\dot{\sigma}_{(1)} + \ldots + p^{(n)}\sigma^{(n)}_{(1)} = q^{(0)}\varepsilon_{(1)} + q^{(1)}\dot{\varepsilon}_{(1)} + \ldots + q^{(n)}\varepsilon^{(n)}_{(1)} \quad (1.70)$$

and (ii) a tensorial differential equation for the deviatoric parts (denoted by primes) of these tensors:

$$\sigma_{ij} + p^{(1)\prime}\dot{\sigma}'_{ij} + \ldots + p^{(n)\prime}\sigma^{(n)\prime}_{ij} = q^{(0)\prime}\varepsilon_{ij} + q^{(1)}\varepsilon_{ij} + \ldots + q^{(n)}\varepsilon^{(n)}_{ij}. \quad (1.71)$$

In (1.70) (respectively, (1.71)), $p^{(1)}, \ldots, q^{(n)}$ $(p^{(1)\prime}, \ldots, q^{(n)\prime})$ are the coefficients of pressure-dilatation (deviatoric) type responses. Adding parabolic-type heat conduction results in extra terms on the right-hand side in (1.70):

$$\begin{aligned}\sigma_{(1)} + p^{(1)}\dot{\sigma}_{(1)} + \cdots + p^{(n)}\sigma^{(n)}_{(1)} &= q^{(0)}\varepsilon_{(1)} + q^{(1)}\dot{\varepsilon}_{(1)} + \ldots + q^{(n)}\varepsilon^{(n)}_{(1)} \\ &\quad + r^{(0)}\,(T - T_0) + r^{(1)}\dot{T} + \cdots + r^{(n)}T^{(n)}.\end{aligned} \quad (1.72)$$

This type of model can also be written for a locally anisotropic material response with all the constitutive coefficients becoming rank 4 tensors. However, introducing a hyperbolic-type heat conduction is not as straightforward.

Hyperbolic formulation of viscothermoelasticity
with one relaxation time

In order to set up the constitutive viscothermoelastic equations generalising those of thermoelasticity with one relaxation time (1.65), we admit one internal variable α_{ij} in the internal energy u (Ostoja-Starzewski 2014). As always, the specific power of deformation is $l = \sigma_{ij}\dot{\varepsilon}_{ij}/\rho$ and the classical relation (1.25) holds, whereby, switching the entropy s to the absolute temperature T, we have $\psi = \psi(T, q_i, \varepsilon_{ij}, \alpha_{ij})$ and $s = s(T, q_i, \varepsilon_{ij}, \alpha_{ij})$. By modifying (1.64), the free energy is now taken in the form (again with $\vartheta = T - T_0$):

$$\text{free energy: } \psi(\varepsilon_{ij}, \alpha_{ij}, q_i, T) = \frac{1}{2}\sum_{r=1}^{2}(\varepsilon_{ij} - \alpha_{ij}^{(r)})\boldsymbol{C}_{ijkl}^{(r)}(\varepsilon_{kl} - \alpha_{kl}^{(r)}) + M_{ij}\varepsilon_{ij}\vartheta - \frac{C_E}{2T_0}\vartheta^2 + \frac{t_0}{2T_0}\lambda_{ij}q_iq_j,$$

where $\alpha_{ij}^{(1)} \equiv \alpha_{ij}$, $\alpha_{ij}^{(2)} = 0$, $\boldsymbol{C}_{ijkl}^{(r)}$ $(r = 1, 2)$ are the stiffness tensors, M_{ij} and C_E are defined as before. This choice of ψ corresponds to a particular setup of the *Zener model*: Maxwell element in parallel with a spring. In effect, $\boldsymbol{C}_{ijkl}^{(1)}$ is the stiffness tensor associated with the Maxwell element while $\boldsymbol{C}_{ijkl}^{(2)}$ is the stiffness tensor of the spring. It follows that

$$\begin{aligned} \sigma_{ij}^{(q)} &= \sum\nolimits_{r=1}^{2} \boldsymbol{C}_{ijkl}^{(r)}(\varepsilon_{kl} - \alpha_{kl}^{(r)}) + M_{ij}\vartheta, \\ \beta_{ij}^{(q)} &= -\boldsymbol{C}_{ijkl}^{(1)}(\varepsilon_{kl} - \alpha_{kl}), \\ s &= -M_{ij}\varepsilon_{ij} + \frac{C_E}{T_0}\vartheta. \end{aligned}$$

Restricting the model to linear dissipative processes, we can satisfy (1.48) and (1.52) by adopting:

$$\rho T s^{*(i)} = \rho\varphi(\mathbf{V}) = \frac{1}{2}\dot{\alpha}_{ij}\boldsymbol{F}_{ijkl}\dot{\alpha}_{kl} + \frac{1}{2T}\lambda_{ij}q_iq_j,$$

which leads to the governing equation

$$\sigma_{mn} + \left(\boldsymbol{C}_{ijkl}^{(1)}\right)^{-1}\boldsymbol{F}_{mnkl}\dot{\sigma}_{ij} = \boldsymbol{C}_{mnpr}^{(2)}\varepsilon_{pr} + \left(\boldsymbol{C}_{ijkl}^{(1)}\right)^{-1}\boldsymbol{F}_{mnkl}\left(\boldsymbol{C}_{ijkl}^{(1)} + \boldsymbol{C}_{ijkl}^{(2)}\right)\dot{\varepsilon}_{kl} + M_{mn}\vartheta + \left(\boldsymbol{C}_{ijkl}^{(1)}\right)^{-1}\boldsymbol{F}_{mnkl}M_{ij}\dot{\vartheta}.$$

The heat conduction is hyperbolic according to the telegraph-like equation for the temperature coupled with the displacement field, where $\hat{\dot{u}}_{i,j} = \dot{u}_{i,j} + t_0\ddot{u}_{i,j}$

$$(k_{ij}\vartheta,_j),_i - C_E\hat{\dot{\vartheta}} + T_0M_{ij}\hat{\dot{u}}_{i,j} = 0.$$

Hyperbolic formulation of viscothermoelasticity with two relaxation times

To generalise the constitutive thermoelastic equations with two relaxation times, the internal energy and dissipation function are taken as $u = u(\varepsilon_{ij}, T, \dot{T})$ and $\phi = \phi(\dot{\varepsilon}_{ij}, q_i, \dot{q}_i)$. By adding the internal strain term to (1.67), the free energy is

$$\rho\psi(\varepsilon_{ij}, \alpha_{ij}, T, \dot{T}) = \frac{1}{2}(\varepsilon_{ij} - \alpha_{ij})\mathbf{C}^{(1)}_{ijkl}(\varepsilon_{kl} - \alpha_{kl}) + \frac{1}{2}\varepsilon_{ij}\mathbf{C}^{(2)}_{ijkl}\varepsilon_{kl} + M_{ij}\varepsilon_{ij}\vartheta - \frac{C_E}{2T_0}\vartheta^2 - \frac{C_E}{\vartheta_0}t_0\vartheta\dot{\vartheta},$$

which leads to

$$\sigma^{(q)}_{ij} = \rho\frac{\partial\psi}{\partial\varepsilon_{ij}} = \mathbf{C}^{(1)}_{ijkl}(\varepsilon_{kl} - \alpha_{kl}) + \mathbf{C}^{(2)}_{ijkl}\varepsilon_{kl} + M_{ij}\vartheta$$

$$s = -\rho\frac{\partial\psi}{\partial T} = -M_{ij}\varepsilon_{ij} + \frac{C_E}{T_0}\vartheta + \frac{C_E}{T_0}t_0\dot{\vartheta}$$

$$\beta^{(q)}_{ij} = \rho\frac{\partial\psi}{\partial\alpha_{ij}} = -\mathbf{C}^{(1)}_{ijkl}(\varepsilon_{kl} - \alpha_{kl}).$$

Next, adopting the dissipation function

$$\rho T s^{*(i)} \equiv \rho\phi(\dot{\varepsilon}_{ij}, \dot{\alpha}_{ij}, q_i) = \dot{\alpha}_{ij}\mathbf{F}_{ijkl}\dot{\alpha}_{kl} + t_1 M_{ij}\dot{\varepsilon}_{ij}\dot{\vartheta} + \frac{1}{T}\lambda_{ij}q_i q_j$$

leads to the constitutive equation linking stress (and its rate) with strain (and its rate) with temperature (and its first and second rates):

$$\sigma_{mn} + \left(\mathbf{C}^{(1)}_{ijkl}\right)^{-1}\mathbf{F}_{mnkl}\dot{\sigma}_{ij} = \mathbf{C}^{(2)}_{mnpr}\varepsilon_{pr} + \left(\mathbf{C}^{(1)}_{ijkl}\right)^{-1}\mathbf{F}_{mnkl}\left(\mathbf{C}^{(1)}_{ijkl} + \mathbf{C}^{(2)}_{ijkl}\right)\dot{\varepsilon}_{kl} + M_{mn}\vartheta + \left[t_1 M_{mn} + M_{ij}\left(\mathbf{C}^{(1)}_{ijkl}\right)^{-1}\mathbf{F}_{mnkl}M\right]\dot{\vartheta} + t_1 M_{ij}\left(\mathbf{C}^{(1)}_{ijkl}\right)^{-1}\mathbf{F}_{mnkl}\ddot{\vartheta}.$$

Again, the heat conduction is hyperbolic according to the telegraph-like equation for the temperature coupled (albeit in a simpler fashion than in the one relaxation time case) with the displacement field

$$(k_{ij}\vartheta_{,j})_{,i} - C_E(\dot{\vartheta} + t_0\ddot{\vartheta}) + T_0 M_{ij}\dot{u}_{i,j} = 0.$$

1.5.3 Piezoelectricity

The field equations of piezoelectro-elastodynamics for a generally anisotropic medium (Nowacki 1975) involve a combination of the strain-displacement relations (1.34) and the equations of motion (1.18), along with:

- the electric field-potential relations

$$E_i = -\Phi_{,i}\,,$$

- the electric induction equation

$$D_{i,i} = 0,$$

- the constitutive relations

$$\sigma_{ij} = \mathsf{C}_{ijkl}\varepsilon_{kl} - \mathsf{e}_{kij}E_k,$$
$$D_i = \mathsf{e}_{ijk}\varepsilon_{jk} + \varepsilon_{ik}E_k.$$

The latter can equivalently be written as

$$\varepsilon_{ij} = \mathsf{S}_{ijpq}\left(\sigma_{pq} + \mathsf{e}_{kpq}\right)E_k,$$
$$D_i = \mathsf{P}_{ipq}\sigma_{pq} + \left(\mathsf{P}_{ipq}\mathsf{e}_{kpq} + \varepsilon_{ik}\right)E_k,$$

where

$$\mathsf{P}_{ipq} = \mathsf{e}_{iab}\mathsf{S}_{abpq}.$$

The triplet $(u_i, E_{ij}, \sigma_{ij})$ represents an elastic motion of the body with the physical properties described by $(\rho, \mathsf{C}_{ijkl})$ that, in general, depend on $\boldsymbol{x} \in \mathcal{B}$. Also, Φ is the electric potential, while E_i and D_i denote the electric and induction vector fields, respectively, that represent a piezoelectric motion of a body with the properties $(\mathsf{e}_{ijk}, \varepsilon_{ij})$ in which e_{ijk} and ε_{ij} are the piezoelectric and dielectric permeability tensor fields, respectively, also generally depending on $\boldsymbol{x} \in \mathcal{B}$. The latter two tensors possess the following symmetries:

$$\mathsf{e}_{kij} = \mathsf{e}_{kji}, \qquad \varepsilon_{ij} = \varepsilon_{ji}.$$

By analogy to elasticity and thermoelasticity, discussed earlier, there are two basic formulations of the field equations: either in terms of the displacement and electric potential (i.e. generalising (1.39))

$$\left(\mathsf{C}_{ijkl}u_{(k,l)}\right)_{,j} + \mathsf{e}_{kij}\Phi_{,kj} + b_i = \rho\ddot{u}_i,$$
$$\mathsf{e}_{ikl}u_{k,li} - \varepsilon_{ik}\Phi_{,ik} = 0,$$

or in terms of the stress and electric potential (i.e. generalising (1.43)):

$$\left(\rho^{-1}\sigma_{(ik,k}\right)_{,j)} - \mathsf{S}_{ijkl}\ddot{\sigma}_{kl} + \mathsf{P}_{sij}\ddot{\Phi}_s = 0,$$
$$\left(\mathsf{P}_{ipq}\sigma_{pq}\right)_{,i} - \left[\left(\mathsf{P}_{ipq}\mathsf{e}_{kpq} + \varepsilon_{ik}\right)\Phi_{,k}\right]_{,i} = 0.$$

The internal energy is

$$u\left(\boldsymbol{\varepsilon}, \boldsymbol{E}\right) = \frac{1}{2}\varepsilon_{ij}\mathsf{C}_{ijkl}\varepsilon_{kl} + \frac{1}{2}E_i\varepsilon_{ik}E_j,$$

which, by a Legendre transformation, leads to an expression for the free energy, then explicitly showing the coupling of mechanical with electrical fields.

The piezoelectricity may be generalised to thermo-piezoelectricity along the lines of thermoelasticity models of either parabolic or hyperbolic types, with a dissipation function appearing accordingly. Thermo-magnetoelasticity and other mechanical-electromagnetic models may be formulated in analogous ways.

1.6 Inelastic Materials

1.6.1 Damage

In order to grasp an anisotropic damage state, the classical (deterministic) continuum damage mechanics (CDM) makes use of fabric tensors in a manner

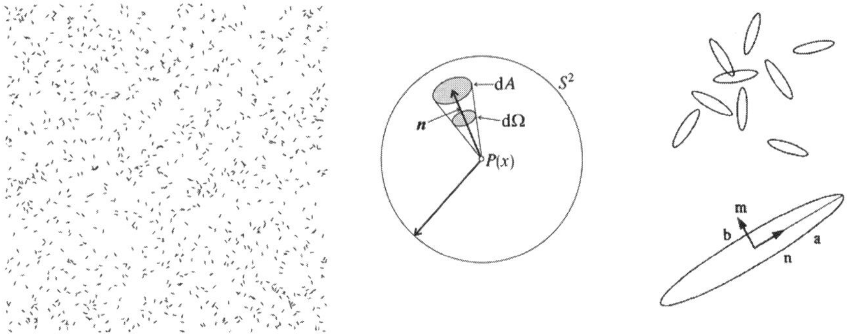

(a) (b) (c)

Figure 1.3 (a) Random crack field; (b) distribution of $\mathbf{n}$ on the unit sphere; (c) random ellipse field with a single ellipse (with semi-axes b, a) showing the $(\mathbf{m}, \mathbf{n})$ pair for determination of the fabric tensor.

borrowed from the mechanics of granular media (Lubarda & Krajcinovic 1993; Murakami 2012; Ganczarski, Egner & Skrzypek 2015). First, there is a random field of cracks (Figure 1.3(a)), each specified by an orientation vector $\boldsymbol{n}$ (i.e. a unit vector normal to the crack), which is analogous to a field of grain-grain contacts, each described by $\boldsymbol{n}$. With reference to Figure 1.3(b), for a system in 3D (2D), one introduces a probability density function of $\boldsymbol{n}$ on a unit sphere S^2 (respectively, on a circle)

$$p(\boldsymbol{n}) = D_0 + D_{ij} f_{ij}(\boldsymbol{n}) + \boldsymbol{D}_{ijkl} \boldsymbol{f}_{ijkl}(\boldsymbol{n}) + \ldots$$

This is a generalised Fourier series with respect to the irreducible tensor bases

$$\begin{aligned} f_{ij}(\boldsymbol{n}) &= n_i n_j - \frac{1}{3}\delta_{ij} \\ f_{ijkl}(\boldsymbol{n}) &= n_i n_j n_k n_l - \frac{1}{7}(\delta_{ij} n_k n_l + \delta_{ik} n_j n_l + \delta_{il} n_j n_k + \delta_{jk} n_i n_l \\ &\quad + \delta_{jl} n_i n_k + \delta_{kl} n_i n_j) + \frac{1}{5 \times 7}(\delta_{ij} n_k n_l + \delta_{ik} n_j n_l + \delta_{il} n_j n_k), \end{aligned}$$

while

$$D_0 = \frac{1}{4\pi} \int_{S^2} p(\boldsymbol{n})\, d\Omega, \tag{1.73a}$$

$$D_{ij} = \frac{1}{4\pi} \frac{3 \times 5}{2} \int_{S^2} p(\boldsymbol{n})\, f_{ij}(\boldsymbol{n})\, d\Omega, \tag{1.73b}$$

$$\boldsymbol{D}_{ijkl} = \frac{1}{4\pi} \frac{3 \times 5 \times 7 \times 9}{2 \times 3 \times 4} \int_{S^2} p(\boldsymbol{n})\, \boldsymbol{f}_{ijkl}(\boldsymbol{n})\, d\Omega, \tag{1.73c}$$

are the scalar, second-order, fourth-order, damage tensors (respectively, fabric tensors), describing the directional distribution of damage state (grain–grain contacts). Thus, the overall directional distribution of damage state of a material is described through the tensors D_0, D_{ij}, D_{ijkl}, ..., just as the overall distribution of contacts is characterised in a granular medium. Depending on the specific choice of the constitutive model, either just one (D_0), or two, or more damage tensors are included in the analysis.

This picture of CDM corresponds to smoothing of the morphology within the RVE (recall Subsection 1.1.1) by assuming a random field of cracks of infinite extent at the continuum point of the macroscopic continuum model. This 'separation of scales' assumption may, and should be, abandoned in favour of treating the continuum point as the SVE: the function $\xi(\mathbf{n})$ possesses mesoscale fluctuations, so that D_0, D_{ij}, D_{ijkl} are rank 0, 2 and 4 TRFs.

In continuum mechanics, damage is interpreted as a decrease in elasticity property consequent to a decrease of the areas transmitting internal forces, through the appearance and subsequent growth of microcracks and microcavities. The oldest measure of damage set up on the scale of RVE is a scalar D accounting for a loss of cross-sectional area carrying the stress:

$$0 \leq D \leq 1,$$

where $D = 0$ corresponds to a virgin element, while $D = 1$ to a fully damaged element. Due to a progressive atomic decohesion and complete loss of a load-carrying ability, in most materials the total damage/fracture occurs at $D < 1$, typically in the range 0.2–0.8. This scalar measure of damage (effectively, D_0) is useful for 1D models and 2D or 3D models where damage is assumed to occur and/or evolve isotropically. However, the stress and strain field are generally anisotropic, so that the evolution of damage is anisotropic. Thus, one has to involve higher-order tensors D_{ij}, D_{ijkl}, ... from (1.73). A chosen damage tensor then appears as an argument of the free energy ψ; other arguments of ψ may include the effects of isotropic/kinematic plastic hardening, e.g. Maugin (1999); Murakami (2012). Proceeding with the TIV formalism, the rate of D_{ij} (or D_{ijkl}, ...) appears as argument of the dissipation function.

However, in order to account for the stochastic, rather than deterministic, character of response and evolution of material damage, one must note the finite-size effects in the actual spatial distribution of cracks, which, in turn, implies that damage tensors are TRFs.

Instead of D_{ij} of (1.73b), one can use formulas that better account for the shape of the crack such as, say, it being an elliptical void:

$$D_{ij} = \frac{\pi}{A} \sum_k (a^2 n_i n_j + b^2 m_i m_j)^{(k)}.$$

Here a and b are the semi-axes of an ellipse and the summation is carried out over all the voids within a given domain, Figure 1.3(c). Counterintuitive properties of the cross-correlation of D_{ij} with a random anti-plane elasticity tensor C_{ij}, both taken on the same mesoscale, have been reported in Ostoja-Starzewski (2008).

1.6.2 Perfect Plasticity

One of the simplest dissipative material models is that of perfect plasticity. In the case of the continuum model set on the scale of a very large (theoretically infinite) number of grains, assuming their statistics possess the statistical spatial homogeneity and isotropy, the Huber–von Mises–Hencky yield condition may be applicable

$$\frac{1}{2}\sigma'_{ij}\sigma'_{ij} = k^2,$$

where k is the *yield stress in shear.* Taking k to be a (scalar) random field is analogous to taking, say, the conductivity tensor to be isotropic (i.e. $C_{ij} = C\delta_{ij}$) everywhere, with C being a scalar random field.

In the case of othrotropy of individual crystals and their yield being independent of the mean stress $\sigma_{(1)}$, the six-parameter *Hill's condition* is applicable

$$\begin{aligned} &F\left(\sigma_{22}-\sigma_{33}\right)^2 + G\left(\sigma_{33}-\sigma_{11}\right)^2 + H\left(\sigma_{11}-\sigma_{22}\right)^2 \\ &+ 2L\sigma_{23}^2 + 2M\sigma_{31}^2 + 2N\sigma_{12}^2 = 1, \end{aligned}$$

for which, in principle, all six coefficients can be determined by experimental (or, using appropriate micromechanics, computational) tests.

The general form of an anisotropic yield condition for perfect plasticity (dating back to von Mises) is known to have the form

$$\sigma_{ij}\boldsymbol{A}_{ijkl}\sigma_{kl} = 1.$$

Here the tensor of plastic moduli $\boldsymbol{A}_{ijkl}$ has the same major and minor symmetries as the stiffness tensor, i.e. $\boldsymbol{A}_{ijkl} = \boldsymbol{A}_{jikl} = \boldsymbol{A}_{ijlk} = \boldsymbol{A}_{klij}$ and, hence, has 21 components. While extensive discussions of these and many other models exist in the literature, e.g. Skrzypek & Ganczarski (1999), this model may apply on a mesoscale (SVE level) involving a finite number of crystal grains, as indicated by Figure 0.1(b), with $\boldsymbol{A}_{ijkl}$ being a TRF. Again, the larger is the SVE, the weaker is the randomness and the closer is the response to that on the macroscopic (RVE) level.

The same considerations apply to soil plasticity (Houlsby & Puzrin 2007), perhaps even more so than to metals, because statistical aspects of soils and foundations are even more pronounced than in metals.

1.7 Generalised Continuum Theories

1.7.1 Basic Concepts of Micropolar Theories

There are many ways in which classical continuum theories may be generalised; see Maugin (2017) for a compact, yet very readable, presentation of multifarious avenues. One of the oldest generalisations comprises so-called micropolar theories. We briefly introduce that theory because tensor-valued random fields of micropolar type are introduced in Chapter 4.

Force transfer and degrees of freedom

Dating back to Cosserat & Cosserat (1909), that approach admits the presence of moment-tractions in addition to force-tractions acting on any infinitesimal surface element. To be fully consistent with kinematics, each continuum point is endowed with three rotations φ_i $(i = 1, \ldots, 3)$ (or $\boldsymbol{\varphi}$) in addition to three displacements u_i $(i = 1, \ldots, 3)$ (or $\boldsymbol{u}$). Thus, each point has six degrees of freedom and the micropolar (Cosserat) theory is seen as a continuum generalisation of a rigid body, just like a classical (Cauchy) theory is a generalisation of points of zero size. With $\mathbf{u}$ and $\boldsymbol{\varphi}$ being, in general, independent functions of position and time, $\boldsymbol{\varphi}$ is not the same as the macro-rotation given by the gradient of $\boldsymbol{u}$

$$\varphi_i \neq \frac{1}{2}\boldsymbol{\varepsilon}_{ijk} u_{k,j} \tag{1.74}$$

Restricting attention to small deformations, the kinematics of a continuum point is described by two tensors

$$\text{strain: } \gamma_{ji} = u_{i,j} - \boldsymbol{\varepsilon}_{kji}\varphi_k, \qquad \text{torsion-curvature: } \kappa_{ji} = \varphi_{i,j}\,, \tag{1.75}$$

which, in general, are non-symmetric.

While the classical (or Cauchy) continuum mechanics relies on the force stress concept as the only interaction between contiguous material particles, the micropolar continuum also involves a couple stress concept. Mathematically, the latter is an adaption of the former, and this is seen as follows. First, upon taking a finite surface area ΔA $(= L^2)$, defined by an outer unit normal $\boldsymbol{n}$, and a force $\Delta \boldsymbol{F}$ acting on ΔA, one takes the limit of the ratio of $\Delta \boldsymbol{F}$ to ΔA

$$\lim_{\Delta A \to 0} \frac{\Delta \boldsymbol{F}^{(\boldsymbol{n})}}{\Delta A} = \boldsymbol{t}^{(\boldsymbol{n})}. \tag{1.76}$$

It is the basic postulate of continuum mechanics that such a limit is well defined, i.e. that it is finite except the singularity points in the body, such as crack tips. Next, considering the force balance on a Cauchy tetrahedron, one finds a *force–stress tensor* $\boldsymbol{\tau}$ as a linear mapping from $\boldsymbol{n}$ into $\boldsymbol{t}^{(\boldsymbol{n})}$

$$\boldsymbol{t}^{(\boldsymbol{n})} = \boldsymbol{\tau} \cdot \boldsymbol{n} \quad \text{or} \quad t_i^{(\boldsymbol{n})} = \tau_{ji} \cdot n_j. \tag{1.77}$$

The symbol τ_{ji} is employed to now indicate a generally non-symmetric Cauchy stress, in contradistinction to the symmetric Cauchy stress σ_{ji} introduced in (1.17).

While the global and local forms of conservation of linear momentum – expressed, respectively, by (1.17) and (1.18) – carry through, the angular momentum equations change. First, in a micropolar (or Cosserat) continuum, one proceeds by analogy to (1.76) to define a moment traction from a surface couple (moment) $\Delta \boldsymbol{M}$ acting on ΔA:

$$\lim_{\Delta A \to 0} \frac{\Delta \boldsymbol{M}^{(\boldsymbol{n})}}{\Delta A} = \boldsymbol{m}^{(\boldsymbol{n})}, \tag{1.78}$$

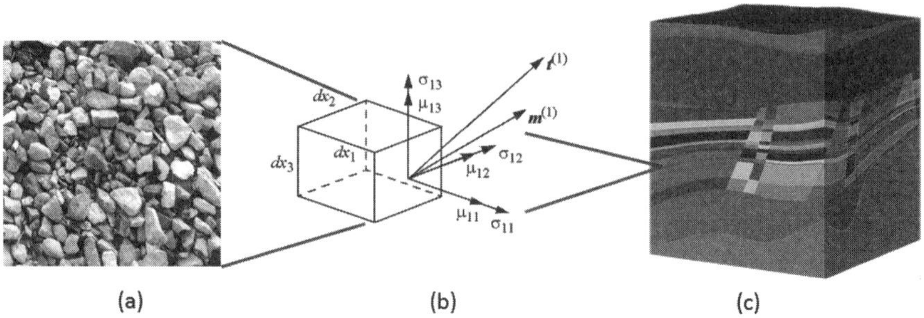

Figure 1.4 A granular medium (a) carrying force and moment tractions, is represented as a micropolar continuum point (b) within a macroscopic model of a geological formation (c)

leading to, now by analogy to (1.77), a *couple-stress tensor* $\boldsymbol{\mu}$ as a linear mapping from $\boldsymbol{n}$ into $\boldsymbol{m}^{(\boldsymbol{n})}$

$$\boldsymbol{m}^{(\boldsymbol{n})} = \boldsymbol{\mu} \cdot \boldsymbol{n} \quad \text{or} \quad m_i^{(\boldsymbol{n})} = \mu_{ji} \cdot n_j. \tag{1.79}$$

In Figure 1.4 both $\boldsymbol{t}^{(\boldsymbol{n})}$ and $\boldsymbol{m}^{(\boldsymbol{n})}$ are shown acting on one face of SVE, a dV element of the macroscopic micropolar (and inhomogeneous) continuum. Then, the conservation of angular momentum (1.19) needs to include the contributions of body and surface couples, leading to the local form

$$\boldsymbol{\varepsilon}_{ijk}\tau_{jk} + \mu_{ji,j} + Y_i = I_{ij}\ \ddot{\varphi}. \tag{1.80}$$

The linear and angular momentum balance laws may also be obtained from the invariance of energy balance under the Galilean group of transformations (Eringen 1999; Nowacki 1975). A modern discussion of the modelling based on generalised continua can be found in dell'Isola, Seppecher & Della Corte (2015).

Micropolar elasticity

Clearly, (τ_{ij}, γ_{ij}) and (μ_{ij}, κ_{ij}) are conjugate pairs. Assuming a micropolar material of linear elastic type, its energy density is given by a quadratic form

$$u = \frac{1}{2}\gamma_{ij}\boldsymbol{C}^{(1)}_{ijkl}\gamma_{kl} + \frac{1}{2}\kappa_{ij}\boldsymbol{C}^{(2)}_{ijkl}\kappa_{kl},$$

so that Hooke's law is generalised from (1.33) to

$$\tau_{ij} = \boldsymbol{C}^{(1)}_{ijkl}\gamma_{kl} \qquad \mu_{ij} = \boldsymbol{C}^{(2)}_{ijkl}\kappa_{kl}. \tag{1.81}$$

Here $\boldsymbol{C}^{(1)}_{ijkl}$ and $\boldsymbol{C}^{(2)}_{ijkl}$ are two micropolar stiffness tensors. Note that, due to the assumption of energy density u, we have the major symmetry of both stiffness (and hence, compliance) tensors

$$\boldsymbol{C}^{(1)}_{ijkl} = \boldsymbol{C}^{(1)}_{klij} \qquad \boldsymbol{C}^{(2)}_{ijkl} = \boldsymbol{C}^{(2)}_{klij}.$$

However, the two minor symmetries of (1.35) ($ijkl \leftrightarrow jikl \leftrightarrow ijlk$) do not hold since the tensors τ_{ij} and γ_{ij} as well as μ_{ij} and κ_{ij} are, in general, non-symmetric. Indeed, this is the reason for calling this theory an *asymmetric elasticity* by Nowacki (1986). The inverse of (1.81) is written as

$$\gamma_{ij} = \mathbf{S}^{(1)}_{ijkl}\tau_{kl} \qquad \kappa_{ij} = \mathbf{S}^{(2)}_{ijkl}\mu_{kl}.$$

In principle, the energy density of linear micropolar elasticity is written most generally as

$$u = \frac{1}{2}\gamma_{ij}\mathbf{C}^{(1)}_{ijkl}\gamma_{kl} + \gamma_{ij}\mathbf{C}^{(3)}_{ijkl}\kappa_{kl} + \frac{1}{2}\kappa_{ij}\mathbf{C}^{(2)}_{ijkl}\kappa_{kl},$$

in which the mixed term accounts for thenon-centrosymmetric (chiral) effects, essentially due to some helical microstructure. This is seen more explicitly from

$$\tau_{ij} = \mathbf{C}^{(1)}_{ijkl}\gamma_{kl} + \mathbf{C}^{(3)}_{ijkl}\kappa_{kl} \qquad \mu_{ij} = \mathbf{C}^{(3)}_{ijkl}\gamma_{kl} + \mathbf{C}^{(2)}_{ijkl}\kappa_{kl},$$

which replaces (1.81).

We end this subsection by noting that, just like in classical elasticity, we can express a micropolar field problem in displacements and rotations or in stresses and couple-stresses (Nowacki 1986).

1.7.2 Local versus Non-Local Continuum Mechanics

Elasticity

Following Bažant & Jirásek (2002), with reference to (1.1), we recall the fundamental equation of the continuum physics theory. Typically, $\mathbf{u}$ and $\mathbf{f}$ are tensor fields over a certain spatio-temporal domain $\mathcal{D} \times \mathcal{T}$. The next step is the *locality condition*: 'The operator $\mathcal{L}$ is called local if, for two functions $\mathbf{u}$ and $\mathbf{v}$ identical on an open set $O \subset \mathcal{D}$, their images $\mathcal{L}\mathbf{u}$ and $\mathcal{L}\mathbf{v}$ are also identical in O.' Equivalently, we can say that whenever $\mathbf{u}(\boldsymbol{x}) = \mathbf{v}(\boldsymbol{x})$ for all $\boldsymbol{x}$ in a neighbourhood of point $\boldsymbol{x}_0$, then $\mathcal{L}\mathbf{u}(\boldsymbol{x}_0) = \mathcal{L}\mathbf{v}(\boldsymbol{x}_0)$. Clearly, differential operators satisfy this condition because the derivatives of an arbitrary order do not change if the differentiated field changes only outside a small neighbourhood of the point where a derivative is taken.

Let us first consider the 1D local elasticity described by

$$-\frac{d}{dx}\left[E(x)\frac{d}{dx}u(x)\right] = f(x). \tag{1.82}$$

It is easily verified that the locality condition is satisfied here. Also note that (1.82) combines three relations:

$$\text{strain-displacement} \qquad \varepsilon(x) = \frac{d}{dx}u(x), \tag{1.83a}$$

$$\text{equilibrium} \qquad \frac{d}{dx}\sigma(x) + f(x) = 0, \tag{1.83b}$$

$$\text{constitutive equation} \qquad \sigma(x) = E(x)\,\varepsilon(x). \tag{1.83c}$$

On the other hand, in non-local elasticity (1.83c) is replaced by

$$\sigma(x) = \int_{-\infty}^{\infty} E(x,\xi)\, u(\xi)\, d\xi, \tag{1.84}$$

where $E(x,\xi)$ is the kernel of the elastic integral operator. It generalises the notion of conventional (local) elastic modulus. The resulting generalisation of (1.82) is

$$-\frac{d}{dx}\left[\int_{-\infty}^{\infty} E(x,\xi)\frac{du}{dx}(\xi)\, d\xi\right] = f(x). \tag{1.85}$$

Due to the presence of spatial integral, the locality condition is violated, unless the elastic kernel has the degenerate form $E(x,\xi) = E(x)\delta(x-\xi)$, whereby the local/classical elasticity is recovered.

This would suggest that the local theories are described by differential equations, whereas the non-local theories by integro-differential ones. In general, continuum models for solids can be divided into:

1. *strictly local models*, encompassing non-polar simple materials;
2. *weakly non-local models*, encompassing polar theories (micropolar, microstretch, micromorphic, multipolar) and gradient theories;
3. *strongly non-local models*, encompassing those of the integral type above.

Before we introduce some of these particular models in some detail, we extend the non-local model from 1D to 3D. Working within a linear small-strain assumption, an elasticity theory can be derived from the hypothesis that the elastic energy is a quadratic functional:

$$W = \frac{1}{2}\int_{\mathcal{D}}\int_{\mathcal{D}} \varepsilon_{ij}(\boldsymbol{x})\, \boldsymbol{C}_{ijkl}(\boldsymbol{x},\boldsymbol{\xi})\, \varepsilon_{kl}(\boldsymbol{\xi})\, d\boldsymbol{x}d\boldsymbol{\xi}. \tag{1.86}$$

Observe that this is in contradistinction to the local elasticity where the total energy can be expressed as a spatial integral depending only on the local value of strain. Thus, only if $\boldsymbol{C}(\boldsymbol{x},\boldsymbol{\xi}) = \boldsymbol{C}(\boldsymbol{x})\,\delta(\boldsymbol{x}-\boldsymbol{\xi})$, does (1.86) reduce to

$$W = \frac{1}{2}\int_{\mathcal{D}} \varepsilon_{ij}(\boldsymbol{x})\, \boldsymbol{C}_{ijkl}(\boldsymbol{x})\, \varepsilon_{kl}(\boldsymbol{x})\, d\boldsymbol{x}.$$

Many particular models as well as more general theories (possibly with dissipation) have been, and continue to be, discussed in the literature.

The ensemble average of a random local medium is non-local

As pointed out in Section 1.1, the exact solution of the average of a field problem governed by a linear random (and local) operator on the domain $\mathcal{B}$ follows from (1.4). The operator $\langle \mathcal{L}^{-1}\rangle^{-1}$ is a deterministic and non-local. This was illustrated in terms of a Fourier-type heat conduction problem, a result which immediately carries over to anti-plane elasticity. Moving to a general setting of linear elastostatics on the random field of a fourth-rank stiffness tensor

$\mathbf{C}_{ijkl} = \{\mathbf{C}_{ijkl}(\omega, \boldsymbol{x}); \omega \in \Omega, \boldsymbol{x} \in \mathcal{B}\}$, the equations governing the average fields in a Cauchy-type continuum are (Beran & McCoy 1970*a*)

$$\langle \sigma_{ij}(\boldsymbol{x}) \rangle_{,j} = 0 \tag{1.87a}$$

$$\langle \sigma_{ij}(\boldsymbol{x}) \rangle = \int_{\mathcal{B}} \mathbf{\Lambda}_{ijkl}(\boldsymbol{x}, \boldsymbol{x}') \langle \varepsilon_{kl}(\boldsymbol{x}') \rangle \, d\boldsymbol{x}' \tag{1.87b}$$

$$\langle \varepsilon_{kl} \rangle = (\langle u_{k,l} \rangle + \langle u_{l,k} \rangle)/2. \tag{1.87c}$$

In (1.87b) $\mathbf{\Lambda}_{ijkl}(\boldsymbol{x}, \boldsymbol{x}')$ is an infinite sum of integro-differential operators, involving the moments of all orders of the random field $\mathbf{C}_{ijkl}$

$$\mathbf{\Lambda}_{ijkl}(\boldsymbol{x}, \boldsymbol{x}') = [\langle \mathbf{C}_{ijkl} \rangle + \mathbf{D}_{ijkl}(\boldsymbol{x}')] \, \delta(\boldsymbol{x} - \boldsymbol{x}') + \mathbf{E}_{ijkl}(\boldsymbol{x}, \boldsymbol{x}') + \ldots, \tag{1.88}$$

where $\mathbf{D}_{ijkl}(\boldsymbol{x}')$ and $\mathbf{E}_{ijkl}(\boldsymbol{x}, \boldsymbol{x}')$ are functions of the statistical properties of $\mathbf{C}_{ijkl}$ and the free-space Green's function of the non-statistical problem. Addition of a deterministic body force field f_i does not change the results. When the fluctuations in $\mathbf{C}_{ijkl}$ are small, $\mathbf{D}_{ijkl}(\boldsymbol{x}')$ and $\mathbf{E}_{ijkl}(\boldsymbol{x}, \boldsymbol{x}')$ may be evaluated explicitly, and this was done by Beran & McCoy (1970*a*) and Beran & McCoy (1970*b*) in the special case of the realisations $\mathbf{C}_{ijkl}(\omega)$ being locally isotropic, i.e. expressed in terms of a "vector" random field of two Lamé coefficients $\{[\lambda(\omega, \boldsymbol{x}), \mu(\omega, \boldsymbol{x})]; \omega \in \Omega, \boldsymbol{x} \in \mathcal{B}\}$; recall the Introduction.

Next, considering this random field to be statistically homogeneous and mean-ergodic, one may disregard the contributions of this operator for $|\boldsymbol{x} - \boldsymbol{x}'| > l_c$ (the correlation length). Thus, since only the neighbourhood within the distance l_c of $\boldsymbol{x}$ has a significant input into the integral (1.87b), one may expand $\langle \varepsilon_{kl}(\boldsymbol{x}') \rangle$ in a power series about $\boldsymbol{x}$

$$\langle \varepsilon_{kl}(\boldsymbol{x}') \rangle = \langle \varepsilon_{kl}(\boldsymbol{x}) \rangle + (x'_m - x_m) \langle \varepsilon_{kl}(\boldsymbol{x}) \rangle_{,m} + \frac{(x'_m - x_m)(x'_n - x_n)}{2} \langle \varepsilon_{kl}(\boldsymbol{x}) \rangle_{,mn} \cdots$$

so as to obtain from (1.87) and (1.88)

$$\langle \sigma_{ij}(\boldsymbol{x}) \rangle = \int_{\mathcal{B}} \mathbf{\Lambda}_{ijkl}(\boldsymbol{x}, \boldsymbol{x}') \, d\boldsymbol{x}' \, \langle \varepsilon_{kl}(\boldsymbol{x}) \rangle + \int_{\mathcal{B}} \mathbf{\Lambda}_{ijkl}(\boldsymbol{x}, \boldsymbol{x}') (x'_m - x_m) \, d\boldsymbol{x}' \, \langle \varepsilon_{kl}(\boldsymbol{x}) \rangle_{,m} + \ldots$$

This, in turn, can be rewritten as a sum of local, plus first gradient, plus higher gradient strain effects

$$\langle \sigma_{ij} \rangle = \mathbf{C}^*_{ijkl} \langle \varepsilon_{kl} \rangle + \mathbf{D}^*_{ijklm} \langle \varepsilon_{kl} \rangle_{,m} + \mathbf{E}^*_{ijklmn} \langle \varepsilon_{kl} \rangle_{,mn} + \ldots$$

Thus, $\int_{\mathcal{B}} \mathbf{\Lambda}_{ijkl}(\boldsymbol{x}, \boldsymbol{x}') \, d\boldsymbol{x}'$ in (1.87) is recognised as the *effective stiffness* $\mathbf{C}^*_{ijkl}$, indeed the stiffness of a single realisation $B(\omega)$ of the random material $\mathcal{B}$. It follows that, for a given random medium $\mathcal{B}$, one may determine the microstructural statistics, and hence the higher-order approximations $\mathbf{D}^*_{ijklm}$, $\mathbf{E}^*_{ijklmn}$, ...

1.8 Concluding Remarks

Using a common perspective of TIV, we have attempted to introduce a range of continuum theories: from classical (Cauchy-type), single- and multi-field, to generalised, such as micropolar (Cosserat-type) and nonlocal. Precisely due to the presence of the dissipation function, this approach can also bridge down to statistical physics involving nanoscale violations of the Second Law of thermodynamics. In all these theories and countless other possibilities, the key role is played by the tensor fields of dependent quantities and constitutive responses, which are generally anisotropic and spatially inhomogeneous. This is the motivation for admitting the spatial randomness of tensor-valued random fields.

2

Mathematical Preliminaries

In this chapter, we review mathematical tools that are necessary to solve our problems. We start from a review of *Linear Algebra*. Two things are important here. First, a specialist in applied mathematics or physics usually understands 'linear space' as a set of rows or columns that contain numbers. We found that it is easier to formulate and to start to solve a problem in a coordinate-free form. Later, one chooses the most convenient coordinate system to write down the answer.

Second, there exist many equivalent definitions of tensors. We found that initially it is convenient to use Definitions 1 and 2 and even the most abstract Theorem 1 to define the tensor product of linear spaces rather than the classical physical Definition 3.

Material in *topology, groups and group actions* is standard. We would like to mention Example 7, where we consider the so-called 'classical groups' that play an important rôle later.

Group representations are familiar to all physicists. We consider *orthogonal* tensor representations of compact topological groups as a particular case of *group actions*. Especially we are interested in their matrix entries and Clebsch–Gordan coefficients.

A connection between the theory of homogeneous and isotropic random fields and classical invariant theory has been known since the 1930s. The reason for that is as follows. Physically interesting random fields take values in subspaces of tensors that are invariant under the natural action of a permutation group in the space of *all* tensors of a fixed rank, in particular, in spaces of *symmetric tensors*. On the other hand, the space of symmetric tensors is naturally isomorphic to the space of homogeneous polynomials; see Equation (2.10) below. Therefore, homogeneous polynomial mappings that intertwine different group representations, are important. It is the classical invariant theory that studies the above described polynomial mappings. We pay special attention to *symmetric isotropic tensors and symmetric covariant tensors*. Many of them were introduced by the authors.

Many integrals that arise in studies of homogeneous and isotropic random fields, contain a function that takes values in a convex compact subset of a finite-dimensional real linear space. The geometry of the above set, in particular, the structure of the set of its *extreme points*, determines the spectral expansion of the field: the simpler the set of extreme points is, the simpler the spectral expansion is. We review the theory of convex compacta in Section 2.8.

In Section 2.9 we review the theory of homogeneous tensor-valued random fields. A special rôle is played by Theorem 10, which completely describes the class of such fields. The above theorem is the starting point of our research in Chapter 3.

The results of integration in the theory of homogeneous and isotropic random fields often contain *special functions*. We describe the necessary functions in Section 2.10.

2.1 Linear Spaces

In this section we review some results in Linear Algebra. Each result appears first in a coordinate-free form, then in a coordinate-dependent setting.

2.1.1 Forms

Let $\mathbb{F}$ be either the field $\mathbb{R}$ of real numbers or the field $\mathbb{C}$ of complex numbers. Let V be a finite-dimensional linear space over $\mathbb{F}$. In particular, let $\mathbb{F}^1$ denote the *one-dimensional* linear space over $\mathbb{F}$. A *linear form* on V is a function $\boldsymbol{v}^*\colon V \to \mathbb{F}^1$ satisfying the following condition

$$\boldsymbol{v}^*(\alpha\boldsymbol{v} + \beta\boldsymbol{w}) = \alpha\boldsymbol{v}^*(\boldsymbol{v}) + \beta\boldsymbol{v}^*(\boldsymbol{w}), \qquad \boldsymbol{v}, \boldsymbol{w} \in V, \quad \alpha, \beta \in \mathbb{F}.$$

The set V^* of all linear forms on V is a linear space (the *dual space* to V). The addition of linear forms is defined as

$$(\boldsymbol{v}^* + \boldsymbol{w}^*)(\boldsymbol{v}) := \boldsymbol{v}^*(\boldsymbol{v}) + \boldsymbol{w}^*(\boldsymbol{v}),$$

while the multiplication of a linear form $\boldsymbol{v}^*$ by a scalar $\alpha \in \mathbb{F}$ is defined as

$$(\alpha\boldsymbol{v}^*)(\boldsymbol{v}) := \alpha\boldsymbol{v}^*(\boldsymbol{v}).$$

The spaces V and V^* have the same dimension and therefore must be isomorphic. However, there exists no *natural isomorphism* between a space and its dual, see Section 2.11.

In what follows we write $\langle\boldsymbol{v}^*, \boldsymbol{v}\rangle = \boldsymbol{v}^*(\boldsymbol{v})$ for $\boldsymbol{v}^* \in V^*$ and $\boldsymbol{v} \in V$.

In a coordinate form, let m be the dimension of the space V, and let $\{\boldsymbol{e}_1, \dots, \boldsymbol{e}_m\}$ be a basis of V. Specify a vector $\boldsymbol{v} \in V$ by its coordinates v^1, …, v^m in the above basis:

$$\boldsymbol{v} = \sum_{i=1}^{m} v^i\boldsymbol{e}_i. \tag{2.1}$$

In what follows, we use the *Einstein summation convention*: whenever an index variable appears twice in a single term it implies summation of that term over all the values of the index. With this convention, Equation (2.1) becomes

$$\boldsymbol{v} = v^i \boldsymbol{e}_i,$$

or just $\boldsymbol{v} = v^i$ when $\boldsymbol{e}_i$ is understood. Let $\boldsymbol{e}^i \in V^*$ be the linear functional acting on the vector (2.1) by $\langle \boldsymbol{e}^i, \boldsymbol{v} \rangle = v^i$. The functionals $\{\boldsymbol{e}^1, \dots, \boldsymbol{e}^m\}$ form a basis of V^*, the so-called *dual basis* with respect to the basis $\{\boldsymbol{e}_1, \dots, \boldsymbol{e}_m\}$. The unique linear mapping $A\colon V \to V^*$ that transforms $\boldsymbol{e}_i$ to $\boldsymbol{e}^i$ is an isomorphism between V and V^*. This isomorphism is not natural but rather *accidental*: it changes if the basis $\{\boldsymbol{e}_1, \dots, \boldsymbol{e}_m\}$ is changed.

Let $V^{**} := (V^*)^*$ be the space of linear forms on V^*, the *second dual* to V. Consider the mapping $\varepsilon_V\colon V \to V^{**}$ that maps a vector $\boldsymbol{v} \in V$ to the linear form on V^* whose value on the element $\boldsymbol{v}^* \in V^*$ is equal to $\langle \boldsymbol{v}^*, \boldsymbol{v} \rangle$:

$$\varepsilon_V : \boldsymbol{v} \to [\boldsymbol{v}^* \mapsto \langle \boldsymbol{v}^*, \boldsymbol{v} \rangle].$$

The mapping ε_V is the natural isomorphism between V and V^{**}.

Assume that V is a complex vector space. Define the *conjugate space* $\overline{V}$ which has the same vector addition as V but scalar-vector multiplication defined by

$$\mathbb{C} \times V \to V, \qquad (z, \boldsymbol{v}) \mapsto \bar{z}\boldsymbol{v}.$$

Let V and W be two linear spaces. A function $A\colon V \times W \to \mathbb{F}^1$ is called a *bilinear form* on $V \times W$ if for all $\boldsymbol{v}$, $\boldsymbol{v}_1$, $\boldsymbol{v}_2 \in V$, $\boldsymbol{w}$, $\boldsymbol{w}_1$, $\boldsymbol{w}_2 \in W$, α, $\beta \in \mathbb{F}$ we have:

$$\begin{aligned} A(\alpha\boldsymbol{v}_1 + \beta\boldsymbol{v}_2, \boldsymbol{w}) &= \alpha A(\boldsymbol{v}_1, \boldsymbol{w}) + \beta A(\boldsymbol{v}_2, \boldsymbol{w}), \\ A(\boldsymbol{v}, \alpha\boldsymbol{w}_1 + \beta\boldsymbol{w}_2) &= \alpha A(\boldsymbol{v}, \boldsymbol{w}_1) + \beta A(\boldsymbol{v}, \boldsymbol{w}_2). \end{aligned} \tag{2.2}$$

The set $L(V, W; \mathbb{F}^1)$ of all bilinear forms on $V \times W$ is a linear space with respect to the obviously defined operations of addition of bilinear forms and multiplication of a bilinear form by a scalar.

In a coordinate form, let n be the dimension of the space W, and let $\{\boldsymbol{f}_1, \dots, \boldsymbol{f}_n\}$ be a basis of W. Denote

$$a_{ij} = A(\boldsymbol{e}_i, \boldsymbol{f}_j).$$

Using (2.2), we obtain

$$A(e^i, f^j) = a_{ij} e^i f^j.$$

The matrix with matrix entries a_{ij} is called the *matrix of the bilinear form* A.

In general, let r be a positive integer, and let V_1, V_2, $\dots$, V_r be finite-dimensional linear spaces over $\mathbb{F}$. A function $A\colon V_1 \times V_2 \times \cdots \times V_r \to \mathbb{F}^1$ is called a *multilinear form* if for all i, $1 \leq i \leq p$ and for all $\boldsymbol{v}_j \in V_j$, $j \neq i$, $\boldsymbol{w}_1$, $\boldsymbol{w}_2 \in V_i$, α, $\beta \in \mathbb{F}$ we have

$$\begin{aligned}
A(\boldsymbol{v}_1, \ldots, \boldsymbol{v}_{i-1}, \alpha\boldsymbol{w}_1 + \beta\boldsymbol{w}_2, \boldsymbol{v}_{i+1}, \ldots, \boldsymbol{v}_r) \\
= \alpha A(\boldsymbol{v}_1, \ldots, \boldsymbol{v}_{i-1}, \boldsymbol{w}_1, \boldsymbol{v}_{i+1}, \ldots, \boldsymbol{v}_r) \\
+ \beta A(\boldsymbol{v}_1, \ldots, \boldsymbol{v}_{i-1}, \boldsymbol{w}_2, \boldsymbol{v}_{i+1}, \ldots, \boldsymbol{v}_r).
\end{aligned}$$

The set $L(V_1, V_2, \ldots, V_r; \mathbb{F}^1)$ of all multilinear forms on $V_1 \times V_2 \times \cdots \times V_r$ is a linear space with respect to the obviously defined operations of addition of bilinear forms and multiplication of a bilinear form by a scalar.

A bilinear form $B\colon V \times V \to \mathbb{F}^1$ is called *non-degenerate* if $B(\boldsymbol{v}, \boldsymbol{w}) = 0$ for all $\boldsymbol{w} \in V$ implies that $\boldsymbol{v} = \mathbf{0}$ and $B(\boldsymbol{v}, \boldsymbol{w}) = 0$ for all $\boldsymbol{v} \in V$ implies that $\boldsymbol{w} = \mathbf{0}$. It is called *symmetric* if $B(\boldsymbol{v}, \boldsymbol{w}) = B(\boldsymbol{w}, \boldsymbol{v})$.

Let V be a vector space over $\mathbb{R}$. A bilinear form $B\colon V \times V \to \mathbb{R}$ is called *positive definite* if $B(\boldsymbol{v}, \boldsymbol{v}) > 0$ for every $\boldsymbol{v} \in V$ with $\boldsymbol{v} \neq \mathbf{0}$. An *inner product* on V is a symmetric, non-degenerate and positive definite bilinear form. The value of the inner product on the vectors $\boldsymbol{v}$ and $\boldsymbol{w}$ is denoted by $(\boldsymbol{v}, \boldsymbol{w})$.

Every inner product determines an isomorphism between V and V^* associating with every vector $\boldsymbol{w} \in V$ a linear form $\boldsymbol{w}^*\colon \boldsymbol{v} \to (\boldsymbol{v}, \boldsymbol{w})$.

Let V be a vector space over $\mathbb{C}$ and let I be the identity operator in V. A *real structure* on V is a mapping $J\colon V \to V$ that satisfies the following conditions:

$$\begin{aligned}
J(\alpha\boldsymbol{v} + \beta\boldsymbol{w}) &= \overline{\alpha}J(\boldsymbol{v}) + \overline{\beta}J(\boldsymbol{w}), \qquad \alpha, \beta \in \mathbb{C}, \quad \boldsymbol{v}, \boldsymbol{w} \in V, \\
J^2 &= I.
\end{aligned}$$

In coordinates, a mapping

$$J(\alpha_1\boldsymbol{e}_1 + \cdots + \alpha_m\boldsymbol{e}_m) := \overline{\alpha_1}\boldsymbol{e}_1 + \cdots + \overline{\alpha_m}\boldsymbol{e}_m$$

is a real structure on V.

Later on we need also a *quaternionic structure*. This is a mapping $J\colon V \to V$ that satisfies the following conditions:

$$\begin{aligned}
J(\alpha\boldsymbol{v} + \beta\boldsymbol{w}) &= \overline{\alpha}J(\boldsymbol{v}) + \overline{\beta}J(\boldsymbol{w}), \qquad \alpha, \beta \in \mathbb{C}, \quad \boldsymbol{v}, \boldsymbol{w} \in V, \\
J^2 &= -I.
\end{aligned}$$

A function $B\colon V \times V \to \mathbb{C}$ is called a *Hermitian form* if for all α, $\beta \in \mathbb{C}$, $\boldsymbol{v}$, $\boldsymbol{w}$, $\boldsymbol{w}_1$, $\boldsymbol{w}_2 \in V$

$$\begin{aligned}
B(\boldsymbol{v}, \alpha\boldsymbol{w}_1 + \beta\boldsymbol{w}_2) &= \alpha B(\boldsymbol{v}, \boldsymbol{w}_1) + \beta B(\boldsymbol{v}, \alpha\boldsymbol{w}_2), \\
B(\boldsymbol{v}, \boldsymbol{w}) &= \overline{B(\boldsymbol{w}, \boldsymbol{v})}.
\end{aligned}$$

A *Hermitian inner product* is a non-degenerate positive definite Hermitian form $(\boldsymbol{v}, \boldsymbol{w})$ on V. For every Hermitian linear product and every real structure J, we have a non-degenerate bilinear form $(J\boldsymbol{v}, \boldsymbol{w})$ that determines an isomorphism between V and V^* associating with every vector $\boldsymbol{w} \in V$ a linear form $\boldsymbol{w}^*\colon \boldsymbol{v} \to (J\boldsymbol{v}, \boldsymbol{w})$.

2.1.2 Tensors

An important particular case of a bilinear form is as follows.

Definition 1. The *tensor product* $V \otimes W$ of linear spaces V and W is the linear space $L(V^*, W^*; \mathbb{F}^1)$. The tensor product $\boldsymbol{v} \otimes \boldsymbol{w}$ of the vectors $\boldsymbol{v} \in V$ and $\boldsymbol{w} \in W$ is the element of the space $V \otimes W$ acting by

$$\boldsymbol{v} \otimes \boldsymbol{w}(\boldsymbol{v}^*, \boldsymbol{w}^*) := \langle \boldsymbol{v}^*, \boldsymbol{v} \rangle \langle \boldsymbol{w}^*, \boldsymbol{w} \rangle, \qquad \boldsymbol{v}^* \in V, \quad \boldsymbol{w}^* \in W.$$

A *rank* 2 *tensor* is an element of the tensor product $V \otimes W$.

In a coordinate form, a rank 2 tensor A is a matrix with elements

$$a^{ij} := A(\boldsymbol{e}^i, \boldsymbol{f}^j).$$

The tensor product of the vectors v^i and f^j is $a^{ij} = v^i f^j$.

Let τ be the linear operator acting from $V \times W$ to $V \otimes W$ by

$$\tau(\boldsymbol{v}, \boldsymbol{w}) := \boldsymbol{v} \otimes \boldsymbol{w}.$$

Theorem 1 (The universal mapping property)**.** *The pair* $(V \otimes W, \tau)$ *satisfies the* universal mapping property*: for any linear space* X *and bilinear map* $\beta \colon V \times W \to X$, *there exists a unique* linear *operator* $B \colon U \otimes V \to X$ *such that* $\beta = B \circ \tau$*:*

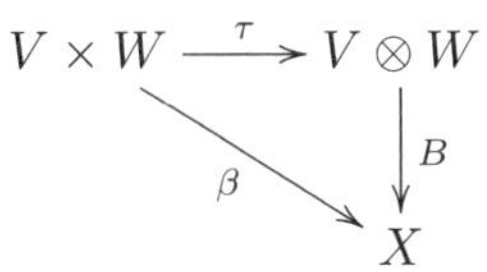

(2.3)

In other words, the diagram (2.3) is *commutative*: all directed paths in the diagram with the same start and endpoints lead to the same result by composition.

Example 1. As a first application of Theorem 1, put $X = L(V, W)$. For any $(\boldsymbol{v}^*, \boldsymbol{w}) \in V^* \times W$, let $B_{\boldsymbol{v}^*, \boldsymbol{w}}$ be the rank-one linear operator:

$$B_{\boldsymbol{v}^*, \boldsymbol{w}}(\boldsymbol{v}) := \langle \boldsymbol{v}^*, \boldsymbol{v} \rangle \boldsymbol{w}, \qquad \boldsymbol{v} \in V.$$

The map $\beta(\boldsymbol{v}^*, \boldsymbol{w}) := B_{\boldsymbol{v}^*, \boldsymbol{w}}$ from $V^* \times W$ to $L(V, W)$ is bilinear. By Theorem 1, there is a unique linear map $B \colon V^* \otimes W \to L(V, W)$ such that $B(\boldsymbol{v}^* \otimes \boldsymbol{w}) = B_{\boldsymbol{v}^*, \boldsymbol{w}}$. In other words, B is a natural isomorphism between the spaces $V^* \otimes W$ and $L(V, W)$.

In a coordinate form, we have $B(\boldsymbol{v}_i \otimes \boldsymbol{w}^j) \boldsymbol{v}^k = \langle \boldsymbol{v}_i, \boldsymbol{v}^k \rangle \boldsymbol{w}^j$.

Let V, W, X and Y be finite-dimensional linear spaces, let $A \in L(V, X)$ and let $B \in L(W, Y)$. The *tensor product of linear operators*, $A \otimes B$, is a unique element of the space $L(V \otimes W, X \otimes Y)$ such that

$$(A \otimes B)(\boldsymbol{v} \otimes \boldsymbol{w}) := A(\boldsymbol{v}) \otimes B(\boldsymbol{w}), \qquad \boldsymbol{v} \in W, \quad \boldsymbol{w} \in W.$$

Definition 2. The *r-fold tensor product* $V_1 \otimes V_2 \otimes \cdots \otimes V_r$ is the linear space $L(V_1^*, V_2^*, \dots, V_r^*; \mathbb{F}^1)$. The r-fold tensor product $\boldsymbol{v}_1 \otimes \boldsymbol{v}_2 \otimes \cdots \otimes \boldsymbol{v}_r$ of the vectors

$\boldsymbol{v}_1 \in V_1$, $\boldsymbol{v}_2 \in V_2$, ..., $\boldsymbol{v}_r \in V_r$ is the element of the space $V_1 \otimes V_2 \otimes \cdots \otimes V_r$ acting by

$$\boldsymbol{v}_1 \otimes \boldsymbol{v}_2 \otimes \cdots \otimes \boldsymbol{v}_r(\boldsymbol{v}_1^*, \boldsymbol{v}_2^*, \ldots, \boldsymbol{v}_r^*) := \langle \boldsymbol{v}_1^*, \boldsymbol{v}_1 \rangle \langle \boldsymbol{v}_2^*, \boldsymbol{v}_2 \rangle \cdots \langle \boldsymbol{v}_r^*, \boldsymbol{v}_r \rangle$$

for all $1 \leq i \leq r$ and for all $\boldsymbol{v}_i^* \in V_i^*$. A *rank r tensor* is an element of the r-fold tensor product $V_1 \otimes V_2 \otimes \cdots \otimes V_r$. By convention, the 0-fold tensor product of the empty family of linear spaces is equal to $\mathbb{F}^1$.

In a coordinate form, a rank r tensor $\boldsymbol{A}$ is an r-dimensional array with elements

$$a^{i_1 i_2 \cdots i_r} := A(\boldsymbol{e}^{i_1}, \boldsymbol{e}^{i_2}, \ldots, \boldsymbol{e}^{i_r}).$$

The *universal mapping property* relative to r-linear maps is as follows: if W is a vector space and $\beta \colon V_1 \times V_2 \times \cdots \times V_r \to W$ is an r-linear map (i.e. linear in each argument), then there exists a unique linear map $B \colon V_1 \otimes V_2 \otimes \cdots \otimes V_r \to W$ such that $B(\boldsymbol{v}_1 \otimes \boldsymbol{v}_2 \otimes \cdots \otimes \boldsymbol{v}_r) = \beta(\boldsymbol{v}_1, \boldsymbol{v}_2, \cdots, \boldsymbol{v}_r)$. In other words: *the construction of the tensor product of linear spaces reduces the study of multilinear mappings to the study of linear ones.*

Let V_1, ..., V_r, W_1, ..., W_r be finite-dimensional linear spaces, and let $A_i \in L(V_i, W_i)$ for $1 \leq i \leq r$. The tensor product of linear operators, $A_1 \otimes \cdots \otimes A_r$, is a unique element of the space $L(V_1 \otimes \cdots \otimes V_r, W_1 \otimes \cdots \otimes W_r)$ such that

$$(A_1 \otimes \cdots \otimes A_r)(\boldsymbol{v}_1 \otimes \cdots \otimes \boldsymbol{v}_r) := A_1(\boldsymbol{v}_1) \otimes \cdots \otimes A_r(\boldsymbol{v}_r), \qquad \boldsymbol{v}_i \in V_i.$$

If all the spaces V_i, $1 \leq i \leq r$, are copies of the same space V, then we write $V^{\otimes r}$ for the r-fold tensor product of V with itself, and $\boldsymbol{v}^{\otimes r}$ for the tensor product of r copies of a vector $\boldsymbol{v} \in V$. Similarly, for $A \in L(V, V)$ we write $A^{\otimes r}$ for the r-fold tensor product of A with itself.

Denote $\mathsf{T}^p_q(V) := V^{\otimes p} \otimes (V^*)^{\otimes q}$ and call this linear space the *space of mixed tensors of type* (p, q) relative to V. By our convention, $\mathsf{T}^0_0(V) := \mathbb{F}^1$.

In a coordinate form, let $\{\boldsymbol{e}_1, \ldots, \boldsymbol{e}_m\}$ be a basis of V, and $\{\boldsymbol{e}^1, \ldots, \boldsymbol{e}^m\}$ be a dual basis of V^*. Let A^i_j be the matrix describing the change of basis in V:

$$\boldsymbol{e}'_j = A^i_j \boldsymbol{e}_j, \tag{2.4}$$

and let B^i_j be the matrix of transformation from the basis $\{\boldsymbol{e}^1, \ldots, \boldsymbol{e}^m\}$ to the basis dual to $\{\boldsymbol{e}'_1, \ldots, \boldsymbol{e}'_m\}$. The tensors

$$\{\, \boldsymbol{e}^{i_1} \otimes \cdots \otimes \boldsymbol{e}^{i_q} \otimes \boldsymbol{e}_{j_1} \otimes \cdots \otimes \boldsymbol{e}_{j_p} \colon 1 \leq i_k, j_\ell \leq m \,\} \tag{2.5}$$

form a basis in the space $\mathsf{T}^p_q(V)$.

Definition 3. A tensor $\boldsymbol{T} \in \mathsf{T}^p_q(V)$ is a mapping which associates with each basis of V a family of scalar components $\boldsymbol{T}^{j_1 \cdots j_p}_{i_1 \cdots i_q}$. Under the transformation (2.4) given by its coordinates in the basis the components of the tensor transform by

$$\boldsymbol{T}'^{j_1 \cdots j_p}_{i_1 \cdots i_q} = A^{k_1}_{i_1} \cdots A^{k_q}_{i_q} \boldsymbol{T}^{\ell_1 \cdots \ell_p}_{k_1 \cdots k_q} B^{j_1}_{\ell_1} \cdots B^{j_p}_{\ell_p}.$$

In particular, the elements of the space $\mathbb{F}^1 = \mathsf{T}^0_0(V)$ are *scalars*, the elements of the space $V = \mathsf{T}^1_0(V)$ are *vectors*, the elements of the space $V^* = \mathsf{T}^0_1(V)$ are *covectors*, the elements of the space $L(V) = \mathsf{T}^1_1(V)$ are *matrices*. In what follows, we denote scalars (rank 0 tensors) by small Greek letters α, β, ..., vectors and covectors (rank 1 tensors) by small bold italic Latin letters $\boldsymbol{v}$, $\boldsymbol{w}$, ..., matrices and other rank 2 tensors by capital Latin letters A, B, ..., tensors of rank 3 or more or of arbitrary rank by capital bold italic sans serif Latin letters $\boldsymbol{\mathsf{S}}$, $\boldsymbol{\mathsf{T}}$, ..., if the above does not conflict with traditional physical notation.

The *tensor contraction* is a unique linear form $C \in (V^* \otimes V)^*$ such that $\langle C, \boldsymbol{v}^* \otimes \boldsymbol{v} \rangle = \langle \boldsymbol{v}^*, \boldsymbol{v} \rangle$. Let $B\colon V^* \otimes V \to L(V)$ be the natural isomorphism between $V^* \otimes V$ and $L(V)$ constructed in Example 1. The *trace* of a linear operator $A \in L(V)$ is defined by $\operatorname{tr} A = CB^{-1}A$, or by a commutative diagram:

$$\begin{array}{ccc} V^* \otimes V & \xrightarrow{C} & \mathbb{F}^1 \\ {\scriptstyle B}\big\downarrow & \nearrow_{\operatorname{tr}} & \\ L(V) & & \end{array} \tag{2.6}$$

In a coordinate form, the matrix entries of the linear operator A are $a^j_i = \langle \boldsymbol{e}_i \otimes \boldsymbol{e}^j, A \rangle$. Then

$$\langle C, A \rangle = a^i_i,$$

the usual definition of the trace.

In what follows, our real linear spaces are always equipped with an inner product, while complex linear spaces are always equipped with a Hermitian inner product and a real structure. Therefore, we have an isomorphism between a linear space V and its dual, V^*, and do not care about the difference between upper and lower indices in tensors. We adopt the following convention: an index is a tensor index if and only if it is lower (if this does not contradict to the traditional physical notation).

If $(\cdot,\cdot)_i$ is a (Hermitian) inner product on the space V_i, $1 \le i \le m$, then there exists a unique (Hermitian) inner product $(\cdot,\cdot)$ on $V_1 \otimes V_2 \otimes \cdots \otimes V_m$ such that for all $\boldsymbol{v}_i$, $\boldsymbol{w}_i \in V_i$, $1 \le i \le m$, we have

$$(\boldsymbol{v}_1 \otimes \cdots \otimes \boldsymbol{v}_m, \boldsymbol{w}_1 \otimes \cdots \otimes \boldsymbol{w}_m) = (\boldsymbol{v}_1, \boldsymbol{w}_1)_1 \cdot \cdots \cdot (\boldsymbol{v}_m, \boldsymbol{w}_m)_m.$$

2.2 Topology

Let X be a non-empty set. A family $\mathcal{O}$ of subsets of X is called a *topology* if $\varnothing \in \mathcal{O}$, $X \in \mathcal{O}$, every union of sets of $\mathcal{O}$ and the intersection of any two sets from $\mathcal{O}$ are sets of $\mathcal{O}$. The pair $(X, \mathcal{O})$ is called a *topological space*. The sets from $\mathcal{O}$ are called *open*.

Example 2. Consider a finite-dimensional linear space V. Let $(\cdot,\cdot)$ be a (Hermitian) inner product in V, and let $\|\cdot\|$ be the corresponding norm:

$$\|\boldsymbol{v}\| := \sqrt{(\boldsymbol{v},\boldsymbol{v})}.$$

Declare a subset $O \subseteq V$ open, if it is either empty or for any $\boldsymbol{v} \in O$ there exists a positive real number $\varepsilon = \varepsilon(\boldsymbol{v})$ such that the *open ball*

$$B(\boldsymbol{v},\varepsilon) := \{\, \boldsymbol{w} \in V : \|\boldsymbol{w} - \boldsymbol{v}\| < \varepsilon \,\}$$

is a subset of O.

A *neighbourhood* of a point $x \in X$ is any subset of X which contains an open set containing x. A subset of X is called *closed* if its complement is open. An intersection of any family of closed sets is closed. The *closure* of a set A is the intersection of all closed sets containing A.

The *interior* of A is the union of all open sets containing in A. The *boundary* of A is the set of all points in the closure of A that do not belong to the interior of A.

A mapping f of a topological space X into a topological space Y is called *continuous* if the inverse image under f of every open subset of Y is an open subset of X.

Let A be a subset of a topological space X. The topology *induced* on A by the topology of X is the family of intersections with A of open sets of X. The set A with this topology is called a *subspace* of X.

A *base* of the topology of a topological space X is any set $\mathcal{C}$ of open sets of X such that every open set in X is the union of sets in $\mathcal{C}$.

Let X and Y be two topological spaces. By definition, the *product topology* on the Cartesian product $X \times Y$ has as a base the set of finite intersections of Cartesian products of an open set of X with an open set in Y.

A topological space X satisfies the *Hausdorff separation axiom* if any two different points in X can be separated by disjoint open sets. For example, the topological space V of Example 2 is Hausdorff. From now on we consider only Hausdorff topological spaces. It is easy to see that a finite topological space X is Hausdorff if and only if the topology of X is *discrete*, i.e. all subsets of X are open.

The following definition is a little bit abstract, therefore we first consider its simplified version. A subspace K of the topological space V of Example 2 is called *compact* if any sequence of points in K contains a subsequence that converges to a point in K.

It is possible to prove that K is compact if and only if every covering of K by open sets contains a finite covering of K. The following general definition follows: a topological space K is compact if every covering of K by open sets contains a finite covering of K. A topological space X is called *locally compact* if any point in X has a neighbourhood with compact closure. The space V of Example 2 is not compact, but is locally compact.

A topological space X is called *connected* if it is not the union of two disjoint non-empty open sets. The connected component of a point x in the space X is the union of all connected subspaces which contain x. The topological space V of Example 1 is connected.

Two particular kinds of topological spaces will play an important rôle later on.

2.2.1 Real-Analytic Manifolds

Let m be a positive integer, and let U and V be two subsets in $\mathbb{R}^m$. Let $\varphi\colon U \to V$. Then φ may be written as

$$y_1 = \varphi_1(x_1, \dots, x_m), \dots, y_m = \varphi_m(x_1, \dots, x_m).$$

Assume that these functions are real-analytic.

Let M be a set. A *chart* in M is a pair (U, φ), where $U \subseteq M$, and φ is a one-to-one map from U to an open set $\varphi(U) \subseteq \mathbb{R}^m$. For any charts $(U_\alpha, \varphi_\alpha)$ and (U_β, φ_β) with $U_\alpha \cap U_\beta \neq \varnothing$, consider the mapping

$$\varphi_\beta \circ \varphi_\alpha^{-1}\colon \varphi_\alpha(U_\alpha \cap U_\beta) \to \varphi_\beta(U_\alpha \cap U_\beta),$$

which is called a *chart change.* A family $\{\, (U_\alpha, \varphi_\alpha)\colon \alpha \in A \,\}$ of charts on M is called an *analytic atlas* if:

- U_α form a covering of M;
- both the domain and the range of any chart change are open subsets of $\mathbb{R}^m$;
- any chart change is real-analytic.

The set of all analytic atlases is partially ordered by inclusion. Any atlas $\mathcal{A}$ is contained in a unique *maximal* atlas $\mathcal{A}_{\max}$ with respect to the above partial ordering. Any maximal atlas $\mathcal{A}_{\max}$ is called an *analytic structure* on M. A pair $(M, \mathcal{A}_{\max})$ is called a *real-analytic manifold.* In what follows we consider only real-analytic manifolds and call them just manifolds.

Introduce a topology in M by declaring a subset $O \subseteq M$ open, if for any chart (U, h) of the maximal atlas the set $h(O \cap U)$ is open in $\mathbb{R}^m$. In what follows we assume that the above topology satisfies the Hausdorff separation axiom.

Let M_1 and M_2 be two manifolds, and let $f\colon M_1 \to M_2$. The mapping f is called *real-analytic* if for any chart (U_1, φ_1) of M_1 and for any chart (U_2, φ_2) of M_2 with $f(M_1) \cap U_2 \neq \varnothing$, the mapping $\varphi_1^{-1} \circ f \circ \varphi_2$ is real-analytic.

Let x be a point in the domain U of a chart (U, φ) of a manifold M, and let $\varphi\colon U \to \mathbb{R}^m$. The number m is called the *local dimension* of the manifold M at x. The local dimension does not depend on the choice of a chart and is constant on each connected component of M. The above constant is called the dimension of the component.

Example 3. The identity matrix in $\mathbb{R}^m$ is an analytic atlas of $\mathbb{R}^m$. The corresponding maximal atlas determines a real-analytic structure on $\mathbb{R}^m$.

Consider the centred sphere $S^{m-1} = \{x_1^2 + \cdots + x_m^2 = 1\}$ of radius 1 in $\mathbb{R}^m$. Two stereographic projections to the planes $x_m = \pm 1$ determine an analytic atlas. The corresponding maximal atlas determines a real-analytic structure on S^{m-1}. In particular, the maximal atlas contains a special chart that is familiar under the name *spherical coordinates*. The domain of the above chart is a dense open subset of the sphere. Nevertheless, most calculations successfully use spherical coordinates as if they would cover all of the sphere.

2.2.2 Stratified Spaces

Let X be a topological space, and let $\{ M_i \colon i \in I \}$ be a partition of X.

Definition 4. The partition $\{ M_i \colon i \in I \}$ is called a *stratification* if it satisfies the following conditions:

- it is *locally finite*, i.e. any compact subset of X intersects only finitely many sets M_i;
- the sets M_i are connected manifolds;
- for each $i \in I$ the closure of M_i is equal to $M_i \cup \bigcup_{j \in I_i} M_j$, where $I_i \subseteq I \setminus \{i\}$, and $\dim M_j < \dim M_i$ for each $j \in M_i$.

The pair $(X, \{ M_i \colon i \in I \})$ is called a *stratified space*. The sets M_i are called *strata*.

Example 4. The sets $\{\mathbf{0}\}$ and $E \setminus \{\mathbf{0}\}$ are strata of a finite-dimensional real linear space E.

The sets $\{0\}$ and $(0, \infty)$ are strata of the topological space $[0, \infty)$.

2.3 Groups

A *group* is a set G together with a product, i.e. a mapping $G \times G \to G$, $(g_1, g_2) \mapsto g_1 g_2$, that satisfies the following conditions.

Associativity: $(g_1 g_2) g_3 = g_1 (g_2 g_3)$, g_1, g_2, $g_3 \in G$.
Identity element: there exist $e \in G$ such that $eg = ge = g$ for all $g \in G$.
Inverse element: for any $g \in G$, there exists $g^{-1} \in G$ with $gg^{-1} = g^{-1} g = e$.

A subgroup of G is a subset H of G such that $e \in H$, g_1, $g_2 \in H$ implies $g_1 g_2 \in H$, and $g \in H$ implies $g^{-1} \in H$.

The set $gH := \{ gh \colon h \in H \}$ is called the *left coset* of H in G with respect to g. Two left cosets either do not intersect or are equal. The set of left cosets is denoted by G/H.

For any $g \in G$, and for any subgroup H of G, the set $gHg^{-1} := \{ ghg^{-1} : h \in H \}$ is a subgroup of G that is called a group *conjugate* to H. As g runs over G, the subgroups gHg^{-1} run over the *conjugacy class* $[H]$ of the subgroup H. A subgroup H is called *normal* if its conjugacy class contains only one element. For each g_1H and $g_2H \in G/H$, define their product as

$$(g_1H)(g_2H) := \{ k_1k_2 : k_1 \in g_1H, k_2 \in g_2H \}.$$

The above operation is well defined, i.e. the result is a left coset and does not depend on the choice of representatives $k_1 \in g_1H$ and $k_2 \in g_2H$, associative, and has identity element H. The inverse of an element $gH \in G/H$ is $g^{-1}H$. The set G/H together with the above-described product is called a *quotient group*.

For any subgroup H of G, the *normaliser* of H in G is

$$N(H) = \{ g \in G : gHg^{-1} = H \}.$$

It is the largest subgroup of G containing H as a normal subgroup.

Let G and H be two groups. A mapping $\rho\colon G \to H$ is called a *group homomorphism* if it respects multiplication, i.e.

$$\rho(g_1g_2) = \rho(g_1)\rho(g_2), \qquad g_1, g_2 \in G.$$

For example, the mapping $\rho\colon G \to G/H$, that maps an element $g \in G$ to the left coset gH of a normal subgroup H, is a group homomorphism.

A *group isomorphism* is a one-to-one homomorphism $\rho\colon G \to H$ such that ρ^{-1} is a group homomorphism.

A group G is called *abelian* if $gh = hg$ for all g, $h \in H$.

Let G and H be two groups. The Cartesian product of the sets $G \times H$ carries the group structure defined by the multiplication:

$$(g_1, h_1)(g_2, h_2) := (g_1g_2, h_1h_2), \qquad g_1, g_2 \in G, \quad h_1, h_2 \in H.$$

With this multiplication, $G \times H$ is the Cartesian product of the groups G and H.

A *topological group* is a set G which is a group and a topological space such that the mapping $(g, h) \mapsto g^{-1}h$ of $G \times G$ into G is continuous.

A *Lie group* is a set G which is a group and a real-analytic manifold such that the mapping $(g, h) \mapsto g^{-1}h$ of $G \times G$ into G is real-analytic.

Example 5 (Symmetric group)**.** Let r be a positive integer, and let $\Omega = \{1, 2, \dots, r\}$. A *permutation* on Ω is a one-to-one mapping of Ω onto itself. The *symmetric group* on r letters is the set Σ_r of all permutations on Ω with composition of mappings as the group multiplication. Equipped with discrete topology, Σ_r becomes a topological group. Any subgroup of the symmetric group is called the *permutation group*.

A permutation $\sigma \in \Sigma_r$ is called a *cycle* of length m if for m distinct numbers $i_1, \dots, i_r$, σ maps i_j to i_{j+1}, $1 \leq j \leq m-1$, maps i_m to i_1, and leaves all other points fixed. This cycle is denoted by $(i_1 \cdots i_r)$. Note that any cycle of length 1 is the identity permutation. A cycle of length 2 is called the *transposition*.

Two cycles are called *disjoint* if they do not move a common point. Each permutation can be written as a product of disjoint cycles. This product is unique up to the order in which the cycles appear in the product and the inclusion or exclusion of cycles of length 1. We exclude all cycles of length 1 from the product.

It is easy to see that

$$(i_1 \cdots i_m) = (i_1 i_m)(i_1 i_{m-1}) \cdots (i_1 i_2).$$

It follows that each cycle, and therefore every permutation, can be written as a product of transpositions. This product is *not* uniquely determined, but the numbers of transpositions in different products are either all even or all odd. The permutation is called *even* (resp. *odd*) if it may be represented as a product of even (resp. odd) number of permutations.

Define the homomorphism ε from Σ_r to the multiplicative group $Z_2 := \{-1, 1\}$ as follows: $\varepsilon(\sigma) = 1$ if σ is even and $\varepsilon(\sigma) = -1$ if σ is odd. The number $\varepsilon(\sigma)$ is called the *sign* of the permutation σ.

Example 6. A linear space V together with the operation of vector addition is an abelian group. It is called the *additive group of* V.

Example 7 (Classical groups)**.** Let V be a finite-dimensional linear space. The set of all invertible linear operators from V to V together with composition of operators is a group with identity element I, where $I\boldsymbol{v} = \boldsymbol{v}$ for all $\boldsymbol{v} \in V$. This group is called the *general linear group* of V and is denoted by $\mathrm{GL}(V)$. The topology on $\mathrm{GL}(V)$ is induced by the topology on $V \otimes V$ described in Example 2.

In a coordinate form, let n be the dimension of V. The operator $A \in \mathrm{GL}(V)$ acts on a vector $\boldsymbol{v} = v^i \boldsymbol{e}_i$ as

$$(Av^i \boldsymbol{e}_i)^j = a^j_i \boldsymbol{v}^i \boldsymbol{e}_j, \qquad 1 \leq j \leq n.$$

The numbers $a^j_i \in \mathbb{F}$ are called the *matrix entries* of A with respect to the basis $\{\boldsymbol{e}_1, \dots, \boldsymbol{e}_n\}$ for V. The group of all $n \times n$ invertible matrices with coefficients in $\mathbb{F}$ is called the *general linear group of rank* n and is denoted by $\mathrm{GL}(n, \mathbb{F})$. The mapping $\mu\colon \mathrm{GL}(V) \to \mathrm{GL}(n, \mathbb{F})$ that maps $A \in \mathrm{GL}(V)$ to the matrix with entries a_{ij} is a group isomorphism.

The group $\mathrm{GL}(n, \mathbb{F})$ has two connected components. The connected component of the identity matrix δ_{ij} is the set of all matrices with positive determinant, while the second connected component is the set of all matrices with negative determinant.

The *special linear group of rank* n, $\mathrm{SL}(n, \mathbb{F})$ is the set of all matrices A such that $\det(A) = 1$. The group $\mathrm{SL}(V) := \mu^{-1}(\mathrm{SL}(n, \mathbb{F}))$ is independent of the choice of basis and is called the *special linear group of* V. Both groups are connected.

Let $\mathbb{F} = \mathbb{R}$. The set $O(V)$ of all $g \in GL(V)$ such that $(g\boldsymbol{v}, g\boldsymbol{w}) = (\boldsymbol{v}, \boldsymbol{w})$ for all $\boldsymbol{v}$, $\boldsymbol{w} \in V$, is a subgroup of $GL(V)$, it is called the *orthogonal group of* V. This group has two connected components. The connected component of the identity operator is the *special orthogonal group of* V, $SO(V) := O(V) \cap SL(V)$. Both $O(V)$ and $SO(V)$ are compact topological groups.

In a coordinate form, $O(n) := \mu(O(V))$ is the set of $n \times n$ matrices A with $AA^\top = I$, the *orthogonal group of rank* n. The elements of $O(n)$ are *orthogonal matrices.* The connected component of the identity matrix, $SO(n) := \mu(SO(V))$, is called the *special orthogonal group of rank* n. The determinant of any matrix in $SO(n)$ is equal to 1, while the determinant of any matrix in the second connected component is equal to -1.

Let $\mathbb{F} = \mathbb{C}$. The set $U(V)$ of all $g \in GL(V)$ such that $(g\boldsymbol{v}, g\boldsymbol{w}) = (\boldsymbol{v}, \boldsymbol{w})$ for all $\boldsymbol{v}$, $\boldsymbol{w} \in V$, is a subgroup of $GL(V)$, it is called the *unitary group of* V. The *special unitary group of* V is $SU(V) := U(V) \cap SL(V)$. Both $U(V)$ and $SU(V)$ are compact topological groups.

In a coordinate form, $U(n) := \mu(U(V))$ is the set of $n \times n$ matrices A with $AA^* = I$, the *unitary group of rank* n. Here A^* is the conjugate transposed matrix to A. The elements of $U(n)$ are *unitary matrices.* The *special unitary group of rank* n is $SU(n) := \mu(SU(V))$.

Other classical groups are described in Weyl (1997) and Goodman & Wallach (2009). All classical groups are locally compact and are Lie groups.

Let M_n be the space of $n \times n$ complex matrices. For a matrix $M \in M_n$, define the *principal minors* by

$$\Delta_i(M) := \det \begin{pmatrix} m_{11} & \cdots & m_{1i} \\ \vdots & \ddots & \vdots \\ m_{i1} & \cdots & m_{ii} \end{pmatrix}, \qquad 1 \leq i \leq n.$$

Theorem 2 (Cholesky decomposition). *Let $M \in M_n$ be a symmetric matrix with $\Delta_i(M) \neq 0$, $1 \leq i \leq n$. There exists an upper-triangular matrix $B \in M_n$ such that $M = B^\top B$. The matrix B is uniquely determined by M up to left multiplication by a diagonal matrix with entries ± 1.*

Hansen (2010) proved that the Cholesky decomposition remains true if the matrix M is infinite.

2.4 Group Actions

2.4.1 Actions, Stabilisers and Orbits

Let X be a non-empty set, and let G be a group with identity element e. A *left action* of the group G on the set X is a mapping $\rho\colon G \times X \to X$ that maps a pair $(g, x) \in G \times X$ to a point $g \cdot x \in X$ such that $e \cdot x = x$ (the identity axiom) and $g \cdot (h \cdot x) = (gh) \cdot x$ (the compatibility axiom) for all g, $h \in G$ and for all $x \in X$.

An action is called *transitive* if for each pair x, $y \in X$ there exists a $g \in G$ such that $g \cdot x = y$. An action is called *faithful* if for any $g \in G$, $g \neq e$ there is an $x \in X$ such that $g \cdot x \neq x$.

The *stabiliser* G_x of a point $x \in X$ is the set of all $g \in G$ such that $g \cdot x = x$. The stabiliser of any point $x \in X$ is a subgroup of G. The *orbit* of x in X is the set $G \cdot x := \{ g \cdot x \colon g \in G \}$. The set of all orbits is denoted by X/G. This set forms a partition of X. The group action is transitive if and only if it has only one orbit. Let $\pi \colon X \to X/G$ be the *orbital mapping* that maps a point $x \in X$ to its orbit.

If G is a topological group and X is a topological space, we consider only *continuous left actions*, i.e. when the mapping from $G \times X$ to X that maps (g, x) to $g \cdot x$ is continuous. If G is a Lie group and X is a manifold, we consider only *real-analytic left actions*, i.e. when the mapping from $G \times X$ to X that maps (g, x) to $g \cdot x$ is real-analytic. Introduce the *quotient topology* on X/G: a subset $O \subset X/G$ is open if and only if the inverse image $\pi^{-1}(O)$ is open in X. The introduced topological space is called the *quotient* of X under the action of G.

A left action is called *proper*, if the map $(g, x) \mapsto (g \cdot x, x)$ from $G \times X$ to $X \times X$ is proper: the inverse image of any compact set is compact. If an action is continuous and proper, then the quotient is Hausdorff.

Two point x, $y \in X$, and two orbits $G \cdot x$, $G \cdot y \in X/G$ are of the same *orbit type*, with notation $x \sim y$, and $G \cdot x \sim G \cdot y$, if G_x is conjugate to G_y within G, that is,

$$G_y = \{ g^{-1} h g \colon h \in G_x \}$$

for some $g \in G$. A point x *dominates* y, and an orbit $G \cdot x$ dominates $G \cdot y$, with notation $y \precsim x$, and $G \cdot y \precsim G \cdot x$, if G_x is conjugate within G to a subgroup of G_y, that is, $g^{-1} G_x g \subset G_y$ for some $g \in G$.

Let H be a subgroup of G. The *set of fixed points for H in X* is

$$X^H = \{ y \in X \colon h \cdot y = y \quad \text{for all} \quad h \in H \}.$$

If H is a compact subgroup of a Lie group G and acts analytically on a manifold M, then each connected component of M^H is a closed set and a manifold.

Example 8 (Affine spaces). Formally, an *affine space* is a triple $(E, V, +)$, where V is an additive group of a linear space described in Example 6, and $+$ is a transitive and faithful action of V on E. The space V is the *underlying space* of E, or the *space domain*. We denote by $A + \boldsymbol{x}$ the result of the action of the vector $\boldsymbol{x} \in V$ on a point $A \in E$. The map $A \mapsto A + \boldsymbol{x}$ is called the *translation* by the vector $\boldsymbol{x}$. The vector in V that translates the element $A \in E$ to the element $B \in E$ is denoted by $B - A$.

Let O be a point in E, and let $\Theta_O \colon E \to V$ be the following map:

$$\Theta_O(B) = B - O.$$

There exists a unique structure of a linear space on E such that Θ_O ia an isomorphism of linear spaces. Denote the set E endowed with the above structure by E_O and call it the *vectorisation* of E at O.

To introduce coordinates into E, choose $d+1$ points O, A_1, ..., A_d such that $\{ A_i - O \colon 1 \le i \le d \}$ is a basis for V, and call this set an *affine frame* with an origin O. The coordinates of $A \in E$ in this frame are the reals λ_1, ..., λ_d such that

$$A - O = \sum_{i=1}^{d} \lambda_i (A_i - O).$$

Example 9 (Actions of classical groups)**.** Any classical group $\mathrm{G}(n, \mathbb{F})$ acts continuously on $\mathbb{F}^n$ by matrix-vector multiplication.

Put $\boldsymbol{x}_\rho := (0, 0, \dots, 0, \rho)^\top \in \mathbb{R}^n$ with $\rho \ge 0$. The $\mathrm{O}(n)$-orbit of the point $\boldsymbol{x}_0$ is $\boldsymbol{x}_0$. The stabiliser $\mathrm{O}(n)_{\boldsymbol{x}_0}$ is $\mathrm{O}(n)$. The orbit of a point $\boldsymbol{x}_\rho$ with $\rho > 0$ is the *centred sphere of radius* ρ:

$$\mathrm{O}(n) \cdot \boldsymbol{x}_\rho = \{ \boldsymbol{x} \in \mathbb{R}^n \colon \|\boldsymbol{x}\| = \rho \}.$$

The stabiliser of any point $\boldsymbol{x}_\rho$ with $\rho > 0$ is the subgroup of matrices of the form

$$\begin{pmatrix} A & \mathbf{0} \\ \mathbf{0}^\top & 1 \end{pmatrix}, \qquad A \in \mathrm{O}(n-1).$$

The quotient $\mathbb{R}^n / \mathrm{O}(n)$ is the interval $[0, \infty)$.

2.4.2 Application: Symmetric and Alternating Tensors

Let V be a linear space of dimension n over the field $\mathbb{F}$ and let r be a positive integer. The symmetric group Σ_r acts on $\tau(V^r)$ by permuting the positions of the factors in the tensor product as

$$\sigma \cdot (\boldsymbol{v}_1 \otimes \cdots \otimes \boldsymbol{v}_r) := \boldsymbol{v}_{\sigma^{-1}(1)} \otimes \cdots \otimes \boldsymbol{v}_{\sigma^{-1}(r)}, \qquad \sigma \in \Sigma_r. \tag{2.7}$$

This action may be extended to $V^{\otimes r}$ by the universal mapping property and therefore defines a representation σ_r of the group Σ_r. Define the linear operator $P^+ \in L(V^{\otimes r})$ by

$$P^+ \boldsymbol{T} := \frac{1}{r!} \sum_{\sigma \in \Sigma_r} \sigma_r(\sigma) \cdot \boldsymbol{T}.$$

The range of the operator P^+ is denoted by $\mathsf{S}^r(V)$ and is called the *space of symmetric tensors of rank r over V*. The elements of this space are called rank r *symmetric tensors*. For example, the space $\mathsf{S}^2(V)$ is spanned by the tensors $\boldsymbol{v} \otimes \boldsymbol{w} + \boldsymbol{w} \otimes \boldsymbol{v}$ for $\boldsymbol{v}$, $\boldsymbol{w} \in V$.

In a coordinate form, the group Σ_r acts on the tensor $\boldsymbol{T}_{i_1\cdots i_r}$ by

$$(\sigma_r(\sigma)\cdot\boldsymbol{T})_{i_1\cdots i_r} := \boldsymbol{T}_{\sigma^{-1}(i_1)\cdots\sigma^{-1}(i_r)}, \qquad \sigma\in\Sigma_r. \tag{2.8}$$

For any $\boldsymbol{v}_1, \dots, \boldsymbol{v}_r \in V$ consider the function

$$\boldsymbol{v}\mapsto\prod_{j=1}^{r}(\boldsymbol{v}_j,\boldsymbol{v}), \qquad \boldsymbol{v}\in V. \tag{2.9}$$

The linear span of the above functions is denoted by $P^r(V)$ and is called the *space of homogeneous polynomials* (*or n-ary forms*) *of degree r on V*. The elements of this space are *homogeneous polynomials* or *n-ary forms* of degree r.

In a coordinate form, let $\boldsymbol{v} = x^i\boldsymbol{e}_i \in V$ and $\boldsymbol{v}_j = v^i_j\boldsymbol{e}_i$. The mapping (2.9) becomes

$$(x_1,\dots,x_n)^\top\mapsto\prod_{j=1}^{r}\overline{v^i_j}x^i.$$

Define $f\colon V^r \to P^r(V)$ as the r-bilinear map that maps an element $(\boldsymbol{v}_1,\dots,\boldsymbol{v}_r)\in V^r$ to the mapping (2.9). By the universal mapping property, there exists a unique map $F\colon \mathsf{S}^r(V)\to P^r(V)$ such that

$$F(P(\boldsymbol{v}_1\otimes\cdots\otimes\boldsymbol{v}_r)) = \prod_{j=1}^{r}(\boldsymbol{v}_j,\boldsymbol{v})$$

with the following commutative diagram:

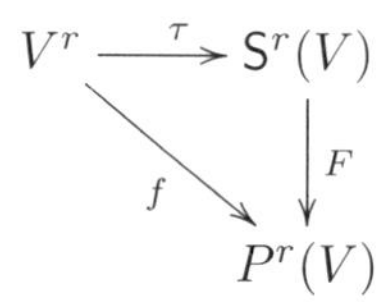

(2.10)

Moreover, F is an isomorphism between $\mathsf{S}^r(V)$ and $P^r(V)$.

Let Σ be a subgroup of the symmetric group Σ_r having $|\Sigma|$ elements. Define the linear operator $P^+_\Sigma \in L(V^{\otimes r})$ by

$$P^+_\Sigma\boldsymbol{T} := \frac{1}{|\Sigma|}\sum_{\sigma\in\Sigma}\sigma_r(\sigma)\cdot\boldsymbol{T}. \tag{2.11}$$

The tensors lying in the range $P^+_\Sigma(V^{\otimes r})$ are important for continuum physics. For example, the tensors in $\mathsf{S}^2(V)$ are called *stress tensors*.

Example 10 (Elasticity tensor)**.** Let $r = 4$, and let $\Sigma\subset\Sigma_4$ be the 8-element subgroup generated by the transpositions (12), (34) and the product (13)(24). We have $P^+_\Sigma(V^{\otimes 4}) = \mathsf{S}^2(\mathsf{S}^2(V))$, and the tensors in this space are called *elasticity tensors*.

Define the linear operator $P^- \in L(V^{\otimes r})$ by

$$P^- \mathsf{T} := \frac{1}{r!} \sum_{\sigma \in \Sigma_r} \varepsilon(\sigma)\sigma \cdot \mathsf{T}.$$

The range of the operator P^- is denoted by $\Lambda^r(V)$ and is called the *space of alternating or skew-symmetric tensors of rank r over V*. The elements of this space are called rank r *alternating tensors* or rank r *skew-symmetric tensors.* For example, the space $\Lambda^2(V)$ is spanned by the tensors $\boldsymbol{v} \otimes \boldsymbol{w} - \boldsymbol{w} \otimes \boldsymbol{v}$ for $\boldsymbol{v}$, $\boldsymbol{w} \in V$.

Let n be the dimension of V. For $0 \leq r \leq n$, the dimension of the space $\Lambda^r(V)$ is equal to $\binom{n}{r}$, and for $r > n$ we have $\Lambda^r(V) = \{\mathbf{0}\}$.

Assume that the group Σ contains odd permutations. Define the linear operator $P^-_\Sigma \in L(V^{\otimes r})$ by

$$P^-_\Sigma \mathsf{T} := \frac{1}{|\Sigma|} \sum_{\sigma \in \Sigma} \varepsilon(\sigma)\sigma \cdot \mathsf{T}. \tag{2.12}$$

2.5 Group Representations

Let G be a topological group. A *representation* of G or a G*-module* is a pair (V, ρ), where V is a linear space over a field $\mathbb{F}$, and $\rho\colon G \times V \to V$ is an action of G on V such that for each $g \in G$ the translation $U(g)\colon \boldsymbol{v} \mapsto \rho(g, \boldsymbol{v})$ is a $\mathbb{F}$-linear map. When V is *finite-dimensional*, we suppose in addition that the action ρ is *continuous.* As usual, we denote $\rho(g, \boldsymbol{v})$ by $g \cdot \boldsymbol{v}$. In this notation, the defining equations of an action take the form

$$e \cdot \boldsymbol{v} = \boldsymbol{v}, \qquad (gh) \cdot \boldsymbol{v} = g \cdot (h \cdot \boldsymbol{v}).$$

In terms of translations, they become

$$U(e) = I, \qquad U(g)U(h) = U(gh).$$

Thus, $U(g)$ is a linear operator in V with inverse $U(g^{-1})$, and the map $g \mapsto U(g)$ is a homomorphism between G and the group $\operatorname{Aut}(V)$ of linear invertible operators in V. Conversely, given any such homomorphism U, we define an action

$$\rho\colon G \times V \to V, \qquad (g, \boldsymbol{v}) \mapsto U(g)\boldsymbol{v}.$$

Moreover, if the space V is finite-dimensional, then ρ is continuous if and only if U is continuous. A choice of a basis for V determines an isomorphism between $\operatorname{Aut}(V)$ and $\mathrm{GL}(n, \mathbb{F})$, where $n = \dim V$. A continuous homomorphism $U\colon G \to \mathrm{GL}(n, \mathbb{F})$ is called a *matrix representation.* In what follows we will denote representations either (V, ρ) or $U(g)$, or even just V, when ρ is thought.

Let (V, ρ) and (W, τ) be representations of G. An operator $A \in L(V, W)$ is called an *intertwining operator* if

$$A(g \cdot \boldsymbol{v}) = g \cdot (A\boldsymbol{v}), \qquad g \in G, \quad \boldsymbol{v} \in V. \tag{2.13}$$

The intertwining operators form a linear space $\mathrm{Hom}_G(V, W)$ over $\mathbb{F}$. Another name for an intertwining operator is an *equivariant map*.

The representations (V, ρ) and (W, τ) are called *equivalent* if the space $\mathrm{Hom}_G(V, W)$ contains an invertible operator.

An inner product $V \times V \to \mathbb{F}$, $(\boldsymbol{u}, \boldsymbol{v}) \mapsto \langle \boldsymbol{u}, \boldsymbol{v} \rangle$ is called *G-invariant* if $\langle g \cdot \boldsymbol{u}, g \cdot \boldsymbol{v} \rangle = \langle \boldsymbol{u}, \boldsymbol{v} \rangle$ for all $g \in G$ and $\boldsymbol{u}$, $\boldsymbol{v} \in V$. A representation (V, ρ) together with a G-invariant inner product is called *orthogonal* if $\mathbb{F} = \mathbb{R}$ and *unitary* if $\mathbb{F} = \mathbb{C}$.

If the space V is finite-dimensional, then a choice of an orthonormal basis for V defines a homomorphism $G \to \mathrm{O}(n)$ if $\mathbb{F} = \mathbb{R}$ or $G \to \mathrm{U}(n)$ if $\mathbb{F} = \mathbb{C}$. Any *finite-dimensional* representation of a *compact* group G possesses a G-invariant inner product.

A subspace $W \subset V$ is called *invariant* or a *submodule* if $g \cdot \boldsymbol{w} \in W$ for $g \in G$ and $\boldsymbol{w} \in W$. A non-zero representation V if called *irreducible* if it has no invariant subspaces other than $\{\mathbf{0}\}$ and V. A representation which is not irreducible is called *reducible*.

Example 11. Put $V = \mathbb{F}^1$. $U(g) = 1$. This representation is called *trivial*.

Let G be one of the groups $\mathrm{GL}(n, \mathbb{F})$, $O(n)$, or $U(n)$. The representation with $V = \mathbb{F}^1$ and $U(g) = \det g$ is called *determinant representation*.

Let G be any classical group of Example 7. The representation with $V = \mathbb{F}^n$ and $U(g) = g$ is called *defining representation*.

The direct sum $V \oplus W$ of two G-modules V and W becomes a G-module under the action $g \cdot (\boldsymbol{v}, \boldsymbol{w}) = (g \cdot \boldsymbol{v}, g \cdot \boldsymbol{w})$. This module is called the *direct sum* of the G-modules V and W. If both V and W are finite-dimensional with $m = \dim V$ and $n = \dim W$, then the direct sum of representations is formed by block matrices:

$$g \mapsto \begin{pmatrix} A(g) & 0 \\ 0 & B(g) \end{pmatrix}.$$

Let G be a compact group. Denote by $\mathrm{Irr}(G, \mathbb{F})$ the set of equivalence classes of irreducible G-modules over $\mathbb{F}$. If W is equivalent to a submodule of V we say that W is *contained* in V.

If W is irreducible then we define the *multiplicity* of W in V as

$$n(W, V) = \dim_{\mathbb{F}} \mathrm{Hom}_G(W, V).$$

To explain this definition, consider the map

$$d_W \colon \mathrm{Hom}_G(W, V) \otimes W \to V, \qquad A \otimes \boldsymbol{w} \to A\boldsymbol{w}.$$

The operator d_W intertwines the representations

$$g \cdot (A \otimes \boldsymbol{w}) = A \otimes (g \cdot \boldsymbol{w}) \tag{2.14}$$

and V. Define the map d as the direct sum of d_W over $W \in \mathrm{Irr}(G, \mathbb{F})$. This map is invertible and establishes an equivalence between the representation V and the direct sum

$$\sum_{W \in \operatorname{Irr}(G,\mathbb{F})} \oplus \operatorname{Hom}_G(W,V) \otimes W$$

of the representations (2.14). The image of the map d_W in V is called the W-*isotypical summand* of V. It is generated by the irreducible submodules of V that are equivalent to W. The multiplicity of W in V is simply the number of copies of W contained in V.

Let V and W be representations of G. The *tensor product* acts in $V \otimes W$ by $g \cdot (\boldsymbol{v} \otimes \boldsymbol{w}) = (g \cdot \boldsymbol{v}) \otimes (d \cdot \boldsymbol{w})$. The group G acts in the linear space $\operatorname{Hom}_G(V,W)$ of intertwining operators by $(g \cdot A)\boldsymbol{v} = g \cdot (Ag^{-1} \cdot \boldsymbol{v})$, where $A \in \operatorname{Hom}_G(V,W)$, $g \in G$ and $\boldsymbol{v} \in V$. In other words, the diagram

$$\begin{array}{ccc} V & \xrightarrow{A} & W \\ {\scriptstyle g}\downarrow & & \downarrow{\scriptstyle g} \\ V & \xrightarrow[g \cdot A]{} & W \end{array}$$

is commutative. In the case when V is complex and $W = \mathbb{C}^1$ is the complex trivial representation, the representation $\operatorname{Hom}_G(V,\mathbb{C}^1) = V^*$ is called the *dual representation*.

If V is finite-dimensional and $\boldsymbol{v}_1, \ldots, \boldsymbol{v}_n$ is a basis of V, then

$$g \cdot \boldsymbol{v}_j = \sum_{i=1}^{n} u_{ij}(g)\boldsymbol{v}_j.$$

If $\boldsymbol{v}_1^*, \ldots, \boldsymbol{v}_n^*$ is a dual basis, then

$$g \cdot \boldsymbol{v}_j^* = \sum_{i=1}^{n} u_{ij}^*(g)\boldsymbol{v}_j$$

and

$$u_{ij}^*(g) = \langle g \cdot \boldsymbol{v}_j^*, \boldsymbol{v}_i \rangle = \langle \boldsymbol{v}_j^*, g^{-1} \cdot \boldsymbol{v}_i \rangle = \left\langle \boldsymbol{v}_j^*, \sum_{k=1}^{n} u_{ki}(g^{-1})\boldsymbol{v}_k \right\rangle = u_{ji}(g^{-1}),$$

that is, g acts via the transpose of the inverse.

The conjugate space $\overline{V}$ is also a G-module, called the *conjugate representation.*

When V is unitary, then the matrix entries of both V^* and $\overline{V}$ are complex conjugate to those of V.

A representation V is called *self-conjugate* if V and $\overline{V}$ are equivalent.

The *character* of a finite-dimensional representation V is defined as

$$\operatorname{ch}_V(g) := \operatorname{tr} U(g), \qquad g \in G.$$

In particular, the character of a one-dimensional representation is this presentation itself. The characters of irreducible unitary representations of an abelian topological group G form a multiplicative group $\hat{G}$, the *character group* of the group G. The base of the topology in $\hat{G}$ consists of finite intersections of the sets

$$\{\rho \in \hat{G} \colon \rho(K) \subset O\},$$

where K runs over all compact subsets of G and O runs over open subsets of $\mathrm{U}(1)$. The above topology is therefore called the *compact-open topology.*

If $V_1, \dots, V_k$ are irreducible and pairwise non-equivalent representations, then their characters are linearly independent. For any representations V and (W) we have

$$\mathrm{ch}_{V \oplus W}(g) = \mathrm{ch}_V(g) + \mathrm{ch}_W(g), \quad \mathrm{ch}_{V \otimes W}(g) = \mathrm{ch}_V(g)\,\mathrm{ch}_W(g).$$

Let $U(g)$ be a representation of a group G, and let $T(h)$ be a representation of a group H. The *outer tensor product* of the above representations is the representation $(U \hat{\otimes} T)$ of the Cartesian product $G \times H$ defined by

$$(U \hat{\otimes} T)(g, h) = U(g) \otimes T(h), \qquad g \in G, \quad h \in H.$$

If the representations $U(g)$ and $(T(h))$ are irreducible, then the representation $U \hat{\otimes} T$ is irreducible, and all irreducible representations of $G \times H$ are of the form $U \hat{\otimes} T$.

Let G be a compact Hausdorff topological group. Let $C(G, \mathbb{F})$ be the space of $\mathbb{F}$-valued continuous functions on G with norm

$$\|f\| := \max_{g \in G} |f(g)|.$$

Consider the following action of the group G on the set $C(G, \mathbb{F})$:

$$(g \cdot f)(h) := f(g^{-1} h), \qquad h \in G.$$

A function $f \in C(G, \mathbb{F})$ is called *almost invariant* if the linear span of the orbit $G \cdot f$ is finite-dimensional. Denote by $R(G, \mathbb{F})$ the set of all almost invariant functions.

Let X be a Hausdorff topological space, and let $\mathfrak{B}(X)$ be the σ-field of Borel sets in X. A measure μ on $\mathfrak{B}(X)$ is called *tight* if, for any Borel set A, $\mu(A)$ is the supremum of $\mu(K)$ over all compact subsets K of A. μ is called *locally finite* if every point of X has a neighbourhood O with $\mu(O) < \infty$. A *Radon measure* is a locally finite tight measure. Every finite measure defined on Borel sets of a finite-dimensional linear space is Radon.

Let G be a compact Hausdorff topological group. There exists a unique measure μ on the Borel σ-field of G satisfying the following conditions.

1. μ is Radon.
2. μ is *left-invariant*: $\mu(g^{-1}A) = \mu(A)$ for any Borel set A and for any $g \in G$.
3. μ is *probabilistic*: $\mu(G) = 1$.

The above measure is called the *Haar measure* on G.

Let $L^2(G, \mathbb{F}, \mathrm{d}\mu)$ be the Hilbert space of $\mathbb{F}$-valued square-integrable functions with respect to the Haar measure. The inner product in the space $L^2(G, \mathbb{F}, \mathrm{d}\mu)$ is defined as

$$(f_1, f_2) := \int_G \overline{f_1(g)} f_2(g) \, \mathrm{d}\mu(g).$$

Theorem 3 (Peter–Weyl). *The set $R(G, \mathbb{F})$ is dense in both $C(G, \mathbb{F})$ and $L^2(G, \mathbb{F}, \mathrm{d}\mu)$.*

To use this theorem in practice, consider the fine structure of the set $R(G, \mathbb{F})$. Let $U(g)$ be either an irreducible orthogonal representation of G in a real linear space or an irreducible unitary representation of G in a complex linear space. Consider the bilinear map from $V \times V$ to $C(G, \mathbb{F})$ that maps a pair $(\boldsymbol{x}, \boldsymbol{y})$ to the function $(U(g)\boldsymbol{x}, \boldsymbol{y})$. By the Universal Mapping Property, there is a linear map $\Phi_V \colon V \otimes V \to G(G, \mathbb{F})$ with $\Phi_V(\boldsymbol{x} \otimes \boldsymbol{y})(g) = (U(g)\boldsymbol{x}, \boldsymbol{y})$. Denote by $R_U(G, \mathbb{F})$ the image of $V \otimes V$ under Φ_V in $R(G, \mathbb{F})$.

Theorem 4 (The Fine Structure Theorem, Hofmann & Morris, 2013). *The set $\{\, R_U(G, \mathbb{F}) \colon U \in \operatorname{Irr}(G, \mathbb{F}) \,\}$ is an orthogonal family of closed vector subspaces of the Hilbert space $L^2(G, \mathbb{F}, \mathrm{d}\mu)$. Its algebraic direct sum is*

$$\sum_{U \in \operatorname{Irr}(G, \mathbb{F})} R_U(G, \mathbb{F}) = R(G, \mathbb{F}),$$

while its orthogonal Hilbert space direct sum is

$$\sum_{U \in \operatorname{Irr}(G, \mathbb{F})} \oplus R_U(G, \mathbb{F}) = L^2(G, \mathbb{F}, \mathrm{d}\mu).$$

The dimension of the space $R_U(G, \mathbb{F})$ is

$$\dim R_U(G, \mathbb{F}) = m \dim U,$$

where $m = \dim U$ if $\mathbb{F} = \mathbb{C}$.

In a coordinate form, it is possible to choose a basis $\{\boldsymbol{e}_1, \dots, \boldsymbol{e}_{\dim U}\}$ of the space V in such a way that the matrix entries

$$\{\, (U(g)\boldsymbol{e}_i, \boldsymbol{e}_j) \colon U \in \operatorname{Irr}(G, \mathbb{F}), 1 \le i \le \dim U, 1 \le j \le m \,\}$$

form an orthogonal basis in the Hilbert space $L^2(G, \mathbb{F}, \mathrm{d}\mu)$, and the set of their finite linear combinations is equal to $R(G, \mathbb{F})$, therefore, is dense in $C(G, \mathbb{F})$.

The Fine Structure Theorem holds true for the case of the spaces $C(G/H, \mathbb{F})$ and $L^2(G/H, \mathbb{F}, \mathrm{d}\mu)$, where H is a closed subgroup of G. The group G acts on the set $C(G/H, \mathbb{F})$ as follows:

$$(g \cdot f)(hH) = f(g^{-1}hH).$$

The matrix entries

$$\{ u_{ij}(g) = (U(g)\boldsymbol{e}_i, \boldsymbol{e}_j) \colon U \in \operatorname{Irr}(G, \mathbb{F}), 1 \leq i \leq \dim U, 1 \leq j \leq m'_U \}$$

form an orthogonal basis in the Hilbert space $L^2(G/H, \mathbb{F}, \mathrm{d}\mu)$, and the set of their finite linear combinations is equal to $R(G, \mathbb{F})$, therefore, is dense in $C(G, \mathbb{F})$. The set $\{ \boldsymbol{e}_j) \colon 1 \leq j \leq m'_U \}$ is a basis of the intersection of the linear span of the first m vectors of the basis and the isotypic subspace of the trivial representation of the group H.

Assume that a compact group G is *easily reducible*. This means that for any three irreducible representation $S(g)$ in the space V, $T(g)$ in the space W, and $U(g)$ in the space X the multiplicity m_U of $U(g)$ in $S \otimes T$ is equal to either 0 or 1. For example, the groups $\mathrm{O}(2)$ and $\mathrm{O}(3)$ are easily reducible. Assume $m_U = 1$. Let $\{ \boldsymbol{e}_i^U \colon 1 \leq i \leq \dim U \}$ be an orthonormal basis in the space X, and similarly for $S(g)$ and $T(g)$. There are two natural bases in the space $V \otimes W$. The *coupled basis* is

$$\{ \boldsymbol{e}_i^S \otimes \boldsymbol{e}_j^T \colon 1 \leq i \leq \dim S, 1 \leq j \leq \dim T \}.$$

The *uncoupled basis* is

$$\{ \boldsymbol{e}_k^U \colon m_U = 1, 1 \leq k \leq \dim U \}.$$

In a coordinate form, the elements of the space $V \otimes W$ are matrices with $\dim S$ rows and $\dim T$ columns. The coupled basis consists of matrices having 1 in the ith row and jth column, and all other entries equal to 0. Denote by $c^{k[i,j]}_{U[S,T]}$ the coefficients of expansion of the vectors of uncoupled basis in the coupled basis:

$$\boldsymbol{e}_k^U = \sum_{i=1}^{\dim S} \sum_{j=1}^{\dim T} c^{k[i,j]}_{U[S,T]} \boldsymbol{e}_i^S \otimes \boldsymbol{e}_j^T. \tag{2.15}$$

The numbers $c^{k[i,j]}_{U[S,T]}$ are called the *Clebsch–Gordan coefficients* of the group G. In the coupled basis, the vectors of the uncoupled basis are matrices $c^k_{U[S,T]}$ with matrix entries $c^{k[i,j]}_{U[S,T]}$, the *Clebsch–Gordan matrices.* The uncoupled basis is orthonormal, therefore we have

$$\sum_{i=1}^{\dim S} \sum_{j=1}^{\dim T} \left| c^{k[i,j]}_{U[S,T]} \right|^2 = 1. \tag{2.16}$$

Denote by $D^S(g)$ the matrix of the representation $S(g)$ in the basis $\{ \boldsymbol{e}_i^S \colon 1 \leq i \leq \dim s \}$, and similarly for $T(g)$ and $U(g)$. Let C be the square matrix whose rows are enumerated by the pairs (U, k), $m_U = 1$, $1 \leq k \leq \dim U$ and whose columns are enumerated by the pairs (i, j), $1 \leq i \leq \dim S$, $1 \leq j \leq \dim T$. The matrix entries of C are as follows:

$$C_{(U,k)(i,j)} := c^{k[i,j]}_{U[S,T]}.$$

According to (2.15), the matrix C is the transition matrix from the coupled basis to the uncoupled one. If the representations $S(g)$ and $T(g)$ are unitary (resp. orthogonal), then the matrix C is unitary (resp. orthogonal). In particular,

$$D^S(g)\otimes D^T(g)=C^{-1}\left(\oplus_{U\colon m_U=1}D^U(g)\right)C,$$

or in a coordinate form:

$$D^S_{ij}(g)D^T_{lm}(g)=\sum_{U\colon m_U=1}\sum_{k,n=1}^{\dim S}\overline{c^{k[i,l]}_{U[S,T]}}D^U_{kn}(g)c^{n[j,m]}_{U[S,T]}. \tag{2.17}$$

Let $\lambda_U\in\mathbb{F}$ with $|\lambda_U|=1$ be a *phase*. Define new Clebsch–Gordan coefficients by

$$(c^{k[i,j]}_{U[S,T]})'=\lambda_U c^{k[i,j]}_{U[S,T]}.$$

New coefficients still satisfy (2.17) and therefore may serve as Clebsch–Gordan coefficients. Multiply both sides of (2.17) by $\overline{D^U_{kn}(g)}$ and integrate over G with respect to the Haar measure. By the Peter–Weyl theorem, we obtain

$$\dim\tau\int_G D^S_{ij}(g)D^T_{lm}(g)\overline{D^U_{kn}(g)}\,\mathrm{d}\mu(g)=\overline{c^{k[i,\ell]}_{U[S,T]}}c^{n[j,m]}_{U[S,T]}. \tag{2.18}$$

Choose numbers i_0 with $1\le i_0\le\dim S$, l_0 with $1\le l_0\le\dim T$ and k_0 with $1\le k_0\le\dim U$, and put $i=j=i_0$, $l=m=l_0$ and $k=n=k_0$. We obtain

$$\dim U\int_G D^S_{i_0i_0}(g)D^T_{\ell_0\ell_0}(g)\overline{D^U_{k_0k_0}(g)}\,\mathrm{d}\mu(g)=\left|c^{k_0[i_0,\ell_0]}_{U[S,T]}\right|^2.$$

By (2.16), we may choose i_0, ℓ_0 and k_0 in such a way that the right-hand side is not equal to 0. Choose the phase λ_U in such a way that

$$c^{k_0[i_0,\ell_0]}_{U[S,T]}=\left(\dim U\int_G D^S_{i_0i_0}(g)D^T_{\ell_0\ell_0}(g)\overline{D^U_{k_0k_0}(g)}\,\mathrm{d}\mu(g)\right)^{1/2} \tag{2.19}$$

and calculate the remaining Clebsch–Gordan coefficients using (2.18):

$$c^{n[j,m]}_{U[S,T]}=\frac{\dim U}{c^{k_0[i_0,\ell_0]}_{U[S,T]}}\int_G D^S_{i_0j}(g)D^T_{l_0m}(g)\overline{D^U_{k_0n}(g)}\,\mathrm{d}\mu(g). \tag{2.20}$$

Consider a representation $U(g)$ of a compact Lie group G in a real finite-dimensional linear space V as a group action. Then, there are only finitely many distinct orbit types. Moreover, the orbit types in V form a stratification of V, and the orbit types in the quotient V/G form a stratification of V/G.

Define the *directed graph* $\mathcal{T}$ of the orbit type stratification as follows. The vertices of $\mathcal{T}$ are orbit types in V/G. One draws an arrow from $A\in\mathcal{T}$ to $B\in\mathcal{T}$, $A\to B$, if B is a subset of the closure of A, that is, elements of A converge to elements of B.

The directed graph has the following properties. If $A\to B$ and $A\ne B$, then $\dim A>\dim B$ and $\dim\pi^{-1}(A)>\dim\pi^{-1}(B)$, where π is the orbital mapping from V to V/G. The relation $A\to B$ defines a *partial ordering* in $\mathcal{T}$, that is: if

$A \to B$ and $B \to C$, then $A \to C$ and $A = B$ if and only if $A \to B$ and $B \to A$. If we define the dimension of V/G as the maximal dimension of its strata, then any chain has at most $1 + \dim V/G$ many elements. A component $A \in \mathcal{T}$ is minimal with respect to the introduced partial ordering if and only if A is closed in V/G and $\pi^{-1}(A)$ is closed in V. We call it the *minimal orbit type*. Similarly, a component $A \in \mathcal{T}$ is maximal if and only if A is open in V/G and $\pi^{-1}(A)$ is open in V. In the last case, A is also connected and dense in V/G, $\pi^{-1}(A)$ is connected and dense in V, and the maximal element is unique. We call it the *principal orbit type*.

An orbit type A determines the conjugacy class $[H]$ of subgroups of G. Moreover, $A \to B$ if and only if any group in the conjugacy class determined by A is a subgroup of some group in the conjugacy class determined by B. This relation defines a partial ordering on the set of conjugacy classes of stabilisers of the representation $U(g)$. Following Golubitsky, Stewart & Schaeffer (1988), we call the corresponding directed graph the *lattice of isotropy subgroups*.

In figures, we only draw the arrows between immediate successors. We will draw two orbit types (and the corresponding conjugacy classes of stabilisers) at thew same horizontal level if and only if they have the same dimension.

Let H be the stabiliser group of a vector $\boldsymbol{x} \in V$, and let A be the orbit type of $\boldsymbol{x}$. The fixed point set V^H is a subspace of V. For any group K of the normaliser $N(H)$ containing H, the space V^H is an invariant subspace of the representation (ρ, V) of the group K. The dimension of V^H is calculated by the *trace formula*:

$$\dim V^H = \int_H \mathrm{ch}_U(h)\,\mathrm{d}h, \tag{2.21}$$

where $\mathrm{d}h$ is the probabilistic Haar measure on H. The dimension of the stratum A is

$$\dim A = \dim V^H + \dim H - \dim N(H),$$

while the dimension of the stratum $\pi^{-1}(A)$ is

$$\dim \pi^{-1}(A) = \dim V^H + \dim G - \dim N(H).$$

From now on, we denote by U with different indices a *unitary* representation of a group G, and by ρ with different indices an *orthogonal* representation of G.

Example 12 (Irreducible unitary representations of SO(2))**.** The group SO(2) is isomorphic to the group U(1) of complex numbers of the form $\mathrm{e}^{\mathrm{i}\varphi}$, $0 \leq \varphi < 2\pi$. The character group of the abelian group U(1) is the additive group $\mathbb{Z}$ of integers. The character U^ℓ is

$$U^\ell(\varphi) = \mathrm{e}^{\mathrm{i}\ell\varphi}, \qquad \ell \in \mathbb{Z}.$$

The Haar measure on U(1) is

$$\mathrm{d}\mu(\varphi) = \frac{1}{2\pi}\mathrm{d}\varphi.$$

The Fine Structure Theorem tells us that the Fourier coefficients of a function $f \in L^2(\mathrm{U}(1), \mathrm{d}\mu)$ is

$$A^\ell(f) = \frac{1}{2\pi} \int_0^{2\pi} \mathrm{e}^{-\mathrm{i}\ell\varphi} f(\varphi)\, \mathrm{d}\varphi,$$

while its Fourier series is

$$f(\varphi) = \sum_{\ell \in \mathbb{Z}} A^\ell(f) \mathrm{e}^{\mathrm{i}\ell\varphi}.$$

The Fourier series of a square integrable function converges in the space $L^2(\mathrm{U}(1), \mathrm{d}\mu)$, while that of a continuous function converges uniformly.

Example 13 (Irreducible unitary representations of O(2))**.** The group $O(2)$ has two connected components. The connected component of the identity matrix is SO(2), the set of matrices

$$g_\varphi = \begin{pmatrix} \cos\varphi & \sin\varphi \\ -\sin\varphi & \cos\varphi \end{pmatrix}, \qquad 0 \leq \varphi < 2\pi.$$

The elements of SO(2) are called *rotations*.

The second component is

$$\mathrm{O}(2) \setminus \mathrm{SO}(2) = \{\, \sigma_x g_\varphi \colon g_\varphi \in \mathrm{SO}(2) \,\},$$

where

$$\sigma_x = \begin{pmatrix} 1 & 0 \\ 0 & -1 \end{pmatrix}$$

is the reflection through the x axis. The elements of this component are called *reflections*.

The irreducible unitary representations of O(2) are as follows: trivial representation $U^+(g) = 1$, determinant representation $U^-(g) = \det g$ acting in $\mathbb{C}^1$ and the representations U^ℓ, $\ell \geq 1$, acting in $\mathbb{C}^2$ as follows:

$$U^\ell(g_\varphi) = \begin{pmatrix} \mathrm{e}^{-\mathrm{i}\ell\varphi} & 0 \\ 0 & \mathrm{e}^{\mathrm{i}\ell\varphi} \end{pmatrix}, \qquad U^\ell(\sigma_x) = \begin{pmatrix} 0 & 1 \\ 1 & 0 \end{pmatrix}.$$

Later on, we will describe the irreducible *orthogonal* representations of the groups O(2) and O(3) using the following algorithm, which is applicable for any compact group G. The set $\mathrm{Irr}(G, \mathbb{C})$ of equivalence classes of irreducible unitary representations of G is the disjoint union of three subsets.

1. $\mathrm{Irr}(G, \mathbb{C})_\mathbb{R}$, the representations of *real type*. There exists a G-invariant real structure on V.
2. $\mathrm{Irr}(G, \mathbb{C})_\mathbb{C}$, the representations of *complex type*. The representation is *not* self-conjugate.
3. $\mathrm{Irr}(G, \mathbb{C})_\mathbb{H}$, the representations of *quaternionic type*. There exists a G-invariant quaternionic structure on V.

An irreducible unitary representation V of real type generates an irreducible orthogonal representation under restriction to the $(+1)$ eigenspace V_+ of the G-invariant real structure. The multiplication by $\mathrm{i} \in \mathbb{C}$ is an invertible intertwining operator between V_+ and V_-, the (-1) eigenspace of the above structure.

Any irreducible unitary representation V of complex type becomes irreducible and orthogonal when V is viewed as a real linear space with the same action of G. Two irreducible orthogonal representations obtained in this way from irreducible unitary representations V_1 and V_2 are equivalent if and only if either V_1 is equivalent to V_2 or V_1 is equivalent to $\overline{V_2}$.

Finally, an irreducible unitary representation V of quaternionic type becomes irreducible and orthogonal when V is viewed as a real linear space with the same action of G. No other irreducible orthogonal representations of G exist.

Example 14 (Irreducible orthogonal representations of SO(2) and O(2))**.** All irreducible unitary representations of the group O(2) are of real type. It follows that the restriction of each of them to the $(+1)$ eigenspace V_+ of the G-invariant real structure defines a one-to-one correspondence between $\mathrm{Irr}(\mathrm{O}(2), \mathbb{C})$ and $\mathrm{Irr}(\mathrm{O}(2), \mathbb{R})$. Following Auffray, Kolev & Olive (2017), consider three different representatives, or *models*, of each equivalence class in $\mathrm{Irr}(\mathrm{O}(2), \mathbb{R})$.

The first model is as follows. The first two representations are $\rho^+(g) = 1$ and $\rho^-(g) = \det g$. The representation $\rho^\ell(g)$, $\ell \geq 1$, are defined in the space $V_\ell = \mathbb{R}^2$ as follows:

$$\rho^\ell(g_\varphi) = \begin{pmatrix} \cos(\ell\varphi) & \sin(\ell\varphi) \\ -\sin(\ell\varphi) & \cos(\ell\varphi) \end{pmatrix}, \qquad \rho^\ell(\sigma_x) = \sigma_x.$$

The subgroup $\mathrm{O}(1) = \{I, \sigma_x\}$ has two irreducible orthogonal representations: $\rho^+(g) = 1$ and $\rho^-(g) = \det g$. The subspace of V^ℓ spanned by the first (resp. second) basis vector, carries the representation ρ^+ (resp. ρ^-), therefore we denote the first (resp. second) basis vector by $\boldsymbol{e}_1$ (resp. $\boldsymbol{e}_{-1}$).

The second model is as follows. Let $P^\ell(\mathbb{R}^2)$ be the real linear space of real-valued homogeneous polynomials of degree ℓ in two variables x and y. The group action

$$(g \cdot P)(\boldsymbol{x}) := P(\rho_\ell(g)^{-1}(\boldsymbol{x})), \qquad P \in P^\ell(\mathbb{R}^2), \quad \boldsymbol{x} = (x, y)^\top \in \mathbb{R}^2 \tag{2.22}$$

is a representation of the group O(2) in the space $P^\ell(\mathbb{R}^2)$. This representation is reducible. The subspace $\mathrm{H}^\ell(\mathbb{R}^2)$ of *harmonic polynomials* with vanishing Laplacian is an invariant subspace of the representation (2.22). It is well known that any real-valued harmonic function on $\mathbb{R}^2$ is either the real or the imaginary part of a holomorphic function. It follows that the two-dimensional subspace $\mathrm{H}^\ell(\mathbb{R}^2)$ is generated by its elements $\mathrm{Re}(x + \mathrm{i}y)^\ell$ and $\mathrm{Im}(x + \mathrm{i}y)^\ell$. Consider the mapping that maps $\boldsymbol{e}_1 = (1, 0)^\top \in \mathbb{R}^2$ to $\mathrm{Re}(x + \mathrm{i}y)^\ell \in \mathrm{H}^\ell(\mathbb{R}^2)$ and $\boldsymbol{e}_2 = (0, 1)^\top \in \mathbb{R}^2$ to $\mathrm{Im}(x + \mathrm{i}y)^\ell \in \mathrm{H}^\ell(\mathbb{R}^2)$. By linearity, it extends to an orthogonal intertwining operator between the representations of the two models.

The third model is as follows. Let r be a non-negative integer. The group action

$$(\rho^{\otimes r}(g))(\boldsymbol{v}_1, \dots, \boldsymbol{v}_r) := g\boldsymbol{v}_1 \otimes \cdots \otimes g\boldsymbol{v}_r$$

of the group $\mathrm{O}(2)$ on $(\mathbb{R}^2)^r$ can be extended by linearity to the orthogonal representation $(U^1)^{\otimes r}$. The representation $(\mathbb{R}^1, \rho^{\otimes 0})$ is trivial. For $r \geq 1$, the action (2.7) can also be extended by linearity to the orthogonal representation $((\mathbb{R}^n)^{\otimes r}, \rho_r)$ of the symmetric group Σ_r. It is clear that $\rho_r(\sigma)$ commutes with $\rho^{\otimes r}(g)$ for all $\sigma \in \Sigma_r$ and for all $g \in \mathrm{O}(n)$. We can therefore define a representation $((\mathbb{R}^n)^{\otimes r}, \tau)$ of the group $\mathrm{O}(n) \times \Sigma_r$ by

$$\tau(g, \sigma)(\boldsymbol{T}) := \rho^{\otimes r}(g)\rho_r(\sigma)(\boldsymbol{T}), \qquad \boldsymbol{T} \in (\mathbb{R}^n)^{\otimes r}.$$

Let G be a closed subgroup of $\mathrm{O}(n)$, and let Σ be a subgroup of Σ_r. The restriction of the representation $(\mathbb{R}^n)^{\otimes r}, \tau)$ to $\Sigma \times G$ is the direct sum of the irreducible representations of the group $\Sigma \times G$. Every such representation has the form $\tau'(\sigma)\hat{\otimes}\tau''(g)$, where $\tau' \in \hat{\Sigma}$, $\tau'' \in \hat{G}$.

Let (τ, ρ^+) be the direct sum of all representations $\tau_+(\sigma)\hat{\otimes}\tau''(g)$, where $(\tau_+, \mathbb{R}^1)$ is the trivial representation of Σ. We have $\rho^+ = P^+_\Sigma((\mathbb{R}^n)^{\otimes r})$, where P^+_Σ is the linear operator (2.11).

In particular, if $\Sigma = \Sigma_r$, then $\rho^+ = \mathsf{S}^r(\mathbb{R}^n)$, the set of all rank r symmetric tensors over $\mathbb{R}^n$, or the *rth symmetric tensor power* of $\mathbb{R}^n$. The representation $(\tau, \mathsf{S}^r(\mathbb{R}^n))$ is called the *rth symmetric tensor power* of the defining representation and is denoted by $(\mathsf{S}^r(g), \mathsf{S}^r(\mathbb{R}^n))$.

Assume that the group Σ contains odd permutations. Let (τ, U^-) be the direct sum of all representations $\varepsilon(\sigma)\hat{\otimes}\tau''(g)$. We have $\rho^- = P^-_\Sigma((\mathbb{R}^n)^{\otimes r})$, where P^-_Σ is the linear operator (2.12).

In particular, if $\Sigma = \Sigma_r$, then $\rho^- = \Lambda^r(\mathbb{R}^n)$, the set of all rank r skew-symmetric tensors over $\mathbb{R}^n$, or the *rth skew-symmetric tensor power* of $\mathbb{R}^n$. The representation $(\tau, \Lambda^r(\mathbb{R}^n))$ is called the *rth skew-symmetric tensor power* of the defining representation and is denoted by $(\Lambda^r(g), \Lambda^r(\mathbb{R}^n))$. When $G = \mathrm{O}(n)$, $n \geq 2$, we have the following decomposition:

$$g \otimes g = \mathsf{S}^2(g) \oplus \Lambda^2(g).$$

Define the *contraction operator* acting from $(\mathbb{R}^n)^{\otimes r}$ to $(\mathbb{R}^n)^{\otimes(r-2)}$ as follows. If $r \leq 1$, let the contraction operator be the zero operator. Otherwise, for any pair $1 \leq i < j \leq r$ the ij-contraction operator is defined on the tensors of the form $\boldsymbol{x}_1 \otimes \cdots \otimes \boldsymbol{x}_r$ by

$$\begin{aligned} C_{ij}(\boldsymbol{x}_1 \otimes \cdots \otimes \boldsymbol{x}_r) = (\boldsymbol{x}_i, \boldsymbol{x}_j)\boldsymbol{x}_1 \otimes \cdots \otimes \boldsymbol{x}_{i-1} \otimes \boldsymbol{x}_{i+1} \otimes \cdots \\ \otimes \boldsymbol{x}_{j-1} \otimes \boldsymbol{x}_{j+1} \otimes \cdots \otimes \boldsymbol{x}_r. \end{aligned}$$

A tensor $\boldsymbol{T} \in (\mathbb{R}^n)^{\otimes r}$ is called *harmonic* if any contraction operator maps it to the zero tensor. In particular, any scalar and any vector are harmonic tensors.

The restriction of the ℓth tensor power of the representation $(\mathbb{R}^2, g)$ to the space $\mathsf{H}^\ell(\mathbb{R}^2)$ of rank ℓ harmonic tensors is equivalent to $(\mathrm{H}^\ell(\mathbb{R}^2), \rho_\ell)$. An intertwining operator $\Phi\colon \mathsf{H}^\ell(\mathbb{R}^2) \to \mathrm{H}^\ell(\mathbb{R}^2)$ acts as follows:

$$(\Phi \boldsymbol{T})(\boldsymbol{x}) = \boldsymbol{T}(\boldsymbol{x}, \dots, \boldsymbol{x}). \tag{2.23}$$

The polynomial $\Phi\boldsymbol{T}$ is indeed harmonic, because

$$(\Delta\Phi\boldsymbol{T})(\boldsymbol{x}) = \ell(\ell-1)(\operatorname{tr}\boldsymbol{T})(\boldsymbol{x}, \dots, \boldsymbol{x}) = \boldsymbol{0}.$$

The inverse operator $\Phi^{-1}\colon \mathrm{H}^\ell(\mathbb{R}^2) \to \mathsf{H}^\ell(\mathbb{R}^2)$ acts by *polarisation*:

$$(\Phi^{-1}P)(\boldsymbol{x}_1, \dots, \boldsymbol{x}_\ell) = \frac{1}{\ell!} \left.\frac{\partial^\ell}{\partial t_1 \cdots \partial t_\ell}\right|_{t_1=\cdots=t_\ell=0} P(t_1\boldsymbol{x}_1 + \cdots + t_\ell\boldsymbol{x}_\ell). \tag{2.24}$$

Following Auffray et al. (2017), put $\mathsf{H}^0(\mathbb{R}^2) = \mathsf{H}^{-1}(\mathbb{R}^2) = \mathbb{R}^1$, and let the representation ρ^+ (resp. ρ^-) acts in $\mathsf{H}^0(\mathbb{R}^2)$ (resp. $\mathsf{H}^{-1}(\mathbb{R}^2)$). The spaces $\mathsf{H}^\ell(\mathbb{R}^2)$, $\ell \geq -1$, have special names. In particular, $\mathsf{H}^{-1}(\mathbb{R}^2)$ is the space of *pseudo-scalars*, or *hemitropic coefficients*. $\mathsf{H}^0(\mathbb{R}^2)$ is the space of *scalars*, or *isotropic coefficients*. $\mathsf{H}^1(\mathbb{R}^2)$ is the space of *vectors*. $\mathsf{H}^2(\mathbb{R}^2)$ is the space of *deviators*. $\mathsf{H}^\ell(\mathbb{R}^2)$, $\ell \geq 3$, is the space of *ℓth-order deviators*.

The characters of the two-dimensional representations ρ^ℓ, $\ell \geq 1$, are as follows

$$\operatorname{ch}_{\rho^\ell}(g_\varphi) = 2\cos(\ell\varphi), \qquad \operatorname{ch}_{\rho_\ell}(kg_\varphi) = 0.$$

Expanding the product of two characters into a sum of characters, we obtain a tensor multiplication table for the irreducible orthogonal representations of O(2) as follows:

	ρ^+	ρ^-	ρ^ℓ
ρ^+	ρ^+		
ρ^-	ρ^-	ρ^+	
ρ^m	ρ^m	ρ^m	$\rho^{\lvert\ell-m\rvert} \oplus \rho^{\ell+m}, m \neq \ell$ $\rho^+ \oplus \rho^- \oplus \rho^{2\ell}, m = \ell$

The Clebsch–Gordan coefficients for the irreducible orthogonal representations of O(2) are easy to calculate using (2.19) and (2.20). For $n \neq m$, the Clebsch–Gordan matrices are as follows

$$c^{-1}_{\lvert m-n\rvert[m,n]} = c^0_{-[m,m]} = \begin{pmatrix} 0 & 1/\sqrt{2} \\ -1/\sqrt{2} & 0 \end{pmatrix},$$

$$c^1_{\lvert m-n\rvert[m,n]} = c^0_{+[m,m]} = \begin{pmatrix} 1/\sqrt{2} & 0 \\ 0 & 1/\sqrt{2} \end{pmatrix},$$

$$c^{-1}_{m+n[m,n]} = \begin{pmatrix} 0 & 1/\sqrt{2} \\ 1/\sqrt{2} & 0 \end{pmatrix}, \qquad c^1_{m+n[m,n]} = \begin{pmatrix} -1/\sqrt{2} & 0 \\ 0 & 1/\sqrt{2} \end{pmatrix}. \tag{2.25}$$

The restrictions of the representations ρ^+ and ρ^- to the subgroup SO(2) are equal to ρ^0, its trivial representation. The representations ρ^ℓ remain irreducible after restriction. No other irreducible orthogonal representations of SO(2) exist.

The characters of the representations ρ^ℓ, $\ell \geq 1$, are

$$\operatorname{ch}_{\rho^\ell}(\varphi) = 2\cos(\ell\varphi).$$

Expanding the product of two characters into a sum of characters, we obtain a tensor multiplication table for the irreducible orthogonal representations of SO(2) as follows:

	ρ^0	$\rho^\ell, \ell \geq 1$
ρ^0	ρ^0	ρ^ℓ
$\rho^m, m \geq 1$	ρ^m	$\rho^{\lvert\ell-m\rvert} \oplus \rho^{\ell+m}, m \neq \ell$
		$2\rho^0 \oplus \rho^{2\ell}, m = \ell$

As we see, the group SO(2) is not easily reducible. A general method of calculating Clebsch–Gordan coefficients for such groups is described in Klimyk (1979). Instead of using the above method, we use the fact that the representations of SO(2) act in the same space as those of O(2) and share the same basis. Formulae for Clebsch–Gordan coefficients follow:

$$\begin{aligned}
c^{-1}_{|m-n|[m,n]} &= c^0_{0,2[m,m]} = \begin{pmatrix} 0 & 1/\sqrt{2} \\ -1/\sqrt{2} & 0 \end{pmatrix}, \\
c^{1}_{|m-n|[m,n]} &= c^0_{0,1[m,m]} = \begin{pmatrix} 1/\sqrt{2} & 0 \\ 0 & 1/\sqrt{2} \end{pmatrix}, \\
c^{-1}_{m+n[m,n]} &= \begin{pmatrix} 0 & 1/\sqrt{2} \\ 1/\sqrt{2} & 0 \end{pmatrix}, \qquad c^{1}_{m+n[m,n]} = \begin{pmatrix} -1/\sqrt{2} & 0 \\ 0 & 1/\sqrt{2} \end{pmatrix},
\end{aligned} \tag{2.26}$$

where the first copy of the trivial representation acts in the linear space generated by the identity operator, while its second copy acts in the space of skew-symmetric matrices.

The directed graph of the orbit type stratification of the representation $(\rho_\ell, \mathbb{R}^2)$ is as follows:

$$\begin{array}{c} \{0\} \\ \uparrow \\ (0,\infty) \end{array} \quad . \tag{2.27}$$

The lattice of isotropy subgroups for O(2) is

$$\begin{array}{c} [\mathrm{O}(2)] \\ \uparrow \\ [Z_2] \end{array} \quad ,$$

while that for SO(2) is

$$\begin{array}{c}[\mathrm{SO}(2)]\,.\\ \uparrow \\ \big| \\ [E]\end{array}$$

In what follows we will draw only the lattice of isotropy subgroups.

Example 15 (Irreducible unitary representations of SU(2))**.** The group SU(2) consists of the matrices

$$g = \begin{pmatrix} \alpha & \beta \\ -\overline{\beta} & \overline{\alpha} \end{pmatrix}, \qquad \alpha, \beta \in \mathbb{C}, \quad |\alpha|^2 + |\beta|^2 = 1. \tag{2.28}$$

The action

$$g \cdot P(u, v) = P(g^{-1}(u, v)^\top)$$

is an irreducible representation $U^\ell(g)$ of SU(2) in the complex linear space $P^{2\ell}(\mathbb{C}^2)$ of complex-valued homogeneous polynomials of degree 2ℓ in two complex variables, where $\ell = 0,\ 1/2,\ 1,\ \dots$. Another common name for this space is the space of 2-ary (or binary) forms of degree 2ℓ on $\mathbb{C}^2$. Note that $U^\ell(-I) = I$ if and only if ℓ is integer.

Introduce an inner product in the space $P^{2\ell}(\mathbb{C}^2)$ in such a way that the binary forms

$$\boldsymbol{e}_m(u, v) := (-1)^{\ell+m} \sqrt{\frac{(2\ell+1)!}{(\ell+m)!(\ell-m)!}} u^{\ell+m} v^{\ell-m}, \tag{2.29}$$

where $m = -2\ell,\ -2\ell+1,\ \dots,\ 2\ell$, constitute an orthonormal basis. The basis (2.29) is called the *Wigner orthonormal basis.*

The matrix entries of the operators $U^\ell(g)$ in the above basis are called *Wigner D functions* and are denoted by $D^\ell_{mn}(g)$. The tensor product $U^{\ell_1} \otimes U^{\ell_2}$ is expanding as follows:

$$U^{\ell_1}(g) \otimes U^{\ell_2}(g) = \sum_{\ell=|\ell_1-\ell_2|}^{\ell_1+\ell_2} \oplus U^\ell(g). \tag{2.30}$$

The standard choice of the phase for the Clebsch–Gordan coefficients is

$$c^{|\ell_1-\ell_2|[\ell_1,-\ell_2]}_{\ell[\ell_1,\ell_2]} > 0.$$

Realise the linear space $\mathbb{R}^3$ with coordinates x_{-1}, x_0 and x_1 as the set of traceless Hermitian matrices over $\mathbb{C}^2$ with entries

$$\begin{pmatrix} x_0 & x_1 + \mathrm{i}x_{-1} \\ x_1 - \mathrm{i}x_{-1} & -x_0 \end{pmatrix}.$$

The matrix (2.28) acts on the so realised $\mathbb{R}^3$ as follows:

$$\pi(g)\begin{pmatrix} x_0 & x_1+\mathrm{i}x_{-1} \\ x_1-\mathrm{i}x_{-1} & -x_0 \end{pmatrix} := g^*\begin{pmatrix} x_0 & x_1+\mathrm{i}x_{-1} \\ x_1-\mathrm{i}x_{-1} & -x_0 \end{pmatrix} g.$$

The mapping π is a homomorphism of SU(2) onto SO(3). The kernel of π is $\pm I$.

The *Cartan map* is the linear map $\psi\colon \mathrm{H}^\ell(\mathbb{C}^3)\to P^{2\ell}(\mathbb{C}^2)$ for *integer* ℓ, where $\mathrm{H}^\ell(\mathbb{C}^3)$ is the space of complex-valued harmonic polynomials of degree ℓ in three complex variables x, y and z. It is given by

$$(\psi(P))(u,v)=P\left(\frac{u^2-v^2}{2},\frac{u^2+v^2}{2\mathrm{i}},uv\right).$$

The inverse map ψ^{-1} is constructed in the following way. Let $P\in P^{2\ell}(\mathbb{C}^2)$ with

$$P(u,v)=\sum_{m=0}^{2\ell}a_m u^k v^{2\ell-k}.$$

For each m, replace the expression $u^m v^{2\ell-m}$ with $z^m(-x+y\mathrm{i})^{\ell-m}$ if $0\le m\le\ell$, and with $z^{2\ell-m}(x+y\mathrm{i})^{\ell-m}$ if $\ell+1\le m\le 2\ell$. Denote the resulting polynomial by $p(x,y,z)$. Let r be the integer part of $\ell/2$. Put

$$p_r(x,y,z)=\begin{cases}\frac{1}{(2r+1)!}\Delta^r p(x,y,z), & \text{if } \ell \text{ is even},\\ \frac{6(r+1)}{(2r+3)!}\Delta^r p(x,y,z), & \text{otherwise.}\end{cases}$$

For $0\le k\le r$ define $p_k(x,y,z)$ recursively by

$$p_k(x,y,z)=\mu_k\Delta^k\left(p(x,y,z)-\sum_{j=k+1}^{r}(x^2+y^2+z^2)^j p_j(x,y,z)\right),$$

where

$$\mu_k=\frac{(2\ell-4k+1)!(\ell-k)!}{(2\ell-2k+1)!k!(\ell-2k)!}.$$

Put

$$(\psi^{-1}P)(x,y,z)=p_0(x,y,z),$$

This map intertwines the representation $U^\ell(g)$ with the representation given by the action

$$(g\cdot P)(x,y,z)=P((\pi(g))^{-1}(x,y,z)^\top),$$

see Olive, Kolev, Desmorat & Desmorat (2017, Theorem 5.1) for a proof.

Example 16 (Irreducible unitary representations of SO(3) and O(3))**.** Assume that U is an irreducible unitary representation of SO(3). Then $U\circ\pi$ is an irreducible unitary representation of SU(2) with kernel $\pm E$. Then we have $U\circ\pi=U^\ell$ for some integer ℓ. In other words, every irreducible unitary representation U^ℓ of SU(2) with integer ℓ gives rise to an irreducible unitary representation of

SO(3), and no other irreducible unitary representations exist. We denote the above representation of SO(3) again by U^ℓ.

Let SO(2) be the subgroup of SO(3) that leaves the vector $(0,0,1)^\top$ fixed. The restriction of U^ℓ to SO(2) is equivalent to the direct sum of irreducible unitary representations $(\mathbb{C}^1, \mathrm{e}^{\mathrm{i}m\varphi})$, $-\ell \le m \le \ell$ of SO(2). Moreover, the space of the representation $(\mathbb{C}^1, \mathrm{e}^{\mathrm{i}m\varphi})$ is spanned by the vector $\boldsymbol{e}_m(u,v)$ of the Wigner basis (2.29).This is where their enumeration comes from.

The group $O(3)$ is the Cartesian product of its normal subgroups SO(3) and $\{I, -I\}$. The elements of SO(3) are rotations, while the elements of the second component are reflections. Therefore, any irreducible unitary representation of $O(3)$ is the outer tensor product of some U^ℓ by an irreducible unitary representation of $\{E, -E\}$. The latter group has two irreducible unitary representation: trivial U^+ and determinant U^-. Denote $U^{\ell g} := U^\ell \hat{\otimes} U^+$ and $U^{\ell u} := U^\ell \hat{\otimes} U^-$ (g by German *gerade*, even and u by *ungerade*, odd). These are all irreducible unitary representations of $O(3)$.

Introduce a chart on SO(3), the *Euler angles*. Any rotation g may be performed by three successive rotations:

- rotation $g_0(\psi)$ about the x_0-axis through an angle ψ, $0 \le \psi < 2\pi$;
- rotation $g_{-1}(\theta)$ about the x_{-1}-axis through an angle θ, $0 \le \theta \le \pi$,
- rotation $g_0(\varphi)$ about the x_0-axis through an angle φ, $0 \le \varphi < 2\pi$.

The angles ψ, θ and φ are the *Euler angles*. The Wigner D functions are $D^\ell_{mn}(\varphi, \theta, \psi)$. The Wigner D functions D^ℓ_{m0} do not depend on ψ and may be written as $D^\ell_{m0}(\varphi, \theta)$. The *spherical harmonics* Y^m_ℓ are defined by

$$Y^m_\ell(\theta, \varphi) := \sqrt{\frac{2\ell+1}{4\pi}} \overline{D^\ell_{m0}(\varphi, \theta)}. \tag{2.31}$$

Let (r, θ, φ) be the spherical coordinates in $\mathbb{R}^3$:

$$\begin{aligned} x_{-1} &= r \sin\theta \sin\varphi, \\ x_0 &= r\cos\theta, \\ x_1 &= r \sin\theta \cos\varphi. \end{aligned} \tag{2.32}$$

The measure $\mathrm{d}S := \sin\theta \,\mathrm{d}\varphi \,\mathrm{d}\theta$ is the Lebesgue measure on the *unit sphere* $S^2 := \{\, \boldsymbol{x} \in \mathbb{R}^3 \colon \|\boldsymbol{x}\| = 1 \,\}$. The spherical harmonics are orthonormal:

$$\int_{S^2} Y^{m_1}_{\ell_1}(\theta, \varphi) \overline{Y^{m_2}_{\ell_2}(\theta, \varphi)} \,\mathrm{d}S = \delta_{\ell_1 \ell_2} \delta_{m_1 m_2}.$$

Example 17 (Irreducible orthogonal representations of SO(3) and O(3))**.** The first model of the above representations is as follows. Let $\mathrm{H}^\ell(\mathbb{R}^3)$ be the space of real-valued homogeneous harmonic polynomials of degree ℓ in three real variables. There exist two irreducible orthogonal representations of thc group O(3) on $\mathrm{H}^\ell(\mathbb{R}^3)$:

$$\rho^\ell(g)P(\boldsymbol{x}) = P(g^{-1}\boldsymbol{x}),$$
$$\rho^{\ell *}(g)P(\boldsymbol{x}) = \det g P(g^{-1}\boldsymbol{x}).$$

We observe that $\rho^\ell(-I)P(\boldsymbol{x}) = P(-\boldsymbol{x})$. When ℓ is even, we have $P(-\boldsymbol{x}) = P(\boldsymbol{x})$, that is, $\rho^\ell(-I)$ is the identity operator in $\mathrm{H}^\ell(\mathbb{R}^3)$. On the other hand, when ℓ is odd, we have $P(-\boldsymbol{x}) = -P(\boldsymbol{x})$, that is, $\rho^\ell(-I)$ multiplies all elements of the space $\mathrm{H}^\ell(\mathbb{R}^3)$ by -1. In other words, the representation ρ^ℓ is the orthogonal version of the unitary *gerade* representation $U^{\ell g}$ for even ℓ and the orthogonal version of the unitary *ungerade* representation $U^{\ell u}$ for odd ℓ. Similarly, the representation $\rho^{\ell *}$ is the orthogonal version of the unitary *ungerade* representation $U^{\ell u}$ for even ℓ and the orthogonal version of the unitary *gerade* representation $U^{\ell g}$ for odd ℓ.

The restriction of both representations to the subgroup SO(3) is the same representation ρ^ℓ, the orthogonal version of the unitary representation U^ℓ. It follows from (2.30) that

$$\begin{aligned}\rho^{\ell_1} \otimes \rho^{\ell_2} = \rho^{\ell_1 *} \otimes \rho^{\ell_2 *} &= \sum_{\ell=|\ell_1-\ell_2|}^{\ell_1+\ell_2} \oplus \rho^\ell,\\ \rho^{\ell_1 *} \otimes \rho^{\ell_2} = \rho^{\ell_1} \otimes \rho^{\ell_2 *} &= \sum_{\ell=|\ell_1-\ell_2|}^{\ell_1+\ell_2} \oplus \rho^{\ell *}.\end{aligned} \tag{2.33}$$

The operators of the second model act in the space $\mathsf{H}^\ell(\mathbb{R}^3)$ of rank ℓ harmonic tensors by

$$\rho^\ell \mathsf{T}(\boldsymbol{x},\dots,\boldsymbol{x}) = \mathsf{T}(g^{-1}\boldsymbol{x},\dots,g^{-1}\boldsymbol{x}),$$
$$\rho^{\ell *} \mathsf{T}(\boldsymbol{x},\dots,\boldsymbol{x}) = \det g \mathsf{T}(g^{-1}\boldsymbol{x},\dots,g^{-1}\boldsymbol{x}).$$

The direct and inverse intertwining operators between the two models are given by Equations (2.23) and (2.24).

The third model is as follows. For any polynomial $P \in P^{2\ell}(\mathbb{C}^2)$ with integer ℓ denote by $\overline{P}$ the polynomial whose coefficients are conjugate to those of P. Define the mapping $J\colon P^{2\ell}(\mathbb{C}^2) \to P^{2\ell}(\mathbb{C}^2)$ as

$$(JP)(u,v) := \overline{P}(-v,u).$$

It is easy to prove that J is a real structure on $P^{2\ell}(\mathbb{C}^2)$ that respects both ρ^ℓ and $\rho^{\ell *}$. The orthonormal basis in the space $P^{2\ell}_{\mathbb{R}}(\mathbb{C}^2)$ of eigenvectors of J with eigenvalue $(-1)^\ell$ was proposed by Gordienko (2002). The vectors of the *Gordienko basis* are as follows ($m \geq 1$):

$$\begin{aligned}\boldsymbol{P}_{-m}(u,v) &:= \frac{(-\mathrm{i})^{\ell-1}}{\sqrt{2}}[(-1)^m \boldsymbol{e}_m(u,v) - \boldsymbol{e}_{-m}(u,v)],\\ \boldsymbol{P}_0(u,v) &:= (-\mathrm{i})^\ell \boldsymbol{e}_0(u,v),\\ \boldsymbol{P}_m(u,v) &:= -\frac{(-\mathrm{i})^\ell}{\sqrt{2}}[(-1)^m \boldsymbol{e}_m(u,v) + \boldsymbol{e}_{-m}(u,v)].\end{aligned}$$

In this basis, the representations ρ^ℓ and $\rho^{\ell *}$ become orthogonal.

Remark 1. The vector $\boldsymbol{P}_0(u,v)$ of the Gordienko basis is proportional to the vector $\boldsymbol{e}_0(u,v)$ of the Wigner basis. It follows that the one-dimensional real subspace generated by $\boldsymbol{P}_0(u,v)$ carries the one-dimensional representation of the subgroup O(2).

Denote the matrix entries of the matrix $\rho_\ell(g)$ by $\rho^\ell_{mn}(\varphi,\theta,\psi)$. If $g=g(\varphi,\theta,\psi)$, then

$$\rho_\ell(g)=\rho_\ell(g_0(\varphi))\rho_\ell(g_{-1}(\theta))\rho_\ell(g_0(\psi))$$

by definition of a representation. Denote the matrix entries of the matrix $\rho_\ell(g_0(\varphi))$ by $\Omega^\ell_{0,m,n}(\varphi)$, where $-\ell\le m,n\le\ell$. The non-zero entries are

$$\Omega^\ell_{0,0,0}(\varphi)=1,\qquad \Omega^\ell_{0,m,m}(\varphi)=\cos(m\varphi),\qquad \Omega^\ell_{0,-m,m}(\varphi)=\sin(m\varphi), \tag{2.34}$$

where $m=\pm1,\pm2,\dots,\pm\ell$. Denote the matrix entries of the matrix $\rho_\ell(g_{-1}(\theta))$ by $\Omega^\ell_{-1,m,n}(\theta)$. Then we have

$$\rho^\ell_{mn}(\varphi,\theta,\psi)=\sum_{p,q=-\ell}^{\ell}\Omega^\ell_{0,m,p}(\varphi)\Omega^\ell_{-1,p,q}(\theta)\Omega^\ell_{0,q,n}(\psi).$$

We are only interested in the matrix entries ρ^ℓ_{m0}. In this case $n=0$. It follows from (2.34) that $q=0$ and $p=\pm m$. The non-zero entries $\Omega^\ell_{-1,\pm m,0}(\theta)$ are

$$\begin{aligned}\Omega^\ell_{-1,0,0}(\theta)&=\frac{(-1)^\ell}{2^\ell\ell!}\frac{\mathrm{d}^\ell}{\mathrm{d}\mu^\ell}(1-\mu^2)^\ell,\\ \Omega^\ell_{-1,m,0}(\theta)&=-\frac{(-1)^\ell}{2^\ell\ell!}\sqrt{\frac{2(\ell+m)!}{(\ell-m)!}}\frac{1}{(1-\mu^2)^{m/2}}\frac{\mathrm{d}^{\ell-m}}{\mathrm{d}\mu^{\ell-m}}(1-\mu^2)^\ell,\end{aligned} \tag{2.35}$$

where $m\ge1$ and $\mu=\cos\theta$. We obtain:

$$\rho^\ell_{m0}(\varphi,\theta)=\begin{cases}\Omega^\ell_{-1,|m|,0}(\theta)\sin(|m|\varphi), & m<0,\\ \Omega^\ell_{-1,m,0}(\theta)\cos(m\varphi), & m\ge0.\end{cases}$$

On the second connected component of O(3), the matrix entries of the representations ρ_ℓ remain the same, and those of the representations ρ^*_ℓ are multiplied by -1.

The Clebsh–Gordan coefficients of the groups SO(3) and O(3) with respect to the Gordienko basis were calculated by Godunov & Gordienko (2004). We call them *Godunov–Gordienko coefficients* and denote them by $g^{m[m_1,m_2]}_{\ell[\ell_1,\ell_2]}$. We describe the algorithm for calculation of the Godunov–Gordienko coefficients, following Selivanova (2014).

First, fix non-negative integers ℓ, ℓ_1 and ℓ_2 with $|\ell_1-\ell_2|\le\ell\le\ell_1+\ell_2$. Calculate the Clebsch–Gordan matrices $C^{\pm\ell}_{\ell[\ell_1,\ell_2]}$. If $\ell\ne1$, calculate also $C^{\pm1}_{1[\ell_1,\ell_1]}$ and $C^{\pm1}_{1[\ell_2,\ell_2]}$. Formulae for calculation the classical Clebsch–Gordan coefficients may be found in, e.g. Varshalovich, Moskalev & Khersonskiĭ (1988).

Second, let V_ℓ be the square matrix with $2\ell+1$ rows and the following non-zero entries:

$$V_\ell = \frac{\mathrm{i}^{\ell-1}}{\sqrt{2}} \begin{pmatrix} -1 & & & & & & (-1)^\ell \\ & \dots & & & & \dots & \\ & & -1 & & (-1)^m & & \\ & & & \mathrm{i}\sqrt{2} & & & \\ & & -\mathrm{i} & & -\mathrm{i}(-1)^m & & \\ & \dots & & & & \dots & \\ -\mathrm{i} & & & & & & -\mathrm{i}(-1)^\ell \end{pmatrix}.$$

Calculate the Godunov–Gordienko matrices $G^{\ell}_{\ell[\ell_1,\ell_2]}$, $G^{-\ell}_{\ell[\ell_1,\ell_2]}$ with $\ell > 0$, and $G^{0}_{0[\ell_1,\ell_1]}$ by the following formulae:

$$\begin{aligned} G^{\ell}_{\ell[\ell_1,\ell_2]} &= \frac{-(-\mathrm{i}^\ell)}{\sqrt{2}} V_{\ell_1}[(-1)^\ell C^{\ell}_{\ell[\ell_1,\ell_2]} + C^{-\ell}_{\ell[\ell_1,\ell_2]}](V_{\ell_2})^\top, \\ G^{-\ell}_{\ell[\ell_1,\ell_2]} &= \frac{-\mathrm{i}^{\ell-1}}{\sqrt{2}} V_{\ell_1}[(-1)^\ell C^{\ell}_{\ell[\ell_1,\ell_2]} - C^{-\ell}_{\ell[\ell_1,\ell_2]}](V_{\ell_2})^\top, \\ G^{0}_{0[\ell_1,\ell_1]} &= V_{\ell_1} C^{0}_{0[\ell_1,\ell_1]}(V_{\ell_1})^\top. \end{aligned}$$

If $\ell \neq 1$, calculate also $G^{\pm 1}_{1[\ell_1,\ell_1]}$ and $G^{\pm 1}_{1[\ell_2,\ell_2]}$ by Godunov & Gordienko (2004, equations (1.27)–(1.29)), that is, let

$$r^n_\ell = \frac{1}{2}\sqrt{\frac{3(\ell+n+1)(\ell-n)}{\ell(\ell+1)(2\ell+1)}}.$$

The matrix $G^{-1}_{1[\ell,\ell]}$ is skew-symmetric. The only diagonal that lies over the main one and may contain non-zero elements is as follows.

$$(g^{-1[-\ell+1,-\ell]}_{1[\ell,\ell]}, \dots, g^{-1[\ell,\ell-1]}_{1[\ell,\ell]}) = (r^{\ell-1}_\ell, \dots, r^1_\ell, 0, \sqrt{2}r^0_\ell, -r^1_\ell, \dots, -r^{\ell-1}_\ell).$$

The non-zero diagonals of the matrix $G^{1}_{1[\ell,\ell]}$ are as follows.

$$\begin{aligned} (g^{1[\ell-1,-\ell]}_{1[\ell,\ell]}, \dots, g^{1[-\ell,\ell-1]}_{1[\ell,\ell]}) &= (-r^{\ell-1}_\ell, \dots, -r^1_\ell, \sqrt{2}r^0_\ell, -\sqrt{2}r^0_\ell, r^1_\ell, \dots, r^{\ell-1}_\ell), \\ (g^{1[\ell,-\ell+1]}_{1[\ell,\ell]}, \dots, g^{1[-\ell,\ell-1]}_{1[\ell,\ell]}) &= (-r^{\ell-1}_\ell, \dots, -r^1_\ell, 0, 0, r^1_\ell, \dots, r^{\ell-1}_\ell). \end{aligned}$$

Third, starting from $m = \ell$, subtracting -1 and up to $m = 2$, calculate the Godunov–Gordienko matrices $G^{-(m-1)}_{\ell[\ell_1,\ell_2]}$ and $G^{m-1}_{\ell[\ell_1,\ell_2]}$ by

$$\begin{aligned} G^{-(m-1)}_{\ell[\ell_1,\ell_2]} = &-\frac{1}{\sqrt{(\ell+m)(\ell-m+1)}}\{-\frac{1}{\sqrt{3}}[\sqrt{\ell_1(\ell_1+1)(2\ell_1+1)} \\ &\times G^{-1}_{1[\ell_1,\ell_1]} G^{-m}_{\ell[\ell_1,\ell_2]} + \sqrt{\ell_2(\ell_2+1)(2\ell_2+1)} G^{-m}_{\ell[\ell_1,\ell_2]}(G^{-1}_{1[\ell_2,\ell_2]})^\top] \\ &- \frac{1}{\sqrt{(\ell+m)(\ell-m+1)}}\{-\frac{1}{\sqrt{3}}[\sqrt{\ell_1(\ell_1+1)(2\ell_1+1)} \\ &\times G^{1}_{1[\ell_1,\ell_1]} G^{m}_{\ell[\ell_1,\ell_2]} + \sqrt{\ell_2(\ell_2+1)(2\ell_2+1)} G^{m}_{\ell[\ell_1,\ell_2]}(G^{1}_{1[\ell_2,\ell_2]})^\top]\} \end{aligned}$$

and

$$
\begin{aligned}
G^{m-1}_{\ell[\ell_1,\ell_2]} = &-\frac{1}{\sqrt{(\ell+m)(\ell-m+1)}}\{\frac{1}{\sqrt{3}}[\sqrt{\ell_1(\ell_1+1)(2\ell_1+1)} \\
&\times G^1_{1[\ell_1,\ell_1]}G^{-m}_{\ell[\ell_1,\ell_2]} + \sqrt{\ell_2(\ell_2+1)(2\ell_2+1)}G^{-m}_{\ell[\ell_1,\ell_2]}(G^1_{1[\ell_2,\ell_2]})^\top] \\
&-\frac{1}{\sqrt{(\ell+m)(\ell-m+1)}}\{-\frac{1}{\sqrt{3}}[\sqrt{\ell_1(\ell_1+1)(2\ell_1+1)} \\
&\times G^{-1}_{1[\ell_1,\ell_1]}G^{m}_{\ell[\ell_1,\ell_2]} + \sqrt{\ell_2(\ell_2+1)(2\ell_2+1)}G^{m}_{\ell[\ell_1,\ell_2]}(G^{-1}_{1[\ell_2,\ell_2]})^\top]\}.
\end{aligned}
$$

Finally, calculate the Godunov–Gordienko matrix $G^0_{\ell[\ell_1,\ell_2]}$ by

$$
\begin{aligned}
G^{0}_{\ell[\ell_1,\ell_2]} = &\frac{1}{\sqrt{2\ell(\ell+1)}}\{\frac{1}{\sqrt{3}}[\sqrt{\ell_1(\ell_1+1)(2\ell_1+1)} \\
&\times G^1_{1[\ell_1,\ell_1]}G^{-1}_{\ell[\ell_1,\ell_2]} + \sqrt{\ell_2(\ell_2+1)(2\ell_2+1)}G^{-1}_{\ell[\ell_1,\ell_2]}(G^1_{1[\ell_2,\ell_2]})^\top] \\
&-\frac{1}{\sqrt{\ell(\ell+1)}}\{-\frac{1}{\sqrt{3}}[\sqrt{\ell_1(\ell_1+1)(2\ell_1+1)} \\
&\times G^{-1}_{1[\ell_1,\ell_1]}G^{1}_{\ell[\ell_1,\ell_2]} + \sqrt{\ell_2(\ell_2+1)(2\ell_2+1)}G^{1}_{\ell[\ell_1,\ell_2]}(G^{-1}_{1[\ell_2,\ell_2]})^\top]\}.
\end{aligned}
$$

If the numbers ℓ_1, ℓ_2 and ℓ_3 are non-negative integers, but do not satisfy the *triangle condition*

$$|\ell_1-\ell_2| \leq \ell_3 \leq \ell_1+\ell_2,$$

then put $g^{m_3[m_1,m_2]}_{\ell_3[\ell_1,\ell_2]} := 0$. The following integral will be important in the sequel.

Theorem 5 (Gaunt integral)**.**

$$
\begin{aligned}
\int_{S^2} S^{m_1}_{\ell_1}(\theta,\varphi)S^{m_2}_{\ell_2}(\theta,\varphi)S^{m_3}_{\ell_3}(\theta,\varphi)\,\mathrm{d}\Omega = &\sqrt{\frac{(2\ell_1+1)(2\ell_2+1)}{4\pi(2\ell_3+1)}} \\
&\times g^{m_3[m_1,m_2]}_{\ell_3[\ell_1,\ell_2]}g^{0[0,0]}_{\ell_3[\ell_1,\ell_2]}.
\end{aligned}
\tag{2.36}
$$

This theorem can be proved in exactly the same way as its complex counterpart; see, for example, Marinucci & Peccati (2011).

The elements of the linear space where the representation ρ_1 acts, are called *axial vectors* or pseudo-vectors. They do not change under rotations and change sign under reflections.

The directed graph of the representations ρ_1^* and ρ_1 is (2.27). The representation ρ_2 of SO(3) may be realised in the five-dimensional space V of 3×3 symmetric traceless matrices by

$$\rho_2(g)A = gAg^{-1}, \qquad g\in \mathrm{SO}(3), \quad A \in E.$$

The lattice of isotropy subgroups is

$$\begin{array}{c}[\mathrm{SO}(3)]\\ \uparrow\\ [\mathrm{O}(2)]\\ \uparrow\\ [D_2]\end{array}$$

where D_2 is the group generated by the matrices

$$\sigma_1 = \begin{pmatrix} -1 & 0 & 0 \\ 0 & -1 & 0 \\ 0 & 0 & 1 \end{pmatrix}, \qquad \sigma_2 = \begin{pmatrix} 1 & 0 & 0 \\ 0 & -1 & 0 \\ 0 & 0 & -1 \end{pmatrix};$$

see Golubitsky et al. (1988).

The intertwining operator between the spaces of the second and third models is the restriction of the Cartan map to the space $\mathrm{H}^{2\ell}(\mathbb{R}^3)$.

Example 18 (Expansions of representations of orthogonal groups)**.** Put $n = 2$, $r = 2$, $G = \mathrm{O}(2)$ and $\Sigma = \Sigma_2$. Then the representation $(\tau, (\mathbb{R}^2)^{\otimes 2})$ of the group $\mathrm{O}(2) \times \Sigma_2$ is the direct sum of three irreducible components

$$\tau = [\rho_+(g)\tau_+(\sigma)] \oplus [\rho_-(g)\varepsilon(\sigma)] \oplus [\rho_2(g)\tau_+(\sigma)].$$

The space of the first component is the span of the identity matrix. The mapping $\alpha I \to \alpha$ is the isomorphism between the above space and the field $\mathbb{R}$ of scalars. The space of the second component consists of 2×2 skew-symmetric matrices. Its elements are *pseudo-scalars* (they do not change under rotations and change sign under reflections, i.e. transform according to the non-trivial representation $\rho_-(g)$ of $\mathrm{O}(2)$). Finally, the space of the third component consists of 2×2 traceless symmetric matrices, or deviators. The second component is $(\Lambda^2(g), \Lambda^2(\mathbb{R}^2))$, and the direct sum of the first and third components is $(\mathsf{S}^2(g), \mathsf{S}^2(\mathbb{R}^2))$.

Put $n = 3$, $r = 2$, $G = \mathrm{O}(3)$ and $\Sigma = \Sigma_2$. Then the representation $(\tau, (\mathbb{R}^3)^{\otimes 2})$ of the group $\mathrm{O}(3) \times \Sigma_3$ is the direct sum of three irreducible components

$$\tau = [\rho_0(g)\tau_+(\sigma)] \oplus [\rho_1(g)\varepsilon(\sigma)] \oplus [\rho_2(g)\tau_+(\sigma)].$$

The one-dimensional space of the first component is the span of the identity matrix and consists of scalars. The three-dimensional space of the second component is the space $\Lambda^2(\mathbb{R}^3)$ of 3×3 skew-symmetric matrices. Its elements are three-dimensional *bivectors* or pseudo-vectors. Finally, the five-dimensional space of the third component consists of 3×3 traceless symmetric matrices (deviators). Again, the second component is $(\Lambda^2(g), \Lambda^2(\mathbb{R}^3))$, and the direct sum of the first and third components is $(\mathsf{S}^2(g), \mathsf{S}^2(\mathbb{R}^3))$.

In general, consider the orthogonal representation $(\tau, (\mathbb{R}^3)^{\otimes r})$ of the group $\mathrm{O}(3) \times \Sigma_r$ acting by

$$\tau(g, \sigma)(\boldsymbol{T}) := (\rho_1)^{\otimes r}(g)\rho_r(\sigma)(\boldsymbol{T}), \qquad \boldsymbol{T} \in (\mathbb{R}^3)^{\otimes r}.$$

This representation is reducible and may be represented as the direct sum of irreducible representations as follows:

$$\tau(g, \sigma) = \sum_{\ell=0}^{r} \sum_{q=1}^{N_r^\ell} \oplus \tilde{\rho}_\ell(g)\rho_q(\sigma),$$

where q is called the *seniority index* of the component $\tilde{\rho}_\ell(g)\rho_q(\sigma)$, see Andrews & Ghoul (1982), and where $\tilde{\rho}_\ell = \rho_\ell$ when ℓ is even and $\tilde{\rho}_\ell = \rho_\ell^*$ otherwise, which easily follows from (2.33). The number N_r^ℓ of copies of the representation $\tilde{\rho}_\ell$ is given by

$$N_r^\ell = \sum_{k=0}^{\lfloor (r-\ell)/3 \rfloor} (-1)^k \binom{r}{k} \binom{2r - 3k - \ell - 2}{r - 2}.$$

Example 19 (Irreducible unitary representations of $\mathbb{R}^n$)**.** Any irreducible unitary representation of the additive group $\mathbb{R}^n$ has the form

$$\boldsymbol{x} \mapsto \mathrm{e}^{\mathrm{i}(\boldsymbol{p}, \boldsymbol{x})}, \qquad \boldsymbol{p} \in \mathbb{R}^n. \tag{2.37}$$

The character group $\hat{\mathbb{R}^n}$ is accidentally isomorphic to $\mathbb{R}^n$. Physicists call $\mathbb{R}^n$ the *space domain* and $\hat{\mathbb{R}^n}$ the *wavenumber domain.* The character (2.37) is called the *plane wave.*

2.6 Point Groups

Let (ρ, V) be an orthogonal representation of the group $\mathrm{O}(3)$ in a real finite-dimensional space. Consider ρ as a group action. Possible orbit types of this action correspond to conjugacy classes of closed subgroups of $\mathrm{O}(3)$. Such subgroups are called the *point groups.* Therefore, a classification of the above classes is of great interest.

Let r be a non-negative integer, and let $(\mathbb{R}^3)^{\otimes r}$ be the tensor product of r copies of the space $\mathbb{R}^3$ with the convention $(\mathbb{R}^3)^{\otimes 0} = \mathbb{R}^1$. Let V be an invariant subspace of the orthogonal representation $g \mapsto g^{\otimes r}$ of the group $\mathrm{O}(3)$. For any tensor $\boldsymbol{V} \in \mathsf{V}$, the group

$$G_{\mathsf{V}} = \{\, g \in \mathrm{O}(3) \colon g^{\otimes r}\boldsymbol{V} = \boldsymbol{V} \,\}$$

is called the *stabiliser* or the *isotropy subgroup* of $\boldsymbol{V}$ by mathematicians and the *symmetry group* of $\boldsymbol{V}$ by physicists. The *orbit* of $\boldsymbol{V}$ is the set $\{\, g^{\otimes r}\boldsymbol{V} \colon g \in \mathrm{O}(3) \,\}$. Two tensors $\boldsymbol{V}_1$ and $\boldsymbol{V}_2$ lie on the same orbit if and only if their stabilisers are *conjugate*, that is, $G_{\boldsymbol{V}_2} = gG_{\boldsymbol{V}_1}g^{-1}$ for some $g \in \mathrm{O}(3)$. As the tensor $\boldsymbol{V}$ runs over its own orbit, the stabiliser $G_{\boldsymbol{V}}$ runs over the *conjugacy class*

$$[G_{\boldsymbol{V}_0}] = \{\, gG_{\boldsymbol{V}_0}g^{-1} : g \in \mathrm{O}(3)\,\}$$

of a fixed element $\boldsymbol{V}_0 \in \mathsf{V}$. The conjugacy class of a stabiliser is called the *isotropy class* or the *symmetry class*.

Theoretically, any conjugacy class of closed subgroups of the group $\mathrm{O}(3)$ can serve as a symmetry class. Therefore, we need to study the classification of conjugacy classes. Let $I \in \mathrm{O}(3)$ be the identity matrix.

Theorem 6. *Every closed subgroup of* $\mathrm{O}(3)$ *is conjugate to precisely one group of the following list:*

- *Type I, closed subgroups of* $\mathrm{SO}(3)$: $\mathrm{SO}(3)$, $\mathrm{O}(2)$, $\mathrm{SO}(2)$, $\mathcal{T}$, $\mathcal{O}$, $\mathcal{I}$, $\{\, D_n : n \geq 2\,\}$, $\{\, Z_n : n \geq 1\,\}$.
- *Type II, Cartesian products* $K \times Z_2^c$, *where* K *is a group of type I and* $Z_2^c = \{\pm I\}$.
- *Type III:* $\mathrm{O}(2)^-$, $\mathcal{O}^-$, $\{\, D_m^v : m \geq 2\,\}$, $\{\, D_{2m}^h : m \geq 2\,\}$, $\{\, Z_{2m}^- : m \geq 1\,\}$.

The groups $\mathcal{T}$ of order 12, $\mathcal{O}$ of order 24 and $\mathcal{I}$ of order 60 are the group of rotations that fix the regular tetrahedron, octahedron (or cube) and icosahedron (or dodecahedron). They are called the *tetrahedral*, *octahedral*, and *icosahedral* groups.

The *dihedral group* D_m of order $2m$ is the group of the symmetries of a regular m-gon lying in the (x, y)-plane. It consists of rotations through $2k\pi/m$ about the z-axis $(0 \leq k \leq m-1)$ and flips about symmetry axes of the m-gon. The rotations form the *cyclic group* Z_m of order m.

Let $\pi\colon \mathrm{O}(3) \to \mathrm{SO}(3)$ be the homomorphism whose kernel is Z_2^c. For each group G of type III there exist two groups $L \subset K \subset \mathrm{SO}(3)$ such that the quotient K/L consists of two cosets, $\pi(G) = K$ and $G \cap \mathrm{SO}(3) = L$. In particular, $\mathrm{O}(2)^-$ corresponds to the choice $L = \mathrm{SO}(2)$ and $K = \mathrm{O}(2)$, $\mathcal{O}^-$ corresponds to $L = \mathcal{T}$ and $K = \mathcal{O}$, D_m^v corresponds to $L = Z_m$ and $K = D_m$, D_{2m}^h corresponds to $L = D_m$, $K = D_{2m}$ and Z_{2m}^- corresponds to $L = Z_m$ and $K = Z_{2m}$. The restriction of π to G is an isomorphism between G and K.

Another classification is by *systems*:

- Triclinic: Z_1 and Z_2^c.
- Monoclinic: Z_2, Z_2^- and $Z_2 \times Z_2^c$.
- Orthotropic: D_2, D_2^v and $D_2 \times Z_2^c$.
- Cubic: $\mathcal{T}$, $\mathcal{T} \times Z_2^c$, $\mathcal{O}$, $\mathcal{O}^-$ and $\mathcal{O} \times Z_2^c$.
- Icosahedral: $\mathcal{I}$ and $\mathcal{I} \times Z_2^c$.
- Transverse isotropic: $\mathrm{SO}(2)$, $\mathrm{SO}(2) \times Z_2^c$, $\mathrm{O}(2)$, $\mathrm{O}(2)^-$ and $\mathrm{O}(2) \times Z_2^c$.
- Isotropic: $\mathrm{SO}(3)$ and $\mathrm{O}(3)$.

Table 2.1 *Normalisers of point groups*

G	$N_{\mathrm{O}(3)}(G)$
Z_1, Z_2^c, SO(3), O(3)	O(3)
D_2, $D_2 \times Z_2^c$, $\mathcal{T}$, $\mathcal{T} \times Z_2^c$, $\mathcal{O}$, $\mathcal{O} \times Z_2^c$	$\mathcal{O} \times Z_2^c$
$\{ Z_n : n \geq 2 \}$, $\{ Z_n \times Z_2^c : n \geq 2 \}$, $\{ Z_{2m}^- : m \geq 1 \}$,	
SO(2), SO(2) $\times Z_2^c$, O(2), O(2)$^-$, O(2) $\times Z_2^c$	O(2) $\times Z_2^c$
D_2^v, D_4^h	$D_4 \times Z_2^c$
$\{ D_n : n \geq 3 \}$, $\{ D_n \times Z_2^c : n \geq 3 \}$, $\{ D_n^v : n \geq 3 \}$	
$\{ D_{2n}^h : n \geq 3 \}$	$D_{2n} \times Z_2^c$
$\mathcal{I}$, $\mathcal{I} \times Z_2^c$	$\mathcal{I} \times Z_2^c$

- n-gonal: $\{ D_n : n \geq 3 \}$, $\{ Z_n : n \geq 3 \}$, $\{ D_n \times Z_2^c : n \geq 3 \}$, $\{ Z_n \times Z_2^c : n \geq 3 \}$, $\{ D_n^v : n \geq 3 \}$.
- $2n$-gonal: $\{ D_{2n}^h : n \geq 2 \}$, $\{ Z_{2n}^- : n \geq 2 \}$.

The triclinic, monoclinic, orthorhombic, cubic, trigonal ($n = 3$), tetragonal ($n = 4$) and hexagonal ($n = 6$) systems are called *crystal systems.* All classes of the above systems are *crystal classes.* For each crystal class, there exist a *lattice* $\{ k\boldsymbol{a} + l\boldsymbol{b} + m\boldsymbol{c} : k, l, m \in \mathbb{Z} \}$ generated by three linearly independent vectors $\boldsymbol{a}$, $\boldsymbol{b}$ and $\boldsymbol{c}$ such that any group of the class maps this lattice into itself.

According to Section 3.1, we consider a homogeneous V-valued random field that is (G, ρ)-isotropic, where the group G lies between the stabiliser $G_{\boldsymbol{\mathsf{V}}}$ of some tensor $\boldsymbol{\mathsf{V}} \in \mathsf{V}$ and the normaliser $N(G_{\boldsymbol{\mathsf{V}}})$. The normalisers of all point groups are given in Table 2.1 adapted from Brock (2014).

2.7 Invariant Theory

Let V and W be two finite-dimensional linear spaces over the same field $\mathbb{F}$. Let (V, ρ) and (W, σ) be two representations of a group G. A mapping $h \colon W \to V$ is called a *covariant* or *form-invariant* or a *covariant tensor* of the pair of representations (V, ρ) and (W, σ), if

$$h(\sigma(g)\boldsymbol{w}) = \rho(g)h(\boldsymbol{w}), \qquad g \in G.$$

In other words, the diagram

$$\begin{array}{ccc} W & \xrightarrow{\;h\;} & V \\ \Big\downarrow{\scriptstyle\sigma} & & \Big\downarrow{\scriptstyle\rho} \\ W & \xrightarrow{\;h\;} & V \end{array}$$

is commutative.

If $V = \mathbb{F}^1$ and ρ is the trivial representation of G, then the corresponding covariant scalars are called *absolute invariants* (or just invariants) of the representation (W, σ), hence the name *Invariant Theory.* Note that the set

$\mathbb{F}[W]^G$ of invariants is an *algebra* over the field $\mathbb{F}$, that is, a linear space over $\mathbb{F}$ with bilinear multiplication operation and the multiplication identity 1. The product of a covariant $h\colon W \to V$ and an invariant $f \in \mathbb{F}[W]^G$ is again a covariant. In other words, the covariant tensors of the pair of representations (V, ρ) and (W, σ) form a *module* over the algebra of invariants of the representation (W, σ).

A mapping $h\colon W \to V$ is called *homogeneous polynomial mapping of degree d* if for any $\boldsymbol{v} \in V$ the mapping $\boldsymbol{w} \mapsto (h(\boldsymbol{w}), \boldsymbol{v})$ lies in $\mathrm{H}^d(W)$. The mapping h is called a *polynomial covariant of degree d* if it is a homogeneous polynomial mapping of degree d and a covariant.

Let G be a closed subgroup of the group $\mathrm{GL}(W, \mathbb{F})$. Assume that its determinant representation ρ is not trivial. The corresponding covariant tensors are called *relative covariant tensors*.

Let (W, σ) be the defining representation of G, and (V, ρ) be the rth tensor power of the defining representation. The corresponding covariant tensors are said to have *an order r*. The covariant tensors of degree 0 and of order r of the group $\mathrm{O}(n)$ are known as *isotropic tensors*.

The algebra of invariants and the module of covariant tensors were an object of intensive research. The first general result was obtained by Gordan (1868). He proved that for any finite-dimensional complex representation of the group $G = \mathrm{SL}(2, \mathbb{C})$ the algebra of invariants and the module of covariant tensors are finitely generated. In other words, there exists an *integrity basis*: a finite set of invariant homogeneous polynomials $I_1, \dots, I_N$ such that every polynomial invariant can be written as a polynomial in $I_1, \dots, I_N$. An integrity basis is called *minimal* if none of its elements can be expressed as a polynomial in the others. A minimal integrity basis is not necessarily unique, but all minimal integrity bases have the same amount of elements of each degree.

The algebra of invariants is not necessarily free. Some polynomial relations between generators, called *syzygies*, may exist.

Hilbert (1890) proved that if a finite-dimensional representation of a group $\mathrm{SL}(n, \mathbb{C})$ or $\mathrm{GL}(2, \mathbb{C})$ is completely reducible then the algebra of invariants is finitely generated. Hilbert's proof was not constructive and was criticised by Gordan. In response to Gordan's criticism, Hilbert (1893) described a way to construct generators for the algebra of invariants. In fact, these two papers formed the foundation of modern commutative algebra.

Hilbert's results remain true for a wide class of groups and their representations, including real finite-dimensional representations of compact groups. Moreover, in this case the elements of an integrity basis *separate the orbits of the corresponding group action*. This means the following: if $I_n(\boldsymbol{x}_1) = I_n(\boldsymbol{x}_2)$ for all n then $\boldsymbol{x}_1$ and $\boldsymbol{x}_2$ lie on the same orbit. This property is so important that it rises a definition. A finite set of invariant (not necessarily polynomial) functions is called a *functional basis* if this set separates the orbits. A functional basis is *minimal* if no proper subset of it is a functional basis.

Any integrity basis is a functional basis. The converse is wrong. For example, let $G = \mathrm{O}(3)$, let $W = \mathbb{R}^3$ and let $\rho(g) = g$. The one-element set $\{\|\boldsymbol{x}\|^2\}$ is an integrity basis. The one-element set $\{\|\boldsymbol{x}\|^4\}$ is a functional basis but not an integrity basis.

The importance of polynomial invariants can be explained by the following result. Let G be a closed subgroup of the group $\mathrm{O}(3)$, the group of symmetries of a material. Let (V, ρ), (V_1, ρ_1), ..., (V_N, ρ_N) be finitely many orthogonal representations of G in real finite-dimensional spaces. Let $\mathsf{T}\colon \mathsf{V}_1 \oplus \cdots \oplus \mathsf{V}_N \to \mathsf{V}$ be an *arbitrary* (say, measurable) covariant of the pair ρ and $\rho_1 \oplus \cdots \oplus \rho_N$. Let $\{\, I_k\colon 1 \leq k \leq K \,\}$ be an integrity basis for *polynomial* invariants of the representation $\rho_1 \oplus \cdots \oplus \rho_N$, and let $\{\, \boldsymbol{\mathsf{T}}_l\colon 1 \leq l \leq L \,\}$ be an integrity basis for *polynomial* covariant tensors of the pair ρ and $\rho_1 \oplus \cdots \oplus \rho_N$. Following Wineman & Pipkin (1964), we call $\boldsymbol{\mathsf{T}}_l$ *basic covariant tensors.*

Theorem 7 (Wineman & Pipkin (1964))**.** *A function* $\mathsf{T}\colon \mathsf{V}_1 \oplus \cdots \oplus \mathsf{V}_N \to \mathsf{V}$ *is a measurable covariant of the pair* ρ *and* $\rho_1 \oplus \cdots \oplus \rho_N$ *if and only if it has the form*

$$\mathsf{T}(\boldsymbol{T}_1, \ldots, \boldsymbol{T}_N) = \sum_{l=1}^{L} \varphi_l(I_1, \ldots, I_K)\mathsf{T}_l(\boldsymbol{T}_1, \ldots, \boldsymbol{T}_N),$$

where φ_l *are real-valued measurable functions of the elements of an integrity basis.*

In 1939 in the first edition of Weyl (1997), Hermann Weyl proved that any polynomial covariant of degree d and of order r of the group $\mathrm{O}(n)$ is a linear combination of products of Kronecker's deltas δ_{ij} and second degree homogeneous polynomials $x_i x_j$.

Example 20 (Ogden tensors)**.** Let ν be a non-negative integer. The Ogden tensor $\boldsymbol{\mathsf{I}}^\nu$ of rank $2\nu + 2$ is determined inductively as

$$\boldsymbol{\mathsf{I}}^0_{ij} := \delta_{ij}, \qquad \boldsymbol{\mathsf{I}}^1_{ijk\ell} := \frac{1}{2}(\delta_{ik}\delta_{j\ell} + \delta_{i\ell}\delta_{jk}),$$
$$\boldsymbol{\mathsf{I}}^\nu_{i_1 \cdots i_{2\nu+2}} := \nu^{-1}(\boldsymbol{\mathsf{I}}^1_{i_1 p i_3 i_4}\boldsymbol{\mathsf{I}}^{\nu-1}_{p i_2 i_5 \cdots i_{2\nu+2}} + \cdots + \boldsymbol{\mathsf{I}}^1_{i_1 p i_{2\nu+1} i_{2\nu+2}}\boldsymbol{\mathsf{I}}^{\nu-1}_{p i_2 \cdots i_{2\nu-1} i_{2\nu}}),$$

where there is a summation over p. In what follows we will omit the upper index.

The above tensors are isotropic tensors of order $2\nu+2$. Moreover, they are basic covariant tensors of degree 0 and of order $2\nu + 2$ of the pair of representations $(\mathsf{S}^{2\nu+2}(\mathbb{R}^n), \mathsf{S}^{2\nu+2}(g))$ and $(\mathbb{R}^n, g)$.

Example 21 (Symmetric isotropic tensors)**.** Let $d = 3$, let $(W, \sigma) = (\mathbb{R}^3, g)$ be the defining representation of the group $\mathrm{O}(3)$, and let $(V, \rho) = ((\mathbb{R}^3)^{\otimes 2r}, g^{\otimes 2r})$ be the $2r$th tensor power of the defining representation. The number of different products of r Kronecker's deltas is

$$N(2r) = \frac{(2r)!}{r!2^r},$$

which is equal to the number of different ways to distribute the $2r$ indices among r Kronecker's deltas. In particular, $N(0) = 1$, the only rank 0 isotropic tensor is 1, and the representation $(\mathbb{R}^1, g^{\otimes 0})$ contains one copy of the trivial representation. Next, $N(2) = 1$, the only rank 2 isotropic tensor is

$$L^1_{ij} = \delta_{ij}, \tag{2.38}$$

and the representation $((\mathbb{R}^3)^{\otimes 2}, g^{\otimes 2})$ contains 1 copy of the trivial representation.

When $r = 2$, we have $N(4) = 3$, the representation $((\mathbb{R}^3)^{\otimes 4}, g^{\otimes 4})$ contains three copies of the trivial representation, and the three rank 4 isotropic tensors are $\delta_{ij}\delta_{kl}$, $\delta_{ik}\delta_{jl}$ and $\delta_{il}\delta_{jk}$. Now, put $V = \mathsf{S}^2(\mathsf{S}^2(\mathbb{R}^3))$ and $\rho = \mathsf{S}^2(\mathsf{S}^2(g))$. We have $\mathsf{S}^2(\mathsf{S}^2(\mathbb{R}^3)) = P^+_\Sigma((\mathbb{R}^3)^{\otimes 4})$, where P^+_Σ is the linear operator (2.11), and where Σ is the group introduced in Example 10. By (2.11), the isotropic tensors for this case are the sums of the above three tensors over the orbits of the action (2.8) of the group Σ on the set of isotropic tensors. It is easy to check that the above action has two orbits, and the isotropic tensors are

$$\mathbf{L}^1_{ijkl} = \delta_{ij}\delta_{kl}, \qquad \mathbf{L}^2_{ijkl} = 2\mathbf{I}_{ijkl}. \tag{2.39}$$

When $r = 3$, we have $N(6) = 15$ and the representation $((\mathbb{R}^3)^{\otimes 3}, g^{\otimes 3})$ contains 15 copies of the trivial representation. To find symmetric isotropic tensors, consider a group Σ_p of order 8, generated by transpositions (ij), $(i'j')$ and $(ii')(jj')(kk')$ of the above indices. The group Σ_p acts on the set of isotropic tensors. The sums of isotropic tensors over each orbit are the symmetric isotropic tensors. The results of calculations are as follows:

$$\begin{aligned}
\mathbf{L}^1 &= \delta_{ij}\delta_{kk'}\delta_{i'j'}, \\
\mathbf{L}^2 &= 2\delta_{kk'}\mathbf{I}_{iji'j'}, \\
\mathbf{L}^3 &= 2(\delta_{ij}\mathbf{I}_{kj'i'k'} + \delta_{i'j'}\mathbf{I}_{ijkk'}), \\
\mathbf{L}^4 &= 2(\delta_{ik}\mathbf{I}_{jj'i'k'} + \delta_{jk}\mathbf{I}_{iji'k'}), \\
\mathbf{L}^5 &= 2(\delta_{kj'}\mathbf{I}_{iji'k'} + \delta_{ki'}\mathbf{I}_{ijj'k'}).
\end{aligned} \tag{2.40}$$

The situation becomes quite different when $r = 4$. We have $N(8) = 105$, but the representation $((\mathbb{R}^3)^{\otimes 8}, g^{\otimes 8})$ contains only 91 copies of the trivial representation. This means that $105 - 91 = 14$ rank 8 isotropic tensors must be linear combinations of 91 linearly independent tensors. In other words, 14 syzygies should exist. Indeed, Kearsley & Fong (1975) listed 14 linear combinations (*reduction equations*) of products which are identically zero.

Consider the subgroup $\Sigma' \subset \Sigma_8$ of order 16 generated by the transpositions (ij), (kl), $(i'j')$ and $(k'l')$. Monchiet & Bonnet (2011) call the elements of this group *minor symmetries*. They reported 17 orbits of the group Σ' acting on the set of 105 isotropic tensors by (2.8), and found one syzygy.

We are interested in another subgroup $\Sigma_e \subset \Sigma_8$ of order 128 generated by the transpositions (ij), (kl), $(i'j')$, $(k'l')$ and products $(ik)(il)$, $(i'k')(j'l')$ and

Table 2.2 Σ_e*-symmetric rank* 8 *isotropic tensors*

Tensor	**Value**
$\boldsymbol{\mathsf{L}}^{\mathbf{1}}_{\boldsymbol{i}\cdots\boldsymbol{l}'}$	$\delta_{ij}\delta_{kl}\delta_{i'j'}\delta_{k'l'}$
$\boldsymbol{\mathsf{L}}^{\mathbf{2}}_{\boldsymbol{i}\cdots\boldsymbol{l}'}$	$2(\delta_{ij}\delta_{kl}\mathsf{I}_{i'j'k'l'}+\delta_{i'j'}\delta_{k'l'}\mathsf{I}_{ijkl})$
$\mathsf{L}^{3}_{i\cdots l'}$	$2(\delta_{ij}(\delta_{i'j'}\mathsf{I}_{klk'l'}+\delta_{k'l'}\mathsf{I}_{kli'j'})+\delta_{kl}(\delta_{i'j'}\mathsf{I}_{ijk'l'}+\delta_{k'l'}\mathsf{I}_{iji'j'}))$
$\boldsymbol{\mathsf{L}}^{\mathbf{4}}_{\boldsymbol{i}\cdots\boldsymbol{l}'}$	$4\mathsf{I}_{ijkl}\mathsf{I}_{i'j'k'l'}$
$\mathsf{L}^{5}_{i\cdots l'}$	$8(\delta_{ij}\mathsf{I}_{kli'j'k'l'}+\delta_{kl}\mathsf{I}_{iji'j'k'l'}+\delta_{i'j'}\mathsf{I}_{ijklk'l'}+\delta_{k'l'}\mathsf{I}_{ijkli'j'})$
$\mathsf{L}^{6}_{i\cdots l'}$	$4(\mathsf{I}_{iji'j'}\mathsf{I}_{klk'l'}+\mathsf{I}_{ijk'l'}\mathsf{I}_{kli'j'})$
$\mathsf{L}^{7}_{i\cdots l'}$	$4(\mathsf{I}_{iji'k'}\mathsf{I}_{klj'l'}+\mathsf{I}_{iji'l'}\mathsf{I}_{klj'k'}+\mathsf{I}_{ijj'k'}\mathsf{I}_{kli'l'}+\mathsf{I}_{ijj'l'}\mathsf{I}_{kli'k'})$
$\mathsf{L}^{30}_{i\cdots l'}$	$4(\mathsf{I}_{ijkk'}\mathsf{I}_{i'j'll'}+\mathsf{I}_{ijll'}\mathsf{I}_{i'j'lk'}+\mathsf{I}_{ijlk'}\mathsf{I}_{i'j'kl'}+\mathsf{I}_{ijll'}\mathsf{I}_{i'j'kk'}$ $+\mathsf{I}_{ijki'}\mathsf{I}_{k'l'lj'}+\mathsf{I}_{ijkj'}\mathsf{I}_{k'l'li'}+\mathsf{I}_{ijli'}\mathsf{I}_{k'l'kj'}+\mathsf{I}_{ijlj'}\mathsf{I}_{k'l'ki'})$

$(ii')(jj')(kk')(ll')$. Eight orbits have been found. The sums of elements over each orbit are Σ_e-symmetric rank 8 isotropic tensors shown in Table 2.2. We use the short notation $i\cdots l' := ijkli'j'k'l'$. Σ_e-symmetric rank 8 isotropic tensors found by Lomakin (1965) are shown in bold. We also found one reduction equation:

$$\mathsf{L}^{30}_{i\cdots l'} = 8\mathsf{L}^{1}_{i\cdots l'} - 4\mathsf{L}^{2}_{i\cdots l'} - 4\mathsf{L}^{3}_{i\cdots l'} + 2\mathsf{L}^{4}_{i\cdots l'} + 2\mathsf{L}^{5}_{i\cdots l'} + 2\mathsf{L}^{6}_{i\cdots l'} - \mathsf{L}^{7}_{i\cdots l'}.$$

The enumeration of isotropic tensors and covariant tensors is so chosen in order to write down Equation (3.98) in a compact form.

Example 22 (Symmetric covariant tensors)**.** Consider covariant tensors of degree 2 and of order 2. For the pair of representations $((\mathbb{R}^3)^{\otimes 2}, g^{\otimes 2})$ and $(\mathbb{R}^3, g)$, the only possible one is

$$\|\boldsymbol{x}\|^2 L^2_{ij}(\boldsymbol{x}) = x_i x_j, \tag{2.41}$$

and the representation $((\mathbb{R}^3)^{\otimes 2}, g^{\otimes 2})$ indeed contains one copy of the irreducible representation ρ^2 of the group $\mathrm{O}(3)$. The same remains true for the pair of representations $(\mathsf{S}^2(\mathbb{R}^3), \mathsf{S}^2(g))$ and $(\mathbb{R}^3, g)$.

In the case of degree 2 and of order 4, for the pair of representations $((\mathbb{R}^3)^{\otimes 4}, g^{\otimes 4})$ and $(\mathbb{R}^3, g)$ we have six covariant tensors:

$$\delta_{il}x_jx_k, \delta_{jk}x_ix_l, \delta_{jl}x_ix_k, \delta_{ik}x_jx_l, \delta_{kl}x_ix_j, \delta_{ij}x_kx_l.$$

The representation $((\mathbb{R}^3)^{\otimes 4}, g^{\otimes 4})$ of the group $\mathrm{O}(3)$ indeed contains six copies of the representation ρ^2.

Now, put $(V, \rho) = (\mathsf{S}^2(\mathsf{S}^2(\mathbb{R}^3)), \mathsf{S}^2(\mathsf{S}^2(g)))$. We have $\mathsf{S}^2(\mathsf{S}^2(\mathbb{R}^3)) = P^+_\Sigma((\mathbb{R}^3)^{\otimes 4})$, where P^+_Σ is the linear operator (2.11), and where Σ is the group introduced in Example 10. By (2.11), the symmetric covariant tensors for this case are the sums of the above six covariant tensors over the orbits of the action

$$\sigma(x_{i_1}x_{i_2}\delta_{i_3i_4}) := x_{\sigma^{-1}(i_1)}x_{\sigma^{-1}(i_2)}\delta_{\sigma^{-1}(i_3)\sigma^{-1}(i_4)}, \qquad \sigma \in \Sigma$$

Table 2.3 Σ_p-symmetric covariant tensors of degree 2 and of order 6.

Tensor	Value
$\mathsf{L}^6(\boldsymbol{x})$	$\delta_{ij}\delta_{i'j'}x_kx_{k'}$
$\mathsf{L}^7(\boldsymbol{x})$	$\delta_{i'j'}\delta_{kk'}x_ix_j + \delta_{ij}\delta_{kk'}x_{i'}x_{j'}$
$\mathsf{L}^8(\boldsymbol{x})$	$2\mathsf{I}_{iji'j'}x_kx_{k'}$
$\mathsf{L}^9(\boldsymbol{x})$	$\delta_{i'j'}(\delta_{jk}x_ix_{k'} + \delta_{ik}x_jx_{k'}) + \delta_{ij}(\delta_{j'k'}x_{i'}x_k + \delta_{i'k'}x_{j'}x_k)$
$\mathsf{L}^{10}(\boldsymbol{x})$	$\delta_{jk}(\delta_{j'k'}x_ix_{i'} + \delta_{i'k'}x_ix_{j'}) + \delta_{ik}(\delta_{j'k'}x_jx_{i'} + \delta_{i'k'}x_jx_{j'})$
$\mathsf{L}^{11}(\boldsymbol{x})$	$\delta_{i'j'}(\delta_{jk'}x_ix_k + \delta_{ik'}x_jx_k) + \delta_{ij}(\delta_{kj'}x_{i'}x_{k'} + \delta_{ki'}x_{j'}x_{k'})$
$\mathsf{L}^{12}(\boldsymbol{x})$	$\delta_{jk'}(\delta_{kj'}x_ix_{i'} + \delta_{ki'}x_ix_{j'}) + \delta_{ik'}(\delta_{kj'}x_jx_{i'} + \delta_{ki'}x_jx_{j'})$
$\mathsf{L}^{13}(\boldsymbol{x})$	$\delta_{kk'}(\delta_{jj'}x_ix_{i'} + \delta_{ji'}x_ix_{j'} + \delta_{ij'}x_jx_{i'} + \delta_{ii'}x_jx_{j'})$
$\mathsf{L}^{14}(\boldsymbol{x})$	$2(\mathsf{I}_{i'j'k'k}x_ix_j + \mathsf{I}_{ijkk'}x_{i'}x_{j'})$
$\mathsf{L}^{15}(\boldsymbol{x})$	$2(\mathsf{I}_{jki'j'}x_ix_{k'} + \mathsf{I}_{iki'j'}x_jx_{k'} + \mathsf{I}_{ijj'k'}x_kx_{i'} + \mathsf{I}_{iji'k'}x_kx_{j'})$
$\mathsf{L}^{22}(\boldsymbol{x})$	$(\delta_{ji'}\delta_{j'k'} + \delta_{jj'}\delta_{i'k'})x_ix_k + (\delta_{ii'}\delta_{j'k'} + \delta_{ij'}\delta_{i'k'})x_jx_k$ $+ 2(\mathsf{I}_{ijkj'}x_{i'}x_{k'} + \mathsf{I}_{ijki'}x_{j'}x_{k'})$

of the group Σ on the set of covariant tensors. It is easy to check that the above action has two orbits, and the symmetric covariant tensors are

$$\begin{aligned}\|\boldsymbol{x}\|^2\mathsf{L}^3_{ijkl}(\boldsymbol{x}) &= \delta_{il}x_jx_k + \delta_{jk}x_ix_l + \delta_{jl}x_ix_k + \delta_{ik}x_jx_l,\\ \|\boldsymbol{x}\|^2\mathsf{L}^4_{ijkl}(\boldsymbol{x}) &= \delta_{kl}x_ix_j + \delta_{ij}x_kx_l.\end{aligned} \tag{2.42}$$

In the case of degree 2 and of order 6, for the pair of representations $((\mathbb{R}^3)^{\otimes 6}, g^{\otimes 6})$ and $(\mathbb{R}^3, g)$ we have 45 covariant tensors. Under the action of the group Σ_p, we obtain the symmetric covariant tensors shown in Table 2.3, where we omitted $\|\boldsymbol{x}\|^2$ in the first column.

We found a syzygy: the symmetric covariant tensor $\mathsf{L}^{22}(\boldsymbol{x})$ is expressed as a linear combination of the remaining tensors as follows:

$$\begin{aligned}\mathsf{L}^{22}(\boldsymbol{x}) = {} & 4\mathsf{L}^1 - 2\mathsf{L}^2 - 2\mathsf{L}^3 + \mathsf{L}^4 + \mathsf{L}^5 - 4\mathsf{L}^6(\boldsymbol{x}) - 4\mathsf{L}^7(\boldsymbol{x}) + 2\mathsf{L}^8(\boldsymbol{x})\\ & + 2\mathsf{L}^9(\boldsymbol{x}) - \mathsf{L}^{10}(\boldsymbol{x}) + 2\mathsf{L}^{11}(\boldsymbol{x}) - \mathsf{L}^{12}(\boldsymbol{x}) + 2\mathsf{L}^{13}(\boldsymbol{x}) + 2\mathsf{L}^{14}(\boldsymbol{x})\\ & - \mathsf{L}^{15}(\boldsymbol{x}).\end{aligned}$$

In the case of degree 2 and of order 8, for the pair of representations $((\mathbb{R}^3)^{\otimes 8}, g^{\otimes 8})$ and $(\mathbb{R}^3, g)$ we have $105 \times 4 = 420$ covariant tensors (in each of the 105 isotropic tensors one can change any term $\delta_{i_1 i_2}$ with $x_{i_1}x_{i_2}$). We have found 13 orbits of the group Σ_e of order 128 of Example 21. The sums of elements on each orbit are shown in Table 2.4. We found three reduction equations:

$$\begin{aligned}\mathsf{L}^{31}_{i\dots l'}(\boldsymbol{x}) = {} & 16\mathsf{L}^1_{i\dots l'} - 4\mathsf{L}^2_{i\dots l'} - 4\mathsf{L}^3_{i\dots l'} + \mathsf{L}^5_{i\dots l'} - 12\mathsf{L}^8_{i\dots l'}(\boldsymbol{x}) + 2\mathsf{L}^9_{i\dots l'}(\boldsymbol{x})\\ & + 4\mathsf{L}^{10}_{i\dots l'}(\boldsymbol{x}) + 4\mathsf{L}^{11}_{i\dots l'}(\boldsymbol{x}) + 2\mathsf{L}^{13}_{i\dots l'}(\boldsymbol{x}) - \mathsf{L}^{14}_{i\dots l'}(\boldsymbol{x}),\\ \mathsf{L}^{32}_{i\dots l'}(\boldsymbol{x}) = {} & 8\mathsf{L}^1_{i\dots l'} - 4\mathsf{L}^2_{i\dots l'} + 2\mathsf{L}^4_{i\dots l'} - 2\mathsf{L}^6_{i\dots l'} + \mathsf{L}^7_{i\dots l'} - 4\mathsf{L}^8_{i\dots l'}(\boldsymbol{x})\\ & + 2\mathsf{L}^9_{i\dots l'}(\boldsymbol{x}) + 2\mathsf{L}^{10}_{i\dots l'}(\boldsymbol{x}) - \mathsf{L}^{12}_{i\dots l'}(\boldsymbol{x}) - 2\mathsf{L}^{13}_{i\dots l'}(\boldsymbol{x}) + \mathsf{L}^{14}_{i\dots l'}(\boldsymbol{x})\\ & + \mathsf{L}^{15}_{i\dots l'}(\boldsymbol{x}) - \frac{1}{2}\mathsf{L}^{16}_{i\dots l'}(\boldsymbol{x}) + 2\mathsf{L}^{17}_{i\dots l'}(\boldsymbol{x}),\end{aligned}$$

Table 2.4 Σ_e-*symmetric basic covariant tensors of degree* 2 *and of order* 8

Covariant	**Value**
$\boldsymbol{L}^{8}_{i\cdots l'}(\boldsymbol{x})$	$\delta_{ij}\delta_{kl}(\delta_{i'j'}x_{k'}x_{l'}+\delta_{k'l'}x_{i'}x_{j'})+\delta_{i'j'}\delta_{k'l'}(\delta_{ij}x_kx_l+\delta_{kl}x_ix_j)$
$\boldsymbol{L}^{9}_{i\cdots l'}(\boldsymbol{x})$	$2(\boldsymbol{I}_{ijkl}(\delta_{i'j'}x_{k'}x_{l'}+\delta_{k'l'}x_{i'}x_{j'})+\boldsymbol{I}_{i'j'k'l'}(\delta_{ij}x_kx_l+\delta_{kl}x_ix_j))$
$\boldsymbol{L}^{10}_{i\cdots l'}(\boldsymbol{x})$	$\delta_{ij}\delta_{kl}(\delta_{i'k'}x_{j'}x_{l'}+\delta_{i'l'}x_{j'}x_{k'}+\delta_{j'k'}x_{i'}x_{l'}+\delta_{j'l'}x_{i'}x_{k'})$
	$+\delta_{i'j'}\delta_{k'l'}(\delta_{ik}x_jx_l+\delta_{il}x_jx_k+\delta_{jk}x_ix_l+\delta_{jl}x_ix_k)$
$\boldsymbol{L}^{11}_{i\cdots l'}(\boldsymbol{x})$	$\delta_{ij}\delta_{i'j'}(\delta_{kk'}x_lx_{l'}+\delta_{kl'}x_lx_{k'}+\delta_{lk'}x_kx_{l'}+\delta_{ll'}x_kx_{k'})$
	$+\delta_{ij}\delta_{k'l'}(\delta_{ki'}x_lx_{j'}+\delta_{kj'}x_lx_{j'}+\delta_{li'}x_kx_{j'}+\delta_{lj'}x_kx_{i'})$
	$+\delta_{kl}\delta_{i'j'}(\delta_{ik'}x_jx_{l'}+\delta_{jk'}x_jx_{l'}+\delta_{il'}x_jx_{k'}+\delta_{jl'}x_ix_{k'})$
	$+\delta_{kl}\delta_{k'l'}(\delta_{ii'}x_jx_{j'}+\delta_{ij'}x_jx_{i'}+\delta_{ji'}x_ix_{j'}+\delta_{jj'}x_ix_{i'})$
$\boldsymbol{L}^{12}_{i\cdots l'}(\boldsymbol{x})$	$2(\boldsymbol{I}_{ijkl}(\delta_{i'k'}x_{j'}x_{l'}+\delta_{i'l'}x_{j'}x_{k'}+\delta_{j'k'}x_{i'}x_{l'}+\delta_{j'l'}x_{i'}x_{k'})$
	$+\boldsymbol{I}_{i'j'k'l'}(\delta_{ik}x_jx_l+\delta_{il}x_jx_k+\delta_{jk}x_ix_l+\delta_{jl}x_ix_k))$
$\boldsymbol{L}^{13}_{i\cdots l'}(\boldsymbol{x})$	$2((\delta_{ij}\boldsymbol{I}_{kli'j'}+\delta_{kl}\boldsymbol{I}_{iji'j'})x_{k'}x_{l'}+(\delta_{ij}\boldsymbol{I}_{klk'l'}+\delta_{kl}\boldsymbol{I}_{ijk'l'})x_{i'}x_{j'}$
	$+(\delta_{i'j'}\boldsymbol{I}_{ijk'l'}+\delta_{k'l'}\boldsymbol{I}_{iji'j'})x_kx_l+(\delta_{i'j'}\boldsymbol{I}_{klk'l'}+\delta_{k'l'}\boldsymbol{I}_{kli'j'})x_ix_j)$
$\boldsymbol{L}^{14}_{i\cdots l'}(\boldsymbol{x})$	$2((\delta_{ij}\boldsymbol{I}_{kli'k'}+\delta_{kl}\boldsymbol{I}_{iji'k'})x_{j'}x_{l'}+(\delta_{ij}\boldsymbol{I}_{kli'l'}+\delta_{kl}\boldsymbol{I}_{iji'l'})x_{j'}x_{k'}$
	$+(\delta_{ij}\boldsymbol{I}_{klj'k'}+\delta_{kl}\boldsymbol{I}_{ijj'k'})x_{i'}x_{l'}+(\delta_{ij}\boldsymbol{I}_{klj'l'}+\delta_{kl}\boldsymbol{I}_{ijj'l'})x_{i'}x_{k'}$
	$+(\delta_{i'j'}\boldsymbol{I}_{ikk'l'}+\delta_{k'l'}\boldsymbol{I}_{iki'j'})x_jx_l+(\delta_{i'j'}\boldsymbol{I}_{ilk'l'}+\delta_{k'l'}\boldsymbol{I}_{ili'j'})x_jx_k$
	$+(\delta_{i'j'}\boldsymbol{I}_{jkk'l'}+\delta_{k'l'}\boldsymbol{I}_{jki'j'})x_ix_l+(\delta_{i'j'}\boldsymbol{I}_{jlk'l'}+\delta_{k'l'}\boldsymbol{I}_{jli'j'})x_ix_k)$
$\boldsymbol{L}^{15}_{i\cdots l'}(\boldsymbol{x})$	$8(\boldsymbol{I}_{ijkli'j'}x_{k'}x_{l'}+\boldsymbol{I}_{ijklk'l'}x_{i'}x_{j'}+\boldsymbol{I}_{iji'j'k'l'}x_kx_l+\boldsymbol{I}_{kli'j'k'l'}x_ix_j)$
$\boldsymbol{L}^{16}_{i\cdots l'}(\boldsymbol{x})$	$8(\boldsymbol{I}_{ijkli'k'}x_{j'}x_{l'}+\boldsymbol{I}_{ijkli'l'}x_{j'}x_{k'}+\boldsymbol{I}_{ijklj'k'}x_{i'}x_{l'}+\boldsymbol{I}_{ijklj'l'}x_{i'}x_{k'}$
	$+\boldsymbol{I}_{iki'j'k'l'}x_jx_l+\boldsymbol{I}_{ili'j'k'l'}x_jx_k+\boldsymbol{I}_{jki'j'k'l'}x_ix_l+\boldsymbol{I}_{jli'j'k'l'}x_ix_k)$
$\boldsymbol{L}^{17}_{i\cdots l'}(\boldsymbol{x})$	$2(\boldsymbol{I}_{iji'j'}(\delta_{kk'}x_lx_{l'}+\delta_{kl'}x_lx_{k'}+\delta_{lk'}x_kx_{l'}+\delta_{ll'}x_kx_{k'})$
	$+\boldsymbol{I}_{ijk'l'}(\delta_{ki'}x_lx_{j'}+\delta_{kj'}x_lx_{i'}+\delta_{li'}x_kx_{j'}+\delta_{lj'}x_kx_{i'}$
	$+\boldsymbol{I}_{kli'j'}(\delta_{ik'}x_jx_{l'}+\delta_{il'}x_jx_{k'}+\delta_{jk'}x_ix_{l'}+\delta_{jl'}x_ix_{k'})$
	$+\boldsymbol{I}_{klk'l'}(\delta_{ii'}x_jx_{j'}+\delta_{ij'}x_jx_{i'}+\delta_{ji'}x_ix_{j'}+\delta_{jj'}x_ix_{i'})$
$\boldsymbol{L}^{31}_{i\cdots l'}(\boldsymbol{x})$	$\delta_{ij}(2\boldsymbol{I}_{ki'k'l'}x_lx_{j'}+2\boldsymbol{I}_{kj'k'l'}x_lx_{i'}+2\boldsymbol{I}_{li'k'l'}x_kx_{j'}+2\boldsymbol{I}_{li'k'l'}x_kx_{i'}$
	$+(\delta_{ki'}\delta_{j'k'}+\delta_{kj'}\delta_{i'k'})x_lx_{l'}+(\delta_{ki'}\delta_{j'l'}+\delta_{kj'}\delta_{i'l'})x_lx_{k'}$
	$+(\delta_{li'}\delta_{j'k'}+\delta_{lj'}\delta_{i'k'})x_kx_{l'}+(\delta_{li'}\delta_{j'l'}+\delta_{lj'}\delta_{i'l'})x_kx_{k'})$
	$+\delta_{kl}(2\boldsymbol{I}_{ii'k'l'}x_jx_{j'}+2\boldsymbol{I}_{ij'k'l'}x_jx_{i'}+2\boldsymbol{I}_{ji'k'l'}x_ix_{j'}+2\boldsymbol{I}_{jj'k'l'}x_ix_{i'}$
	$+(\delta_{ii'}\delta_{j'l'}+\delta_{ij'}\delta_{i'k'})x_jx_{l'}+(\delta_{ii'}\delta_{j'l'}+\delta_{ij'}\delta_{i'l'})x_jx_{k'}$
	$+(\delta_{ji'}\delta_{j'k'}+\delta_{jj'}\delta_{i'k'})x_ix_{l'}+(\delta_{ji'}\delta_{j'l'}+\delta_{jj'}\delta_{i'l'})x_ix_{k'})$
	$+\delta_{i'j'}(2\boldsymbol{I}_{ijkk'}x_lx_{l'}+2\boldsymbol{I}_{ijkl'}x_lx_{k'}+2\boldsymbol{I}_{ijlk'l'}x_kx_{l'}+2\boldsymbol{I}_{ijll'}x_kx_{k'}$
	$+(\delta_{ik}\delta_{lk'}+\delta_{il}\delta_{k'k'})x_jx_{l'}+(\delta_{ik}\delta_{ll'}+\delta_{il}\delta_{kl'})x_jx_{k'}$
	$+(\delta_{jk}\delta_{lk'}+\delta_{jl}\delta_{k'k'})x_ix_{l'}+(\delta_{jk}\delta_{ll'}+\delta_{jl}\delta_{kl'})x_ix_{k'})$
	$+\delta_{k'l'}(2\boldsymbol{I}_{ijki'}x_lx_{j'}+2\boldsymbol{I}_{ijkj'}x_lx_{i'}+2\boldsymbol{I}_{jili'}x_kx_{j'}+2\boldsymbol{I}_{jjlj'}x_kx_{i'}$
	$+(\delta_{ik}\delta_{lj'}+\delta_{il}\delta_{kj'})x_jx_{i'}+(\delta_{ik}\delta_{lk'}+\delta_{il}\delta_{kk'})x_jx_{l'}$
	$+(\delta_{jk}\delta_{li'}+\delta_{jl}\delta_{ki'})x_ix_{j'}+(\delta_{jk}\delta_{lj'}+\delta_{jl}\delta_{kj'})x_ix_{i'})$
$\boldsymbol{L}^{32}_{i\cdots l'}(\boldsymbol{x})$	$2((\boldsymbol{I}_{iji'k'}\delta_{kj'}+\boldsymbol{I}_{ijj'k'}\delta_{ki'})x_lx_{l'}+(\boldsymbol{I}_{iji'l'}\delta_{kj'}+\boldsymbol{I}_{ijj'l'}\delta_{ki'})x_lx_{k'}$
	$+(\boldsymbol{I}_{iji'k'}\delta_{kl'}+\boldsymbol{I}_{iji'l'}\delta_{kk'})x_lx_{j'}+(\boldsymbol{I}_{ijj'k'}\delta_{kl'}+\boldsymbol{I}_{ijj'l'}\delta_{kk'})x_lx_{i'}$
	$+(\boldsymbol{I}_{iji'k'}\delta_{lj'}+\boldsymbol{I}_{iji'k'}\delta_{li'})x_kx_{l'}+(\boldsymbol{I}_{iji'l'}\delta_{lj'}+\boldsymbol{I}_{ijj'l'}\delta_{li'})x_kx_{k'}$
	$+(\boldsymbol{I}_{iji'k'}\delta_{ll'}+\boldsymbol{I}_{iji'l'}\delta_{lk'})x_kx_{j'}+(\boldsymbol{I}_{ijj'k'}\delta_{ll'}+\boldsymbol{I}_{ijj'l'}\delta_{lk'})x_kx_{i'}$
	$+(\boldsymbol{I}_{kli'k'}\delta_{jj'}+\boldsymbol{I}_{klj'k'}\delta_{ii'})x_jx_{l'}+(\boldsymbol{I}_{kli'l'}\delta_{ij'}+\boldsymbol{I}_{klj'l'}\delta_{ii'})x_jx_{k'}$
	$+(\boldsymbol{I}_{kli'k'}\delta_{il'}+\boldsymbol{I}_{kli'l'}\delta_{ik'})x_jx_{j'}+(\boldsymbol{I}_{klj'k'}\delta_{il'}+\boldsymbol{I}_{klj'l'}\delta_{ik'})x_jx_{i'}$
	$+(\boldsymbol{I}_{kli'k'}\delta_{jj'}+\boldsymbol{I}_{klj'k'}\delta_{ji'})x_ix_{l'}+(\boldsymbol{I}_{kli'l'}\delta_{jj'}+\boldsymbol{I}_{klj'l'}\delta_{ji'})x_ix_{k'}$
	$+(\boldsymbol{I}_{kli'k'}\delta_{jl'}+\boldsymbol{I}_{kli'l'}\delta_{jk'})x_ix_{j'}+(\boldsymbol{I}_{klj'k'}\delta_{jl'}+\boldsymbol{I}_{klj'l'}\delta_{jk'})x_ix_{i'})$

Table 2.4 (*Cont.*)

Covariant	**Value**
$\boldsymbol{L}^{33}_{i\cdots l'}(\boldsymbol{x})$	$2(\delta_{ik}(\boldsymbol{I}_{ji'k'l'}x_lx_{j'} + \boldsymbol{I}_{jj'i'k'}x_lx_{l'} + \boldsymbol{I}_{jj'i'l'}x_lx_{k'} + \boldsymbol{I}_{jj'k'l'}x_lx_{i'}$ $+\boldsymbol{I}_{li'j'k'}x_jx_{l'} + \boldsymbol{I}_{li'k'l'}x_jx_{j'} + \boldsymbol{I}_{li'j'l'}x_jx_{k'} + \boldsymbol{I}_{lj'k'l'}x_jx_{i'})$ $+\delta_{il}(\boldsymbol{I}_{ji'k'l'}x_kx_{j'} + \boldsymbol{I}_{jj'i'k'}x_kx_{l'} + \boldsymbol{I}_{jj'i'l'}x_kx_{k'} + \boldsymbol{I}_{jj'k'l'}x_kx_{i'}$ $+\boldsymbol{I}_{ki'k'l'}x_jx_{j'} + \boldsymbol{I}_{kj'i'k'}x_jx_{l'} + \boldsymbol{I}_{ki'j'l'}x_jx_{k'} + \boldsymbol{I}_{kj'k'l'}x_jx_{i'})$ $+\delta_{jk}(\boldsymbol{I}_{ij'i'k'}x_lx_{l'} + \boldsymbol{I}_{ii'k'l'}x_lx_{j'} + \boldsymbol{I}_{ij'i'l'}x_lx_{k'} + \boldsymbol{I}_{ij'k'l'}x_lx_{i'}$ $+\boldsymbol{I}_{li'j'k'}x_ix_{l'} + \boldsymbol{I}_{li'k'l'}x_ix_{j'} + \boldsymbol{I}_{li'j'l'}x_ix_{k'} + \boldsymbol{I}_{lj'k'l'}x_ix_{i'})$ $+\delta_{jl}(\boldsymbol{I}_{ii'k'l'}x_kx_{j'} + \boldsymbol{I}_{ij'i'k'}x_kx_{l'} + \boldsymbol{I}_{ij'i'l'}x_kx_{k'} + \boldsymbol{I}_{ij'k'l'}x_kx_{i'}$ $+\boldsymbol{I}_{ki'k'l'}x_ix_{j'} + \boldsymbol{I}_{kj'i'k'}x_jx_{l'} + \boldsymbol{I}_{ki'j'l'}x_lx_{k'} + \boldsymbol{I}_{kj'k'l'}x_ix_{i'}))$

$$\begin{aligned}\boldsymbol{L}^{33}_{i\cdots l'}(\boldsymbol{x}) &= 24\boldsymbol{L}^{1}_{i\cdots l'} - 4\boldsymbol{L}^{2}_{i\cdots l'} - 8\boldsymbol{L}^{3}_{i\cdots l'} - 2\boldsymbol{L}^{4}_{i\cdots l'} + 2\boldsymbol{L}^{5}_{i\cdots l'} + 2\boldsymbol{L}^{6}_{i\cdots l'} - \boldsymbol{L}^{7}_{i\cdots l'}\\ &\quad - 12\boldsymbol{L}^{8}_{i\cdots l'}(\boldsymbol{x}) - 2\boldsymbol{L}^{9}_{i\cdots l'}(\boldsymbol{x}) + 2\boldsymbol{L}^{10}_{i\cdots l'}(\boldsymbol{x}) + 4\boldsymbol{L}^{11}_{i\cdots l'}(\boldsymbol{x}) + 3\boldsymbol{L}^{12}_{i\cdots l'}(\boldsymbol{x})\\ &\quad + 2\boldsymbol{L}^{13}_{i\cdots l'}(\boldsymbol{x}) - \boldsymbol{L}^{14}_{i\cdots l'}(\boldsymbol{x}) + \boldsymbol{L}^{15}_{i\cdots l'}(\boldsymbol{x}) - \frac{1}{2}\boldsymbol{L}^{16}_{i\cdots l'}(\boldsymbol{x}).\end{aligned}$$

In the case of degree 4 and of order 4 we have only one covariant:

$$\|\boldsymbol{x}\|^4\boldsymbol{L}^{5}_{ijkl}(\boldsymbol{x}) = x_ix_jx_kx_l. \tag{2.43}$$

In the case of degree 4 and of order 6, for the pair of representations $((\mathbb{R}^3)^{\otimes 6}, g^{\otimes 6})$ and $(\mathbb{R}^3, g)$ we have 15 covariant tensors. Under Σ_p, we found the following symmetric covariant tensors.

$$\|\boldsymbol{x}\|^4\boldsymbol{L}^{16}(\boldsymbol{x}) = \delta_{kk'}x_ix_jx_{i'}x_{j'},$$

$$\begin{aligned}\|\boldsymbol{x}\|^4\boldsymbol{L}^{17}(\boldsymbol{x}) &= \delta_{ij}x_kx_{i'}x_{j'}x_{k'} + \delta_{i'j'}x_ix_jx_kx_{k'},\\ \|\boldsymbol{x}\|^4\boldsymbol{L}^{18}(\boldsymbol{x}) &= \delta_{ii'}x_jx_kx_{j'}x_{k'} + \delta_{ij'}x_jx_kx_{i'}x_{k'} + \delta_{ji'}x_ix_kx_{j'}x_{k'}\\ &\quad + \delta_{jj'}x_ix_kx_{i'}x_{k'},\\ \|\boldsymbol{x}\|^4\boldsymbol{L}^{19}(\boldsymbol{x}) &= \delta_{ik'}x_jx_kx_{i'}x_{j'} + \delta_{jk'}x_ix_kx_{i'}x_{j'} + \delta_{ki'}x_ix_jx_{j'}x_{k'}\\ &\quad + \delta_{kj'}x_ix_jx_{i'}x_{k'},\\ \|\boldsymbol{x}\|^4\boldsymbol{L}^{20}(\boldsymbol{x}) &= \delta_{ik}x_jx_{i'}x_{j'}x_{k'} + \delta_{jk}x_ix_{i'}x_{j'}x_{k'} + \delta_{i'k'}x_ix_jx_kx_{j'}\\ &\quad + \delta_{j'k'}x_ix_jx_kx_{i'}.\end{aligned} \tag{2.44}$$

In the case of degree 4 and of order 8, we have $6 \times 35 = 210$ covariant tensors (each of $N(4) = 6$ products of 2 Kronecker's deltas can be combined with $\binom{8}{4} = 35$ products of the form $x_{i_1}x_{i_2}x_{i_3}x_{i_4}$). We found 10 orbits of the group Σ. The sums of elements on each orbit are shown in Table 2.5.

We found two reduction equations:

$$\begin{aligned}\boldsymbol{L}^{34}_{i\cdots l'}(\boldsymbol{x}) &= -4\boldsymbol{L}^{8}_{i\cdots l'}(\boldsymbol{x}) + 2\boldsymbol{L}^{9}_{i\cdots l'}(\boldsymbol{x}) + 2\boldsymbol{L}^{10}_{i\cdots l'}(\boldsymbol{x}) - \boldsymbol{L}^{12}_{i\cdots l'}(\boldsymbol{x}) + 2\boldsymbol{L}^{13}_{i\cdots l'}(\boldsymbol{x})\\ &\quad - \boldsymbol{L}^{14}_{i\cdots l'}(\boldsymbol{x}) - \boldsymbol{L}^{15}_{i\cdots l'}(\boldsymbol{x}) + \frac{1}{2}\boldsymbol{L}^{16}_{i\cdots l'}(\boldsymbol{x}) + 8\boldsymbol{L}^{19}_{i\cdots l'}(\boldsymbol{x}) - 4\boldsymbol{L}^{21}_{i\cdots l'}(\boldsymbol{x})\\ &\quad + 2\boldsymbol{L}^{23}_{i\cdots l'}(\boldsymbol{x}) - 4\boldsymbol{L}^{24}_{i\cdots l'}(\boldsymbol{x}) + 2\boldsymbol{L}^{25}_{i\cdots l'}(\boldsymbol{x}),\end{aligned}$$

Table 2.5 *Σ-symmetric basic covariant tensors of degree 4 and of order 8*

Symbol	**Value**
$\boldsymbol{L}^{18}_{i\cdots l'}(\boldsymbol{x})$	$\delta_{ij}\delta_{kl}x_{i'}x_{j'}x_{k'}x_{l'} + \delta_{i'j'}\delta_{k'l'}x_i x_j x_k x_l$
$\boldsymbol{L}^{19}_{i\cdots l'}(\boldsymbol{x})$	$(\delta_{ij}x_k x_l + \delta_{kl}x_i x_j)(\delta_{i'j'}x_{k'}x_{l'} + \delta_{k'l'}x_{i'}x_{j'})$
$\boldsymbol{L}^{20}_{i\cdots l'}(\boldsymbol{x})$	$2(\boldsymbol{I}_{ijkl}x_{i'}x_{j'}x_{k'}x_{l'} + \boldsymbol{I}_{i'j'k'l'}x_i x_j x_k x_l)$
$\boldsymbol{L}^{21}_{i\cdots l'}(\boldsymbol{x})$	$(\delta_{ij}x_k x_l + \delta_{kl}x_i x_j)(\delta_{i'k'}x_{j'}x_{l'} + \delta_{i'l'}x_{j'}x_{k'} + \delta_{j'k'}x_{i'}x_{l'} + \delta_{j'l'}x_{i'}x_{k'})$
	$+(\delta_{i'j'}x_{k'}x_{l'} + \delta_{k'l'}x_{i'}x_{j'})(\delta_{ik}x_j x_l + \delta_{il}x_j x_k + \delta_{jk}x_i x_l + \delta_{jl}x_i x_k)$
$\boldsymbol{L}^{22}_{i\cdots l'}(\boldsymbol{x})$	$\delta_{ij}(\delta_{ki'}x_l x_{j'}x_{k'}x_{l'} + \delta_{kj'}x_l x_{i'}x_{k'}x_{l'} + \delta_{kk'}x_l x_{i'}x_{j'}x_{l'} + \delta_{kl'}x_l x_{i'}x_{j'}x_{k'}$
	$+\delta_{li'}x_k x_{j'}x_{k'}x_{l'} + \delta_{lj'}x_k x_{i'}x_{k'}x_{l'} + \delta_{lk'}x_k x_{i'}x_{j'}x_{l'} + \delta_{ll'}x_k x_{i'}x_{j'}x_{k'})$
	$+\delta_{kl}(\delta_{ii'}x_j x_{j'}x_{k'}x_{l'} + \delta_{ij'}x_j x_{i'}x_{k'}x_{l'} + \delta_{ik'}x_j x_{i'}x_{j'}x_{l'} + \delta_{il'}x_j x_{i'}x_{j'}x_{k'}$
	$+\delta_{ji'}x_i x_{j'}x_{k'}x_{l'} + \delta_{jj'}x_i x_{i'}x_{k'}x_{l'} + \delta_{jk'}x_i x_{i'}x_{j'}x_{l'} + \delta_{jl'}x_i x_{i'}x_{j'}x_{k'})$
	$+\delta_{i'j'}(\delta_{ik'}x_j x_k x_l x_{l'} + \delta_{jk'}x_i x_k x_l x_{l'} + \delta_{kk'}x_i x_j x_l x_{l'} + \delta_{lk'}x_i x_j x_k x_{l'}$
	$+\delta_{il'}x_k x_{l'}x_{j'}x_{k'} + \delta_{jl'}x_k x_{l'}x_{i'}x_{k'} + \delta_{kl'}x_i x_j x_{k'}x_{l'} + \delta_{ll'}x_i x_j x_k x_{k'})$
	$+\delta_{k'l'}(\delta_{ii'}x_j x_k x_l x_{j'} + \delta_{ji'}x_i x_k x_{j'}x_{l'} + \delta_{ki'}x_j x_j x_l x_{j'} + \delta_{li'}x_i x_j x_k x_{j'}$
	$+\delta_{ij'}x_j x_k x_l x_{i'} + \delta_{jj'}x_i x_k x_l x_{i'} + \delta_{kj'}x_i x_j x_l x_{i'} + \delta_{lj'}x_i x_j x_k x_{i'})$
$\boldsymbol{L}^{23}_{i\cdots l'}(\boldsymbol{x})$	$(\delta_{ik}x_j x_l + \delta_{il}x_j x_k + \delta_{jk}x_i x_l + \delta_{jl}x_i x_k)$
	$\times(\delta_{i'k'}x_{j'}x_{l'} + \delta_{i'l'}x_{j'}x_{k'} + \delta_{j'k'}x_{i'}x_{l'} + \delta_{j'l'}x_{i'}x_{k'})$
$\boldsymbol{L}^{24}_{i\cdots l'}(\boldsymbol{x})$	$2(\boldsymbol{I}_{iji'j'}x_k x_l x_{k'}x_{l'} + \boldsymbol{I}_{ijk'l'}x_k x_l x_{i'}x_{j'}$
	$+\boldsymbol{I}_{kli'j'}x_i x_j x_{k'}x_{l'} + \boldsymbol{I}_{klk'l'}x_i x_j x_{i'}x_{j'})$
$\boldsymbol{L}^{25}_{i\cdots l'}(\boldsymbol{x})$	$2[(\boldsymbol{I}_{iji'k'}x_{j'}x_{l'} + \boldsymbol{I}_{iji'l'}x_{j'}x_{k'} + \boldsymbol{I}_{ijj'k'}x_{i'}x_{l'} + \boldsymbol{I}_{ijj'l'}x_{i'}x_{k'})x_k x_l$
	$+(\boldsymbol{I}_{iki'j'}x_{k'}x_{l'} + \boldsymbol{I}_{ikk'l'}x_{i'}x_{j'})x_j x_l + (\boldsymbol{I}_{ili'j'}x_{k'}x_{l'} + \boldsymbol{I}_{ilk'l'}x_{i'}x_{j'})x_j x_k$
	$+(\boldsymbol{I}_{jki'j'}x_{k'}x_{l'} + \boldsymbol{I}_{jkk'l'}x_{i'}x_{j'})x_i x_l + (\boldsymbol{I}_{jli'j'}x_{k'}x_{l'} + \boldsymbol{I}_{jlk'l'}x_{i'}x_{j'})x_j x_k$
	$+(\boldsymbol{I}_{kli'k'}x_{j'}x_{l'} + \boldsymbol{I}_{kli'l'}x_{j'}x_{k'} + \boldsymbol{I}_{klj'k'}x_{i'}x_{l'} + \boldsymbol{I}_{klj'l'}x_{i'}x_{k'})x_i x_j]$
$\boldsymbol{L}^{34}_{i\cdots l'}(\boldsymbol{x})$	$2[(\boldsymbol{I}_{iki'k'}x_{j'}x_{l'} + \boldsymbol{I}_{iki'l'}x_{j'}x_{k'} + \boldsymbol{I}_{ikj'k'}x_{i'}x_{l'} + \boldsymbol{I}_{ikj'l'}x_{i'}x_{k'})x_j x_l$
	$+(\boldsymbol{I}_{ili'k'}x_{j'}x_{l'} + \boldsymbol{I}_{ili'l'}x_{j'}x_{k'} + \boldsymbol{I}_{ilj'k'}x_{i'}x_{l'} + \boldsymbol{I}_{ilj'l'}x_{i'}x_{k'})x_j x_k$
	$+(\boldsymbol{I}_{jki'k'}x_{j'}x_{l'} + \boldsymbol{I}_{jki'l'}x_{j'}x_{k'} + \boldsymbol{I}_{jkj'k'}x_{i'}x_{l'} + \boldsymbol{I}_{jkj'l'}x_{i'}x_{k'})x_i x_l$
	$+(\boldsymbol{I}_{jli'k'}x_{j'}x_{l'} + \boldsymbol{I}_{jli'l'}x_{j'}x_{k'} + \boldsymbol{I}_{jlj'k'}x_{i'}x_{l'} + \boldsymbol{I}_{jlj'l'}x_{i'}x_{k'})x_i x_k$
$\boldsymbol{L}^{35}_{i\cdots l'}(\boldsymbol{x})$	$2[\boldsymbol{I}_{lj'k'l'}x_i x_j x_k x_{i'} + (\delta_{jk}\delta_{ll'} + \delta_{jl}\delta_{kl'})x_i x_{i'}x_{j'}x_{k'}$
	$+\boldsymbol{I}_{li'k'l'}x_i x_j x_k x_{j'} + (\delta_{ik}\delta_{ll'} + \delta_{il}\delta_{kl'})x_j x_{i'}x_{j'}x_{k'}$
	$+\boldsymbol{I}_{ijll'}x_k x_{i'}x_{j'}x_{k'} + (\delta_{li'}\delta_{j'l'} + \delta_{lj'}\delta_{i'l'})x_i x_j x_k x_{k'}$
	$+\boldsymbol{I}_{ijkl'}x_l x_{i'}x_{j'}x_{k'} + (\delta_{li'}\delta_{j'k'} + \delta_{lj'}\delta_{i'k'})x_i x_j x_k x_{l'}$
	$+\boldsymbol{I}_{kjk'l'}x_i x_j x_l x_{i'} + (\delta_{jk}\delta_{lk'} + \delta_{jl}\delta_{kk'})x_i x_{i'}x_{j'}x_{l'}$
	$+\boldsymbol{I}_{ki'k'l'}x_i x_j x_l x_{j'} + (\delta_{ik}\delta_{lk'} + \delta_{il}\delta_{kk'})x_j x_{i'}x_{j'}x_{l'}$
	$+\boldsymbol{I}_{ijlk'l'}x_k x_{i'}x_{j'}x_{l'} + (\delta_{ki'}\delta_{j'l'} + \delta_{kj'}\delta_{i'l'})x_i x_j x_l x_{k'}$
	$+\boldsymbol{I}_{ijkk'}x_l x_{i'}x_{j'}x_{l'} + (\delta_{ki'}\delta_{j'l'} + \delta_{kj'}\delta_{i'l'})x_i x_j x_l x_{l'}$
	$+\boldsymbol{I}_{ji'k'l'}x_i x_k x_l x_{i'} + (\delta_{jk}\delta_{lj'} + \delta_{jl}\delta_{kj'})x_i x_{i'}x_{k'}x_{l'}$
	$+\boldsymbol{I}_{ji'k'l'}x_i x_k x_l x_{j'} + (\delta_{ik}\delta_{lj'} + \delta_{il}\delta_{kj'})x_j x_{i'}x_{k'}x_{l'}$
	$+\boldsymbol{I}_{ijlj'}x_k x_{i'}x_{k'}x_{l'} + (\delta_{ji'}\delta_{j'l'} + \delta_{jj'}\delta_{i'l'})x_i x_k x_l x_{k'}$
	$+\boldsymbol{I}_{ijkj'}x_l x_{i'}x_{k'}x_{l'} + (\delta_{ji'}\delta_{j'l'} + \delta_{jj'}\delta_{i'k'})x_i x_k x_l x_{l'}$
	$+\boldsymbol{I}_{ij'k'l'}x_j x_k x_l x_{i'} + (\delta_{ik}\delta_{li'} + \delta_{jl}\delta_{ki'})x_i x_{j'}x_{k'}x_{l'}$
	$+\boldsymbol{I}_{ii'k'l'}x_j x_k x_l x_{j'} + (\delta_{ik}\delta_{li'} + \delta_{il}\delta_{ki'})x_j x_{j'}x_{k'}x_{l'}$
	$+\boldsymbol{I}_{ijli'}x_k x_{j'}x_{k'}x_{l'} + (\delta_{ii'}\delta_{j'l'} + \delta_{ij'}\delta_{i'l'})x_j x_k x_l x_{k'}$
	$+\boldsymbol{I}_{ijki'}x_l x_{j'}x_{k'}x_{l'} + (\delta_{ii'}\delta_{j'k'} + \delta_{ij'}\delta_{i'k'})x_j x_k x_{l'}x_{l'}]$

Table 2.6 Σ_e-symmetric basic covariant tensors of degree 6 and of order 8

Symbol	Value
$\mathsf{L}^{26}_{i\cdots l'}(\boldsymbol{x})$	$(\delta_{ij}x_kx_l+\delta_{kl}x_ix_j)x_{i'}x_{j'}x_{k'}x_{l'}$ $+(\delta_{i'j'}x_{k'}x_{l'}+\delta_{k'l'}x_{i'}x_{j'})x_ix_jx_kx_l$
$\mathsf{L}^{27}_{i\cdots l'}(\boldsymbol{x})$	$(\delta_{ik}x_jx_l+\delta_{il}x_jx_k+\delta_{jk}x_ix_l+\delta_{jl}x_ix_k)x_{i'}x_{j'}x_{k'}x_{l'}$ $+(\delta_{i'k'}x_{j'}x_{l'}+\delta_{i'l'}x_{j'}x_{k'}+\delta_{j'k'}x_{i'}x_{l'}+\delta_{j'l'}x_{i'}x_{k'})x_ix_jx_kx_l$
$\mathsf{L}^{28}_{i\cdots l'}(\boldsymbol{x})$	$(\delta_{ii'}x_{j'}x_{k'}x_{l'}+\delta_{ij'}x_{i'}x_{k'}x_{l'}+\delta_{ik'}x_{i'}x_{j'}x_{l'}$ $+\delta_{il'}x_{i'}x_{j'}x_{k'})x_jx_kx_l+(\delta_{ji'}x_{j'}x_{k'}x_{l'}+\delta_{jj'}x_{i'}x_{k'}x_{l'}$ $+\delta_{jk'}x_{i'}x_{j'}x_{l'}+\delta_{jl'}x_{i'}x_{j'}x_{k'})x_ix_kx_l$ $+(\delta_{ki'}x_{j'}x_{k'}x_{l'}+\delta_{kj'}x_{i'}x_{k'}x_{l'}+\delta_{kk'}x_{i'}x_{j'}x_{l'}$ $+\delta_{kl'}x_{i'}x_{j'}x_{k'})x_ix_jx_l+(\delta_{li'}x_{j'}x_{k'}x_{l'}+\delta_{lj'}x_{i'}x_{k'}x_{l'}$ $+\delta_{lk'}x_{i'}x_{j'}x_{l'}+\delta_{ll'}x_{i'}x_{j'}x_{k'})x_ix_jx_k$

$$\begin{aligned}\mathsf{L}^{35}_{i\cdots l'}(\boldsymbol{x}) &= 4\mathsf{L}^{8}_{i\cdots l'}(\boldsymbol{x})-2\mathsf{L}^{9}_{i\cdots l'}(\boldsymbol{x})-2\mathsf{L}^{13}_{i\cdots l'}(\boldsymbol{x})+\mathsf{L}^{15}_{i\cdots l'}(\boldsymbol{x})-8\mathsf{L}^{18}_{i\cdots l'}(\boldsymbol{x})\\&\quad-8\mathsf{L}^{19}_{i\cdots l'}(\boldsymbol{x})+4\mathsf{L}^{20}_{i\cdots l'}(\boldsymbol{x})+2\mathsf{L}^{21}_{i\cdots l'}(\boldsymbol{x})+2\mathsf{L}^{22}_{i\cdots l'}(\boldsymbol{x})+4\mathsf{L}^{24}_{i\cdots l'}(\boldsymbol{x})\\&\quad-\mathsf{L}^{25}_{i\cdots l'}(\boldsymbol{x}).\end{aligned}$$

In the case of degree 6 and of order 6, we have only one symmetric covariant tensor:

$$\|\boldsymbol{x}\|^4\mathsf{L}^{21}(\boldsymbol{x}) = x_ix_jx_kx_{i'}x_{j'}x_{k'}. \tag{2.45}$$

In the case of degree 6 and of order 8, we have $\binom{8}{2}=28$ covariant tensors. We found 3 orbits of the group Σ_e and no reduction equations. The sums of elements on each orbit are shown in Table 2.6.

Finally, the only covariant of degree 8 and of order 8 is

$$\|\boldsymbol{x}\|^8\mathsf{L}^{29}_{i\cdots l'}(\boldsymbol{x}) = x_ix_jx_kx_lx_{i'}x_{j'}x_{k'}x_{l'}. \tag{2.46}$$

2.8 Convex Compacta

Let V be a real vector space. A subset $K\subset V$ is called *convex* if for all $\boldsymbol{u}$, $\boldsymbol{v}\in K$ and for all $\theta\in(0,1)$ the point

$$\boldsymbol{w} := (1-\theta)\boldsymbol{u}+\theta\boldsymbol{v} \tag{2.47}$$

lies in K.

An *affine subspace* of the space V is the set

$$G := \{\,\boldsymbol{v}_0+\boldsymbol{w}\colon \boldsymbol{w}\in W\,\},$$

where $\boldsymbol{v}_0\in V$, and W is a subspace of V. The *dimension* of G is the dimension of W. The dimension of a convex set K is the dimension of the smallest affine subspace that contains K.

Let $\boldsymbol{v}_1, \dots, \boldsymbol{v}_m$ be vectors in V. A *convex combination* of the vectors $\boldsymbol{v}_1, \dots, \boldsymbol{v}_m$ is

$$\boldsymbol{v} := \sum_{i=1}^{m} \theta_i \boldsymbol{v}_i, \qquad \theta_i \geq 0, \quad \sum_{i=1}^{m} \theta_i = 1.$$

The *convex hull* of the set F is the set of all convex combinations of vectors of F. The *closed convex hull* of the set F is the closure of its convex hull in the smallest affine subspace that contains K.

A point $\boldsymbol{w} \in K$ is called an *extreme point* of K if the representation (2.47) is possible only if $\boldsymbol{u} = \boldsymbol{v}$.

A *simplex* is the closed convex hull of $n+1$ points that do not lie in any affine subspace of dimension $n-1$.

Theorem 8 (Minkowski). *Any compact convex set coincides with the closed convex hull of the set of its extreme points.*

Theorem 9 (Carathéodory). *Each point of a compact convex set K of dimension n can be represented as a convex combination of at most $n+1$ extreme points of K. The above representation is unique if and only if K is a simplex.*

2.9 Random Fields

Let $(\Omega, \mathfrak{F}, \mathsf{P})$ be a *probability space*, i.e. Ω is a set, $\mathfrak{F}$ is a σ-field of subsets of Ω, and P is a probability measure on $\mathfrak{F}$. Let V be a finite-dimensional linear space consisting of tensors, and let $\mathfrak{B}(\mathsf{V})$ be the σ-field of Borel sets of V. A mapping $\boldsymbol{\mathsf{T}} \colon \Omega \to \mathsf{V}$ is called a *random tensor* if it is measurable, i.e. for any Borel set B we have $\boldsymbol{\mathsf{T}}^{-1}(B) \in \mathfrak{B}(\mathsf{V})$. If V consists of scalars, the term *random variable* is then used instead of random scalar.

The *expected value* of a random tensor $\boldsymbol{\mathsf{T}}$ is the integral

$$\mathsf{E}[\boldsymbol{\mathsf{T}}] := \int_{\Omega} \boldsymbol{\mathsf{T}}(\omega) \, \mathrm{d}\mathsf{P}(\omega),$$

if it exists.

A *random field* on a real finite-dimensional affine space E is a function of two variables $\boldsymbol{\mathsf{T}} \colon E \times \Omega \to \mathsf{V}$ such that for any $A \in E$ the function $\boldsymbol{\mathsf{T}}(A, \omega)$ is a random tensor.

Let V be the underlying space of E. Let $\|\cdot\|_V$ (resp. $\|\cdot\|_{\mathsf{V}}$) be the norm on V (resp. V) generated by an inner product. A random field $\boldsymbol{\mathsf{T}}(A)$ is called *second-order* if for any $A \in E$ we have

$$\mathsf{E}\|\boldsymbol{\mathsf{T}}(A)\|_{\mathsf{V}}^2 < \infty.$$

A second-order random field is called *mean-square continuous* if for any $O \in E$ we have

$$\lim_{\|A-O\|_V \to 0} \mathsf{E}\|\boldsymbol{\mathsf{T}}(A) - \boldsymbol{\mathsf{T}}(O)\|_{\mathsf{V}}^2 = 0.$$

In what follows we consider only second-order mean-square continuous random fields.

Consider the following quantity:

$$\langle \mathsf{T}(A)\rangle := \mathsf{E}[\mathsf{T}(A)].$$

Mathematicians call this quantity the *mean value* of the random field $\mathsf{T}(A)$, physicists call it the *one-point correlation tensor*. If V is a one-dimensional space, we use the term *one-point correlation function* instead. In what follows we will use physical terms and physical notation, $\langle \mathsf{T}(A)\rangle$.

A random field $\mathsf{T}(A)$ is called *centred* if $\langle \mathsf{T}(A)\rangle = \mathbf{0}$.

Let J be the identity operator in V if V is real, and a real structure on V if V is complex. The *two-point correlation tensor* (*two-point correlation function* if V is a one-dimensional space) is the following quantity:

$$\langle \mathsf{T}(A), \mathsf{T}(B)\rangle := \mathsf{E}[J(\mathsf{T}(A) - \langle \mathsf{T}(A)\rangle) \otimes (\mathsf{T}(B) - \langle \mathsf{T}(B)\rangle)].$$

A random field $\mathsf{T}(A)$ is called *homogeneous* if its one-point correlation tensor is a constant, and its two-point correlation tensor $\langle \mathsf{T}(A), \mathsf{T}(B)\rangle$ depends only on the vector $B-A$. The physical sense of this definition is as follows: the characteristics of the corresponding physical system do not depend on the location of the origin O of the affine frame.

Let V be a complex linear space, let $\hat{V}$ be the wavenumber domain, and let $\mathfrak{B}(\hat{V})$ be the σ-field of Borel sets of the wavenumber domain. A mapping μ that maps $\mathfrak{B}(\hat{V})$ to the set of all Hermitian non-negative-definite operators on V is called a Radon measure if for any $\mathsf{V} \in \mathsf{V}$ the mapping $(\mu(A)\boldsymbol{v}, \boldsymbol{v})\colon \mathfrak{B}(\hat{V}) \to \mathbb{R}^1$ is a finite Radon measure.

Let J be a real structure on a *complex* linear space W, and let V^+ be the real linear span of the eigenvectors of J with eigenvalue 1. The linear space of all Hermitian operators on V is isomorphic to $\mathsf{V}^+ \otimes \mathsf{V}^+ = \mathsf{S}^2(\mathsf{V}^+) \oplus \Lambda^2(\mathsf{V}^+)$. Let $\top$ be the linear operator in $\mathsf{V}^+ \otimes \mathsf{V}^+$ for which $\mathsf{S}^2(\mathsf{V}^+)$ is the set of eigenvectors with eigenvalue 1, and $\Lambda^2(\mathsf{V}^+)$ is the set of eigenvectors with eigenvalue -1 (this is just the coordinate-free definition of the transposed matrix).

Theorem 10. *Formula*

$$\langle \mathsf{T}(A), \mathsf{T}(B)\rangle = \int_{\hat{V}} \mathrm{e}^{\mathrm{i}(\boldsymbol{p}, B-A)}\, \mathrm{d}F(\boldsymbol{p}) \tag{2.48}$$

establishes a one-to-one correspondence between the set of two-point correlation tensors of homogeneous mean-square continuous W-valued random fields on E and the set of Radon measures F on $\mathfrak{B}(\hat{V})$ taking values in the set of all Hermitian non-negative-definite operators on W. If the random field takes values in W_+, then the measure F satisfies the following condition

$$F(-A) = F(A)^\top, \qquad A \in \mathfrak{B}(\hat{V}), \tag{2.49}$$

where $-A := \{\, -\boldsymbol{p}\colon \boldsymbol{p} \in A\,\}$.

Let X be a set, and let $\boldsymbol{v}(x)$ be a centred random field on X taking values in a complex finite-dimensional linear space V. Let $(\cdot,\cdot)$ be an inner product in V, and let J be a real structure on V. Let Λ be a set, $\mathfrak{L}$ be a σ-field of subsets of Λ, and let F be a measure on $\mathfrak{L}$ taking values in the set of Hermitian non-negative-definite operators on V. Let F_0 be the following measure:

$$F_0(A) := \operatorname{tr} F(A), \qquad A \in \mathfrak{L}.$$

Let $f(x,\lambda)$ be a complex-valued function such that for any fixed $x_0 \in X$ the function $f(x_0,\lambda)$ is square integrable with respect to F_0. Let $L^2(f)$ be the closed linear span of the set $\{\, f(x,\lambda)\colon x \in X \,\}$ in the Hilbert space $L^2(Y,F_0)$ of all square integrable functions. The set $\{\, f(x,\lambda)\colon x \in X \,\}$ is called *total* in $L^2(Y,F_0)$ if $L^2(f) = L^2(Y,F_0)$.

Theorem 11 (Karhunen, 1947). *Assume that*

$$\langle \boldsymbol{v}(x), \boldsymbol{v}(y) \rangle = \int_\Lambda f(x,\lambda)\overline{f(y,\lambda)}\,\mathrm{d}F(\lambda) \tag{2.50}$$

and the set $\{\, f(x,\lambda)\colon x \in X \,\}$ *is total in* $L^2(\Lambda,F_0)$. *Then*

$$\boldsymbol{v}(\boldsymbol{x}) = \int_\Lambda f(x,\lambda)\,\mathrm{d}Z(\lambda), \tag{2.51}$$

where Z is a measure on $\mathfrak{L}$ with values in the Hilbert space of all V-valued random vectors $\boldsymbol{w}$ with $\mathsf{E}[\|\boldsymbol{w}\|^2] < \infty$. Moreover, we have

$$\mathsf{E}[JZ(A) \otimes Z(B)] = F(A \cap B), \qquad A, B \in \mathfrak{L}. \tag{2.52}$$

The measure Z is called an *orthogonal scattered random measure*, while the measure F satisfying (2.52) is called the *control measure* of the orthogonal scattered random measure Z.

Is it possible to apply Theorem 11 to Equation 2.48? The answer is negative, because the function $\mathrm{e}^{\mathrm{i}(\boldsymbol{p},B-A)}$ cannot be written in the form $f(B,\lambda)\overline{f(A,\lambda)}$. To overcome this difficulty, fix an origin $O \in E$, write the vector $B - A \in V$ as $B - A = (B - O) - (A - O)$, and finally obtain

$$\mathrm{e}^{\mathrm{i}(\boldsymbol{p},B-A)} = \mathrm{e}^{\mathrm{i}(\boldsymbol{p},B-O)}\overline{\mathrm{e}^{\mathrm{i}(\boldsymbol{p},A-O)}}.$$

By Theorem 11, we obtain

$$\boldsymbol{v}(A) = \int_{\hat{V}} \mathrm{e}^{\mathrm{i}(\boldsymbol{p},A-O)}\,\mathrm{d}Z(\boldsymbol{p}).$$

To avoid frequent repetitions of the same words, we vectorise the affine space E by a choice of the origin $O \in E$ once and forever, and denote the vector space E_O by $\mathbb{R}^d$.

Let G be a closed subgroup of the group $\mathrm{O}(d)$, and let (ρ, W) be an orthogonal representation of the group G in a real finite-dimensional vector space.

Definition 5. A random field $\boldsymbol{T}(\boldsymbol{x})$ is called (G,ρ)-*isotropic* if

$$\begin{aligned}\langle \boldsymbol{T}(g\boldsymbol{x})\rangle &= \rho(g)\langle \boldsymbol{T}(\boldsymbol{x})\rangle,\\ \langle \boldsymbol{T}(g\boldsymbol{x}),\boldsymbol{T}(g\boldsymbol{y})\rangle &= (\rho\otimes\rho)(g)\langle \boldsymbol{T}(\boldsymbol{x}),\boldsymbol{T}(\boldsymbol{y})\rangle, \qquad g\in G.\end{aligned} \tag{2.53}$$

A physical motivation of Definition 5 will be given in Section 3.1.

2.10 Special Functions

The *Gamma function* is

$$\Gamma(z) := \int_0^\infty t^{z-1}\mathrm{e}^{-t}\,\mathrm{d}t, \qquad \operatorname{Re} z > 0.$$

The Gamma function satisfies the recurrence relation

$$\Gamma(z+1) = z\Gamma(z). \tag{2.54}$$

The Lebesgue measure of the sphere od radius r in a d-dimensional space is

$$|S_{d-1}(r)| = \frac{2\pi^{d/2}}{\Gamma(d/2)} r^{d-1}.$$

The *associated Legendre polynomials* are

$$P_\ell^m(x) := \frac{(-1)^m}{2^\ell \ell!}(1-x^2)^{m/2}\frac{\mathrm{d}^{\ell+m}}{\mathrm{d}x^{\ell+m}}(x^2-1)^\ell, \qquad -\ell\le m\le \ell.$$

The factor $(-1)^m$ is called the *Condon–Shortley phase.* We have

$$P_\ell^{-m}(x) = (-1)^m\frac{(\ell-m)!}{(\ell+m)!}P_\ell^m(x), \qquad m\ge 1.$$

The Wigner D functions $D^\ell_{m0}(\varphi,\theta)$ have the form:

$$D^\ell_{m0}(\varphi,\theta) = \sqrt{\frac{(\ell-m)!}{(\ell+m)!}}\mathrm{e}^{-\mathrm{i}m\varphi}P_\ell^m(\cos\theta).$$

In terms of the associated Legendre polynomials, Equation 2.35 takes the form:

$$\begin{aligned}\Omega^\ell_{-1,0,0}(\theta) &= P_\ell^0(\cos\theta),\\ \Omega^\ell_{-1,m,0}(\theta) &= (-1)^m\sqrt{\frac{2(\ell-m)!}{(\ell+m)!}}P_\ell^m(\cos\theta),\end{aligned}$$

Define the *real-valued spherical harmonics* S_ℓ^m by a formula similar to (2.31):

$$S_\ell^m(\theta,\varphi) := \sqrt{\frac{2\ell+1}{4\pi}}T^\ell_{m0}(\varphi,\theta). \tag{2.55}$$

The real-valued spherical harmonics are orthogonal:

$$\int_{S^2} S_{\ell_1}^{m_1}(\theta,\varphi)S_{\ell_2}^{m_2}(\theta,\varphi)\,\mathrm{d}S = \delta_{\ell_1\ell_2}\delta_{m_1m_2}.$$

The *Gegenbauer polynomials* $C_n^\alpha(x)$ are defined by

$$C_n^\alpha(x) := \begin{cases} \frac{1}{\Gamma(\alpha)} \sum_{m=0}^{[n/2]} (-1)^m \frac{\Gamma(\alpha+n-m)}{m!(n-2m)!} (2x)^{n-2m}, & \alpha > -1/2, \alpha \neq 0, \\ \sum_{m=0}^{[n/2]} (-1)^m \frac{(n-m-1)!}{m!(n-2m)!}, & \alpha = 0, n \neq 0, \\ 1, & \alpha = n = 0. \end{cases}$$

Similarly, one can define complex-valued spherical harmonics Y_ℓ^m; see Erdélyi, Magnus, Oberhettinger & Tricomi (1981). The domain of the function Y_ℓ^m is the $(d-1)$-dimensional sphere S^{d-1}; the number of spherical harmonics of degree ℓ is

$$h(d, \ell) = \frac{(2\ell + d - 2)(d + \ell - 3)!}{(d-2)!\ell!}.$$

Let m_0, m_1, ..., m_{d-2} be integers satisfying the condition

$$\ell = m_0 \geq m_1 \geq \cdots \geq m_{d-2} \geq 0 .$$

Let $\boldsymbol{t} = (t_1, t_2, \ldots, t_d)$ be a point in the space $\mathbb{R}^d$. Let

$$r_k := \sqrt{t_{k+1}^2 + t_{k+2}^2 + \cdots + t_d^2} ,$$

where $k = 0$, 1, ..., $d-2$. Consider the functions

$$H(m_k, \pm, \boldsymbol{t}) := \left(\frac{t_{d-1} + \mathrm{i} t_d}{r_{d-2}} \right)^{\pm m_{d-2}} r_{d-2}^{m_{d-2}} \prod_{k=0}^{d-3} r_k^{m_k - m_{k+1}} \times C_{m_k - m_{k+1}}^{m_{k+1} + (N-k-2)/2} \left(\frac{t_{k+1}}{r_k} \right),$$

and denote

$$Y(m_k, \pm, \boldsymbol{t}) := r_0^{-m} H(m_k, \pm, \boldsymbol{t}).$$

The functions $Y(m_k, \pm, \boldsymbol{t})$ are orthogonal in the Hilbert space $L^2(S^{d-1})$ of the square-integrable functions on the unit sphere S^{d-1}, and the square of the length of the vector $Y(m_k, \pm, t)$ is

$$L(m_k) = 2\pi \prod_{k=1}^{d-2} \frac{\pi 2^{k - 2m_k - d + 2} \Gamma(m_{k-1} + m_k + d - 1 - k)}{(m_{k-1} + \frac{d-1-k}{2}(m_{k-1} - m_k)! [\Gamma(m_k + \frac{d-1-k}{2})]^2}.$$

The functions $Y(m_k, \pm, \boldsymbol{t}) / \sqrt{L(m_k)}$ are complex-valued spherical harmonics.

Let $m = m(m_k, \pm)$ be the number of the symbol $(m_0, m_1, \ldots, m_{N-2}, \pm)$ in the lexicographic ordering. The *real-valued spherical harmonics*, $S_\ell^m(\boldsymbol{t})$, can be defined as

$$S_\ell^m(\boldsymbol{t}) := \begin{cases} Y(m_k, +, \boldsymbol{t}) / \sqrt{L(m_k)} & m_{d-2} = 0, \\ \sqrt{2} \operatorname{Re} Y(m_k, +, \boldsymbol{t}) / \sqrt{L(m_k)} , & m_{d-2} > 0, m = m(m_k, +), \\ -\sqrt{2} \operatorname{Im} Y(m_k, -, \boldsymbol{t}) / \sqrt{L(m_k)} , & m_{d-2} > 0, m = m(m_k, -). \end{cases}$$

These harmonics are orthonormal:

$$\int_{S^{d-1}} S^{m_1}_{\ell_1}(\boldsymbol{t}) S^{m_2}_{\ell_2}(\boldsymbol{t})\,\mathrm{d}S = \delta_{m_1 m_2}\delta_{n_1 n_2}. \tag{2.56}$$

It follows that

$$S^0_0(\boldsymbol{t}) = \sqrt{\frac{\Gamma(d/2)}{2\pi^{d/2}}}. \tag{2.57}$$

The *addition theorem for spherical harmonics*, see e.g. Erdélyi et al. (1981, Chap. XI, Sect. 4, Theorem 4) reads

$$\frac{C^{(d-2)/2}_m(\cos\theta)}{C^{(d-2)/2}_m(1)} = \frac{2\pi^{d/2}}{\Gamma(d/2)h(d,\ell)} \sum_{m=1}^{h(d,\ell)} S^m_\ell\left(\frac{\boldsymbol{x}}{\|\boldsymbol{x}\|}\right) S^m_\ell\left(\frac{\boldsymbol{y}}{\|\boldsymbol{y}\|}\right), \tag{2.58}$$

where θ is the angle between the vectors $\boldsymbol{x}$ and $\boldsymbol{y}$.

The *Bessel function of the first kind* of order ν is defined as

$$J_\nu(z) = (z/2)^\nu \sum_{m=0}^{\infty} \frac{(-1)^m}{2^{2m} m!\Gamma(\nu+m+1)} z^{2m}.$$

The *spherical Bessel function* is

$$j_\nu(z) = \sqrt{\pi/(2z)}J_{\nu+1/2}(z).$$

The spherical Bessel functions have the following property:

$$\frac{j_1(x)}{x} = \frac{1}{3}(j_0(x) + j_2(x)) \tag{2.59}$$

The plane wave, $\mathrm{e}^{\mathrm{i}(\boldsymbol{p},\boldsymbol{x})}$, has the following expansion:

$$\mathrm{e}^{\mathrm{i}(\boldsymbol{p},\boldsymbol{x})} = \Gamma(\frac{d-2}{2})2^{\frac{d-2}{2}} \sum_{\ell=0}^{\infty} \mathrm{i}^\ell(\ell + \frac{d-2}{2}) \frac{J_{\ell+\frac{d-2}{2}}(\|\boldsymbol{p}\|\cdot\|\boldsymbol{x}\|)}{(\|\boldsymbol{p}\|\cdot\|\boldsymbol{x}\|)^{\frac{d-2}{2}}} C_\ell^{\frac{d-2}{2}}(\cos\theta),$$

where θ is the angle between the vectors $\boldsymbol{p}$ and $\boldsymbol{x}$. Combining this expansion with (2.58) and taking into account (2.54) and the value

$$C^{(d-2)/2}_\ell(1) = \frac{(\ell+d-3)!}{\ell!(d-3)!},$$

we obtain the *expansion of the plane wave in spherical harmonics*:

$$\begin{aligned} \mathrm{e}^{\mathrm{i}(\boldsymbol{p},\boldsymbol{x})} &= (2\pi)^{d/2} \sum_{\ell=0}^{\infty} \mathrm{i}^\ell(\ell + (d-2)/2) \frac{J_{\ell+(d-2)/2}(\|\boldsymbol{p}\|\cdot\|\boldsymbol{x}\|)}{(\|\boldsymbol{p}\|\cdot\|\boldsymbol{x}\|)^{(d-2)/2}} \\ &\times \sum_{m=1}^{h(d,\ell)} S^m_\ell\left(\frac{\boldsymbol{p}}{\|\boldsymbol{p}\|}\right) S^m_\ell\left(\frac{\boldsymbol{x}}{\|\boldsymbol{x}\|}\right). \end{aligned} \tag{2.60}$$

In particular, for $d=2$ we obtain the *Jacobi–Anger expansion*

$$\mathrm{e}^{\mathrm{i}(\boldsymbol{p},\boldsymbol{x})} = \sum_{\ell=-\infty}^{\infty} \mathrm{i}^\ell J_\ell(\|\boldsymbol{p}\|\cdot\|\boldsymbol{x}\|)\mathrm{e}^{\mathrm{i}\ell(\varphi_{\boldsymbol{p}}-\varphi_{\boldsymbol{x}})}, \tag{2.61}$$

where $\varphi_{\boldsymbol{p}}$ (resp. $\varphi_{\boldsymbol{x}}$) is the polar angle of the vector $\boldsymbol{p}$ (resp. $\boldsymbol{x}$). For $d = 3$ we have the (real version of) *Rayleigh expansion*:

$$\mathrm{e}^{\mathrm{i}(\boldsymbol{p},\boldsymbol{x})} = 4\pi \sum_{\ell=0}^{\infty} \sum_{m=-\ell}^{\ell} \mathrm{i}^{\ell} j_{\ell}(\|\boldsymbol{p}\| \cdot \|\boldsymbol{x}\|) S_{\ell}^{m}(\theta_{\boldsymbol{p}}, \varphi_{\boldsymbol{p}}) S_{\ell}^{m}(\theta_{\boldsymbol{x}}, \varphi_{\boldsymbol{x}}). \tag{2.62}$$

2.11 Bibliographical Remarks

Two terms, *vector space* and *linear space*, are widely used in the literature. The former stresses that the space consists of vectors, while the latter stresses that one can perform linear operations in the space. We prefer the latter term just because most of our spaces consist of tensors rather than vectors.

A preliminary definition of a natural isomorphism is as follows: it is an isomorphism that does not depend on any arbitrary choices (such as the choice of a basis). A precise definition can be given using the language of categories; see, for example, Riehl (2016, Definition 1.4.1). Eilenberg & MacLane (1945) were the first who proved the non-existence of a natural isomorphism between a finite-dimensional space and its dual.

There exist several equivalent definitions of a tensor. We approve Definition 1 and Definition 2, which are situated somewhere between the most beautiful and abstract definition through the universal mapping property due to Bourbaki (1998) (Theorem 1), and classical coordinate-dependent Definition 3.

Note that we use physical convention: a Hermitian form is linear in its *second* argument.

Equation (2.6) is the coordinate-free definition of the trace due to Bourbaki (1998).

For standard references in general topology, we mention monographs by Engelking (1989), Kelley (1975), Kuratowski (1966) and Kuratowski (1968), and textbooks by Conway (2014), Dixmier (1984) and Nagata (1985).

Permutation groups are considered in textbooks by Biggs & White (1979), Cameron (1999), Dixon & Mortimer (1996) and Passman (2012).

The family of classical groups has been defined by Hermann Weyl in 1939 in the first edition of Weyl (1997). See also modern treatment in Goodman & Wallach (2009).

Many books are devoted to *group actions*. We mention three classical books by Bredon (1972), Montgomery & Zippin (1974) and Pontryagin (1966).

There exists an advanced theory of orbifolds based on a sophisticated axiomatic definition; see Boileau, Maillot & Porti (2003).

For the history of representation theory, see Curtis (1999). Good introductory books are Adams (1969), Goodman & Wallach (2009), Fulton & Harris (1991), Murnaghan (1963), Procesi (2007) and Rossmann (2002). In our exposition, we followed mostly Bröcker & tom Dieck (1995). We also mention a classical book by Weyl (1997).

The Haar measure exists on all locally compact Hausdorff topological groups. It is unique up to a positive multiplicative constant. Its existence was proved by Haar (1933).

The Peter–Weyl theorem has been proved by Peter & Weyl (1927). Material about group actions is taken from Duistermaat & Kolk (2000).

The Clebsch–Gordan coefficients were introduced by Clebsch (1872) and Gordan (1875). The history of their calculation for the case of unitary representations of the group SU(2) can be found in Biedenharn & Louck (1981). Numerous formulae are collected in Varshalovich et al. (1988).

The choice of the phase in (2.31) is due to Condon & Shortley (1935).

The Gaunt integral has been calculated by Gaunt (1929).

An introduction to invariant theory may be found in Spencer (1971), Smith (1994) and Deville & Gatski (2012).

The symmetric covariant tensors and reduction equations were found by the authors using the MATLAB Symbolic Math Toolbox.

The Minkowski theorem was proved by Minkowski (1897), while the Carathéodory theorem was proved in Carathéodory (1907). See also a classical review paper by Danzer et al. (1963). Krein & Milman (1940) proved an infinite-dimensional version of the Minkowski theorem, while Choquet (1956) and Bishop & de Leeuw (1959) proved that of the Carathéodory theorem.

Random fields appeared in applied physical papers by Friedmann & Keller (1924), von Kármán & Howarth (1938), Obukhov (1941*a*) and Obukhov (1941*b*), among others.

The first examples of special functions appeared in the early eighteenth century as solutions to differential equations of mathematical physics and results of integration. We use notation by Abramowitz & Stegun (1964), except for notation for real-valued spherical harmonics taken from Erdélyi et al. (1981).

The Rayleigh expansion was in fact proved by Bauer (1859), and the general expansion (2.60) by Gegenbauer (1873), Gegenbauer (1877*a*), Gegenbauer (1877*b*).

3
Mathematical Results

We start from the exact formulation of the problem: what we mean by a homogeneous and isotropic tensor-valued random field and its correlation tensors, and how to describe the field in terms of more simple objects. We illustrate our methods by considering a simple but non-trivial example when the field is defined on the plane and has rank 1. After that, we prove a general result for fields of arbitrary rank. To check that our methods work correctly, we use them to prove again the results of our predecessors. It turns out that the list of these results is impressively short. Finally, we use the general result and solve the above-formulated problem for physically interesting fields of ranks up to 4. Even more results and proofs may be found on this book's website.

3.1 The Problem

In order to motivate introducing of isotropic tensor-valued random fields, consider the following models.

Let $\tau(\boldsymbol{x})$ be the temperature at the point $\boldsymbol{x}$ in the space domain V. Assume that $\tau(\boldsymbol{x})$ is a second-order mean-square continuous random field. If one shifts the origin of a coordinate system, the scalar $\tau(\boldsymbol{x})$ does not change value. It follows that the random field $\tau(\boldsymbol{x})$ is homogeneous.

Moreover, the scalar $\tau(\boldsymbol{x})$ does not change value under rotations and reflections of the coordinate system. It follows that

$$
\begin{aligned}
\langle \tau(g\boldsymbol{x}) \rangle &= \langle \tau(\boldsymbol{x}) \rangle, \\
\langle \tau(g\boldsymbol{x}), \tau(g\boldsymbol{y}) \rangle &= \langle \tau(\boldsymbol{x}), \tau(\boldsymbol{y}) \rangle, \qquad g \in \mathrm{O}(V).
\end{aligned}
$$

This is a particular case of Definition 5 when $G = \mathrm{O}(V)$ and ρ is the trivial representation of G in a real one-dimensional space W.

Let $\boldsymbol{v}(\boldsymbol{x})$ be the velocity of a turbulent fluid at the point $\boldsymbol{x}$ in the space domain V. Assume that $\boldsymbol{v}(\boldsymbol{x})$ is a second-order mean-square continuous random field. It is then homogeneous by the same reasons as in the previous model.

Apply an arbitrary orthogonal transformation $g \in \mathrm{O}(V)$ to the vector field $\boldsymbol{v}(\boldsymbol{x})$. After the transformation g the point $\boldsymbol{x}$ becomes the point $g\boldsymbol{x}$. Evidently the vector $\boldsymbol{v}(\boldsymbol{x})$ is transformed into $g\boldsymbol{v}(\boldsymbol{x})$. The one-point correlation functions of both fields must be equal:

$$\langle \boldsymbol{v}(g\boldsymbol{x})\rangle = \langle g\boldsymbol{v}(\boldsymbol{x})\rangle = g\langle \boldsymbol{v}(\boldsymbol{x})\rangle.$$

By homogeneity, $\langle \boldsymbol{v}(g\boldsymbol{x})\rangle = \langle \boldsymbol{v}(\boldsymbol{x})\rangle$, and we obtain

$$\langle \boldsymbol{v}(g\boldsymbol{x})\rangle = g\langle \boldsymbol{v}(\boldsymbol{x})\rangle.$$

The two-point correlation functions of both fields are equal as well:

$$\langle \boldsymbol{v}(g\boldsymbol{x}), \boldsymbol{v}(g\boldsymbol{y})\rangle = \langle g\boldsymbol{v}(\boldsymbol{x}), g\boldsymbol{v}(\boldsymbol{y})\rangle = (g \otimes g)\langle \boldsymbol{v}(\boldsymbol{x}), \boldsymbol{v}(\boldsymbol{y})\rangle.$$

It follows that the random field $\boldsymbol{v}(\boldsymbol{x})$ is homogeneous and isotropic with respect to the group $G = \mathrm{O}(V)$ and its defining representation.

Identify the space V with the space $\mathbb{R}^d$ by introducing the Cartesian coordinates. The action of the group $\mathrm{O}(d)$ in the space $\mathbb{R}^d$ by the matrix-vector multiplication has the quotient $\mathbb{R}^d/\mathrm{O}(d) = [0,\infty)$. The orbit type stratification is $[0,\infty) = \{0\} \cup (0,\infty)$. The conjugacy class of the stratum $\{0\}$ is $[\mathrm{O}(d)]$, and the corresponding fixed point set is $\{\mathbf{0}\}$. The conjugacy class of the stratum $(0,\infty)$ is $[\mathrm{O}(d-1)]$, and the corresponding fixed point set is $\mathbb{R}^1$, the linear span of the y-axis when $d = 2$ and the linear span of the z-axis when $d = 3$.

Let $E(\boldsymbol{x})$ be the strain tensor of a deformable body. Assume that $E(\boldsymbol{x})$ is a second-order mean-square continuous random field taking values in the space $\mathsf{S}^2(V)$ of symmetric rank 2 tensors over V. Similar considerations prove that the field $E(\boldsymbol{x})$ is homogeneous and isotropic with respect to the group $G = \mathrm{O}(V)$ and its orthogonal representation $\rho(g) = \mathsf{S}^2(g)$.

The orbit types of the above action for the case of $d = 3$ are described in Golubitsky et al. (1988). The group $\mathrm{O}(3)$ acts in the space $\mathsf{S}^2(\mathbb{R}^3)$ by $g \cdot A = gAg^{-1}$, $g \in \mathrm{O}(3)$, $A \in \mathbb{R}^3$. Under this action, the matrix A can be diagonalised. If A has three distinct eigenvalues, then the corresponding conjugacy class is $[G_0] = [D_2 \times Z_2^c]$. If two of the eigenvalues are equal, then the conjugacy class is $[G_1] = [\mathrm{O}(2) \times Z_2^c]$. Finally, when all three eigenvalues are equal, then the conjugacy class is $[G_2] = [\mathrm{O}(3)]$.

Let V^{G_i} be the corresponding fixed point set. For any group G lying between G_i and $N_{\mathrm{O}(3)}(G_i)$, the space V^{G_i} is an invariant subspace of the representation $\mathsf{S}^2(g)$ of the group G. Denote by $\rho(g)$ the restriction of the representation $\mathsf{S}^2(g)$ to the space V^G. When $G = G_i$, the representation ρ is trivial, otherwise it is not trivial. One may consider a homogeneous V^{G_i}-valued random field $E(\boldsymbol{x})$ which is (G,ρ)-isotropic.

Let $\mathsf{e}(\boldsymbol{x})$ be the piezoelectricity tensor (recall Subsection 1.5.3). Assume that $\mathsf{e}(\boldsymbol{x})$ is a second-order mean-square continuous random field taking values in the space $\mathsf{S}^2(V) \otimes V$ of rank 3 tensors over V symmetric in the first two indices. Similar considerations prove that the field $\mathsf{E}(\boldsymbol{x})$ is homogeneous and isotropic with respect to the group $G = \mathrm{O}(E)$ and its orthogonal representation $\rho(g) = \mathsf{S}^2(g) \otimes g$.

The orbit types of the above action for the case of $d = 3$ are described in Geymonat & Weller (2002). There are 15 orbit types; we will discuss them in Section 3.7. For each class $[G_i]$, we may consider a group G lying between G_i and $N_{\mathrm{O}(3)}(G_i)$, and the restriction ρ of the representation $\mathsf{S}^2(g) \otimes g$ of the group G to the invariant subspace V^{G_i}. We consider a homogeneous and (G, ρ)-isotropic random field.

Finally, let $\boldsymbol{C}(\boldsymbol{x})$ be the elasticity tensor of a deformable body. Assume that $\boldsymbol{C}(\boldsymbol{x})$ is a second-order mean-square continuous random field taking values in the space $\mathsf{S}^2(\mathsf{S}^2(V))$ of symmetric rank 2 tensors over $\mathsf{S}^2(V)$. Similar considerations prove that the field $\boldsymbol{C}(\boldsymbol{x})$ is homogeneous and isotropic with respect to the group $G = \mathrm{O}(E)$ and its orthogonal representation $\rho(g) = \mathsf{S}^2(\mathsf{S}^2(g))$.

The orbit types of the above action for the case of $d = 3$ are described in Forte & Vianello (1996). There are eight orbit types; we will discuss them in Section 3.8. Again, it is interesting to obtain a description of a homogeneous and (G, ρ)-isotropic random field.

We arrive at the following general formulation. Let V be a real finite-dimensional space of dimension d. We will be interested in the cases of $d = 2$ and $d = 3$. The former case is closely connected with *plane problems* of continuum physics, while the latter case is connected with its *space problems*. Let $x_1, \dots, x_d$ be the Cartesian coordinates in V, let

$$(\boldsymbol{x}, \boldsymbol{y}) = x_1 y_1 + \cdots + x_d y_d$$

be the standard inner product in $V = \mathbb{R}^d$, and let $\mathrm{O}(d)$ be the orthogonal group of the space $\mathbb{R}^d$.

Let G be a closed subgroup of the group $\mathrm{O}(d)$, and let ρ be an orthogonal representation of the group G in a real finite-dimensional space V. In physically interesting cases, V is a subspace of the tensor power $(\mathbb{R}^d)^{\otimes r}$. We would like to find the general form of the one- and two-point correlation tensors of a homogeneous and (G, ρ)-isotropic random field as well as its spectral expansion.

3.2 An Example

Before proving general theorems, consider a simple but non-trivial example that demonstrates our methods.

Example 23. Consider a mean-square continuous homogeneous random field $\boldsymbol{v}(\boldsymbol{x})\colon \mathbb{R}^2 \to \mathbb{R}^2$. Assume it is $(\mathrm{O}(2), \rho^1)$-isotropic; that is, for any $g \in \mathrm{O}(2)$, we have

$$\begin{aligned} \langle \boldsymbol{v}(g\boldsymbol{x}) \rangle &= g \langle \boldsymbol{v}(\boldsymbol{x}) \rangle, \\ \langle \boldsymbol{v}(g\boldsymbol{x}), \boldsymbol{v}(g\boldsymbol{y}) \rangle &= (g \otimes g) \langle \boldsymbol{v}(\boldsymbol{x}), \boldsymbol{v}(\boldsymbol{y}) \rangle. \end{aligned} \tag{3.1}$$

We have $\langle \boldsymbol{v}(g\boldsymbol{x})\rangle = \langle \boldsymbol{v}(\boldsymbol{x})\rangle$, because the one-point correlation tensor of a homogeneous random field is constant. Then we obtain $\langle \boldsymbol{v}(\boldsymbol{x})\rangle = g\langle \boldsymbol{v}(\boldsymbol{x})\rangle$ for all $g \in \mathrm{O}(2)$. The only vector which is not changing under rotations is $\mathbf{0}$. It follows that $\langle \boldsymbol{v}(\boldsymbol{x})\rangle = \mathbf{0}$.

By Theorem 10, the two-point correlation tensor of a homogeneous $\mathbb{C}^2$-*valued* random field has the form

$$\langle \boldsymbol{v}(\boldsymbol{x}), \boldsymbol{v}(\boldsymbol{y})\rangle = \int_{\hat{\mathbb{R}}^2} e^{\mathrm{i}(\boldsymbol{p},\boldsymbol{y}-\boldsymbol{x})}\,dF(\boldsymbol{p}), \tag{3.2}$$

where $F(\boldsymbol{p})$ is a measure on the *wavenumber domain* $\hat{\mathbb{R}}^2$ taking values in the set of Hermitian non-negative-definite matrices on $\mathbb{C}^2$. If the field takes values in $\mathbb{R}^2$, then

$$F(-A) = F^{\top}(A), \qquad A \in \mathfrak{B}(\hat{\mathbb{R}}^2), \tag{3.3}$$

where $-A = \{\,-\boldsymbol{p} \colon \boldsymbol{p} \in A\,\}$. Note that the set H of *all* Hermitian matrices on $\mathbb{C}^2$ is a four-dimensional *real* linear space.

Consider the group $Z_2 = \{I, -I\}$, where I is the 2×2 identity matrix, and its orthogonal representation ρ acting in H by

$$\rho(I)F = F, \quad \rho(-I)F = F^{\top}, \qquad F \in H.$$

Then Equation (3.3) is equivalent to

$$F(gA) = \rho(g)F(A), \qquad g \in Z_2.$$

On the other hand, it is easy to prove (see details in Section 3.3) that the second equation in (3.1) is equivalent to the following condition:

$$F(gA) = (g \otimes g)F(A), \qquad g \in \mathrm{O}(2). \tag{3.4}$$

The two last display formulae are compatible if and only if $\rho(g)$ and $g \otimes g$ are the same representations of the group Z_2. The transposition does not change the 2×2 symmetric part of a Hermitian matrix F, and multiplies by -1 its skew-symmetric part. In other words, the representation $\rho(g)$ is the direct sum $3A \oplus B$ of three copies of the trivial representation $A(\pm I) = 1$ of the group Z_2 acting in the three-dimensional linear space $\mathsf{S}^2(\mathbb{R}^2)$ of symmetric 2×2 matrices, and one copy of its non-trivial representation $B(\pm I) = \pm 1$ acting in the one-dimensional space $\Lambda^2(\mathbb{R}^2)$ of skew-symmetric 2×2 matrices.

On the other hand, the representation g of the group Z_2 is $2B$, the direct sum of two copies of B. Its tensor square is $4A \neq 3A \oplus B$. However, the restrictions of both representations to the space $\mathsf{S}^2(\mathbb{R}^2)$ are equal. It follows that the measure F in fact takes values in the space $\mathsf{S}^2(\mathbb{R}^2)$. Condition (3.4) takes the form

$$F(gA) = \mathsf{S}^2(g)F(A), \qquad g \in \mathrm{O}(2). \tag{3.5}$$

The next step is to write Equation (3.2) in the form

$$\langle \boldsymbol{v}(\boldsymbol{x}), \boldsymbol{v}(\boldsymbol{y})\rangle = \int_{\hat{\mathbb{R}}^2} e^{\mathrm{i}(\boldsymbol{p},\boldsymbol{y}-\boldsymbol{x})} f(\boldsymbol{p})\,d\mu(\boldsymbol{p}),$$

where $\mu(A) = \operatorname{tr} F(A)$, and where $f(\boldsymbol{p})$ is a measurable function on $\hat{\mathbb{R}}^2$ taking values in the set of symmetric non-negative-definite 2×2 matrices with unit trace. It is easy to see that condition (3.5) is equivalent to the following two conditions:

$$f(g\boldsymbol{p}) = \mathsf{S}^2(g)f(\boldsymbol{p}), \quad \mu(gA) = \mu(A), \qquad g \in O(2). \tag{3.6}$$

Let $(\lambda, \varphi_{\boldsymbol{p}})$ be the polar coordinates in the wavenumber domain. A measure μ is O(2)-invariant if and only if it has the form

$$d\mu = \frac{1}{2\pi} d\varphi_{\boldsymbol{p}}\, d\Phi(\lambda),$$

where Φ is a finite measure on $[0, \infty)$. The two-point correlation tensor of the field takes the form

$$\langle \boldsymbol{v}(\boldsymbol{x}), \boldsymbol{v}(\boldsymbol{y}) \rangle = \frac{1}{2\pi} \int_0^\infty \int_0^{2\pi} e^{\mathrm{i}(\boldsymbol{p}, \boldsymbol{y}-\boldsymbol{x})} f(\boldsymbol{p}) d\varphi_{\boldsymbol{p}}\, d\Phi(\lambda).$$

Consider a point $(\lambda, 0) \in \hat{\mathbb{R}}^2$ with $\lambda > 0$. The stationary subgroup of this point is the group O(1) that contains two elements:

$$E = \begin{pmatrix} 1 & 0 \\ 0 & 1 \end{pmatrix}, \qquad i = \begin{pmatrix} 1 & 0 \\ 0 & -1 \end{pmatrix}.$$

The representation $\mathsf{S}^2(g)$ of the group O(2) is the direct sum of two irreducible components. The first component is the trivial representation ρ^+ and acts in the one-dimensional space generated by the matrix

$$T^0 = \frac{1}{\sqrt{2}} I.$$

The second component is the representation ρ^2 (see Example 14). It acts in the two-dimensional space of symmetric traceless 2×2 matrices (deviators). The restriction of this representation to the subgroup $O(1)$ is as follows:

$$E \mapsto I, \qquad i \mapsto i,$$

that is, the direct sum $A_g \oplus A_u$ of the trivial representation, A_g, and the non-trivial, A_u. The matrix

$$T^{2,+} = \frac{1}{\sqrt{2}} \begin{pmatrix} -1 & 0 \\ 0 & 1 \end{pmatrix}$$

generates the space when the component A_g acts, indeed, $iT^{2,+}i^{-1} = T^{2,+}$. The matrix

$$T^{2,-} = \frac{1}{\sqrt{2}} \begin{pmatrix} 0 & 1 \\ -1 & 0 \end{pmatrix}$$

generates the space when the component A_u acts, indeed, $iT^{2,-}i^{-1} = -T^{2,-}$.

restriction of the function $f(\boldsymbol{p})$ to this stratum, say $f_1(\boldsymbol{p})$, takes values in the convex compact set $\mathcal{C}_1$. The 3×3 Hermitian non-negative-definite matrices with unit trace belonging to $\mathcal{C}_1$ satisfy the following conditions:

$$f^7(\boldsymbol{p}) = f^8(\boldsymbol{p}) = f^9(\boldsymbol{p}) = 0.$$

The contribution of this stratum to the two-point correlation tensor of the field is

$$\frac{1}{6}\sum_{m=0}^{5}\int_{(\hat{\mathbb{R}}^3/\mathcal{T}\times Z_2^c)_1} \operatorname{Re} j_m(\boldsymbol{p}, \boldsymbol{y}-\boldsymbol{x}) f_1^{Sm}(\boldsymbol{p})\,\mathrm{d}\Phi(\boldsymbol{p}).$$

Consider the stratum $(\hat{\mathbb{R}}^3/\mathcal{T}\times Z_2^c)_2$ with stationary subgroup $H_2 = Z_3$. The restrictions of A_g and A_u to H_2 are trivial. The restrictions of the remaining representations to H_2 do not contain trivial components. It follows that the restriction of the function $f(\boldsymbol{p})$ to this stratum, say $f_2(\boldsymbol{p})$, takes values in the convex compact set $\mathcal{C}_2$. The 3×3 Hermitian non-negative-definite matrices with unit trace belonging to $\mathcal{C}_2$ satisfy the following conditions:

$$f^3(\boldsymbol{p}) = f^4(\boldsymbol{p}) = f^5(\boldsymbol{p}) = f^6(\boldsymbol{p}) = f^8(\boldsymbol{p}) = f^9(\boldsymbol{p}) = 0.$$

The contribution of this stratum to the two-point correlation tensor of the field is

$$\begin{aligned}&\frac{1}{6}\sum_{m=0}^{5}\int_{(\hat{\mathbb{R}}^3/\mathcal{T}\times Z_2^c)_2} \operatorname{Re} j_m(\boldsymbol{p}, \boldsymbol{y}-\boldsymbol{x}) f_2^{Sm}(\boldsymbol{p})\,\mathrm{d}\Phi(\boldsymbol{p})\\ &\quad+\frac{1}{6}\sum_{m=0}^{5}\int_{(\hat{\mathbb{R}}^3/\mathcal{T}\times Z_2^c)_2} \operatorname{Im} j_m(\boldsymbol{p}, \boldsymbol{y}-\boldsymbol{x}) f_2^{Am}(\boldsymbol{p})\,\mathrm{d}\Phi(\boldsymbol{p}).\end{aligned}$$

Finally, consider the stratum $(\hat{\mathbb{R}}^3/\mathcal{T}\times Z_2^c)_3$ and $(\hat{\mathbb{R}}^3/\mathcal{T}\times Z_2^c)_0$ with stationary subgroups $H_3 = Z_2^-(\sigma_y)$ and $H_0 = \mathcal{T}\times Z_2^c$. The restrictions of A_g to both H_3 and H_0 are trivial. The restrictions of the remaining representations to both H_3 and H_0 do not contain trivial components. It follows that the restriction of the function $f(\boldsymbol{p})$ to this stratum, say $f_3(\boldsymbol{p})$, takes values in the convex compact set $\mathcal{C}_3$. The 3×3 Hermitian non-negative-definite matrices with unit trace belonging to $\mathcal{C}_3$ satisfy the following conditions:

$$f^3(\boldsymbol{p}) = f^4(\boldsymbol{p}) = \cdots = f^9(\boldsymbol{p}) = 0.$$

It is easy to see that $\mathcal{C}_3$ is an interval with extreme points $\boldsymbol{S}^1$ and $\frac{1}{\sqrt{2}}\boldsymbol{S}^2$. Then we have

$$f_3(\boldsymbol{p}) = C_1(\boldsymbol{p})\boldsymbol{S}^1 + C_2(\boldsymbol{p})\frac{1}{\sqrt{2}}\boldsymbol{S}^2,$$

where $C_1(\boldsymbol{p})$ and $C_2(\boldsymbol{p})$ are the barycentric coordinates of the point $f_3(\boldsymbol{p})$. Denote

$$\Phi_q(\boldsymbol{p}) = C_q(\boldsymbol{p})\,\mathrm{d}\Phi(\boldsymbol{p}), \qquad q = 1, 2.$$

These strata contribute to the two-point correlation tensor of the field by

$$\begin{aligned}&\frac{1}{6}\sum_{m=0}^{5}\int_{(\hat{\mathbb{R}}^3/\mathcal{T}\times Z_2^c)_{0,3}}\operatorname{Re}j_m(\boldsymbol{p},\boldsymbol{y}-\boldsymbol{x})\,\mathrm{d}\Phi_1(\boldsymbol{p})\boldsymbol{S}^1\\&+\frac{1}{6}\sum_{m=0}^{5}\int_{(\hat{\mathbb{R}}^3/\mathcal{T}\times Z_2^c)_{0,3}}\operatorname{Re}j_m(\boldsymbol{p},\boldsymbol{y}-\boldsymbol{x})\,\mathrm{d}\Phi_2(\boldsymbol{p})\frac{1}{\sqrt{2}}\boldsymbol{S}^2.\end{aligned}$$

Let $u_{0n}(\boldsymbol{p},\boldsymbol{x})$ (resp. $u_{1n}(\boldsymbol{p},\boldsymbol{x})$, resp. $u_{2n}(\boldsymbol{p},\boldsymbol{x})$, be different combinations of cosines and sines of p_1x_1, p_2x_2 and p_3x_3 (resp. p_1x_2, p_3x_1 and p_2x_3, resp. p_2x_1, p_1x_3 and p_3x_2). Enumerate them in such a way that the first four combinations contain either 3 or 1 cosine.

Combining everything together, we obtain the following result.

Theorem 31. *The one-point correlation tensor of a homogeneous and $(\mathcal{T}, A\oplus E)$-isotropic random field $\mathbf{e}(\boldsymbol{x})$ is*

$$\langle\mathbf{e}(\boldsymbol{x})\rangle=C\boldsymbol{T}^1,\qquad C\in\mathbb{R}.$$

Its two-point correlation tensor is:

$$\begin{aligned}\langle\mathbf{e}(\boldsymbol{x}),\mathbf{e}(\boldsymbol{y})\rangle&=\frac{1}{6}\sum_{m=0}^{2}\int_{(\hat{\mathbb{R}}^3/\mathcal{T}\times Z_2^c)_{4,5}}\operatorname{Re}j_m(\boldsymbol{p},\boldsymbol{y}-\boldsymbol{x})f_0^{Sm}(\boldsymbol{p})\,\mathrm{d}\Phi(\boldsymbol{p})\\&+\frac{1}{6}\sum_{m=0}^{2}\int_{(\hat{\mathbb{R}}^3/\mathcal{T}\times Z_2^c)_{4,5}}\operatorname{Im}j_m(\boldsymbol{p},\boldsymbol{y}-\boldsymbol{x})f_0^{\Lambda m}(\boldsymbol{p})\,\mathrm{d}\Phi(\boldsymbol{p})\\&+\frac{1}{6}\sum_{m=0}^{2}\int_{(\hat{\mathbb{R}}^3/\mathcal{T}\times Z_2^c)_{1}}\operatorname{Re}j_m(\boldsymbol{p},\boldsymbol{y}-\boldsymbol{x})f_1^{Sm}(\boldsymbol{p})\,\mathrm{d}\Phi(\boldsymbol{p})\\&+\frac{1}{6}\sum_{m=0}^{2}\int_{(\hat{\mathbb{R}}^3/\mathcal{T}\times Z_2^c)_{2}}\operatorname{Re}j_m(\boldsymbol{p},\boldsymbol{y}-\boldsymbol{x})f_2^{Sm}(\boldsymbol{p})\,\mathrm{d}\Phi(\boldsymbol{p})\\&+\frac{1}{6}\sum_{m=0}^{2}\int_{(\hat{\mathbb{R}}^3/\mathcal{T}\times Z_2^c)_{2}}\operatorname{Im}j_m(\boldsymbol{p},\boldsymbol{y}-\boldsymbol{x})f_2^{\Lambda m}(\boldsymbol{p})\,\mathrm{d}\Phi(\boldsymbol{p})\\&+\frac{1}{6}\sum_{m=0}^{2}\int_{(\hat{\mathbb{R}}^3/\mathcal{T}\times Z_2^c)_{0,3}}\operatorname{Re}j_m(\boldsymbol{p},\boldsymbol{y}-\boldsymbol{x})\,\mathrm{d}\Phi_1(\boldsymbol{p})\boldsymbol{S}^1\\&+\frac{1}{6}\sum_{m=0}^{2}\int_{(\hat{\mathbb{R}}^3/\mathcal{T}\times Z_2^c)_{0,3}}\operatorname{Re}j_m(\boldsymbol{p},\boldsymbol{y}-\boldsymbol{x})\,\mathrm{d}\Phi_2(\boldsymbol{p})\frac{1}{\sqrt{2}}\boldsymbol{S}^2.\end{aligned}$$

The field has the form:

$$\begin{aligned}\mathbf{e}(\boldsymbol{x})=C\boldsymbol{T}^1&+\frac{1}{\sqrt{6}}\sum_{m=0}^{2}\sum_{n=1}^{4}\int_{(\hat{\mathbb{R}}^3/\mathcal{T}\times Z_2^c)_{4,5}}u_{mn}(\boldsymbol{p},\boldsymbol{x})\,\mathrm{d}Z_{m0}^{Sn}(\boldsymbol{p})\boldsymbol{T}^{m+1}\\&+\frac{1}{\sqrt{6}}\sum_{m=0}^{2}\sum_{n=5}^{8}\int_{(\hat{\mathbb{R}}^3/\mathcal{T}\times Z_2^c)_{4,5}}u_{mn}(\boldsymbol{p},\boldsymbol{x})\,\mathrm{d}Z_{m0}^{\Lambda n}(\boldsymbol{p})\boldsymbol{T}^{m+1}\end{aligned}$$

restriction of the function $f(\boldsymbol{p})$ to this stratum, say $f_1(\boldsymbol{p})$, takes values in the convex compact set $\mathcal{C}_1$. The 3×3 Hermitian non-negative-definite matrices with unit trace belonging to $\mathcal{C}_1$ satisfy the following conditions:

$$f^7(\boldsymbol{p}) = f^8(\boldsymbol{p}) = f^9(\boldsymbol{p}) = 0.$$

The contribution of this stratum to the two-point correlation tensor of the field is

$$\frac{1}{6}\sum_{m=0}^{5}\int_{(\hat{\mathbb{R}}^3/\mathcal{T}\times Z_2^c)_1} \operatorname{Re} j_m(\boldsymbol{p}, \boldsymbol{y}-\boldsymbol{x}) f_1^{Sm}(\boldsymbol{p})\,\mathrm{d}\Phi(\boldsymbol{p}).$$

Consider the stratum $(\hat{\mathbb{R}}^3/\mathcal{T} \times Z_2^c)_2$ with stationary subgroup $H_2 = Z_3$. The restrictions of A_g and A_u to H_2 are trivial. The restrictions of the remaining representations to H_2 do not contain trivial components. It follows that the restriction of the function $f(\boldsymbol{p})$ to this stratum, say $f_2(\boldsymbol{p})$, takes values in the convex compact set $\mathcal{C}_2$. The 3×3 Hermitian non-negative-definite matrices with unit trace belonging to $\mathcal{C}_2$ satisfy the following conditions:

$$f^3(\boldsymbol{p}) = f^4(\boldsymbol{p}) = f^5(\boldsymbol{p}) = f^6(\boldsymbol{p}) = f^8(\boldsymbol{p}) = f^9(\boldsymbol{p}) = 0.$$

The contribution of this stratum to the two-point correlation tensor of the field is

$$\begin{aligned}&\frac{1}{6}\sum_{m=0}^{5}\int_{(\hat{\mathbb{R}}^3/\mathcal{T}\times Z_2^c)_2} \operatorname{Re} j_m(\boldsymbol{p}, \boldsymbol{y}-\boldsymbol{x}) f_2^{Sm}(\boldsymbol{p})\,\mathrm{d}\Phi(\boldsymbol{p})\\ &\quad+\frac{1}{6}\sum_{m=0}^{5}\int_{(\hat{\mathbb{R}}^3/\mathcal{T}\times Z_2^c)_2} \operatorname{Im} j_m(\boldsymbol{p}, \boldsymbol{y}-\boldsymbol{x}) f_2^{Am}(\boldsymbol{p})\,\mathrm{d}\Phi(\boldsymbol{p}).\end{aligned}$$

Finally, consider the stratum $(\hat{\mathbb{R}}^3/\mathcal{T} \times Z_2^c)_3$ and $(\hat{\mathbb{R}}^3/\mathcal{T} \times Z_2^c)_0$ with stationary subgroups $H_3 = Z_2^-(\sigma_y)$ and $H_0 = \mathcal{T} \times Z_2^c$. The restrictions of A_g to both H_3 and H_0 are trivial. The restrictions of the remaining representations to both H_3 and H_0 do not contain trivial components. It follows that the restriction of the function $f(\boldsymbol{p})$ to this stratum, say $f_3(\boldsymbol{p})$, takes values in the convex compact set $\mathcal{C}_3$. The 3×3 Hermitian non-negative-definite matrices with unit trace belonging to $\mathcal{C}_3$ satisfy the following conditions:

$$f^3(\boldsymbol{p}) = f^4(\boldsymbol{p}) = \cdots = f^9(\boldsymbol{p}) = 0.$$

It is easy to see that $\mathcal{C}_3$ is an interval with extreme points $\boldsymbol{S}^1$ and $\frac{1}{\sqrt{2}}\boldsymbol{S}^2$. Then we have

$$f_3(\boldsymbol{p}) = C_1(\boldsymbol{p})\boldsymbol{S}^1 + C_2(\boldsymbol{p})\frac{1}{\sqrt{2}}\boldsymbol{S}^2,$$

where $C_1(\boldsymbol{p})$ and $C_2(\boldsymbol{p})$ are the barycentric coordinates of the point $f_3(\boldsymbol{p})$. Denote

$$\Phi_q(\boldsymbol{p}) = C_q(\boldsymbol{p})\,\mathrm{d}\Phi(\boldsymbol{p}), \qquad q = 1, 2.$$

These strata contribute to the two-point correlation tensor of the field by

$$\frac{1}{6}\sum_{m=0}^{5}\int_{(\hat{\mathbb{R}}^3/\mathcal{T}\times Z_2^c)_{0,3}}\operatorname{Re} j_m(\boldsymbol{p},\boldsymbol{y}-\boldsymbol{x})\,\mathrm{d}\Phi_1(\boldsymbol{p})\mathsf{S}^1 + \frac{1}{6}\sum_{m=0}^{5}\int_{(\hat{\mathbb{R}}^3/\mathcal{T}\times Z_2^c)_{0,3}}\operatorname{Re} j_m(\boldsymbol{p},\boldsymbol{y}-\boldsymbol{x})\,\mathrm{d}\Phi_2(\boldsymbol{p})\frac{1}{\sqrt{2}}\mathsf{S}^2.$$

Let $u_{0n}(\boldsymbol{p},\boldsymbol{x})$ (resp. $u_{1n}(\boldsymbol{p},\boldsymbol{x})$, resp. $u_{2n}(\boldsymbol{p},\boldsymbol{x})$, be different combinations of cosines and sines of p_1x_1, p_2x_2 and p_3x_3 (resp. p_1x_2, p_3x_1 and p_2x_3, resp. p_2x_1, p_1x_3 and p_3x_2). Enumerate them in such a way that the first four combinations contain either 3 or 1 cosine.

Combining everything together, we obtain the following result.

Theorem 31. *The one-point correlation tensor of a homogeneous and $(\mathcal{T}, A\oplus E)$-isotropic random field $\mathsf{e}(\boldsymbol{x})$ is*

$$\langle\mathsf{e}(\boldsymbol{x})\rangle = C\mathsf{T}^1,\qquad C\in\mathbb{R}.$$

Its two-point correlation tensor is:

$$\begin{aligned}\langle\mathsf{e}(\boldsymbol{x}),\mathsf{e}(\boldsymbol{y})\rangle &= \frac{1}{6}\sum_{m=0}^{2}\int_{(\hat{\mathbb{R}}^3/\mathcal{T}\times Z_2^c)_{4,5}}\operatorname{Re} j_m(\boldsymbol{p},\boldsymbol{y}-\boldsymbol{x})f_0^{\mathsf{S}m}(\boldsymbol{p})\,\mathrm{d}\Phi(\boldsymbol{p})\\ &+\frac{1}{6}\sum_{m=0}^{2}\int_{(\hat{\mathbb{R}}^3/\mathcal{T}\times Z_2^c)_{4,5}}\operatorname{Im} j_m(\boldsymbol{p},\boldsymbol{y}-\boldsymbol{x})f_0^{\Lambda m}(\boldsymbol{p})\,\mathrm{d}\Phi(\boldsymbol{p})\\ &+\frac{1}{6}\sum_{m=0}^{2}\int_{(\hat{\mathbb{R}}^3/\mathcal{T}\times Z_2^c)_{1}}\operatorname{Re} j_m(\boldsymbol{p},\boldsymbol{y}-\boldsymbol{x})f_1^{\mathsf{S}m}(\boldsymbol{p})\,\mathrm{d}\Phi(\boldsymbol{p})\\ &+\frac{1}{6}\sum_{m=0}^{2}\int_{(\hat{\mathbb{R}}^3/\mathcal{T}\times Z_2^c)_{2}}\operatorname{Re} j_m(\boldsymbol{p},\boldsymbol{y}-\boldsymbol{x})f_2^{\mathsf{S}m}(\boldsymbol{p})\,\mathrm{d}\Phi(\boldsymbol{p})\\ &+\frac{1}{6}\sum_{m=0}^{2}\int_{(\hat{\mathbb{R}}^3/\mathcal{T}\times Z_2^c)_{2}}\operatorname{Im} j_m(\boldsymbol{p},\boldsymbol{y}-\boldsymbol{x})f_2^{\Lambda m}(\boldsymbol{p})\,\mathrm{d}\Phi(\boldsymbol{p})\\ &+\frac{1}{6}\sum_{m=0}^{2}\int_{(\hat{\mathbb{R}}^3/\mathcal{T}\times Z_2^c)_{0,3}}\operatorname{Re} j_m(\boldsymbol{p},\boldsymbol{y}-\boldsymbol{x})\,\mathrm{d}\Phi_1(\boldsymbol{p})\mathsf{S}^1\\ &+\frac{1}{6}\sum_{m=0}^{2}\int_{(\hat{\mathbb{R}}^3/\mathcal{T}\times Z_2^c)_{0,3}}\operatorname{Re} j_m(\boldsymbol{p},\boldsymbol{y}-\boldsymbol{x})\,\mathrm{d}\Phi_2(\boldsymbol{p})\frac{1}{\sqrt{2}}\mathsf{S}^2.\end{aligned}$$

The field has the form:

$$\begin{aligned}\mathsf{e}(\boldsymbol{x}) = C\mathsf{T}^1 &+ \frac{1}{\sqrt{6}}\sum_{m=0}^{2}\sum_{n=1}^{4}\int_{(\hat{\mathbb{R}}^3/\mathcal{T}\times Z_2^c)_{4,5}} u_{mn}(\boldsymbol{p},\boldsymbol{x})\,\mathrm{d}Z_{m0}^{\mathsf{S}n}(\boldsymbol{p})\mathsf{T}^{m+1}\\ &+\frac{1}{\sqrt{6}}\sum_{m=0}^{2}\sum_{n=5}^{8}\int_{(\hat{\mathbb{R}}^3/\mathcal{T}\times Z_2^c)_{4,5}} u_{mn}(\boldsymbol{p},\boldsymbol{x})\,\mathrm{d}Z_{m0}^{\Lambda n}(\boldsymbol{p})\mathsf{T}^{m+1}\end{aligned}$$

$$
\begin{aligned}
&+\frac{1}{\sqrt{6}}\sum_{m=0}^{2}\sum_{n=1}^{8}\int_{(\hat{\mathbb{R}}^3/\mathcal{T}\times Z_2^c)_1} u_{mn}(\boldsymbol{p},\boldsymbol{x})\,\mathrm{d}Z^{Sn}_{m1}(\boldsymbol{p})\boldsymbol{T}^{m+1}\\
&+\frac{1}{\sqrt{6}}\sum_{m=0}^{2}\sum_{n=1}^{4}\int_{(\hat{\mathbb{R}}^3/\mathcal{T}\times Z_2^c)_2} u_{mn}(\boldsymbol{p},\boldsymbol{x})\,\mathrm{d}Z^{Sn}_{m2}(\boldsymbol{p})\boldsymbol{T}^{m+1}\\
&+\frac{1}{\sqrt{6}}\sum_{m=0}^{2}\sum_{n=5}^{8}\int_{(\hat{\mathbb{R}}^3/\mathcal{T}\times Z_2^c)_2} u_{mn}(\boldsymbol{p},\boldsymbol{x})\,\mathrm{d}Z^{\Lambda n}_{m2}(\boldsymbol{p})\boldsymbol{T}^{m+1}\\
&+\frac{1}{\sqrt{6}}\sum_{m=0}^{2}\sum_{n=1}^{8}\int_{(\hat{\mathbb{R}}^3/\mathcal{T}\times Z_2^c)_{0,3}} u_{mn}(\boldsymbol{p},\boldsymbol{x})\,\mathrm{d}Z^{Sn}_{m3}(\boldsymbol{p})\boldsymbol{T}^{m+1}\\
&+\frac{1}{\sqrt{6}}\sum_{m=0}^{2}\sum_{n=1}^{8}\int_{(\hat{\mathbb{R}}^3/\mathcal{T}\times Z_2^c)_{0,3}} u_{mn}(\boldsymbol{p},\boldsymbol{x})\,\mathrm{d}Z^{\Lambda n}_{m4}(\boldsymbol{p})\boldsymbol{T}^{m+1},
\end{aligned}
$$

where

$$
\begin{aligned}
\boldsymbol{Z}^{Sn}_0(\boldsymbol{p})&=(Z^{Sn}_{00}(\boldsymbol{p}),\dots,Z^{Sn}_{20}(\boldsymbol{p}))^\top,\\
\boldsymbol{Z}^{\Lambda n}_0(\boldsymbol{p})&=(Z^{\Lambda n}_{00}(\boldsymbol{p}),\dots,Z^{\Lambda n}_{20}(\boldsymbol{p}))^\top,\\
\boldsymbol{Z}^{Sn}_1(\boldsymbol{p})&=(Z^{Sn}_{01}(\boldsymbol{p}),\dots,Z^{Sn}_{21}(\boldsymbol{p}))^\top,\\
\boldsymbol{Z}^{Sn}_2(\boldsymbol{p})&=(Z^{Sn}_{02}(\boldsymbol{p}),\dots,Z^{Sn}_{22}(\boldsymbol{p}))^\top,\\
\boldsymbol{Z}^{\Lambda n}_2(\boldsymbol{p})&=(Z^{\Lambda n}_{02}(\boldsymbol{p}),\dots,Z^{\Lambda n}_{22}(\boldsymbol{p}))^\top,\\
\boldsymbol{Z}^{Sn}_3(\boldsymbol{p})&=(Z^{Sn}_{03}(\boldsymbol{p}),\dots,Z^{Sn}_{23}(\boldsymbol{p}))^\top,\\
\boldsymbol{Z}^{Sn}_4(\boldsymbol{p})&=(Z^{Sn}_{04}(\boldsymbol{p}),\dots,Z^{Sn}_{24}(\boldsymbol{p}))^\top
\end{aligned}
$$

are centred random measures on the corresponding sets with the corresponding control measures and non-zero cross-correlations similar to (3.86)*:*

$$
\mathsf{E}[\boldsymbol{Z}^{Sn}_i(A)\otimes\boldsymbol{Z}^{\Lambda n}_i(B)]=\int_{A\cap B} f^{\Lambda m}_i(\boldsymbol{p})\,\mathrm{d}\Phi(\boldsymbol{p})
$$

for $i=0,\ 2$.

Proof. The spectral expansion of the field follows from Karhunen's theorem. □

Similar cases, when $j=10$, 15, 27 and 43, may be left to the reader.

Consider a new case, when $j=5$. The representation $\rho=10A'$ is trivial, but the group $G=Z_2^-$ is of type III. We have $\pi(G)=Z_2$, $\rho^\pi(g)=10A$, the trivial representation of Z_2. The representation $\widehat{\mathsf{S}^2(\rho^\pi)}$ is trivial on $\pi(G)\cap G$, that is, on the identity matrix, and trivial times -1 on $G\setminus\pi(G)$, that is, on the non-identity element of Z_2^-. This gives $\widehat{\mathsf{S}^2(\rho^\pi)}=45B$, where B is the non-trivial irreducible representation of Z_2^-. The representation $\tilde{\rho}$ of the group $\tilde{G}=Z_2\times Z_2^c$ is

$$
\tilde{\rho}=55A\hat{\otimes}A_g\oplus 45B\hat{\otimes}A_u=55A_g\oplus 45B_u.
$$

The representation A_g is trivial. Under the action of $\tilde{\rho}$, the matrix $f^S(\boldsymbol{p})$ does not change. The operators $45B_u(E)$ and $45B_u(\sigma_h)$ do not change $f^\Lambda(\boldsymbol{p})$, while the operators $45B_u(C_2)$ and $45B_u(i)$ multiply it by -1, see Altmann & Herzig (1994, Table 60.4). The corresponding parts of the spherical Bessel function are

$$\begin{aligned} j_1(\boldsymbol{p}, \boldsymbol{x}) &= \frac{1}{4}(\mathrm{e}^{\mathrm{i}(E\boldsymbol{p},\boldsymbol{x})} + \mathrm{e}^{\mathrm{i}(\sigma_h\boldsymbol{p},\boldsymbol{x})}) = \frac{1}{2}\mathrm{e}^{\mathrm{i}(p_1x_1+p_2x_2)}\cos(p_3x_3), \\ j_2(\boldsymbol{p}, \boldsymbol{x}) &= \frac{1}{4}(\mathrm{e}^{\mathrm{i}(C_2\boldsymbol{p},\boldsymbol{x})} + \mathrm{e}^{\mathrm{i}(i\boldsymbol{p},\boldsymbol{x})}) = \frac{1}{2}\mathrm{e}^{-\mathrm{i}(p_1x_1+p_2x_2)}\cos(p_3x_3). \end{aligned}$$

This result was obtained as follows. The group $\tilde{G}$ is a subgroup of the group $\mathcal{O} \times Z_2^c$. The matrix entries of the orthogonal representation T_{1u} of this group have been calculated while proving Theorem 22. The elements of $\tilde{G}$ act on the vector $\boldsymbol{p}$ by the corresponding matrices.

The structure of the orbit space $\hat{\mathbb{R}}^3/\tilde{G}$ is given by Equation (3.34). The restriction of the representation $55A_g$ to any stationary subgroup is trivial. The restrictions of the representation $45B_u$ to all stationary subgroups except Z_2 and $Z_2 \times Z_2^c$ are trivial. For Z_2, this restriction is $45B$, while for $Z_2 \times Z_2^c$ it is $45B_u$ itself. Both latter restrictions do not contain trivial components. The two-point correlation tensor takes the form

$$\begin{aligned} \langle \boldsymbol{e}(\boldsymbol{x}), \boldsymbol{e}(\boldsymbol{y}) \rangle &= \int_{\hat{\mathbb{R}}^3/Z_2 \times Z_2^c} \cos(p_1z_1 + p_2z_2)\cos(p_3z_3) f^S(\boldsymbol{p})\, \mathrm{d}\Phi(\boldsymbol{p}) \\ &\quad + \int_{(\hat{\mathbb{R}}^3/Z_2 \times Z_2^c)_{1-3,6,7}} \sin(p_1z_1 + p_2z_2)\cos(p_3z_3) f^A(\boldsymbol{p})\, \mathrm{d}\Phi(\boldsymbol{p}). \end{aligned} \tag{3.89}$$

We proceed to find an orthonormal basis in the space $\mathsf{V}^{Z_2^-}$ that respects the representation ρ. By Altmann & Herzig (1994, Table 12.5), the restriction of the representation T_{1u} of the group $\mathcal{O} \times Z_2^c$ to the subgroup Z_2^- is $2A' \oplus A''$. Moreover, two copies of the trivial representation A' act on the x- and y-axes, while the non-trivial representation A'' acts on the z-axis. The symmetric tensor square of this representation is

$$\mathsf{S}^2(2A' \oplus A'') = 4A' \oplus 2A''.$$

Moreover, the four trivial components act in the one-dimensional spaces generated by the matrices $T^1 = \begin{pmatrix} 1&0&0 \\ 0&0&0 \\ 0&0&0 \end{pmatrix}$, $T^2 = \begin{pmatrix} 0&0&0 \\ 0&1&0 \\ 0&0&0 \end{pmatrix}$, $T^3 = \begin{pmatrix} 0&0&0 \\ 0&0&0 \\ 0&0&1 \end{pmatrix}$ and $T^4 = \frac{1}{\sqrt{2}}\begin{pmatrix} 0&0&1 \\ 0&0&0 \\ 1&0&0 \end{pmatrix}$. The two non-trivial components act in the spaces generated by the matrices $T^5 = \frac{1}{\sqrt{2}}\begin{pmatrix} 0&1&0 \\ 1&0&0 \\ 0&0&0 \end{pmatrix}$ and $T^6 = \frac{1}{\sqrt{2}}\begin{pmatrix} 0&0&0 \\ 0&0&1 \\ 0&1&0 \end{pmatrix}$. After tensor multiplication by $2A' \oplus A''$, the 10 copies of the irreducible trivial representation A' act in the spaces generated by the rank 3 tensors

$$\mathsf{T}^m = \begin{cases} T^m \otimes (1,0,0)^\top, & \text{if } 1 \le m \le 4, \\ T^{m-4} \otimes (0,0,1)^\top, & \text{if } 5 \le m \le 8, \\ T^{m-4} \otimes (0,1,0)^\top, & \text{otherwise.} \end{cases}$$

Let $u_n(\boldsymbol{p},\boldsymbol{x})$ be four different combinations of sines and cosines of $p_1x_1+p_2x_2$ and p_3x_3. Enumerate them in such order that the third and fourth combinations contain both sines and cosines.

Theorem 32. *The one-point correlation tensor of a homogeneous and $(Z_2^-, 10A')$-isotropic random field $\mathsf{e}(\boldsymbol{x})$ has the form*

$$\langle \mathsf{e}(\boldsymbol{x})\rangle = \sum_{m=1}^{10} C_m \boldsymbol{T}^m.$$

Its two-point correlation tensor has the form (3.89)*. The field has the form*

$$\begin{aligned}\mathsf{e}(\boldsymbol{x}) &= \sum_{m=1}^{10} C_m \boldsymbol{T}^m + \sum_{n=1}^{2}\int_{\hat{\mathbb{R}}^3/Z_2\times Z_2^c} u_n(\boldsymbol{p},\boldsymbol{x})\,\mathrm{d}Z_m^{Sn}(\boldsymbol{p})\boldsymbol{T}^m\\ &\quad+\sum_{n=3}^{4}\int_{(\hat{\mathbb{R}}^3/Z_2\times Z_2^c)1-3,6,7} u_n(\boldsymbol{p},\boldsymbol{x})\,\mathrm{d}Z_m^{\Lambda n}(\boldsymbol{p})\boldsymbol{T}^m,\end{aligned}$$

where

$$\begin{aligned}\boldsymbol{Z}_m^S(\boldsymbol{p}) &= (Z^{S1}(\boldsymbol{p}),\dots,Z^{S4}(\boldsymbol{p}))^\top,\\ \boldsymbol{Z}_m^\Lambda(\boldsymbol{p}) &= (Z^{\Lambda 1}(\boldsymbol{p}),\dots,Z^{\Lambda 4}(\boldsymbol{p}))^\top\end{aligned}$$

are uncorrelated random measures with control measure Φ and cross-correlation (3.86)*.*

Proof. As usual, the last part follows from Karhunen's theorem. □

The cases of $j = 19$, 24, 31, 36, 38, 40, 48 are similar and may be left to the reader.

Consider the case when G is of type III and ρ is non-trivial. Put $j = 34$. That is, $G = D_6^h$ and $\rho = 3A_1 \oplus B_1$. Then $\tilde{\rho} = D_6 \times Z_2^c$. By Lemma 1, we have

$$\tilde{\rho} = \mathsf{S}^2(\rho^\pi)\hat{\otimes}(10A_g) \oplus \widehat{\Lambda^2(\rho^\pi)}\hat{\otimes}(6A_u).$$

Furthermore, $\pi(G) = D_6$ and $\rho^\pi = 3A_1 \oplus B_2$. Then we obtain $\mathsf{S}^2(\rho^\pi) = 7A_1 \oplus 3B_2$ and $\Lambda^2(\rho^\pi) = 3A_1 \oplus 3B_2$. Because $\pi(G) \cap G = Z_6$, the representation $\widehat{\Lambda^2(\rho^\pi)}$ is equal to $3A_1 \oplus 3B_2$ on Z_6 and $-(3A_1 \oplus 3B_2)$ on $D_6^h \backslash Z_6$. that is, $\widehat{\Lambda^2(\rho^\pi)} = 3A_2 \oplus B_1$. Finally,

$$\tilde{\rho} = (7A_{1g} \oplus 3B_{2g}) \oplus (3A_{2u} \oplus B_{1u}).$$

The next step is to define the basis in the space $\mathsf{V}^{D_3^v}$. The restriction of the representation ρ_1 of the group O(3) to the subgroup D_3^v is $A_1 \oplus E$ by Altmann & Herzig (1994, Table 51.5A or 51.5B), where the representation A_1 acts on the z-axis and the representation E acts in the xy-plane. By Altmann & Herzig (1994, Table 51.8), the symmetric tensor square of the representation $A_1 \oplus E$ is $\mathsf{S}^2(A_1 \oplus E) = 2A_1 \oplus 2E$. The first copy of A_1 acts in the one-dimensional space generated by the matrix $T^1 = \begin{pmatrix}0&0&0\\0&1&0\\0&0&0\end{pmatrix}$, the second copy in the space generated by the matrix $T^2 = \frac{1}{\sqrt{2}}\begin{pmatrix} 1 & 0 & 0 \\ 0 & 0 & 0 \\ 0 & 0 & 1 \end{pmatrix}$. The first copy of E acts in the two-dimensional space generated by the matrices $T^3 = \frac{1}{\sqrt{2}}\begin{pmatrix} 0 & 1 & 0 \\ 1 & 0 & 0 \\ 0 & 0 & 0 \end{pmatrix}$ and $T^4 = \begin{pmatrix} 0 & 0 & 0 \\ 0 & 0 & 1 \\ 0 & 1 & 0 \end{pmatrix}$, the second copy in the space generated by the matrices $T^5 = \frac{1}{\sqrt{2}}\begin{pmatrix} -1 & 0 & 0 \\ 0 & 0 & 0 \\ 0 & 0 & 1 \end{pmatrix}$ and $T^6 = \frac{1}{\sqrt{2}}\begin{pmatrix} 0 & 0 & 1 \\ 0 & 0 & 0 \\ 1 & 0 & 0 \end{pmatrix}$. After tensor multiplication by $A_1 \oplus E$, we obtain four copies of the trivial representation A_1 acting in the one-dimensional spaces generated by the rank 3 tensors $\boldsymbol{T}^1 = T^1 \otimes (0,1,0)^\top$, $\boldsymbol{T}^2 = T^2 \otimes (0,1,0)^\top$, $\boldsymbol{T}^3 = \frac{1}{\sqrt{2}}(T^3 \otimes (1,0,0)^\top + T^4 \otimes (0,0,1)^\top$ and $\boldsymbol{T}^4 = \frac{1}{\sqrt{2}}(T^5 \otimes (1,0,0)^\top + T^6 \otimes (0,0,1)^\top$.

On the other hand, by Altmann & Herzig (1994, Table 54.5), the restriction of the representation ρ_1 of the group O(3) to the subgroup D_6^v is $A_1 \oplus E_1$, where the representation A_1 acts on the z-axis and the representation E_1 acts in the xy-plane. By Altmann & Herzig (1994, Table 54.8), The symmetric tensor square of the representation $A_1 \oplus E_1$ is $\mathsf{S}^2(A_1 \oplus E_1) = 2A_1 \oplus E_1 \oplus E_2$. The first copy of A_1 acts in the one-dimensional space generated by the matrix T^1, the second copy in the space generated by the matrix T^2. The copy of E_1 acts in the two-dimensional space generated by the matrices T^3 and T^4, while the copy of E_2 acts in the space generated by the matrices T^5 and T^6. After tensor multiplication by $A_1 \oplus E_1$, we obtain three copies of the trivial representation A_1 acting in the one-dimensional spaces generated by the rank 3 tensors $\boldsymbol{T}^1$, $\boldsymbol{T}^2$ and $\boldsymbol{T}^3$, and a copy of B_1 acting in the space generated by $\boldsymbol{T}^4$.

Next, we divide the group $\tilde{G} = D_6 \times Z_2^c$ into four subsets:

$$\begin{aligned} G_1 &= \{E, C_3^+, C_3^-, \sigma_{v1}, \sigma_{v2}, \sigma_{v3}\}, \\ G_2 &= \{C_6^+, C_6^-, C_2, \sigma_{d1}, \sigma_{d2}, \sigma_{d3}\}, \\ G_3 &= \{C_{21}'', C_{22}'', C_{23}'', i, S_6^-, S_6^+\}, \\ G_4 &= \{C_{21}', C_{22}', C_{23}', S_3^-, S_3^+, \sigma_h\}. \end{aligned}$$

The four irreducible components A_{1g}, B_{2g}, $3A_{2u}$ and B_{1u} of the representation $\tilde{\rho}$ take the same values on the elements of each group. We write the spherical Bessel function in the form

$$j(\boldsymbol{p}, \boldsymbol{y} - \boldsymbol{x}) = \frac{1}{24} \sum_{m=1}^{4} \sum_{g \in G_m} e^{i(g\boldsymbol{p}, \boldsymbol{y} - \boldsymbol{x})}.$$

and calculate the matrices of the representation $A_{2u} \oplus E_{1u}$, using the Euler angles from Altmann & Herzig (1994, Table 35.1). The above representation is the restriction of the representation ρ_1 of the group O(3) to the subgroup $\tilde{G}$ by Altmann & Herzig (1994, Table 35.10). Then, we calculate the functions

$$j_m(\boldsymbol{p}, \boldsymbol{y} - \boldsymbol{x}) = \sum_{g \in G_m} e^{i(g\boldsymbol{p}, \boldsymbol{y} - \boldsymbol{x})}$$

and obtain

$$\begin{aligned}
j_1(\boldsymbol{p}, \boldsymbol{y}-\boldsymbol{x}) &= 2\exp(\mathrm{i}(p_1z_1+p_3z_3))\cos(p_2z_2)\\
&\quad+2\exp\left(\mathrm{i}\left(\frac{p_1}{2}(-z_1+\sqrt{3}z_2)+p_3z_3\right)\right)\cos\left(\frac{p_2}{2}(\sqrt{3}z_1+z_2)\right)\\
&\quad+2\exp\left(\mathrm{i}\left(\frac{p_1}{2}(-z_1-\sqrt{3}z_2)+p_3z_3\right)\right)\cos\left(\frac{p_2}{2}(\sqrt{3}z_1-z_2)\right),\\
j_2(\boldsymbol{p}, \boldsymbol{y}-\boldsymbol{x}) &= 2\exp(\mathrm{i}(-p_1z_1+p_3z_3))\cos(p_2z_2)\\
&\quad+2\exp\left(\mathrm{i}\left(\frac{p_1}{2}z_1+\frac{p_2}{2}z_2+p_3z_3\right)\right)\cos\left(\frac{p_1}{2}\sqrt{3}z_2-\frac{p_2}{2}\sqrt{3}z_1\right)\\
&\quad+2\exp\left(\mathrm{i}\left(\frac{p_1}{2}z_1-\frac{p_2}{2}z_2+p_3z_3\right)\right)\cos\left(\frac{p_1}{2}\sqrt{3}z_2+\frac{p_2}{2}\sqrt{3}z_1\right),
\end{aligned}$$

and $j_{m+2}(\boldsymbol{p}, \boldsymbol{y}-\boldsymbol{x}) = \overline{j_m(\boldsymbol{p}, \boldsymbol{y}-\boldsymbol{x})}$ for $m = 1, 2$.

Define the matrices $f_0^S(\boldsymbol{p})$ and $f_0^A(\boldsymbol{p})$ by Equation (3.88). The restrictions of all four irreducible components A_{1g}, B_{2g}, A_{2u} and B_{1u} to the stationary subgroups of the strata $(\hat{\mathbb{R}}^3/D_6\times Z_2^c)_5$ and $(\hat{\mathbb{R}}^3/D_6\times Z_2^c)_7$ are trivial. Exactly as in Lemma 2, we prove that the contribution of the above strata to the two-point correlation tensor of the field is as follows:

$$\begin{aligned}
&\frac{1}{2}\sum_{m=1}^{4}\int_{(\hat{\mathbb{R}}^3/D_6\times Z_2^c)_{5,7}}\operatorname{Re} j_m(\boldsymbol{p}, \boldsymbol{y}-\boldsymbol{x}) f_0^{Sm}(\boldsymbol{p})\,\mathrm{d}\Phi(\boldsymbol{p})\\
&\quad+\frac{1}{2}\sum_{m=1}^{4}\int_{(\hat{\mathbb{R}}^3/D_6\times Z_2^c)_{5,7}}\operatorname{Im} j_m(\boldsymbol{p}, \boldsymbol{y}-\boldsymbol{x}) f_0^{Am}(\boldsymbol{p})\,\mathrm{d}\Phi(\boldsymbol{p}),
\end{aligned}$$

where $f_0^{Sm}(\boldsymbol{p})$ (resp. $f_0^{Am}(\boldsymbol{p})$) is the result of action of the operator $\tilde{\rho}(h)$ on the matrix $f_0^S(\boldsymbol{p})$ (resp. $f_0^A(\boldsymbol{p})$) with $h \in G_m$. By Altmann & Herzig (1994, Table 35.4), $f_0^{S1}(\boldsymbol{p}) = f_0^S(\boldsymbol{p})$ and $f_0^{A1}(\boldsymbol{p}) = f_0^A(\boldsymbol{p})$, any element of G_2 multiplies the matrix entries

$$\begin{aligned}
&(f_0^S)_{14}(\boldsymbol{p}), (f_0^S)_{24}(\boldsymbol{p}), (f_0^S)_{34}(\boldsymbol{p}), (f_0^S)_{41}(\boldsymbol{p}), (f_0^S)_{42}(\boldsymbol{p}), (f_0^S)_{43}(\boldsymbol{p}),\\
&(f_0^A)_{14}(\boldsymbol{p}), (f_0^A)_{24}(\boldsymbol{p}), (f_0^A)_{34}(\boldsymbol{p}), (f_0^A)_{41}(\boldsymbol{p}), (f_0^A)_{42}(\boldsymbol{p}), (f_0^A)_{43}(\boldsymbol{p})
\end{aligned}$$

by -1, any element of G_{m+2} multiplies the matrix $f_0^{Am}(\boldsymbol{p})$ by -1, $m = 1$, 2. The matrix $f_0(\boldsymbol{p})$ takes values in the convex compact set $\mathcal{C}_0$ of Hermitian non-negative-definite 4×4 matrices with unit trace.

The restrictions of the irreducible components A_{1g} and A_{2u} to the stationary subgroup of the stratum $(\hat{\mathbb{R}}^3/D_6\times Z_2^c)_6$ are trivial, while the restrictions of the two remaining components are not. The set $\mathcal{C}_1$ is the set of Hermitian non-negative-definite 4×4 matrices with unit trace $f_1(\boldsymbol{p})$ with the following restrictions:

$$\begin{aligned}
(f_1)_{14}(\boldsymbol{p}) &= (f_1)_{24}(\boldsymbol{p}) = (f_1)_{34}(\boldsymbol{p})\\
&= (f_1)_{41}(\boldsymbol{p}) = (f_1)_{42}(\boldsymbol{p}) = (f_1)_{43}(\boldsymbol{p}) = 0.
\end{aligned}$$

The stratum $(\hat{\mathbb{R}}^3/D_6 \times Z_2^c)_6$ contributes to the two-point correlation tensor of the field as follows:

$$\frac{1}{4}\sum_{m=1}^{2}\int_{(\hat{\mathbb{R}}^3/D_6\times Z_2^c)_6} \operatorname{Re} j_m(\boldsymbol{p},\boldsymbol{y}-\boldsymbol{x}) f_1^{Sm}(\boldsymbol{p})\,\mathrm{d}\Phi(\boldsymbol{p})$$
$$+\frac{1}{4}\sum_{m=1}^{2}\int_{(\hat{\mathbb{R}}^3/D_6\times Z_2^c)_6} \operatorname{Im} j_m(\boldsymbol{p},\boldsymbol{y}-\boldsymbol{x}) f_1^{\Lambda m}(\boldsymbol{p})\,\mathrm{d}\Phi(\boldsymbol{p}).$$

The restrictions of the irreducible components A_{1g} and B_{1u} to the stationary subgroup of the stratum $(\hat{\mathbb{R}}^3/D_6 \times Z_2^c)_4$ are trivial, while the restrictions of the two remaining components are not. The set $\mathcal{C}_2$ is the set of Hermitian non-negative-definite 4×4 matrices with unit trace $f_2(\boldsymbol{p})$ with the following restrictions:

$$(f_2^S)_{14}(\boldsymbol{p}) = (f_2^S)_{24}(\boldsymbol{p}) = (f_2^S)_{34}(\boldsymbol{p}) = (f_2^S)_{41}(\boldsymbol{p}) = (f_2^S)_{42}(\boldsymbol{p}) = (f_2^S)_{43}(\boldsymbol{p}) = 0,$$
$$(f_2^\Lambda)_{12}(\boldsymbol{p}) = (f_2^\Lambda)_{13}(\boldsymbol{p}) = (f_2^\Lambda)_{23}(\boldsymbol{p}) = (f_2^\Lambda)_{21}(\boldsymbol{p}) = (f_2^\Lambda)_{31}(\boldsymbol{p}) = (f_2^\Lambda)_{32}(\boldsymbol{p}) = 0.$$

The stratum $(\hat{\mathbb{R}}^3/D_6 \times Z_2^c)_4$ contributes to the two-point correlation tensor of the field as follows:

$$\frac{1}{2}\sum_{m=1}^{4}\int_{(\hat{\mathbb{R}}^3/D_6\times Z_2^c)_4} \operatorname{Re} j_m(\boldsymbol{p},\boldsymbol{y}-\boldsymbol{x}) f_2^{Sm}(\boldsymbol{p})\,\mathrm{d}\Phi(\boldsymbol{p})$$
$$+\frac{1}{4}\sum_{m=1}^{2}\int_{(\hat{\mathbb{R}}^3/D_6\times Z_2^c)_4} \operatorname{Im} j_m(\boldsymbol{p},\boldsymbol{y}-\boldsymbol{x}) f_2^{\Lambda m}(\boldsymbol{p})\,\mathrm{d}\Phi(\boldsymbol{p}).$$

Finally, the restrictions of all non-trivial irreducible components to the stationary subgroups of the remaining strata are non-trivial. The set $\mathcal{C}_3$ is the set of Hermitian non-negative-definite 4×4 matrices with unit trace $f_3(\boldsymbol{p})$ with the following restrictions:

$$(f_3^S)_{14}(\boldsymbol{p}) = (f_3^S)_{24}(\boldsymbol{p}) = (f_3^S)_{34}(\boldsymbol{p}) = (f_3^S)_{41}(\boldsymbol{p}) = (f_3^S)_{42}(\boldsymbol{p}) = (f_3^S)_{43}(\boldsymbol{p}) = 0,$$
$$f_3^\Lambda(\boldsymbol{p}) = 0.$$

The contribution of the remaining strata is

$$\frac{1}{4}\sum_{m=1}^{2}\int_{(\hat{\mathbb{R}}^3/D_6\times Z_2^c)_{0-3}} \operatorname{Re} j_m(\boldsymbol{p},\boldsymbol{y}-\boldsymbol{x}) f_3^{Sm}(\boldsymbol{p})\,\mathrm{d}\Phi(\boldsymbol{p}).$$

Let $u_{1n}(\boldsymbol{p},\boldsymbol{x})$ be four different combinations of sines and cosines of $p_1x_1+p_3x_3$ and p_2x_2, four different combinations of sines and cosines of $\frac{p_1}{2}(-x_1+\sqrt{3}x_2)+p_3x_3$ and $\frac{p_2}{2}(\sqrt{3}x_1+x_2)$ plus four different combinations of sines and cosines of $\frac{p_1}{2}(-x_1-\sqrt{3}x_2)+p_3x_3$ and $\frac{p_2}{2}(\sqrt{3}x_1-x_2)$. Enumerate them in such order that the combinations from seventh till twelfth contain both sines and cosines. Similarly, let $u_{2n}(\boldsymbol{p},\boldsymbol{x})$ be four different combinations of sines and cosines of $-p_1x_1+p_3x_3$ and p_2x_2 plus four different combinations of sines and cosines of $\frac{p_1}{2}(-x_1+\sqrt{3}x_2)+p_3x_3$ and $\frac{p_2}{2}(\sqrt{3}x_1+x_2)$ plus four different combinations of sines and cosines of $\frac{p_1}{2}(-x_1-\sqrt{3}x_2)+p_3x_3$ and $\frac{p_2}{2}(\sqrt{3}x_1-x_2)$. Enumerate them in the similar order.

Theorem 33. *The one-point correlation tensor of a homogeneous and $(D^h_6, 3A_1 \oplus B_1)$-isotropic random field $\mathsf{e}(\boldsymbol{x})$ is*

$$\langle \mathsf{e}(\boldsymbol{x}) \rangle = \sum_{m=1}^{3} C_m \boldsymbol{T}^m, \qquad C_m \in \mathbb{R}.$$

Its two-point correlation tensor has the form

$$\begin{aligned}
\langle \mathsf{e}(\boldsymbol{x}), \mathsf{e}(\boldsymbol{y}) \rangle &= \frac{1}{4}\sum_{m=1}^{2} \int_{(\hat{\mathbb{R}}^3/D_6\times Z_2^c)_{5,7}} \operatorname{Re} j_m(\boldsymbol{p}, \boldsymbol{y}-\boldsymbol{x}) f_0^{Sm}(\boldsymbol{p})\, \mathrm{d}\Phi(\boldsymbol{p}) \\
&+ \frac{1}{4}\sum_{m=1}^{2} \int_{(\hat{\mathbb{R}}^3/D_6\times Z_2^c)_{5,7}} \operatorname{Im} j_m(\boldsymbol{p}, \boldsymbol{y}-\boldsymbol{x}) f_0^{\Lambda m}(\boldsymbol{p})\, \mathrm{d}\Phi(\boldsymbol{p}) \\
&+ \frac{1}{4}\sum_{m=1}^{2} \int_{(\hat{\mathbb{R}}^3/D_6\times Z_2^c)_{6}} \operatorname{Re} j_m(\boldsymbol{p}, \boldsymbol{y}-\boldsymbol{x}) f_1^{Sm}(\boldsymbol{p})\, \mathrm{d}\Phi(\boldsymbol{p}) \\
&+ \frac{1}{4}\sum_{m=1}^{2} \int_{(\hat{\mathbb{R}}^3/D_6\times Z_2^c)_{6}} \operatorname{Im} j_m(\boldsymbol{p}, \boldsymbol{y}-\boldsymbol{x}) f_1^{\Lambda m}(\boldsymbol{p})\, \mathrm{d}\Phi(\boldsymbol{p}) \\
&+ \frac{1}{4}\sum_{m=1}^{2} \int_{(\hat{\mathbb{R}}^3/D_6\times Z_2^c)_{4}} \operatorname{Re} j_m(\boldsymbol{p}, \boldsymbol{y}-\boldsymbol{x}) f_2^{Sm}(\boldsymbol{p})\, \mathrm{d}\Phi(\boldsymbol{p}) \\
&+ \frac{1}{4}\sum_{m=1}^{2} \int_{(\hat{\mathbb{R}}^3/D_6\times Z_2^c)_{4}} \operatorname{Im} j_m(\boldsymbol{p}, \boldsymbol{y}-\boldsymbol{x}) f_2^{\Lambda m}(\boldsymbol{p})\, \mathrm{d}\Phi(\boldsymbol{p}) \\
&+ \frac{1}{4}\sum_{m=1}^{2} \int_{(\hat{\mathbb{R}}^3/D_6\times Z_2^c)_{0-3}} \operatorname{Re} j_m(\boldsymbol{p}, \boldsymbol{y}-\boldsymbol{x}) f_3^{Sm}(\boldsymbol{p})\, \mathrm{d}\Phi(\boldsymbol{p}).
\end{aligned}$$

The field has the form

$$\begin{aligned}
\mathsf{e}(\boldsymbol{x}) &= \sum_{m=1}^{3} C_m \boldsymbol{T}^m + \frac{1}{2}\sum_{m=1}^{2}\sum_{n=1}^{6} \int_{(\hat{\mathbb{R}}^3/D_6\times Z_2^c)_{5,7}} u_{mn}(\boldsymbol{p}, \boldsymbol{x})\, \mathrm{d}Z_{0m}^{Sn}(\boldsymbol{p}) \boldsymbol{T}^m \\
&+ \frac{1}{2}\sum_{m=1}^{2}\sum_{n=7}^{12} \int_{(\hat{\mathbb{R}}^3/D_6\times Z_2^c)_{5,7}} u_{mn}(\boldsymbol{p}, \boldsymbol{x})\, \mathrm{d}Z_{0m}^{\Lambda n}(\boldsymbol{p}) \boldsymbol{T}^m \\
&+ \frac{1}{2}\sum_{m=}^{2}\sum_{n=1}^{6} \int_{(\hat{\mathbb{R}}^3/D_6\times Z_2^c)_{6}} u_{mn}(\boldsymbol{p}, \boldsymbol{x})\, \mathrm{d}Z_{1m}^{Sn}(\boldsymbol{p}) \boldsymbol{T}^m \\
&+ \frac{1}{2}\sum_{m=1}^{2}\sum_{n=7}^{12} \int_{(\hat{\mathbb{R}}^3/D_6\times Z_2^c)_{6}} u_{mn}(\boldsymbol{p}, \boldsymbol{x})\, \mathrm{d}Z_{1m}^{\Lambda n}(\boldsymbol{p}) \boldsymbol{T}^m \\
&+ \frac{1}{2}\sum_{m=1}^{2}\sum_{n=1}^{6} \int_{(\hat{\mathbb{R}}^3/D_6\times Z_2^c)_{4}} u_{mn}(\boldsymbol{p}, \boldsymbol{x})\, \mathrm{d}Z_{2m}^{Sn}(\boldsymbol{p}) \boldsymbol{T}^m
\end{aligned}$$

$$+\frac{1}{2}\sum_{m=1}^{2}\sum_{n=7}^{12}\int_{(\hat{\mathbb{R}}^3/D_6\times Z_2^c)_4} u_{mn}(\boldsymbol{p},\boldsymbol{x})\,\mathrm{d}Z^{\Lambda n}_{2m}(\boldsymbol{p})\boldsymbol{T}^m$$
$$+\frac{1}{2}\sum_{m=1}^{2}\sum_{n=1}^{12}\int_{(\hat{\mathbb{R}}^3/D_6\times Z_2^c)_{0-3}} u_{mn}(\boldsymbol{p},\boldsymbol{x})\,\mathrm{d}Z^{Sn}_{3m}(\boldsymbol{p})\boldsymbol{T}^m,$$

where

$$\begin{aligned}
\boldsymbol{Z}^{Sn}_0(\boldsymbol{p}) &= (Z^{Sn}_{01}(\boldsymbol{p}),\dots,Z^{Sn}_{04}(\boldsymbol{p}))^\top,\\
\boldsymbol{Z}^{\Lambda n}_0(\boldsymbol{p}) &= (Z^{\Lambda n}_{01}(\boldsymbol{p}),\dots,Z^{\Lambda n}_{04}(\boldsymbol{p}))^\top,\\
\boldsymbol{Z}^{Sn}_1(\boldsymbol{p}) &= (Z^{Sn}_{11}(\boldsymbol{p}),\dots,Z^{Sn}_{14}(\boldsymbol{p}))^\top,\\
\boldsymbol{Z}^{\Lambda n}_1(\boldsymbol{p}) &= (Z^{\Lambda n}_{11}(\boldsymbol{p}),\dots,Z^{\Lambda n}_{14}(\boldsymbol{p}))^\top,\\
\boldsymbol{Z}^{\Lambda n}_2(\boldsymbol{p}) &= (Z^{\Lambda n}_{02}(\boldsymbol{p}),\dots,Z^{\Lambda n}_{22}(\boldsymbol{p}))^\top,\\
\boldsymbol{Z}^{Sn}_2(\boldsymbol{p}) &= (Z^{Sn}_{21}(\boldsymbol{p}),\dots,Z^{Sn}_{24}(\boldsymbol{p}))^\top,\\
\boldsymbol{Z}^{\Lambda n}_2(\boldsymbol{p}) &= (Z^{\Lambda n}_{21}(\boldsymbol{p}),\dots,Z^{\Lambda n}_{24}(\boldsymbol{p}))^\top,\\
\boldsymbol{Z}^{Sn}_3(\boldsymbol{p}) &= (Z^{Sn}_{31}(\boldsymbol{p}),\dots,Z^{Sn}_{34}(\boldsymbol{p}))^\top
\end{aligned}$$

are centred random measures on the corresponding sets with the corresponding control measures and non-zero cross-correlations similar to (3.86)*:*

$$\mathsf{E}[\boldsymbol{Z}^{Sn}_i(A)\otimes\boldsymbol{Z}^{\Lambda n}_i(B)] = \int_{A\cap B} f^{\Lambda m}_i(\boldsymbol{p})\,\mathrm{d}\Phi(\boldsymbol{p})$$

for $0\le i\le 2$.

The cases of $j=12$, 17, 21, 22, 29, 33, 44 are similar and may be left to the reader.

The cases when G is of type II, when $j=4$, 6, 8, 11, 14, 16, 18, 20, 23, 25, 28, 30, 32, 35, 37, 39, 41, 45, 47 and 49, are similar to those considered in Section 3.6 and also may be left to the reader.

Finally, consider the most complicated case of $j=2$. First, determine the structure of the representation $\mathsf{S}^2(\rho)$. Using the Clebsch–Gordan rule, we obtain:

$$\mathsf{S}^2(2\rho_1\oplus\rho_2\oplus\rho_3) = 5\rho_0\oplus\rho_1\oplus 10\rho_2\oplus 5\rho_3\oplus 5\rho_4\oplus\rho_5\oplus\rho_6.$$

The symmetric covariant tensors and their syzygies have been calculated in Chapter 2, Equations (2.40), (2.44) and (2.45), Table 2.3.

We determine the polar tensors of the uncoupled basis of the space $\mathsf{S}^2(\mathsf{V}^{Z_1})$. The results are shown in Table 3.5.

Equation (3.28), which determines the $\boldsymbol{M}$-functions, takes the form

$$\boldsymbol{M}^{\rho_{2\ell},n_\ell}_{ijki'jk'}(\boldsymbol{p}) = \sum_{m=-2\ell}^{2\ell}\mathsf{T}^{\rho_{2\ell},n_\ell,m}_{ijki'j'k'}\rho^{2\ell g}_{m0}(\boldsymbol{p}),$$

where $0\le\ell\le 3$, $1\le n_0\le q_0=5$, $1\le n_1\le q_1=10$, $1\le n_2\le q_2=5$ and $n_3=q_3=1$.

Table 3.5 *The polar tensors of the uncoupled basis of the space* $\mathsf{S}^2(\mathsf{V}^{Z_1})$.

Tensor	Value
$T^{\rho_0,1,0}_{ijki'j'k'}$	$\frac{1}{3\sqrt{3}}\delta_{ij}\delta_{i'j'}\delta_{kk'}$
$T^{\rho_0,2,0}_{ijki'j'k'}$	$\frac{1}{3\sqrt{2}}\sum_{l=-1}^{1}\left(\delta_{ij}\delta_{kl}\sum_{m'=-2}^{2} g^{l[m',k']}_{1[2,1]} g^{m'[i',j']}_{2[1,1]} + \delta_{i'j'}\delta_{k'l}\sum_{m=-2}^{2} g^{l[m,k]}_{1[2,1]} g^{m[i,j]}_{2[1,1]}\right)$
$T^{\rho_0,3,0}_{ijki'j'k'}$	$\frac{1}{\sqrt{3}}\sum_{l=-1}^{1}\sum_{m,m'=-2}^{2} g^{l[m,k]}_{1[2,1]} g^{m[i,j]}_{2[1,1]} g^{l[m',k']}_{1[2,1]} g^{m'[i',j']}_{2[1,1]}$
$T^{\rho_0,4,0}_{ijki'j'k'}$	$\frac{1}{\sqrt{5}}\sum_{l,m,m'=-2}^{2} g^{l[m,k]}_{2[2,1]} g^{m[i,j]}_{2[1,1]} g^{l[m',k']}_{2[2,1]} g^{m'[i',j']}_{2[1,1]}$
$T^{\rho_0,5,0}_{ijki'j'k'}$	$\frac{1}{\sqrt{7}}\sum_{l=-3}^{3}\sum_{m,m'=-2}^{2} g^{l[m,k]}_{3[2,1]} g^{m[i,j]}_{2[1,1]} g^{l[m',k']}_{3[2,1]} g^{m'[i',j']}_{2[1,1]}$
$T^{\rho_2,1,m}_{ijki'j'k'}$	$\frac{1}{3}\delta_{ij}\delta_{i'j'} g^{m[k,k']}_{2[1,1]}$
$T^{\rho_2,2,m}_{ijki'j'k'}$	$\frac{1}{\sqrt{6}}\left(\delta_{ij}\sum_{l'=-1}^{1}\sum_{n'=-2}^{2} g^{m[k,l']}_{2[1,1]} g^{l'[n',k']}_{1[2,1]} g^{n'[i',j']}_{2[1,1]} + \delta_{i'j'}\sum_{l=-1}^{1}\sum_{n=-2}^{2} g^{m'[k',l]}_{2[1,1]} g^{l[n,k]}_{1[2,1]} g^{n[i,j]}_{2[1,1]}\right)$
$T^{\rho_2,3,m}_{ijki'j'k'}$	$\sum_{l,l'=-1}^{1}\sum_{n,n'=-2}^{2} g^{m[l,l']}_{2[1,1]} g^{l[n,k]}_{1[2,1]} g^{n[i,j]}_{2[1,1]} g^{l'[n',k']}_{1[2,1]} g^{n'[i',j']}_{2[1,1]}$
$T^{\rho_2,4,m}_{ijki'j'k'}$	$\frac{1}{\sqrt{6}}\left(\delta_{ij}\sum_{l',n'=-2}^{2} g^{m[k,l']}_{2[1,1]} g^{l'[k',n']}_{2[1,2]} g^{n'[i',j']}_{2[1,1]} + \delta_{i'j'}\sum_{l,n=-2}^{2} g^{m'[k',l]}_{2[1,1]} g^{l[k,n]}_{2[1,2]} g^{n[i,j]}_{2[1,1]}\right)$
$T^{\rho_2,5,m}_{ijki'j'k'}$	$\frac{1}{\sqrt{6}}\left(\delta_{ij}\sum_{l'=-3}^{3}\sum_{n'=-2}^{2} g^{m[k,l']}_{2[1,3]} g^{l'[n',k']}_{3[2,1]} g^{n'[i',j']}_{2[1,1]} + \delta_{i'j'}\sum_{l=-3}^{3}\sum_{n=-2}^{2} g^{m'[k',l]}_{2[1,3]} g^{l[n,k]}_{3[2,1]} g^{n[i,j]}_{2[1,1]}\right)$
$T^{\rho_2,6,m}_{ijki'j'k'}$	$\sum_{l,l',n,n'=-2}^{2} g^{m[l,l']}_{2[2,2]} g^{l[n,k]}_{2[2,1]} g^{n[i,j]}_{2[1,1]} g^{l'[n',k']}_{2[2,1]} g^{n'[i',j']}_{2[1,1]}$
$T^{\rho_2,7,m}_{ijki'j'k'}$	$\frac{1}{\sqrt{2}}\left(\sum_{l'=-1}^{1}\sum_{l',n,n'=-2}^{2} g^{m[l,l']}_{2[1,2]} g^{l[n,k]}_{1[2,1]} g^{n[i,j]}_{2[1,1]} g^{l'[n',k']}_{2[2,1]} g^{n'[i',j']}_{2[1,1]} + \sum_{l=-1}^{1}\sum_{l,n,n'=-2}^{2} g^{m'[l',l]}_{2[1,2]} g^{l'[n',k']}_{1[2,1]} g^{n'[i',j']}_{2[1,1]} g^{l[n,k]}_{2[2,1]} g^{n[i,j]}_{2[1,1]}\right)$
$T^{\rho_2,8,m}_{ijki'j'k'}$	$\frac{1}{\sqrt{2}}\left(\sum_{l=-1}^{1}\sum_{n,n'=-2}^{2}\sum_{l'=-3}^{3} g^{m[l,l']}_{2[1,3]} g^{l[n,k]}_{1[2,1]} g^{n[i,j]}_{2[1,1]} g^{l'[n',k']}_{3[2,1]} g^{n'[i',j']}_{2[1,1]} + \sum_{l'=-1}^{1}\sum_{n,n'=-2}^{2}\sum_{l=-3}^{3} g^{m'[l',l]}_{2[1,3]} g^{l'[n',k']}_{1[2,1]} g^{n'[i',j']}_{2[1,1]} g^{l[n,k]}_{3[2,1]} g^{n[i,j]}_{2[1,1]}\right)$
$T^{\rho_2,9,m}_{ijki'j'k'}$	$\frac{1}{\sqrt{2}}\left(\sum_{l,n,n'=-2}^{2}\sum_{l'=-3}^{3} g^{m[l,l']}_{2[2,3]} g^{l[n,k]}_{2[2,1]} g^{n[i,j]}_{2[1,1]} g^{l'[n',k']}_{3[2,1]} g^{n'[i',j']}_{2[1,1]} + \sum_{l',n,n'=-2}^{2}\sum_{l=-3}^{3} g^{m'[l',l]}_{2[2,3]} g^{l'[n',k']}_{2[2,1]} g^{n'[i',j']}_{2[1,1]} g^{l[n,k]}_{3[2,1]} g^{n[i,j]}_{2[1,1]}\right)$
$T^{\rho_2,10,m}_{ijki'j'k'}$	$\sum_{n,n'=-2}^{2}\sum_{l,l'=-3}^{3} g^{m[l,l']}_{2[3,3]} g^{l[n,k]}_{3[2,1]} g^{n[i,j]}_{2[1,1]} g^{l'[n',k']}_{3[2,1]} g^{n'[i',j']}_{2[1,1]}$
$T^{\rho_4,1,m}_{ijki'j'k'}$	$\frac{1}{\sqrt{6}}\left(\delta_{ij}\sum_{l'=-3}^{3}\sum_{n'=-2}^{2} g^{m[k,l']}_{4[1,3]} g^{l'[n',k']}_{3[2,1]} g^{n'[i',j']}_{2[1,1]} + \delta_{i'j'}\sum_{l=-3}^{3}\sum_{n=-2}^{2} g^{m'[k',l]}_{4[1,3]} g^{l[n,k]}_{3[2,1]} g^{n[i,j]}_{2[1,1]}\right)$
$T^{\rho_4,2,m}_{ijki'j'k'}$	$\sum_{l,l',n,n'=-2}^{2} g^{m[l,l']}_{4[2,2]} g^{l[n,k]}_{2[2,1]} g^{n[i,j]}_{2[1,1]} g^{l'[n',k']}_{2[2,1]} g^{n'[i',j']}_{2[1,1]}$
$T^{\rho_4,3,m}_{ijki'j'k'}$	$\frac{1}{\sqrt{2}}\left(\sum_{l=-1}^{1}\sum_{n,n'=-2}^{2}\sum_{l'=-3}^{3} g^{m[l,l']}_{4[1,3]} g^{l[n,k]}_{1[2,1]} g^{n[i,j]}_{2[1,1]} g^{l'[n',k']}_{3[2,1]} g^{n'[i',j']}_{2[1,1]} + \sum_{l'=-1}^{1}\sum_{n,n'=-2}^{2}\sum_{l=-3}^{3} g^{m'[l',l]}_{4[1,3]} g^{l'[n',k']}_{1[2,1]} g^{n'[i',j']}_{2[1,1]} g^{l[n,k]}_{3[2,1]} g^{n[i,j]}_{2[1,1]}\right)$

Table 3.5 (*Cont.*)

Tensor	Value
$\boldsymbol{T}^{\rho_4,4,m}_{ijki'j'k'}$	$\frac{1}{\sqrt{2}}\Big(\sum_{l,n,n'=-2}^{2}\sum_{l'=-3}^{3} g^{m[l,l']}_{4[2,3]} g^{l[n,k]}_{2[2,1]} g^{n[i,j]}_{2[1,1]} g^{l'[n',k']}_{3[2,1]} g^{n'[i',j']}_{2[1,1]}$
	$+\sum_{l',n,n'=-2}^{2}\sum_{l=-3}^{3} g^{m'[l',l]}_{4[2,3]} g^{l'[n',k']}_{2[2,1]} g^{n'[i',j']}_{2[1,1]} g^{l[n,k]}_{3[2,1]} g^{n[i,j]}_{2[1,1]}\Big)$
$\boldsymbol{T}^{\rho_4,5,m}_{ijki'j'k'}$	$\sum_{n,n'=-2}^{2}\sum_{l,l'=-3}^{3} g^{m[l,l']}_{4[3,3]} g^{l[n,k]}_{3[2,1]} g^{n[i,j]}_{2[1,1]} g^{l'[n',k']}_{3[2,1]} g^{n'[i',j']}_{2[1,1]}$
$\boldsymbol{T}^{\rho_6,m}_{ijki'j'k'}$	$\sum_{n,n'=-2}^{2}\sum_{l,l'=-3}^{3} g^{m[l,l']}_{6[3,3]} g^{l[n,k]}_{3[2,1]} g^{n[i,j]}_{2[1,1]} g^{l'[n',k']}_{3[2,1]} g^{n'[i',j']}_{2[1,1]}$

The complicated $\boldsymbol{M}$-functions are symmetric covariant tensor-valued functions. They must express as linear combinations of more simple $\boldsymbol{L}$-functions. To find the coefficients of the above combinations, we use MATLAB Symbolic Math Toolbox. The results of calculations are given in Table 3.6.

The next step is to determine the two-point correlation tensor of the random field $\boldsymbol{e}(\boldsymbol{x})$. We follow Section 3.6 and prove the following result. The two-point correlation tensor has the form

$$\langle \boldsymbol{e}(\boldsymbol{x}), \boldsymbol{e}(\boldsymbol{x})\rangle = \frac{1}{4\pi}\int_0^\infty \int_0^{2\pi}\int_0^\pi \mathrm{e}^{\mathrm{i}(\boldsymbol{p},\boldsymbol{y}-\boldsymbol{x})} f(\boldsymbol{p})\sin(\theta_{\boldsymbol{p}})\,\mathrm{d}\theta_{\boldsymbol{p}}\,\mathrm{d}\varphi_{\boldsymbol{p}}\,\mathrm{d}\Phi(\lambda),$$

where $f(\boldsymbol{p})$ is a measurable function on $\hat{\mathbb{R}}^3$ taking values in the set of all symmetric non-negative-definite operators on $\mathsf{S}^2(\mathbb{R}^3)\otimes\mathbb{R}^3$ with unit trace and satisfying

$$f(g\boldsymbol{p}) = \mathsf{S}^2(\rho(g))f(\boldsymbol{p}), \qquad \boldsymbol{p}\in\hat{\mathbb{R}}^3, \quad g\in \mathrm{O}(3),$$

and Φ is a finite measure on $[0,\infty)$. Exactly as in Section 3.6, we prove that

$$f(\mathbf{0}) = \sum_{n_0=1}^{5} \boldsymbol{T}^{\rho_0,n_0,0} f_{0,n_0}(0).$$

The results of calculations are given in Table 3.7.

Introduce the following notation.

$$\begin{aligned}
u_1(0) &= \frac{1}{\sqrt{3}}f_{0,1}(0) + \frac{2\sqrt{2}}{\sqrt{15}}f_{0,2}(0) + \frac{4}{5\sqrt{3}}f_{0,3}(0) + \frac{6}{5\sqrt{7}}f_{0,5}(0),\\
u_2(0) &= \frac{2}{\sqrt{3}}f_{0,1}(0) - \frac{2\sqrt{2}}{\sqrt{15}}f_{0,2}(0) + \frac{2}{5\sqrt{3}}f_{0,3}(0) + \frac{2}{\sqrt{5}}f_{0,4}(0)\\
&\quad + \frac{8}{5\sqrt{7}}f_{0,5}(0),\\
u_3(0) &= \frac{3\sqrt{3}}{5}f_{0,3}(0) + \frac{1}{\sqrt{5}}f_{0,4}(0) + \frac{16}{5\sqrt{7}}f_{0,5}(0),\\
u_4(0) &= \frac{2}{\sqrt{5}}f_{0,4}(0) + \frac{1}{\sqrt{7}}f_{0,5}(0),\\
u_5(0) &= \frac{\sqrt{3}}{10}f_{0,3}(0) - \frac{1}{6\sqrt{5}}f_{0,4}(0) - \frac{2}{15\sqrt{7}}f_{0,5}(0).
\end{aligned}$$

Table 3.6 *$\boldsymbol{M}$-functions expressed as linear combinations of $\boldsymbol{L}$-functions.*

$\boldsymbol{M}$-function	Expresses as
$\boldsymbol{M}^{\rho_0,1}_{ijki'j'k'}$	$\frac{1}{3\sqrt{3}}\boldsymbol{L}^1$
$\boldsymbol{M}^{\rho_0,2}_{ijki'j'k'}$	$-\frac{\sqrt{2}}{3\sqrt{15}}\boldsymbol{L}^1+\frac{1}{2\sqrt{30}}\boldsymbol{L}^3$
$\boldsymbol{M}^{\rho_0,3}_{ijki'j'k'}$	$\frac{1}{15\sqrt{3}}\boldsymbol{L}^1-\frac{1}{10\sqrt{3}}\boldsymbol{L}^3+\frac{\sqrt{3}}{20}\boldsymbol{L}^4$
$\boldsymbol{M}^{\rho_0,4}_{ijki'j'k'}$	$-\frac{1}{3\sqrt{5}}\boldsymbol{L}^1+\frac{1}{3\sqrt{5}}\boldsymbol{L}^2+\frac{1}{6\sqrt{5}}\boldsymbol{L}^3-\frac{1}{12\sqrt{5}}\boldsymbol{L}^4-\frac{1}{6\sqrt{5}}\boldsymbol{L}^5$
$\boldsymbol{M}^{\rho_0,5}_{ijki'j'k'}$	$-\frac{1}{15\sqrt{7}}\boldsymbol{L}^1+\frac{1}{6\sqrt{7}}\boldsymbol{L}^2-\frac{1}{15\sqrt{7}}\boldsymbol{L}^3-\frac{1}{15\sqrt{7}}\boldsymbol{L}^4+\frac{1}{6\sqrt{7}}\boldsymbol{L}^5$
$\boldsymbol{M}^{\rho_2,1}_{ijki'j'k'}(\boldsymbol{p})$	$-\frac{1}{3\sqrt{6}}\boldsymbol{L}^1+\frac{1}{\sqrt{6}\lVert\boldsymbol{p}\rVert^2}\boldsymbol{L}^6(\boldsymbol{p})$
$\boldsymbol{M}^{\rho_2,2}_{ijki'j'k'}(\boldsymbol{p})$	$\frac{1}{3\sqrt{15}}\boldsymbol{L}^1-\frac{1}{4\sqrt{15}}\boldsymbol{L}^3-\frac{1}{\sqrt{15}\lVert\boldsymbol{p}\rVert^2}\boldsymbol{L}^6(\boldsymbol{p})+\frac{\sqrt{3}}{4\sqrt{5}\lVert\boldsymbol{p}\rVert^2}\boldsymbol{L}^9(\boldsymbol{p})$
$\boldsymbol{M}^{\rho_2,3}_{ijki'j'k'}(\boldsymbol{p})$	$\frac{1}{15\sqrt{6}}\boldsymbol{L}^1+\frac{1}{10\sqrt{6}}\boldsymbol{L}^3-\frac{\sqrt{3}}{20\sqrt{2}}\boldsymbol{L}^4+\frac{1}{5\sqrt{6}\lVert\boldsymbol{p}\rVert^2}\boldsymbol{L}^6(\boldsymbol{p})$
	$-\frac{\sqrt{3}}{10\sqrt{2}\lVert\boldsymbol{p}\rVert^2}\boldsymbol{L}^9(\boldsymbol{p})+\frac{3\sqrt{3}}{20\sqrt{2}\lVert\boldsymbol{p}\rVert^2}\boldsymbol{L}^{10}(\boldsymbol{p})$
$\boldsymbol{M}^{\rho_2,4}_{ijki'j'k'}(\boldsymbol{p})$	$-\frac{1}{3}\boldsymbol{L}^1+\frac{1}{12}\boldsymbol{L}^3+\frac{1}{3\lVert\boldsymbol{p}\rVert^2}\boldsymbol{L}^6(\boldsymbol{p})+\frac{1}{3\lVert\boldsymbol{p}\rVert^2}\boldsymbol{L}^7(\boldsymbol{p})-\frac{1}{12\lVert\boldsymbol{p}\rVert^2}\boldsymbol{L}^9(\boldsymbol{p})-\frac{1}{6\lVert\boldsymbol{p}\rVert^2}\boldsymbol{L}^{11}(\boldsymbol{p})$
$\boldsymbol{M}^{\rho_2,5}_{ijki'j'k'}(\boldsymbol{p})$	$-\frac{1}{3\sqrt{35}}\boldsymbol{L}^1-\frac{1}{6\sqrt{35}}\boldsymbol{L}^3-\frac{2}{3\sqrt{35}\lVert\boldsymbol{p}\rVert^2}\boldsymbol{L}^6(\boldsymbol{p})+\frac{\sqrt{5}}{6\sqrt{7}\lVert\boldsymbol{p}\rVert^2}\boldsymbol{L}^7(\boldsymbol{p})$
	$-\frac{1}{3\sqrt{35}\lVert\boldsymbol{p}\rVert^2}\boldsymbol{L}^9(\boldsymbol{p})+\frac{\sqrt{5}}{6\sqrt{7}\lVert\boldsymbol{p}\rVert^2}\boldsymbol{L}^{11}(\boldsymbol{p})$
$\boldsymbol{M}^{\rho_2,6}_{ijki'j'k'}(\boldsymbol{p})$	$-\frac{1}{3\sqrt{14}}\boldsymbol{L}^1+\frac{1}{3\sqrt{14}}\boldsymbol{L}^2+\frac{1}{6\sqrt{14}}\boldsymbol{L}^3-\frac{1}{12\sqrt{14}}\boldsymbol{L}^4-\frac{1}{6\sqrt{14}}\boldsymbol{L}^5$
	$+\frac{1}{\sqrt{14}\lVert\boldsymbol{p}\rVert^2}\boldsymbol{L}^6(\boldsymbol{p})-\frac{1}{\sqrt{14}\lVert\boldsymbol{p}\rVert^2}\boldsymbol{L}^8(\boldsymbol{p})-\frac{1}{2\sqrt{14}\lVert\boldsymbol{p}\rVert^2}\boldsymbol{L}^9(\boldsymbol{p})$
	$+\frac{1}{4\sqrt{14}\lVert\boldsymbol{p}\rVert^2}\boldsymbol{L}^{10}(\boldsymbol{p})-\frac{1}{2\sqrt{14}\lVert\boldsymbol{p}\rVert^2}\boldsymbol{L}^{12}(\boldsymbol{p})+\frac{1}{2\sqrt{14}\lVert\boldsymbol{p}\rVert^2}\boldsymbol{L}^{15}(\boldsymbol{p})$
$\boldsymbol{M}^{\rho_2,7}_{ijki'j'k'}(\boldsymbol{p})$	$-\frac{2}{3\sqrt{5}}\boldsymbol{L}^1+\frac{1}{2\sqrt{5}}\boldsymbol{L}^2+\frac{1}{6\sqrt{5}}\boldsymbol{L}^3-\frac{1}{4\sqrt{5}}\boldsymbol{L}^5$
	$+\frac{2}{3\sqrt{5}\lVert\boldsymbol{p}\rVert^2}\boldsymbol{L}^6(\boldsymbol{p})+\frac{2}{3\sqrt{5}\lVert\boldsymbol{p}\rVert^2}\boldsymbol{L}^7(\boldsymbol{p})-\frac{1}{2\sqrt{5}\lVert\boldsymbol{p}\rVert^2}\boldsymbol{L}^8(\boldsymbol{p})$
	$-\frac{1}{6\sqrt{5}\lVert\boldsymbol{p}\rVert^2}\boldsymbol{L}^9(\boldsymbol{p})-\frac{1}{3\sqrt{5}\lVert\boldsymbol{p}\rVert^2}\boldsymbol{L}^{11}(\boldsymbol{p})+\frac{1}{4\sqrt{5}\lVert\boldsymbol{p}\rVert^2}\boldsymbol{L}^{12}(\boldsymbol{p})$
	$-\frac{1}{2\sqrt{5}\lVert\boldsymbol{p}\rVert^2}\boldsymbol{L}^{13}(\boldsymbol{p})+\frac{1}{4\sqrt{5}\lVert\boldsymbol{p}\rVert^2}\boldsymbol{L}^{15}(\boldsymbol{p})$
$\boldsymbol{M}^{\rho_2,8}_{ijki'j'k'}(\boldsymbol{p})$	$\frac{16}{15\sqrt{7}}\boldsymbol{L}^1-\frac{1}{2\sqrt{7}}\boldsymbol{L}^2-\frac{31}{60\sqrt{7}}\boldsymbol{L}^3+\frac{3}{20\sqrt{7}}\boldsymbol{L}^4+\frac{1}{4\sqrt{7}}\boldsymbol{L}^5$
	$-\frac{13}{15\sqrt{7}\lVert\boldsymbol{p}\rVert^2}\boldsymbol{L}^6(\boldsymbol{p})-\frac{\sqrt{7}}{6\lVert\boldsymbol{p}\rVert^2}\boldsymbol{L}^7(\boldsymbol{p})+\frac{1}{2\sqrt{7}\lVert\boldsymbol{p}\rVert^2}\boldsymbol{L}^8(\boldsymbol{p})$
	$+\frac{\sqrt{7}}{15\lVert\boldsymbol{p}\rVert^2}\boldsymbol{L}^9(\boldsymbol{p})-\frac{9}{20\sqrt{7}\lVert\boldsymbol{p}\rVert^2}\boldsymbol{L}^{10}(\boldsymbol{p})+\frac{1}{3\sqrt{7}\lVert\boldsymbol{p}\rVert^2}\boldsymbol{L}^{11}(\boldsymbol{p})$
	$-\frac{1}{4\sqrt{7}\lVert\boldsymbol{p}\rVert^2}\boldsymbol{L}^{12}(\boldsymbol{p})+\frac{1}{2\sqrt{7}\lVert\boldsymbol{p}\rVert^2}\boldsymbol{L}^{13}(\boldsymbol{p})+\frac{3}{4\sqrt{7}\lVert\boldsymbol{p}\rVert^2}\boldsymbol{L}^{14}(\boldsymbol{p})-\frac{1}{4\sqrt{7}\lVert\boldsymbol{p}\rVert^2}\boldsymbol{L}^{15}(\boldsymbol{p})$
$\boldsymbol{M}^{\rho_2,9}_{ijki'j'k'}(\boldsymbol{p})$	$-\frac{4\sqrt{2}}{3\sqrt{35}}\boldsymbol{L}^1+\frac{\sqrt{7}}{6\sqrt{10}}\boldsymbol{L}^2+\frac{13}{12\sqrt{70}}\boldsymbol{L}^3-\frac{\sqrt{5}}{12\sqrt{14}}\boldsymbol{L}^4-\frac{\sqrt{7}}{12\sqrt{10}}\boldsymbol{L}^5$
	$+\frac{1}{\sqrt{70}\lVert\boldsymbol{p}\rVert^2}\boldsymbol{L}^6(\boldsymbol{p})+\frac{\sqrt{7}}{2\sqrt{10}\lVert\boldsymbol{p}\rVert^2}\boldsymbol{L}^7(\boldsymbol{p})+\frac{1}{2\sqrt{70}\lVert\boldsymbol{p}\rVert^2}\boldsymbol{L}^8(\boldsymbol{p})$
	$-\frac{3}{2\sqrt{70}\lVert\boldsymbol{p}\rVert^2}\boldsymbol{L}^9(\boldsymbol{p})+\frac{\sqrt{5}}{4\sqrt{14}\lVert\boldsymbol{p}\rVert^2}\boldsymbol{L}^{10}(\boldsymbol{p})-\frac{1}{2\sqrt{70}\lVert\boldsymbol{p}\rVert^2}\boldsymbol{L}^{11}(\boldsymbol{p})$
	$-\frac{1}{4\sqrt{70}\lVert\boldsymbol{p}\rVert^2}\boldsymbol{L}^{12}(\boldsymbol{p})-\frac{\sqrt{2}}{\sqrt{35}\lVert\boldsymbol{p}\rVert^2}\boldsymbol{L}^{13}(\boldsymbol{p})-\frac{\sqrt{5}}{4\sqrt{14}\lVert\boldsymbol{p}\rVert^2}\boldsymbol{L}^{14}(\boldsymbol{p})+\frac{1}{\sqrt{70}\lVert\boldsymbol{p}\rVert^2}\boldsymbol{L}^{15}(\boldsymbol{p})$
$\boldsymbol{M}^{\rho_2,10}_{ijki'j'k'}(\boldsymbol{p})$	$-\frac{11}{10\sqrt{21}}\boldsymbol{L}^1+\frac{1}{4\sqrt{21}}\boldsymbol{L}^2+\frac{3\sqrt{3}}{10\sqrt{7}}\boldsymbol{L}^3-\frac{1}{10\sqrt{21}}\boldsymbol{L}^4-\frac{\sqrt{3}}{4\sqrt{7}}\boldsymbol{L}^5$
	$+\frac{13}{10\sqrt{21}\lVert\boldsymbol{p}\rVert^2}\boldsymbol{L}^6(\boldsymbol{p})+\frac{1}{\sqrt{21}\lVert\boldsymbol{p}\rVert^2}\boldsymbol{L}^7(\boldsymbol{p})-\frac{1}{4\sqrt{21}\lVert\boldsymbol{p}\rVert^2}\boldsymbol{L}^8(\boldsymbol{p})$

Table 3.6 (*Cont.*)

$\boldsymbol{M}$-function	Expresses as
	$-\frac{\sqrt{7}}{10\sqrt{3}\|\boldsymbol{p}\|^2}\boldsymbol{L}^{9}(\boldsymbol{p}) + \frac{\sqrt{3}}{10\sqrt{7}\|\boldsymbol{p}\|^2}\boldsymbol{L}^{10}(\boldsymbol{p}) - \frac{1}{\sqrt{21}\|\boldsymbol{p}\|^2}\boldsymbol{L}^{11}(\boldsymbol{p})$
	$+\frac{\sqrt{3}}{4\sqrt{7}\|\boldsymbol{p}\|^2}\boldsymbol{L}^{12}(\boldsymbol{p}) - \frac{1}{4\sqrt{21}\|\boldsymbol{p}\|^2}\boldsymbol{L}^{13}(\boldsymbol{p}) - \frac{1}{\sqrt{21}\|\boldsymbol{p}\|^2}\boldsymbol{L}^{14}(\boldsymbol{p}) + \frac{\sqrt{3}}{4\sqrt{7}\|\boldsymbol{p}\|^2}\boldsymbol{L}^{15}(\boldsymbol{p})$
$\boldsymbol{M}^{\rho_4,1}_{ijki'j'k'}(\boldsymbol{p})$	$\frac{1}{2\sqrt{105}}\boldsymbol{L}^{1} + \frac{1}{4\sqrt{105}}\boldsymbol{L}^{3} - \frac{\sqrt{5}}{2\sqrt{21}\|\boldsymbol{p}\|^2}\boldsymbol{L}^{6}(\boldsymbol{p}) - \frac{\sqrt{5}}{4\sqrt{21}\|\boldsymbol{p}\|^2}\boldsymbol{L}^{7}(\boldsymbol{p})$
	$-\frac{\sqrt{5}}{4\sqrt{21}\|\boldsymbol{p}\|^2}\boldsymbol{L}^{9}(\boldsymbol{p}) - \frac{\sqrt{5}}{4\sqrt{21}\|\boldsymbol{p}\|^2}\boldsymbol{L}^{11}(\boldsymbol{p}) + \frac{\sqrt{35}}{4\sqrt{3}\|\boldsymbol{p}\|^4}\boldsymbol{L}^{17}(\boldsymbol{p})$
$\boldsymbol{M}^{\rho_4,2}_{ijki'j'k'}(\boldsymbol{p})$	$\frac{2\sqrt{2}}{3\sqrt{35}}\boldsymbol{L}^{1} - \frac{2\sqrt{2}}{3\sqrt{35}}\boldsymbol{L}^{2} - \frac{\sqrt{2}}{3\sqrt{35}}\boldsymbol{L}^{3} + \frac{1}{3\sqrt{70}}\boldsymbol{L}^{4} + \frac{\sqrt{2}}{3\sqrt{35}}\boldsymbol{L}^{5}$
	$-\frac{\sqrt{5}}{3\sqrt{14}\|\boldsymbol{p}\|^2}\boldsymbol{L}^{6}(\boldsymbol{p}) + \frac{\sqrt{5}}{3\sqrt{14}\|\boldsymbol{p}\|^2}\boldsymbol{L}^{8}(\boldsymbol{p}) + \frac{\sqrt{5}}{6\sqrt{14}\|\boldsymbol{p}\|^2}\boldsymbol{L}^{9}(\boldsymbol{p})$
	$-\frac{\sqrt{5}}{12\sqrt{14}\|\boldsymbol{p}\|^2}\boldsymbol{L}^{10}(\boldsymbol{p}) - \frac{5\sqrt{5}}{12\sqrt{14}\|\boldsymbol{p}\|^2}\boldsymbol{L}^{12}(\boldsymbol{p}) + \frac{\sqrt{35}}{12\sqrt{2}\|\boldsymbol{p}\|^2}\boldsymbol{L}^{13}(\boldsymbol{p})$
	$-\frac{\sqrt{5}}{6\sqrt{14}\|\boldsymbol{p}\|^2}\boldsymbol{L}^{15}(\boldsymbol{p}) - \frac{\sqrt{35}}{3\sqrt{2}\|\boldsymbol{p}\|^4}\boldsymbol{L}^{16}(\boldsymbol{p}) - \frac{\sqrt{35}}{12\sqrt{2}\|\boldsymbol{p}\|^4}\boldsymbol{L}^{18}(\boldsymbol{p}) + \frac{\sqrt{35}}{6\sqrt{2}\|\boldsymbol{p}\|^4}\boldsymbol{L}^{19}(\boldsymbol{p})$
$\boldsymbol{M}^{\rho_4,3}_{ijki'j'k'}(\boldsymbol{p})$	$-\frac{8}{5\sqrt{21}}\boldsymbol{L}^{1} + \frac{\sqrt{3}}{4\sqrt{7}}\boldsymbol{L}^{2} + \frac{31}{40\sqrt{21}}\boldsymbol{L}^{3} - \frac{3\sqrt{3}}{40\sqrt{7}}\boldsymbol{L}^{4} - \frac{\sqrt{3}}{8\sqrt{7}}\boldsymbol{L}^{5}$
	$+\frac{2}{\sqrt{21}\|\boldsymbol{p}\|^2}\boldsymbol{L}^{6}(\boldsymbol{p}) + \frac{\sqrt{7}}{4\sqrt{3}\|\boldsymbol{p}\|^2}\boldsymbol{L}^{7}(\boldsymbol{p}) - \frac{\sqrt{3}}{4\sqrt{7}\|\boldsymbol{p}\|^2}\boldsymbol{L}^{8}(\boldsymbol{p})$
	$-\frac{\sqrt{7}}{8\sqrt{3}\|\boldsymbol{p}\|^2}\boldsymbol{L}^{9}(\boldsymbol{p}) - \frac{\sqrt{3}}{8\sqrt{7}\|\boldsymbol{p}\|^2}\boldsymbol{L}^{10}(\boldsymbol{p}) - \frac{1}{2\sqrt{21}\|\boldsymbol{p}\|^2}\boldsymbol{L}^{11}(\boldsymbol{p})$
	$+\frac{\sqrt{3}}{8\sqrt{7}\|\boldsymbol{p}\|^2}\boldsymbol{L}^{12}(\boldsymbol{p}) - \frac{\sqrt{3}}{4\sqrt{7}\|\boldsymbol{p}\|^2}\boldsymbol{L}^{13}(\boldsymbol{p}) - \frac{3\sqrt{3}}{8\sqrt{7}\|\boldsymbol{p}\|^2}\boldsymbol{L}^{14}(\boldsymbol{p})$
	$+\frac{\sqrt{3}}{8\sqrt{7}\|\boldsymbol{p}\|^2}\boldsymbol{L}^{15}(\boldsymbol{p}) - \frac{\sqrt{7}}{4\sqrt{3}\|\boldsymbol{p}\|^4}\boldsymbol{L}^{17}(\boldsymbol{p}) + \frac{\sqrt{21}}{8\|\boldsymbol{p}\|^4}\boldsymbol{L}^{20}(\boldsymbol{p})$
$\boldsymbol{M}^{\rho_4,4}_{ijki'j'k'}(\boldsymbol{p})$	$-\frac{4}{3\sqrt{7}}\boldsymbol{L}^{1} + \frac{\sqrt{7}}{12}\boldsymbol{L}^{2} + \frac{13}{24\sqrt{7}}\boldsymbol{L}^{3} - \frac{5}{24\sqrt{7}}\boldsymbol{L}^{4} - \frac{\sqrt{7}}{24}\boldsymbol{L}^{5}$
	$+\frac{5}{3\sqrt{7}\|\boldsymbol{p}\|^2}\boldsymbol{L}^{6}(\boldsymbol{p}) + \frac{\sqrt{7}}{4\|\boldsymbol{p}\|^2}\boldsymbol{L}^{7}(\boldsymbol{p}) - \frac{11}{12\sqrt{7}\|\boldsymbol{p}\|^2}\boldsymbol{L}^{8}(\boldsymbol{p})$
	$-\frac{11}{24\sqrt{7}\|\boldsymbol{p}\|^2}\boldsymbol{L}^{9}(\boldsymbol{p}) + \frac{1}{24\sqrt{7}\|\boldsymbol{p}\|^2}\boldsymbol{L}^{10}(\boldsymbol{p}) - \frac{1}{4\sqrt{7}\|\boldsymbol{p}\|^2}\boldsymbol{L}^{11}(\boldsymbol{p})$
	$+\frac{11}{24\sqrt{7}\|\boldsymbol{p}\|^2}\boldsymbol{L}^{12}(\boldsymbol{p}) - \frac{5}{12\sqrt{7}\|\boldsymbol{p}\|^2}\boldsymbol{L}^{13}(\boldsymbol{p}) - \frac{5}{8\sqrt{7}\|\boldsymbol{p}\|^2}\boldsymbol{L}^{14}(\boldsymbol{p})$
	$+\frac{5}{24\sqrt{7}\|\boldsymbol{p}\|^2}\boldsymbol{L}^{15}(\boldsymbol{p}) - \frac{\sqrt{7}}{3\|\boldsymbol{p}\|^4}\boldsymbol{L}^{16}(\boldsymbol{p}) - \frac{\sqrt{7}}{4\|\boldsymbol{p}\|^4}\boldsymbol{L}^{17}(\boldsymbol{p})$
	$+\frac{\sqrt{7}}{6\|\boldsymbol{p}\|^4}\boldsymbol{L}^{18}(\boldsymbol{p}) - \frac{\sqrt{7}}{12\|\boldsymbol{p}\|^4}\boldsymbol{L}^{19}(\boldsymbol{p}) + \frac{\sqrt{7}}{8\|\boldsymbol{p}\|^4}\boldsymbol{L}^{20}(\boldsymbol{p})$
$\boldsymbol{M}^{\rho_4,5}_{ijki'j'k'}(\boldsymbol{p})$	$\frac{13}{30\sqrt{154}}\boldsymbol{L}^{1} + \frac{1}{6\sqrt{154}}\boldsymbol{L}^{2} - \frac{17}{30\sqrt{154}}\boldsymbol{L}^{3} - \frac{1}{15\sqrt{154}}\boldsymbol{L}^{4} + \frac{\sqrt{2}}{3\sqrt{77}}\boldsymbol{L}^{5}$
	$+\frac{2\sqrt{2}}{3\sqrt{77}\|\boldsymbol{p}\|^2}\boldsymbol{L}^{6}(\boldsymbol{p}) - \frac{1}{2\sqrt{154}\|\boldsymbol{p}\|^2}\boldsymbol{L}^{7}(\boldsymbol{p}) - \frac{2\sqrt{2}}{3\sqrt{77}\|\boldsymbol{p}\|^2}\boldsymbol{L}^{8}(\boldsymbol{p})$
	$+\frac{\sqrt{7}}{3\sqrt{22}\|\boldsymbol{p}\|^2}\boldsymbol{L}^{9}(\boldsymbol{p}) + \frac{\sqrt{11}}{6\sqrt{14}\|\boldsymbol{p}\|^2}\boldsymbol{L}^{10}(\boldsymbol{p}) + \frac{1}{2\sqrt{154}\|\boldsymbol{p}\|^2}\boldsymbol{L}^{11}(\boldsymbol{p})$
	$-\frac{\sqrt{11}}{6\sqrt{14}\|\boldsymbol{p}\|^2}\boldsymbol{L}^{12}(\boldsymbol{p}) - \frac{2\sqrt{2}}{3\sqrt{77}\|\boldsymbol{p}\|^2}\boldsymbol{L}^{13}(\boldsymbol{p}) + \frac{1}{2\sqrt{154}\|\boldsymbol{p}\|^2}\boldsymbol{L}^{14}(\boldsymbol{p})$
	$-\frac{\sqrt{11}}{6\sqrt{14}\|\boldsymbol{p}\|^2}\boldsymbol{L}^{15}(\boldsymbol{p}) + \frac{5\sqrt{7}}{6\sqrt{22}\|\boldsymbol{p}\|^4}\boldsymbol{L}^{16}(\boldsymbol{p}) - \frac{\sqrt{7}}{\sqrt{22}\|\boldsymbol{p}\|^4}\boldsymbol{L}^{17}(\boldsymbol{p})$
	$+\frac{5\sqrt{7}}{6\sqrt{22}\|\boldsymbol{p}\|^4}\boldsymbol{L}^{18}(\boldsymbol{p}) + \frac{5\sqrt{7}}{6\sqrt{22}\|\boldsymbol{p}\|^4}\boldsymbol{L}^{19}(\boldsymbol{p}) - \frac{\sqrt{7}}{\sqrt{22}\|\boldsymbol{p}\|^4}\boldsymbol{L}^{20}(\boldsymbol{p})$
$\boldsymbol{M}^{\rho_6,1}_{ijki'j'k'}(\boldsymbol{p})$	$\frac{3\sqrt{3}}{4\sqrt{77}}\boldsymbol{L}^{1} - \frac{5}{4\sqrt{231}}\boldsymbol{L}^{2} - \frac{5}{4\sqrt{231}}\boldsymbol{L}^{3} + \frac{1}{2\sqrt{231}}\boldsymbol{L}^{4} + \frac{1}{2\sqrt{231}}\boldsymbol{L}^{5}$
	$-\frac{\sqrt{7}}{4\sqrt{33}\|\boldsymbol{p}\|^2}\boldsymbol{L}^{6}(\boldsymbol{p}) - \frac{\sqrt{7}}{4\sqrt{33}\|\boldsymbol{p}\|^2}\boldsymbol{L}^{7}(\boldsymbol{p}) + \frac{\sqrt{7}}{4\sqrt{33}\|\boldsymbol{p}\|^2}\boldsymbol{L}^{8}(\boldsymbol{p})$
	$+\frac{\sqrt{7}}{4\sqrt{33}\|\boldsymbol{p}\|^2}\boldsymbol{L}^{9}(\boldsymbol{p}) + \frac{\sqrt{7}}{4\sqrt{33}\|\boldsymbol{p}\|^2}\boldsymbol{L}^{11}(\boldsymbol{p}) + \frac{\sqrt{7}}{4\sqrt{33}\|\boldsymbol{p}\|^2}\boldsymbol{L}^{13}(\boldsymbol{p})$
	$+\frac{\sqrt{7}}{4\sqrt{33}\|\boldsymbol{p}\|^2}\boldsymbol{L}^{14}(\boldsymbol{p}) - \frac{\sqrt{21}}{4\sqrt{11}\|\boldsymbol{p}\|^4}\boldsymbol{L}^{16}(\boldsymbol{p}) - \frac{\sqrt{21}}{4\sqrt{11}\|\boldsymbol{p}\|^4}\boldsymbol{L}^{17}(\boldsymbol{p})$
	$-\frac{\sqrt{21}}{4\sqrt{11}\|\boldsymbol{p}\|^4}\boldsymbol{L}^{18}(\boldsymbol{p}) - \frac{\sqrt{21}}{4\sqrt{11}\|\boldsymbol{p}\|^4}\boldsymbol{L}^{19}(\boldsymbol{p}) - \frac{\sqrt{21}}{4\sqrt{11}\|\boldsymbol{p}\|^4}\boldsymbol{L}^{20}(\boldsymbol{p}) + \frac{\sqrt{231}}{4\|\boldsymbol{p}\|^6}\boldsymbol{L}^{21}(\boldsymbol{p})$

Table 3.7 *The non-zero elements of the matrix* $f(\mathbf{0})$.

ijk	$i'j'k'$	$f_{ijki'j'k'}(\mathbf{0})$
$-1-1-1$	$-1-1-1$	$\frac{1}{3\sqrt{3}}f_{0,1}(0)+\frac{2\sqrt{2}}{3\sqrt{15}}f_{0,2}(0)+\frac{4}{15\sqrt{3}}f_{0,3}(0)+\frac{2}{5\sqrt{7}}f_{0,5}(0)$
111	111	
000	000	
$-1-11$	$-1-11$	$\frac{1}{3\sqrt{3}}f_{0,1}(0)\quad\frac{\sqrt{2}}{3\sqrt{15}}f_{0,2}(0)+\frac{1}{15\sqrt{3}}f_{0,3}(0)$
$11-1$	$11-1$	$+\frac{1}{3\sqrt{5}}f_{0,4}(0)+\frac{4}{15\sqrt{7}}f_{0,5}(0)$
$00-1$	$00-1$	
001	001	
$-1-10$	$-1-10$	
110	110	
-100	-100	$\frac{\sqrt{3}}{10}f_{0,3}(0)+\frac{1}{6\sqrt{5}}f_{0,4}(0)+\frac{8}{15\sqrt{7}}f_{0,5}(0)$
010	010	
$-11-1$	$-11-1$	
-111	-111	
$-10-1$	$-10-1$	
011	011	
-101	-101	$\frac{2}{3\sqrt{5}}f_{0,4}(0)+\frac{1}{3\sqrt{7}}f_{0,5}(0)$
$01-1$	$01-1$	
-110	-110	
$-1-1-1$	$11-1$	$\frac{1}{3\sqrt{3}}f_{0,1}(0)+\frac{1}{3\sqrt{30}}f_{0,2}(0)-\frac{2}{15\sqrt{3}}f_{0,3}(0)-\frac{1}{5\sqrt{7}}f_{0,5}(0)$
$-1-11$	111	
$-1-1-1$	$00-1$	
001	111	
$-1-10$	000	
000	110	
$-1-1-1$	-100	$\frac{1}{2\sqrt{15}}f_{0,2}(0)+\frac{\sqrt{2}}{5\sqrt{3}}f_{0,3}(0)-\frac{\sqrt{2}}{5\sqrt{7}}f_{0,5}(0)$
010	111	
$-1-1-1$	-111	
$-11-1$	111	
$-10-1$	000	
000	011	
$-1-11$	010	$\frac{1}{2\sqrt{15}}f_{0,2}(0)-\frac{1}{5\sqrt{6}}f_{0,3}(0)+\frac{1}{3\sqrt{10}}f_{0,4}(0)$
-100	$11-1$	$-\frac{\sqrt{2}}{15\sqrt{7}}f_{0,5}(0)$
$-11-1$	001	
-111	$00-1$	
$-1-10$	011	
$-10-1$	110	
$-1-11$	$-11-1$	$\frac{1}{2\sqrt{15}}f_{0,2}(0)-\frac{1}{5\sqrt{6}}f_{0,3}(0)-\frac{1}{3\sqrt{10}}f_{0,4}(0)+\frac{4\sqrt{2}}{15\sqrt{7}}f_{0,5}(\lambda)$
-111	$11-1$	
-100	$00-1$	

Table 3.7 (*Cont.*)

ijk	$i'j'k'$	$f_{ijki'j'k'}(\mathbf{0})$
001	010	
$-1-10$	$-10-1$	
011	110	
$-1-11$	001	$\frac{1}{3\sqrt{3}}f_{0,1}(0)-\frac{\sqrt{2}}{3\sqrt{15}}f_{0,2}(0)+\frac{1}{15\sqrt{3}}f_{0,3}(0)$
$00-1$	$11-1$	$-\frac{1}{3\sqrt{5}}f_{0,4}(0)-\frac{1}{15\sqrt{7}}f_{0,5}(0)$
$-1-10$	110	
-100	-111	$\frac{\sqrt{3}}{10}f_{0,3}(0)-\frac{1}{6\sqrt{5}}f_{0,4}(0)-\frac{2}{15\sqrt{7}}f_{0,5}(0)$
$-11-1$	010	
$-10-1$	011	
-101	-110	$-\frac{1}{3\sqrt{5}}f_{0,4}(0)+\frac{1}{3\sqrt{7}}f_{0,5}(0)$
-110	$01-1$	
-101	$01-1$	

It is easy to check that the matrix $f(\mathbf{0})$ is non-negative-definite with unit trace if and only if $u_1(0)\geq 0,\dots,u_4(0)\geq 0$, $u_1(0)+\cdots+u_4(0)=1$ and $|u_5(0)|\leq u_3(0)$. The convex compact $\mathcal{C}_1$, where the matrix $f(\mathbf{0})$ can take values, is a simplex with five vertices $A_1=\{u_1(0)=1\}$, $A_2=\{u_2(0)=1\}$, $A_{3,4}=\{u_3(0)=1,u_5(0)=\pm 1\}$ and $A_5=\{u_4(0)=1\}$. The corresponding non-zero values of the functions $f_{0,m}(0)$ are as follows:

$$A_1\colon f_{0,1}(0)=\frac{1}{\sqrt{3}},\, f_{0,2}(0)=\frac{\sqrt{5}}{\sqrt{6}}$$

$$A_2\colon f_{0,1}(0)=\frac{1}{\sqrt{3}},\, f_{0,2}(0)=-\frac{\sqrt{5}}{2\sqrt{6}},$$

$$A_3\colon f_{0,1}(0)=-\frac{4}{3\sqrt{3}},\, f_{0,2}(0)=\frac{\sqrt{5}}{3\sqrt{6}},\, f_{0,3}(0)=\frac{5}{6\sqrt{3}},\, f_{0,4}(0)=\frac{\sqrt{5}}{2},$$

$$A_4\colon f_{0,2}(0)=-\frac{\sqrt{5}}{\sqrt{6}},\, f_{0,3}(0)=-\frac{1}{2\sqrt{3}},\, f_{0,4}(0)=-\frac{5\sqrt{5}}{6},\, f_{0,5}(0)=\frac{2\sqrt{7}}{3},$$

$$A_5\colon f_{0,1}(0)=-\frac{2}{3\sqrt{3}},\, f_{0,2}(0)=\frac{\sqrt{10}}{3\sqrt{3}},\, f_{0,3}(0)=\frac{2}{3\sqrt{3}},\, f_{0,4}(0)=\frac{2\sqrt{5}}{3},\, f_{0,5}(0)=-\frac{\sqrt{7}}{3}.$$

Using Table 3.7, the reader can easily calculate all matrix entries of the matrices A_1–A_5.

Exactly as in Section 3.6, we prove that

$$f(\lambda,0,0)=\sum_{\ell=0}^{3}\sum_{n=1}^{q_\ell}\mathsf{T}^{\rho_{2\ell,n},0}f_{2\ell,n}(\lambda). \tag{3.90}$$

The results of calculations are given in Table 3.8.

Table 3.8 *The non-zero elements of the matrix* $f(\lambda, 0, 0)$.

ijk	$i'j'k'$	$f_{ijki'j'k'}(\lambda, 0, 0)$
$-1-1-1$	$-1-1-1$	$\frac{1}{3\sqrt{3}} f_{0,1}(\lambda) + \frac{2\sqrt{2}}{3\sqrt{15}} f_{0,2}(\lambda) + \frac{4}{15\sqrt{3}} f_{0,3}(\lambda)$
111	111	$+\frac{2}{5\sqrt{7}} f_{0,5}(\lambda) - \frac{1}{3\sqrt{6}} f_{2,1}(\lambda) - \frac{2}{3\sqrt{15}} f_{2,2}(\lambda)$
		$-\frac{2\sqrt{2}}{15\sqrt{3}} f_{2,3}(\lambda) - \frac{1}{\sqrt{35}} f_{2,5}(\lambda) - \frac{2}{5\sqrt{7}} f_{2,8}(\lambda)$
		$-\frac{2}{5\sqrt{21}} f_{2,10}(\lambda) + \frac{\sqrt{3}}{2\sqrt{35}} f_{4,1}(\lambda) + \frac{\sqrt{3}}{5\sqrt{7}} f_{4,3}(\lambda)$
		$+\frac{9}{10\sqrt{154}} f_{4,5}(\lambda) - \frac{5}{4\sqrt{231}} f_{6,1}(\lambda)$
$-1-11$	$-1-11$	$\frac{1}{3\sqrt{3}} f_{0,1}(\lambda) - \frac{\sqrt{2}}{3\sqrt{15}} f_{0,2}(\lambda) + \frac{1}{15\sqrt{3}} f_{0,3}(\lambda)$
$11-1$	$11-1$	$+\frac{1}{3\sqrt{5}} f_{0,4}(\lambda) + \frac{4}{15\sqrt{7}} f_{0,5}(\lambda) - \frac{1}{3\sqrt{6}} f_{2,1}(\lambda)$
		$+\frac{1}{3\sqrt{15}} f_{2,2}(\lambda) - \frac{1}{15\sqrt{6}} f_{2,3}(\lambda) - \frac{1}{3} f_{2,4}(\lambda)$
		$-\frac{1}{3\sqrt{35}} f_{2,5}(\lambda) + \frac{1}{3\sqrt{14}} f_{2,6}(\lambda) + \frac{1}{3\sqrt{5}} f_{2,7}(\lambda)$
		$+\frac{1}{15\sqrt{7}} f_{2,8}(\lambda) - \frac{1}{3\sqrt{70}} f_{2,9}(\lambda) - \frac{\sqrt{3}}{5\sqrt{7}} f_{2,10}(\lambda)$
		$+\frac{1}{2\sqrt{105}} f_{4,1}(\lambda) - \frac{2\sqrt{2}}{3\sqrt{35}} f_{4,2}(\lambda) - \frac{1}{10\sqrt{21}} f_{4,3}(\lambda)$
		$-\frac{1}{6\sqrt{7}} f_{4,4}(\lambda) + \frac{23}{30\sqrt{154}} f_{4,5}(\lambda) - \frac{1}{4\sqrt{231}} f_{6,1}(\lambda)$
-100	-100	$\frac{\sqrt{3}}{10} f_{0,3}(\lambda) + \frac{1}{6\sqrt{5}} f_{0,4}(\lambda) + \frac{8}{15\sqrt{7}} f_{0,5}(\lambda)$
010	010	$-\frac{\sqrt{3}}{10\sqrt{2}} f_{2,3}(\lambda) + \frac{1}{6\sqrt{14}} f_{2,6}(\lambda) - \frac{1}{2\sqrt{5}} f_{2,7}(\lambda)$
		$+\frac{4}{5\sqrt{7}} f_{2,8}(\lambda) + \frac{2\sqrt{2}}{3\sqrt{35}} f_{2,9}(\lambda) + \frac{4}{5\sqrt{21}} f_{2,10}(\lambda)$
		$-\frac{\sqrt{2}}{3\sqrt{35}} f_{4,2}(\lambda) - \frac{2\sqrt{3}}{5\sqrt{7}} f_{4,3}(\lambda) + \frac{2}{3\sqrt{7}} f_{4,4}(\lambda)$
		$+\frac{4\sqrt{2}}{15\sqrt{77}} f_{4,5}(\lambda) - \frac{4}{\sqrt{231}} f_{6,1}(\lambda)$
$-11-1$	$-11-1$	$\frac{\sqrt{3}}{10} f_{0,3}(\lambda) + \frac{1}{6\sqrt{5}} f_{0,4}(\lambda) + \frac{8}{15\sqrt{7}} f_{0,5}(\lambda)$
-111	-111	$-\frac{\sqrt{3}}{10\sqrt{2}} f_{2,3}(\lambda) + \frac{1}{6\sqrt{14}} f_{2,6}(\lambda) + \frac{1}{2\sqrt{5}} f_{2,7}(\lambda) - \frac{1}{5\sqrt{7}} f_{2,8}(\lambda)$
		$+\frac{1}{3\sqrt{70}} f_{2,9}(\lambda) - \frac{2\sqrt{3}}{5\sqrt{7}} f_{2,10}(\lambda) - \frac{\sqrt{2}}{3\sqrt{35}} f_{4,2}(\lambda)$
		$+\frac{\sqrt{3}}{10\sqrt{7}} f_{4,3}(\lambda) + \frac{1}{6\sqrt{7}} f_{4,4}(\lambda) + \frac{23}{15\sqrt{154}} f_{4,5}(\lambda) - \frac{1}{2\sqrt{231}} f_{6,1}(\lambda)$
$00-1$	$00-1$	$\frac{1}{3\sqrt{3}} f_{0,1}(\lambda) - \frac{\sqrt{2}}{3\sqrt{15}} f_{0,2}(\lambda) + \frac{1}{15\sqrt{3}} f_{0,3}(\lambda)$
001	001	$+\frac{1}{3\sqrt{5}} f_{0,4}(\lambda) + \frac{4}{15\sqrt{7}} f_{0,5}(\lambda) - \frac{1}{3\sqrt{6}} f_{2,1}(\lambda)$
		$+\frac{1}{3\sqrt{15}} f_{2,2}(\lambda) - \frac{1}{15\sqrt{6}} f_{2,3}(\lambda) + \frac{1}{3} f_{2,4}(\lambda)$
		$+\frac{4}{3\sqrt{35}} f_{2,5}(\lambda) + \frac{1}{3\sqrt{14}} f_{2,6}(\lambda) - \frac{1}{3\sqrt{5}} f_{2,7}(\lambda)$
		$-\frac{4}{15\sqrt{7}} f_{2,8}(\lambda) - \frac{2\sqrt{2}}{3\sqrt{35}} f_{2,9}(\lambda) + \frac{2}{5\sqrt{21}} f_{2,10}(\lambda)$
		$-\frac{2}{\sqrt{105}} f_{4,1}(\lambda) - \frac{2\sqrt{2}}{3\sqrt{35}} f_{4,2}(\lambda) + \frac{2}{5\sqrt{21}} f_{4,3}(\lambda)$
		$-\frac{2}{3\sqrt{7}} f_{4,4}(\lambda) + \frac{2\sqrt{2}}{15\sqrt{77}} f_{4,5}(\lambda) - \frac{2}{\sqrt{231}} f_{6,1}(\lambda)$
$-1-10$	$-1-10$	$\frac{1}{3\sqrt{3}} f_{0,1}(\lambda) - \frac{\sqrt{2}}{3\sqrt{15}} f_{0,2}(\lambda) + \frac{1}{15\sqrt{3}} f_{0,3}(\lambda)$
110	110	$+\frac{1}{3\sqrt{5}} f_{0,4}(\lambda) + \frac{4}{15\sqrt{7}} f_{0,5}(\lambda) + \frac{\sqrt{2}}{3\sqrt{3}} f_{2,1}(\lambda)$
		$-\frac{2}{3\sqrt{15}} f_{2,2}(\lambda) + \frac{\sqrt{2}}{15\sqrt{3}} f_{2,3}(\lambda) - \frac{1}{\sqrt{35}} f_{2,5}(\lambda)$

Table 3.8 (*Cont.*)

ijk	$i'j'k'$	$f_{ijki'j'k'}(\boldsymbol{\lambda},\mathbf{o},\mathbf{o})$
		$-\frac{\sqrt{2}}{3\sqrt{7}}f_{2,6}(\lambda)+\frac{1}{5\sqrt{7}}f_{2,8}(\lambda)+\frac{\sqrt{5}}{3\sqrt{14}}f_{2,9}(\lambda)$
		$+\frac{1}{5\sqrt{21}}f_{2,10}(\lambda)-\frac{2}{\sqrt{105}}f_{4,1}(\lambda)+\frac{1}{3\sqrt{70}}f_{4,2}(\lambda)$
		$+\frac{2}{5\sqrt{21}}f_{4,3}(\lambda)-\frac{1}{3\sqrt{7}}f_{4,4}(\lambda)-\frac{17}{30\sqrt{154}}f_{4,5}(\lambda)+\frac{\sqrt{3}}{2\sqrt{77}}f_{6,1}(\lambda)$
$-10-1$	$-10-1$	$\frac{\sqrt{3}}{10}f_{0,3}(\lambda)+\frac{1}{6\sqrt{5}}f_{0,4}(\lambda)+\frac{8}{15\sqrt{7}}f_{0,5}(\lambda)$
011	011	$+\frac{\sqrt{3}}{5\sqrt{2}}f_{2,3}(\lambda)-\frac{1}{3\sqrt{14}}f_{2,6}(\lambda)-\frac{3}{5\sqrt{7}}f_{2,8}(\lambda)$
		$-\frac{\sqrt{5}}{3\sqrt{14}}f_{2,9}(\lambda)+\frac{2}{5\sqrt{21}}f_{2,10}(\lambda)+\frac{1}{6\sqrt{70}}f_{4,2}(\lambda)$
		$-\frac{2\sqrt{3}}{5\sqrt{7}}f_{4,3}(\lambda)+\frac{1}{3\sqrt{7}}f_{4,4}(\lambda)-\frac{17}{15\sqrt{154}}f_{4,5}(\lambda)+\frac{\sqrt{3}}{\sqrt{77}}f_{6,1}(\lambda)$
000	000	$\frac{1}{3\sqrt{3}}f_{0,1}(\lambda)+\frac{2\sqrt{2}}{3\sqrt{15}}f_{0,2}(\lambda)+\frac{4}{15\sqrt{3}}f_{0,3}(\lambda)$
		$+\frac{2}{5\sqrt{7}}f_{0,5}(\lambda)+\frac{\sqrt{2}}{3\sqrt{3}}f_{2,1}(\lambda)+\frac{4}{3\sqrt{15}}f_{2,2}(\lambda)$
		$+\frac{4\sqrt{2}}{15\sqrt{3}}f_{2,3}(\lambda)+\frac{2}{\sqrt{35}}f_{2,5}(\lambda)+\frac{4}{5\sqrt{7}}f_{2,8}(\lambda)$
		$+\frac{4}{5\sqrt{21}}f_{2,10}(\lambda)+\frac{4}{\sqrt{105}}f_{4,1}(\lambda)+\frac{8}{5\sqrt{21}}f_{4,3}(\lambda)$
		$+\frac{6\sqrt{2}}{5\sqrt{77}}f_{4,5}(\lambda)+\frac{4}{\sqrt{231}}f_{6,1}(\lambda)$
-101	-101	$\frac{2}{3\sqrt{5}}f_{0,4}(\lambda)+\frac{1}{3\sqrt{7}}f_{0,5}(\lambda)+\frac{\sqrt{2}}{3\sqrt{7}}f_{2,6}(\lambda)$
$01-1$	$01-1$	$-\frac{\sqrt{5}}{3\sqrt{14}}f_{2,9}(\lambda)+\frac{19}{6\sqrt{70}}f_{4,2}(\lambda)+\frac{1}{3\sqrt{7}}f_{4,4}(\lambda)$
		$-\frac{\sqrt{7}}{3\sqrt{22}}f_{4,5}(\lambda)+\frac{1}{\sqrt{231}}f_{6,1}(\lambda)$
-110	-110	$\frac{2}{3\sqrt{5}}f_{0,4}(\lambda)+\frac{1}{3\sqrt{7}}f_{0,5}(\lambda)-\frac{2\sqrt{2}}{3\sqrt{7}}f_{2,6}(\lambda)+\frac{\sqrt{10}}{3\sqrt{7}}f_{2,9}(\lambda)$
		$+\frac{\sqrt{2}}{3\sqrt{35}}f_{4,2}(\lambda)-\frac{2}{3\sqrt{7}}f_{4,4}(\lambda)-\frac{\sqrt{7}}{3\sqrt{22}}f_{4,5}(\lambda)+\frac{1}{\sqrt{231}}f_{6,1}(\lambda)$
$-1-1-1$	$11-1$	$\frac{1}{3\sqrt{3}}f_{0,1}(\lambda)+\frac{1}{3\sqrt{30}}f_{0,2}(\lambda)-\frac{2}{15\sqrt{3}}f_{0,3}(\lambda)$
$-1-11$	111	$-\frac{1}{5\sqrt{7}}f_{0,5}(\lambda)-\frac{1}{3\sqrt{6}}f_{2,1}(\lambda)-\frac{1}{6\sqrt{15}}f_{2,2}(\lambda)$
		$+\frac{\sqrt{2}}{15\sqrt{3}}f_{2,3}(\lambda)-\frac{1}{6}f_{2,4}(\lambda)-\frac{2}{3\sqrt{35}}f_{2,5}(\lambda)$
		$-\frac{1}{3\sqrt{5}}f_{2,7}(\lambda)+\frac{1}{30\sqrt{7}}f_{2,8}(\lambda)-\frac{1}{2\sqrt{70}}f_{2,9}(\lambda)$
		$+\frac{\sqrt{7}}{10\sqrt{3}}f_{2,10}(\lambda)+\frac{1}{\sqrt{105}}f_{4,1}(\lambda)-\frac{1}{20\sqrt{21}}f_{4,3}(\lambda)$
		$-\frac{1}{4\sqrt{7}}f_{4,4}(\lambda)-\frac{\sqrt{7}}{10\sqrt{22}}f_{4,5}(\lambda)-\frac{1}{4\sqrt{231}}f_{6,1}(\lambda)$
$-1-1-1$	-100	$\frac{1}{2\sqrt{15}}f_{0,2}(\lambda)+\frac{\sqrt{2}}{5\sqrt{3}}f_{0,3}(\lambda)-\frac{\sqrt{2}}{5\sqrt{7}}f_{0,5}(\lambda)$
010	111	$-\frac{1}{2\sqrt{30}}f_{2,2}(\lambda)-\frac{1}{5\sqrt{3}}f_{2,3}(\lambda)-\frac{1}{6\sqrt{2}}f_{2,4}(\lambda)$
		$+\frac{2\sqrt{2}}{3\sqrt{35}}f_{2,5}(\lambda)-\frac{1}{3\sqrt{10}}f_{2,7}(\lambda)+\frac{\sqrt{7}}{30\sqrt{2}}f_{2,8}(\lambda)$
		$-\frac{1}{4\sqrt{35}}f_{2,9}(\lambda)-\frac{\sqrt{3}}{5\sqrt{14}}f_{2,10}(\lambda)-\frac{\sqrt{2}}{\sqrt{105}}f_{4,1}(\lambda)$
		$-\frac{\sqrt{7}}{20\sqrt{6}}f_{4,3}(\lambda)-\frac{1}{4\sqrt{14}}f_{4,4}(\lambda)-\frac{1}{5\sqrt{77}}f_{4,5}(\lambda)+\frac{\sqrt{3}}{\sqrt{154}}f_{6,1}(\lambda)$
$-1-1-1$	-111	$\frac{1}{2\sqrt{15}}f_{0,2}(\lambda)+\frac{\sqrt{2}}{5\sqrt{3}}f_{0,3}(\lambda)-\frac{\sqrt{2}}{5\sqrt{7}}f_{0,5}(\lambda)$

Table 3.8 (*Cont.*)

ijk	$i'j'k'$	$f_{ijki'j'k'}(\boldsymbol{\lambda},\mathbf{o},\mathbf{o})$
$-11-1$	111	$-\frac{1}{2\sqrt{30}}f_{2,2}(\lambda)-\frac{1}{5\sqrt{3}}f_{2,3}(\lambda)+\frac{1}{6\sqrt{2}}f_{2,4}(\lambda)$
		$-\frac{1}{3\sqrt{70}}f_{2,5}(\lambda)+\frac{1}{3\sqrt{10}}f_{2,7}(\lambda)-\frac{13}{30\sqrt{14}}f_{2,8}(\lambda)$
		$+\frac{1}{4\sqrt{35}}f_{2,9}(\lambda)+\frac{\sqrt{7}}{5\sqrt{6}}f_{2,10}(\lambda)+\frac{1}{2\sqrt{210}}f_{4,1}(\lambda)$
		$+\frac{13}{20\sqrt{42}}f_{4,3}(\lambda)+\frac{1}{4\sqrt{14}}f_{4,4}(\lambda)-\frac{\sqrt{7}}{10\sqrt{11}}f_{4,5}(\lambda)-\frac{1}{2\sqrt{462}}f_{6,1}(\lambda)$
$-1-1-1$	$00-1$	$\frac{1}{3\sqrt{3}}f_{0,1}(\lambda)+\frac{1}{3\sqrt{30}}f_{0,2}(\lambda)-\frac{2}{15\sqrt{3}}f_{0,3}(\lambda)$
001	111	$-\frac{1}{5\sqrt{7}}f_{0,5}(\lambda)-\frac{1}{3\sqrt{6}}f_{2,1}(\lambda)-\frac{1}{6\sqrt{15}}f_{2,2}(\lambda)$
		$+\frac{\sqrt{2}}{15\sqrt{3}}f_{2,3}(\lambda)+\frac{1}{6}f_{2,4}(\lambda)+\frac{1}{6\sqrt{35}}f_{2,5}(\lambda)+\frac{1}{3\sqrt{5}}f_{2,7}(\lambda)$
		$+\frac{11}{30\sqrt{7}}f_{2,8}(\lambda)+\frac{1}{2\sqrt{70}}f_{2,9}(\lambda)-\frac{\sqrt{3}}{10\sqrt{7}}f_{2,10}(\lambda)-\frac{1}{4\sqrt{105}}f_{4,1}(\lambda)$
		$-\frac{11}{20\sqrt{21}}f_{4,3}(\lambda)+\frac{1}{4\sqrt{7}}f_{4,4}(\lambda)-\frac{1}{5\sqrt{154}}f_{4,5}(\lambda)+\frac{\sqrt{3}}{2\sqrt{77}}f_{6,1}(\lambda)$
$-1-11$	010	$\frac{1}{2\sqrt{15}}f_{0,2}(\lambda)-\frac{1}{5\sqrt{6}}f_{0,3}(\lambda)+\frac{1}{3\sqrt{10}}f_{0,4}(\lambda)$
-100	$11-1$	$-\frac{\sqrt{2}}{15\sqrt{7}}f_{0,5}(\lambda)-\frac{1}{2\sqrt{30}}f_{2,2}(\lambda)+\frac{1}{10\sqrt{3}}f_{2,3}(\lambda)$
		$-\frac{1}{6\sqrt{2}}f_{2,4}(\lambda)+\frac{2\sqrt{2}}{3\sqrt{35}}f_{2,5}(\lambda)+\frac{1}{6\sqrt{7}}f_{2,6}(\lambda)$
		$-\frac{1}{3\sqrt{10}}f_{2,7}(\lambda)-\frac{11}{30\sqrt{14}}f_{2,8}(\lambda)+\frac{\sqrt{7}}{12\sqrt{5}}f_{2,9}(\lambda)$
		$-\frac{1}{5\sqrt{42}}f_{2,10}(\lambda)-\frac{\sqrt{2}}{\sqrt{105}}f_{4,1}(\lambda)-\frac{2}{3\sqrt{35}}f_{4,2}(\lambda)$
		$+\frac{11}{20\sqrt{42}}f_{4,3}(\lambda)+\frac{\sqrt{7}}{12\sqrt{2}}f_{4,4}(\lambda)-\frac{1}{15\sqrt{77}}f_{4,5}(\lambda)+\frac{1}{\sqrt{462}}f_{6,1}(\lambda)$
$-1-11$	$-11-1$	$\frac{1}{2\sqrt{15}}f_{0,2}(\lambda)-\frac{1}{5\sqrt{6}}f_{0,3}(\lambda)-\frac{1}{3\sqrt{10}}f_{0,4}(\lambda)$
-111	$11-1$	$+\frac{4\sqrt{2}}{15\sqrt{7}}f_{0,5}(\lambda)-\frac{1}{2\sqrt{30}}f_{2,2}(\lambda)+\frac{1}{10\sqrt{3}}f_{2,3}(\lambda)$
		$+\frac{1}{6\sqrt{2}}f_{2,4}(\lambda)-\frac{1}{3\sqrt{70}}f_{2,5}(\lambda)-\frac{1}{6\sqrt{7}}f_{2,6}(\lambda)$
		$-\frac{\sqrt{2}}{3\sqrt{5}}f_{2,7}(\lambda)-\frac{1}{30\sqrt{14}}f_{2,8}(\lambda)-\frac{1}{12\sqrt{35}}f_{2,9}(\lambda)$
		$-\frac{\sqrt{6}}{5\sqrt{7}}f_{2,10}(\lambda)+\frac{1}{2\sqrt{210}}f_{4,1}(\lambda)+\frac{2}{3\sqrt{35}}f_{4,2}(\lambda)$
		$+\frac{1}{20\sqrt{42}}f_{4,3}(\lambda)-\frac{1}{12\sqrt{14}}f_{4,4}(\lambda)+\frac{23}{30\sqrt{77}}f_{4,5}(\lambda)-\frac{1}{2\sqrt{462}}f_{6,1}(\lambda)$
$-1-11$	001	$\frac{1}{3\sqrt{3}}f_{0,1}(\lambda)-\frac{\sqrt{2}}{3\sqrt{15}}f_{0,2}(\lambda)+\frac{1}{15\sqrt{3}}f_{0,3}(\lambda)$
$00-1$	$11-1$	$-\frac{1}{3\sqrt{5}}f_{0,4}(\lambda)-\frac{1}{15\sqrt{7}}f_{0,5}(\lambda)-\frac{1}{3\sqrt{6}}f_{2,1}(\lambda)$
		$+\frac{1}{3\sqrt{15}}f_{2,2}(\lambda)-\frac{1}{15\sqrt{6}}f_{2,3}(\lambda)+\frac{1}{2\sqrt{35}}f_{2,5}(\lambda)$
		$-\frac{1}{3\sqrt{14}}f_{2,6}(\lambda)-\frac{1}{10\sqrt{7}}f_{2,8}(\lambda)+\frac{\sqrt{5}}{6\sqrt{14}}f_{2,9}(\lambda)$
		$-\frac{1}{10\sqrt{21}}f_{2,10}(\lambda)\frac{\sqrt{3}}{4\sqrt{35}}f_{4,1}(\lambda)+\frac{2\sqrt{2}}{3\sqrt{35}}f_{4,2}(\lambda)$
		$+\frac{\sqrt{3}}{20\sqrt{7}}f_{4,3}(\lambda)+\frac{5}{12\sqrt{7}}f_{4,4}(\lambda)-\frac{1}{15\sqrt{154}}f_{4,5}(\lambda)+\frac{1}{2\sqrt{231}}f_{6,1}(\lambda)$
-100	-111	$\frac{\sqrt{3}}{10}f_{0,3}(\lambda)-\frac{1}{6\sqrt{5}}f_{0,4}(\lambda)-\frac{2}{15\sqrt{7}}f_{0,5}(\lambda)$
$-11-1$	010	$-\frac{\sqrt{3}}{10\sqrt{2}}f_{2,3}(\lambda)-\frac{1}{6\sqrt{14}}f_{2,6}(\lambda)+\frac{3}{10\sqrt{7}}f_{2,8}(\lambda)$
		$-\frac{\sqrt{5}}{6\sqrt{14}}f_{2,9}(\lambda)-\frac{1}{5\sqrt{21}}f_{2,10}(\lambda)+\frac{\sqrt{2}}{3\sqrt{35}}f_{4,2}(\lambda)$
		$-\frac{3\sqrt{3}}{20\sqrt{7}}f_{4,3}(\lambda)-\frac{5}{12\sqrt{7}}f_{4,4}(\lambda)-\frac{\sqrt{2}}{15\sqrt{77}}f_{4,5}(\lambda)+\frac{1}{\sqrt{231}}f_{6,1}(\lambda)$

Table 3.8 (*Cont.*)

ijk	$i'j'k'$	$f_{ijki'j'k'}(\boldsymbol{\lambda}, \mathbf{o}, \mathbf{o})$
-100	$00-1$	$\frac{1}{2\sqrt{15}} f_{0,2}(\lambda) - \frac{1}{5\sqrt{6}} f_{0,3}(\lambda) - \frac{1}{3\sqrt{10}} f_{0,4}(\lambda)$
001	010	$+\frac{4\sqrt{2}}{15\sqrt{7}} f_{0,5}(\lambda) - \frac{1}{2\sqrt{30}} f_{2,2}(\lambda) + \frac{1}{10\sqrt{3}} f_{2,3}(\lambda)$
		$-\frac{1}{6\sqrt{2}} f_{2,4}(\lambda) + \frac{2\sqrt{2}}{3\sqrt{35}} f_{2,5}(\lambda) - \frac{1}{6\sqrt{7}} f_{2,6}(\lambda)$
		$+\frac{\sqrt{2}}{3\sqrt{5}} f_{2,7}(\lambda) + \frac{\sqrt{2}}{15\sqrt{7}} f_{2,8}(\lambda) - \frac{1}{3\sqrt{35}} f_{2,9}(\lambda)$
		$+\frac{2\sqrt{2}}{5\sqrt{21}} f_{2,10}(\lambda) - \frac{\sqrt{2}}{\sqrt{105}} f_{4,1}(\lambda) + \frac{2}{3\sqrt{35}} f_{4,2}(\lambda)$
		$-\frac{1}{5\sqrt{42}} f_{4,3}(\lambda) - \frac{1}{3\sqrt{14}} f_{4,4}(\lambda) + \frac{4}{15\sqrt{77}} f_{4,5}(\lambda) - \frac{2\sqrt{2}}{\sqrt{231}} f_{6,1}(\lambda)$
$-11-1$	001	$\frac{1}{2\sqrt{15}} f_{0,2}(\lambda) - \frac{1}{5\sqrt{6}} f_{0,3}(\lambda) + \frac{1}{3\sqrt{10}} f_{0,4}(\lambda)$
-111	$00-1$	$-\frac{\sqrt{2}}{15\sqrt{7}} f_{0,5}(\lambda) - \frac{1}{2\sqrt{30}} f_{2,2}(\lambda) + \frac{1}{10\sqrt{3}} f_{2,3}(\lambda)$
		$+\frac{1}{6\sqrt{2}} f_{2,4}(\lambda) - \frac{1}{3\sqrt{70}} f_{2,5}(\lambda) + \frac{1}{6\sqrt{7}} f_{2,6}(\lambda)$
		$+\frac{1}{3\sqrt{10}} f_{2,7}(\lambda) + \frac{\sqrt{7}}{15\sqrt{2}} f_{2,8}(\lambda) - \frac{1}{6\sqrt{35}} f_{2,9}(\lambda)$
		$-\frac{1}{5\sqrt{42}} f_{2,10}(\lambda) + \frac{1}{2\sqrt{210}} f_{4,1}(\lambda) - \frac{2}{3\sqrt{35}} f_{4,2}(\lambda)$
		$-\frac{\sqrt{7}}{10\sqrt{6}} f_{4,3}(\lambda) - \frac{1}{6\sqrt{14}} f_{4,4}(\lambda) - \frac{1}{15\sqrt{77}} f_{4,5}(\lambda) + \frac{1}{\sqrt{462}} f_{6,1}(\lambda)$
$-1-10$	$-10-1$	$\frac{1}{2\sqrt{15}} f_{0,2}(\lambda) - \frac{1}{5\sqrt{6}} f_{0,3}(\lambda) - \frac{1}{3\sqrt{10}} f_{0,4}(\lambda)$
011	110	$+\frac{4\sqrt{2}}{15\sqrt{7}} f_{0,5}(\lambda) + \frac{1}{\sqrt{30}} f_{2,2}(\lambda) - \frac{1}{5\sqrt{3}} f_{2,3}(\lambda)$
		$-\frac{1}{\sqrt{70}} f_{2,5}(\lambda) + \frac{1}{3\sqrt{7}} f_{2,6}(\lambda) - \frac{1}{10\sqrt{14}} f_{2,8}(\lambda)$
		$+\frac{\sqrt{5}}{12\sqrt{7}} f_{2,9}(\lambda) + \frac{\sqrt{2}}{5\sqrt{21}} f_{2,10}(\lambda) - \frac{\sqrt{2}}{\sqrt{105}} f_{4,1}(\lambda)$
		$-\frac{1}{6\sqrt{35}} f_{4,2}(\lambda) - \frac{1}{5\sqrt{42}} f_{4,3}(\lambda) - \frac{1}{6\sqrt{14}} f_{4,4}(\lambda)$
		$-\frac{17}{30\sqrt{77}} f_{4,5}(\lambda) + \frac{\sqrt{3}}{\sqrt{154}} f_{6,1}(\lambda)$
$-1-10$	000	$\frac{1}{3\sqrt{3}} f_{0,1}(\lambda) + \frac{1}{3\sqrt{30}} f_{0,2}(\lambda) - \frac{2}{15\sqrt{3}} f_{0,3}(\lambda)$
000	110	$-\frac{1}{5\sqrt{7}} f_{0,5}(\lambda) + \frac{\sqrt{2}}{3\sqrt{3}} f_{2,1}(\lambda) + \frac{1}{3\sqrt{15}} f_{2,2}(\lambda)$
		$-\frac{2\sqrt{2}}{15\sqrt{3}} f_{2,3}(\lambda) + \frac{1}{2\sqrt{35}} f_{2,5}(\lambda) - \frac{2}{5\sqrt{7}} f_{2,8}(\lambda)$
		$-\frac{2}{5\sqrt{21}} f_{2,10}(\lambda) + \frac{1}{\sqrt{105}} f_{4,1}(\lambda) - \frac{4}{5\sqrt{21}} f_{4,3}(\lambda)$
		$-\frac{3\sqrt{2}}{5\sqrt{77}} f_{4,5}(\lambda) - \frac{2}{\sqrt{231}} f_{6,1}(\lambda)$
$-1-10$	011	$\frac{1}{2\sqrt{15}} f_{0,2}(\lambda) - \frac{1}{5\sqrt{6}} f_{0,3}(\lambda) + \frac{1}{3\sqrt{10}} f_{0,4}(\lambda)$
$-10-1$	110	$-\frac{\sqrt{2}}{15\sqrt{7}} f_{0,5}(\lambda) + \frac{1}{\sqrt{30}} f_{2,2}(\lambda) - \frac{1}{5\sqrt{3}} f_{2,3}(\lambda)$
		$-\frac{1}{\sqrt{70}} f_{2,5}(\lambda) - \frac{1}{3\sqrt{7}} f_{2,6}(\lambda) - \frac{1}{10\sqrt{14}} f_{2,8}(\lambda)$
		$-\frac{\sqrt{5}}{12\sqrt{7}} f_{2,9}(\lambda) + \frac{\sqrt{2}}{5\sqrt{21}} f_{2,10}(\lambda) - \frac{\sqrt{2}}{\sqrt{105}} f_{4,1}(\lambda)$
		$+\frac{1}{6\sqrt{35}} f_{4,2}(\lambda) - \frac{1}{5\sqrt{42}} f_{4,3}(\lambda) + \frac{1}{6\sqrt{14}} f_{4,4}(\lambda)$
		$+\frac{53}{30\sqrt{77}} f_{4,5}(\lambda) + \frac{1}{\sqrt{462}} f_{6,1}(\lambda)$
$-1-10$	110	$\frac{1}{3\sqrt{3}} f_{0,1}(\lambda) - \frac{\sqrt{2}}{3\sqrt{15}} f_{0,2}(\lambda) + \frac{1}{15\sqrt{3}} f_{0,3}(\lambda)$
		$-\frac{1}{3\sqrt{5}} f_{0,4}(\lambda) - \frac{1}{15\sqrt{7}} f_{0,5}(\lambda) + \frac{\sqrt{2}}{3\sqrt{3}} f_{2,1}(\lambda)$

Table 3.8 (*Cont.*)

$\boldsymbol{ijk}$	$\boldsymbol{i'j'k'}$	$\boldsymbol{f_{ijki'j'k'}(\lambda, o, o)}$
		$-\frac{2}{3\sqrt{15}}f_{2,2}(\lambda) + \frac{\sqrt{2}}{15\sqrt{3}}f_{2,3}(\lambda) - \frac{1}{\sqrt{35}}f_{2,5}(\lambda)$
		$+\frac{\sqrt{2}}{3\sqrt{7}}f_{2,6}(\lambda) + \frac{1}{5\sqrt{7}}f_{2,8}(\lambda) - \frac{\sqrt{5}}{3\sqrt{14}}f_{2,9}(\lambda)$
		$+\frac{1}{5\sqrt{21}}f_{2,10}(\lambda) - \frac{2}{\sqrt{105}}f_{4,1}(\lambda) - \frac{1}{3\sqrt{70}}f_{4,2}(\lambda)$
		$+\frac{2}{5\sqrt{21}}f_{4,3}(\lambda) + \frac{1}{3\sqrt{7}}f_{4,4}(\lambda) + \frac{53}{30\sqrt{154}}f_{4,5}(\lambda) + \frac{1}{2\sqrt{231}}f_{6,1}(\lambda)$
$-10-1$	000	$\frac{1}{2\sqrt{15}}f_{0,2}(\lambda) + \frac{\sqrt{2}}{5\sqrt{3}}f_{0,3}(\lambda) - \frac{\sqrt{2}}{5\sqrt{7}}f_{0,5}(\lambda)$
000	011	$+\frac{1}{\sqrt{30}}f_{2,2}(\lambda) + \frac{2}{5\sqrt{3}}f_{2,3}(\lambda) - \frac{1}{\sqrt{70}}f_{2,5}(\lambda)$
		$+\frac{1}{5\sqrt{14}}f_{2,8}(\lambda) - \frac{2\sqrt{2}}{5\sqrt{21}}f_{2,10}(\lambda) - \frac{\sqrt{2}}{\sqrt{105}}f_{4,1}(\lambda)$
		$+\frac{\sqrt{2}}{5\sqrt{21}}f_{4,3}(\lambda) - \frac{6}{5\sqrt{77}}f_{4,5}(\lambda) - \frac{2\sqrt{2}}{\sqrt{231}}f_{6,1}(\lambda)$
$-10-1$	011	$\frac{\sqrt{3}}{10}f_{0,3}(\lambda) - \frac{1}{6\sqrt{5}}f_{0,4}(\lambda) - \frac{2}{15\sqrt{7}}f_{0,5}(\lambda)$
		$+\frac{\sqrt{3}}{5\sqrt{2}}f_{2,3}(\lambda) + \frac{1}{3\sqrt{14}}f_{2,6}(\lambda) - \frac{3}{5\sqrt{7}}f_{2,8}(\lambda)$
		$+\frac{\sqrt{5}}{3\sqrt{14}}f_{2,9}(\lambda) + \frac{2}{5\sqrt{21}}f_{2,10}(\lambda) - \frac{1}{6\sqrt{70}}f_{4,2}(\lambda)$
		$-\frac{2\sqrt{3}}{5\sqrt{7}}f_{4,3}(\lambda) - \frac{1}{3\sqrt{7}}f_{4,4}(\lambda) + \frac{53}{15\sqrt{154}}f_{4,5}(\lambda) + \frac{1}{\sqrt{231}}f_{6,1}(\lambda)$
-101	-110	$-\frac{1}{3\sqrt{5}}f_{0,4}(\lambda) + \frac{1}{3\sqrt{7}}f_{0,5}(\lambda) + \frac{\sqrt{2}}{3\sqrt{7}}f_{2,6}(\lambda)$
-110	$01-1$	$+\frac{\sqrt{5}}{6\sqrt{14}}f_{2,9}(\lambda) - \frac{1}{3\sqrt{70}}f_{4,2}(\lambda) - \frac{1}{6\sqrt{7}}f_{4,4}(\lambda) - \frac{\sqrt{7}}{3\sqrt{22}}f_{4,5}(\lambda) + \frac{1}{\sqrt{231}}f_{6,1}(\lambda)$
-101	$01-1$	$-\frac{1}{3\sqrt{5}}f_{0,4}(\lambda) + \frac{1}{3\sqrt{7}}f_{0,5}(\lambda) - \frac{2\sqrt{2}}{3\sqrt{7}}f_{2,6}(\lambda) - \frac{\sqrt{5}}{3\sqrt{14}}f_{2,9}(\lambda)$
		$-\frac{17}{6\sqrt{70}}f_{4,2}(\lambda) + \frac{1}{3\sqrt{7}}f_{4,4}(\lambda) - \frac{\sqrt{7}}{3\sqrt{22}}f_{4,5}(\lambda) + \frac{1}{\sqrt{231}}f_{6,1}(\lambda)$

Introduce the following notation:

$$u_1(\lambda) = 2f_{-1-1-1-1-1-1}(\lambda), \qquad u_2(\lambda) = 2f_{-1-11-1-11}(\lambda),$$
$$u_3(\lambda) = 2f_{-100-100}(\lambda), \qquad u_4(\lambda) = 2f_{-11-1-11-1}(\lambda),$$
$$u_5(\lambda) = 2f_{00-100-1}(\lambda), \qquad u_6(\lambda) = 2f_{-1-10-1-10}(\lambda),$$
$$u_7(\lambda) = 2(f_{-10-1-10-1}(\lambda), \qquad u_8(\lambda) = f_{000000}(\lambda),$$
$$u_9(\lambda) = 2f_{-101-101}(\lambda) \qquad u_{10}(\lambda) = u_{-110-110}(\lambda),$$
$$u_{11}(\lambda) = f_{-1-1-1-100}(\lambda), \qquad u_{12}(\lambda) = f_{-1-1-1-111}(\lambda),$$
$$u_{13}(\lambda) = f_{-1-1-100-1}(\lambda), \qquad u_{14}(\lambda) = f_{-1-11010}(\lambda),$$
$$u_{15}(\lambda) = f_{-1-1-111-1}(\lambda), \qquad u_{16}(\lambda) = f_{-10000-1}(\lambda),$$
$$u_{17}(\lambda) = f_{-1-10-10-1}(\lambda), \qquad u_{18}(\lambda) = f_{-1-10000}(\lambda),$$
$$u_{19}(\lambda) = f_{-1-10011}(\lambda), \qquad u_{20}(\lambda) = f_{-10-1000}(\lambda),$$
$$u_{21}(\lambda) = f_{-10-1011}(\lambda).$$

Then we have

$$\begin{aligned}
f_{-1-1-111-1}(\lambda) &= \frac{1}{4}u_1(\lambda) + \frac{1}{4}u_2(\lambda) - \frac{1}{2}u_4(\lambda),\\
f_{-1-11-11-1}(\lambda) &= \frac{1}{2\sqrt{2}}u_1(\lambda) - \frac{1}{2\sqrt{2}}u_2(\lambda) - u_{12}(\lambda),\\
f_{-100-111}(\lambda) &= \frac{1}{\sqrt{2}}(u_{11}(\lambda) - u_{14}(\lambda)),\\
f_{-11-1001}(\lambda) &= \frac{1}{\sqrt{2}}(u_{13}(\lambda) - u_{15}(\lambda)),\\
f_{-1-10110}(\lambda) &= \frac{1}{2}u_6(\lambda) - 4u_{10}(\lambda),\\
f_{-101-110}(\lambda) &= \frac{1}{\sqrt{2}}(u_{17}(\lambda) - u_{19}(\lambda),\\
f_{-10101-1}(\lambda) &= \frac{1}{2}u_7(\lambda) - \frac{1}{2}u_9(\lambda) - u_{21}(\lambda).
\end{aligned}$$

Put

$$v_i(\lambda) = \begin{cases} \frac{u_i(\lambda)}{u_1(\lambda)+\cdots+u_5(\lambda)}, & \text{if } 1 \le i \le 4\\ \frac{u_{i+6}(\lambda)}{u_1(\lambda)+\cdots+u_5(\lambda)}, & \text{if } 5 \le i \le 10\\ \frac{u_{i-5}(\lambda)}{u_6(\lambda)+u_7(\lambda)+u_8(\lambda)}, & \text{if } 11 \le i \le 12\\ \frac{u_{i+4}(\lambda)}{u_6(\lambda)+u_7(\lambda)+u_8(\lambda)}, & \text{if } 13 \le i \le 17\\ \frac{u_9(\lambda)}{u_9(\lambda)+u_{10}(\lambda)}, & \text{if } i = 18. \end{cases} \tag{3.91}$$

We see that the set of extreme points of the convex compact $\mathcal{C}_0$ where the function $f(\lambda)$ may take values, consists of three connected components. The functions $v_1(\lambda), \dots, v_{10}(\lambda)$ (resp. $v_{11}(\lambda), \dots, v_{17}(\lambda)$, resp. $v_{18}(\lambda)$) are the coordinates on the first (resp. second, resp. third) connected component.

The functions $f_{i,j}(\lambda)$ are expressed in terms of the functions $u_k(\lambda)$ according to Table 3.9.

Substitute these values in (3.90) and take into account (3.20). We obtain the matrix entries $f_{ijki'j'k'}(\boldsymbol{p})$ expressed in terms of $v_i(\lambda)$ and $M^{\rho_{2\ell},n_\ell}(\boldsymbol{p})$:

$$f_{iiji'j'k'}(\boldsymbol{p}) = \sum_{\ell=0}^{3}\sum_{n_\ell=1}^{q_\ell}\sum_{m=-2\ell}^{2\ell} M^{\rho_{2\ell},n_\ell}(\boldsymbol{p})\sum_{r=0}^{21} a_{\ell,n_\ell,r}v_r(\lambda), \tag{3.92}$$

where we put $v_0(\lambda) = 1$. Using Table 3.9, we express $f_{i\cdots\ell'}(\boldsymbol{p})$ in terms of $u_i(\lambda)$ and $L^q_{i\cdots\ell'}(\boldsymbol{p})$. Combining (2.48) and (3.22) and taking into account that the O(3)-invariant measure on $\hat{\mathbb{R}}^3$ has the form

$$\mathrm{d}\mu(\boldsymbol{p}) = \frac{1}{4\pi}\,\mathrm{d}S\,\mathrm{d}\nu(\lambda),$$

where $\mathrm{d}S$ is the Lebesgue measure on the unit sphere S^2, we obtain

$$\langle \boldsymbol{e}(\boldsymbol{x}), \boldsymbol{e}(\boldsymbol{y}) = \frac{1}{4\pi}\int_0^\infty \int_{S^2} \mathrm{e}^{\mathrm{i}(\boldsymbol{p},\boldsymbol{y}-\boldsymbol{x})} f(\boldsymbol{p})\,\mathrm{d}S\,\mathrm{d}\nu(\lambda).$$

Table 3.9 *The functions $f_{i,j}(\lambda)$ expressed as linear combinations of the functions $u_k(\lambda)$.*

$f_{i,j}(\lambda)$	Expresses as
$f_{0,1}(\lambda)$	$\frac{2}{3\sqrt{3}}u_1(\lambda) + \frac{2}{3\sqrt{3}}u_2(\lambda) - \frac{8}{3\sqrt{3}}u_4(\lambda) + \frac{1}{3\sqrt{3}}u_5(\lambda) + \frac{\sqrt{2}}{3\sqrt{3}}u_6(\lambda)$ $+\frac{1}{3\sqrt{3}}u_8(\lambda) - \frac{8}{3\sqrt{3}}u_{10}(\lambda) + \frac{4}{3\sqrt{3}}u_{13}(\lambda) + \frac{4\sqrt{2}}{3\sqrt{3}}u_{15}(\lambda) + \frac{4}{3\sqrt{3}}u_{18}(\lambda)$
$f_{0,2}(\lambda)$	$\frac{4\sqrt{2}}{3\sqrt{3}}u_1(\lambda) - \frac{2\sqrt{2}}{3\sqrt{3}}u_2(\lambda) - \frac{4\sqrt{2}}{3\sqrt{3}}u_4(\lambda) - \frac{\sqrt{2}}{3\sqrt{3}}u_5(\lambda) - \frac{2\sqrt{2}}{3\sqrt{3}}u_6(\lambda)$ $+\frac{2\sqrt{2}}{3\sqrt{3}}u_8(\lambda) + \frac{8\sqrt{2}}{3\sqrt{3}}u_{10}(\lambda) + \frac{4}{\sqrt{15}}u_{11}(\lambda) + \frac{\sqrt{10}}{3\sqrt{3}}u_{13}(\lambda) + \frac{4}{\sqrt{15}}u_{14}(\lambda)$ $-\frac{14}{3\sqrt{15}}u_{15}(\lambda) + \frac{4}{\sqrt{15}}u_{16}(\lambda) + \frac{4}{\sqrt{15}}u_{17}(\lambda) + \frac{2\sqrt{2}}{3\sqrt{3}}u_{18}(\lambda) + \frac{4}{\sqrt{15}}u_{19}(\lambda)$ $+\frac{4}{\sqrt{15}}u_{20}(\lambda)$
$f_{0,3}(\lambda)$	$-\frac{1}{15\sqrt{3}}u_1(\lambda) + \frac{2}{15\sqrt{3}}u_2(\lambda) + \frac{2\sqrt{3}}{5}u_3(\lambda) + \frac{34}{15\sqrt{3}}u_4(\lambda) + \frac{1}{15\sqrt{3}}u_5(\lambda)$ $+\frac{2}{15\sqrt{3}}u_6(\lambda) + \frac{2\sqrt{3}}{5}u_7(\lambda) + \frac{4}{15\sqrt{3}}u_8(\lambda) - \frac{8}{15\sqrt{3}}u_{10}(\lambda) + \frac{14\sqrt{2}}{5\sqrt{3}}u_{11}(\lambda)$ $+\frac{4\sqrt{3}}{5}u_{12}(\lambda) - \frac{14}{15\sqrt{3}}u_{13}(\lambda) - \frac{2\sqrt{2}}{\sqrt{3}}u_{14}(\lambda) + \frac{2\sqrt{2}}{3\sqrt{3}}u_{15}(\lambda) - \frac{4\sqrt{2}}{5\sqrt{3}}u_{16}(\lambda)$ $-\frac{4\sqrt{2}}{5\sqrt{3}}u_{17}(\lambda) - \frac{8}{15\sqrt{3}}u_{18}(\lambda) - \frac{4\sqrt{2}}{5\sqrt{3}}u_{19}(\lambda) + \frac{8\sqrt{2}}{5\sqrt{3}}u_{20}(\lambda) + \frac{4\sqrt{3}}{5}u_{21}(\lambda)$
$f_{0,4}(\lambda)$	$-\frac{1}{3\sqrt{5}}u_1(\lambda) + \frac{2}{3\sqrt{5}}u_2(\lambda) + \frac{2}{3\sqrt{5}}u_3(\lambda) + \frac{2}{3\sqrt{5}}u_4(\lambda) + \frac{1}{3\sqrt{5}}u_5(\lambda)$ $-\frac{2}{3\sqrt{5}}u_7(\lambda) + \frac{4}{\sqrt{5}}u_9(\lambda) + \frac{16}{3\sqrt{5}}u_{10}(\lambda) - \frac{2\sqrt{2}}{3\sqrt{5}}u_{11}(\lambda) + \frac{4}{3\sqrt{5}}u_{12}(\lambda)$ $+\frac{2}{3\sqrt{5}}u_{13}(\lambda) + \frac{2\sqrt{2}}{\sqrt{5}}u_{14}(\lambda) - \frac{2\sqrt{2}}{\sqrt{5}}u_{15}(\lambda) - \frac{4\sqrt{2}}{3\sqrt{5}}u_{16}(\lambda) - \frac{4\sqrt{2}}{3\sqrt{35}}u_{17}(\lambda)$ $-\frac{8\sqrt{2}}{3\sqrt{5}}u_{18}(\lambda) + \frac{8\sqrt{2}}{3\sqrt{5}}u_{20}(\lambda) + \frac{4}{3\sqrt{5}}u_{21}(\lambda)$
$f_{0,5}(\lambda)$	$\frac{11}{15\sqrt{7}}u_1(\lambda) - \frac{\sqrt{7}}{15}u_2(\lambda) + \frac{32}{15\sqrt{7}}u_3(\lambda) + \frac{8\sqrt{7}}{15}u_4(\lambda) + \frac{4}{15\sqrt{7}}u_5(\lambda)$ $+\frac{1}{5\sqrt{7}}u_6(\lambda) + \frac{52}{15\sqrt{7}}u_7(\lambda) + \frac{2}{5\sqrt{7}}u_8(\lambda) + \frac{4\sqrt{7}}{15}u_{10}(\lambda) - \frac{32\sqrt{2}}{15\sqrt{7}}u_{11}(\lambda)$ $-\frac{8\sqrt{7}}{15}u_{12}(\lambda) - \frac{16}{15\sqrt{7}}u_{13}(\lambda) + \frac{32\sqrt{2}}{15\sqrt{7}}u_{16}(\lambda) + \frac{52\sqrt{2}}{15\sqrt{7}}u_{17}(\lambda) - \frac{4}{5\sqrt{7}}u_{18}(\lambda)$ $-\frac{4\sqrt{14}}{15}u_{19}(\lambda) - \frac{8\sqrt{2}}{5\sqrt{7}}u_{20}(\lambda) - \frac{8\sqrt{7}}{15}u_{21}(\lambda)$
$f_{2,1}(\lambda)$	$-\frac{\sqrt{2}}{3\sqrt{3}}u_1(\lambda) - \frac{\sqrt{2}}{3\sqrt{3}}u_2(\lambda) + \frac{4\sqrt{2}}{3\sqrt{3}}u_4(\lambda) - \frac{1}{3\sqrt{6}}u_5(\lambda) + \frac{2\sqrt{2}}{3\sqrt{3}}u_6(\lambda)$ $+\frac{\sqrt{2}}{3\sqrt{3}}u_8(\lambda) - \frac{8\sqrt{2}}{3\sqrt{3}}u_{10}(\lambda) - \frac{2\sqrt{2}}{3\sqrt{3}}u_{13}(\lambda) - \frac{4\sqrt{2}}{3\sqrt{3}}u_{15}(\lambda) + \frac{4\sqrt{2}}{3\sqrt{3}}u_{18}(\lambda)$
$f_{2,2}(\lambda)$	$-\frac{4}{3\sqrt{15}}u_1(\lambda) + \frac{2}{3\sqrt{15}}u_2(\lambda) + \frac{4}{3\sqrt{15}}u_4(\lambda) + \frac{1}{3\sqrt{15}}u_5(\lambda) - \frac{4}{3\sqrt{15}}u_6(\lambda)$ $+\frac{4}{3\sqrt{15}}u_8(\lambda) + \frac{16}{3\sqrt{15}}u_{10}(\lambda) - \frac{2\sqrt{2}}{\sqrt{15}}u_{11}(\lambda) - \frac{\sqrt{5}}{3\sqrt{3}}u_{13}(\lambda) - \frac{2\sqrt{2}}{\sqrt{15}}u_{14}(\lambda)$ $+\frac{7\sqrt{2}}{3\sqrt{15}}u_{15}(\lambda) - \frac{2\sqrt{2}}{\sqrt{55}}u_{17}(\lambda) + \frac{4\sqrt{2}}{\sqrt{15}}u_{18}(\lambda) + \frac{4}{3\sqrt{15}}u_{19}(\lambda) + \frac{4\sqrt{2}}{\sqrt{15}}u_{20}(\lambda)$
$f_{2,3}(\lambda)$	$\frac{1}{15\sqrt{6}}u_1(\lambda) - \frac{\sqrt{2}}{15\sqrt{3}}u_2(\lambda) - \frac{\sqrt{6}}{5}u_3(\lambda) - \frac{17\sqrt{2}}{15\sqrt{3}}u_4(\lambda) - \frac{1}{15\sqrt{6}}u_5(\lambda)$ $+\frac{2\sqrt{2}}{15\sqrt{3}}u_6(\lambda) + \frac{2\sqrt{6}}{5}u_7(\lambda) + \frac{4\sqrt{2}}{15\sqrt{3}}u_8(\lambda) - \frac{8\sqrt{2}}{15\sqrt{3}}u_{10}(\lambda) - \frac{14}{5\sqrt{3}}u_{11}(\lambda)$ $-\frac{2\sqrt{6}}{5}u_{12}(\lambda) + \frac{7\sqrt{2}}{15\sqrt{3}}u_{13}(\lambda) + \frac{3}{\sqrt{3}}u_{15}(\lambda) - \frac{3}{3\sqrt{3}}u_{16}(\lambda) + \frac{4}{5\sqrt{3}}u_{17}(\lambda)$ $-\frac{8}{5\sqrt{3}}u_{18}(\lambda) - \frac{8\sqrt{2}}{15\sqrt{3}}u_{19}(\lambda) + \frac{16}{5\sqrt{3}}u_{20}(\lambda) + \frac{4\sqrt{6}}{5}u_{21}(\lambda)$
$f_{2,4}(\lambda)$	$-\frac{2}{3}u_2(\lambda) + \frac{4}{3}u_4(\lambda) + \frac{1}{3}u_5(\lambda) - \frac{2\sqrt{2}}{3}u_{11}(\lambda) + u_{13}(\lambda)$ $-\frac{2\sqrt{2}}{3}u_{14}(\lambda) - \frac{\sqrt{2}}{3}u_{15}(\lambda) - \frac{2\sqrt{2}}{3}u_{16}(\lambda)$
$f_{2,5}(\lambda)$	$-\frac{2}{\sqrt{35}}u_1(\lambda) - \frac{2}{3\sqrt{35}}u_2(\lambda) + \frac{16}{3\sqrt{35}}u_4(\lambda) + \frac{4}{3\sqrt{35}}u_5(\lambda) - \frac{2}{\sqrt{35}}u_6(\lambda)$ $+\frac{2}{\sqrt{35}}u_8(\lambda) + \frac{8}{\sqrt{35}}u_{10}(\lambda) + \frac{16\sqrt{2}}{3\sqrt{35}}u_{11}(\lambda) + \frac{16\sqrt{2}}{3\sqrt{35}}u_{14}(\lambda) - \frac{8}{5\sqrt{21}}u_{15}(\lambda)$ $+\frac{16}{\sqrt{105}}u_{16}(\lambda) + \frac{8\sqrt{2}}{3\sqrt{35}}u_{15}(\lambda) + \frac{16\sqrt{2}}{3\sqrt{35}}u_{10}(\lambda) - \frac{4\sqrt{2}}{\sqrt{35}}u_{17}(\lambda) + \frac{2}{\sqrt{35}}u_{18}(\lambda)$ $-\frac{4\sqrt{2}}{\sqrt{35}}u_{19}(\lambda) - \frac{4\sqrt{2}}{\sqrt{35}}u_{20}(\lambda)$

Table 3.9 (*Cont.*)

$f_{i,j}(\lambda)$	Expresses as
$f_{2,6}(\lambda)$	$-\frac{1}{3\sqrt{14}}u_1(\lambda)+\frac{\sqrt{2}}{3\sqrt{7}}u_2(\lambda)+\frac{\sqrt{2}}{3\sqrt{7}}u_3(\lambda)+\frac{\sqrt{2}}{3\sqrt{7}}u_4(\lambda)+\frac{1}{3\sqrt{14}}u_5(\lambda)$ $-\frac{10\sqrt{2}}{3\sqrt{7}}u_7(\lambda)+\frac{4\sqrt{2}}{\sqrt{7}}u_9(\lambda)-\frac{16\sqrt{2}}{3\sqrt{7}}u_{10}(\lambda)-\frac{2}{3\sqrt{7}}u_{11}(\lambda)+\frac{2\sqrt{2}}{3\sqrt{7}}u_{12}(\lambda)$ $+\frac{\sqrt{2}}{3\sqrt{7}}u_{13}(\lambda)+\frac{2}{\sqrt{7}}u_{14}(\lambda)-\frac{2}{\sqrt{7}}u_{15}(\lambda)-\frac{4}{3\sqrt{7}}u_{16}(\lambda)+\frac{16}{3\sqrt{7}}u_{17}(\lambda)$ $-\frac{16}{3\sqrt{7}}u_{19}(\lambda)+\frac{20\sqrt{2}}{3\sqrt{7}}u_{21}(\lambda)$
$f_{2,7}(\lambda)$	$-\frac{1}{\sqrt{5}}u_1(\lambda)+\frac{2}{3\sqrt{5}}u_2(\lambda)-\frac{2}{\sqrt{5}}u_3(\lambda)+\frac{14}{3\sqrt{5}}u_4(\lambda)-\frac{1}{3\sqrt{5}}u_5(\lambda)$ $-\frac{4\sqrt{2}}{3\sqrt{5}}u_{11}(\lambda)+\frac{4}{\sqrt{5}}u_{12}(\lambda)+\frac{2}{\sqrt{5}}u_{13}(\lambda)-\frac{4\sqrt{2}}{3\sqrt{5}}u_{14}(\lambda)-\frac{2\sqrt{2}}{3\sqrt{5}}u_{15}(\lambda)$ $+\frac{8\sqrt{2}}{3\sqrt{5}}u_{16}(\lambda)$
$f_{2,8}(\lambda)$	$-\frac{2}{5\sqrt{7}}u_1(\lambda)+\frac{2}{15\sqrt{7}}u_2(\lambda)+\frac{16}{5\sqrt{7}}u_3(\lambda)-\frac{15}{15\sqrt{7}}u_4(\lambda)-\frac{4}{15\sqrt{7}}u_5(\lambda)$ $+\frac{2}{5\sqrt{7}}u_6(\lambda)-\frac{12}{5\sqrt{7}}u_7(\lambda)+\frac{4}{5\sqrt{7}}u_8(\lambda)-\frac{8}{5\sqrt{7}}u_{10}(\lambda)+\frac{32\sqrt{2}}{15\sqrt{7}}u_{11}(\lambda)$ $-\frac{8}{5\sqrt{7}}u_{12}(\lambda)+\frac{12}{5\sqrt{7}}u_{13}(\lambda)-\frac{8\sqrt{2}}{3\sqrt{7}}u_{14}(\lambda)-\frac{4\sqrt{2}}{3\sqrt{7}}u_{15}(\lambda)+\frac{8\sqrt{2}}{15\sqrt{7}}u_{16}(\lambda)$ $-\frac{2\sqrt{2}}{5\sqrt{7}}u_{17}(\lambda)-\frac{8}{5\sqrt{7}}u_{18}(\lambda)-\frac{2\sqrt{2}}{5\sqrt{7}}u_{19}(\lambda)+\frac{4\sqrt{2}}{5\sqrt{7}}u_{20}(\lambda)-\frac{24}{5\sqrt{7}}u_{21}(\lambda)$
$f_{2,9}(\lambda)$	$-\frac{\sqrt{2}}{3\sqrt{35}}u_1(\lambda)-\frac{\sqrt{2}}{3\sqrt{35}}u_2(\lambda)+\frac{8\sqrt{2}}{3\sqrt{35}}u_3(\lambda)+\frac{8\sqrt{2}}{3\sqrt{35}}u_4(\lambda)-\frac{2\sqrt{2}}{3\sqrt{35}}u_5(\lambda)$ $-\frac{4\sqrt{10}}{3\sqrt{7}}u_7(\lambda)+\frac{8\sqrt{10}}{3\sqrt{7}}u_{10}(\lambda)-\frac{16}{3\sqrt{35}}u_{11}(\lambda)+\frac{4\sqrt{2}}{3\sqrt{35}}u_{12}(\lambda)+\frac{2\sqrt{2}}{3\sqrt{35}}u_{13}(\lambda)$ $+\frac{8}{\sqrt{35}}u_{14}(\lambda)+\frac{4}{\sqrt{35}}u_{15}(\lambda)-\frac{8}{3\sqrt{35}}u_{16}(\lambda)-\frac{4\sqrt{5}}{3\sqrt{7}}u_{17}(\lambda)-\frac{4\sqrt{5}}{3\sqrt{7}}u_{19}(\lambda)$ $+\frac{8\sqrt{10}}{3\sqrt{7}}u_{21}(\lambda)$
$f_{2,10}(\lambda)$	$-\frac{3\sqrt{3}}{10\sqrt{7}}u_1(\lambda)+\frac{13}{10\sqrt{21}}u_2(\lambda)+\frac{16}{5\sqrt{21}}u_3(\lambda)-\frac{52}{5\sqrt{21}}u_4(\lambda)+\frac{2}{5\sqrt{21}}u_5(\lambda)$ $+\frac{2}{5\sqrt{21}}u_6(\lambda)+\frac{8}{5\sqrt{21}}u_7(\lambda)+\frac{4}{5\sqrt{21}}u_8(\lambda)-\frac{8}{5\sqrt{21}}u_{10}(\lambda)-\frac{16\sqrt{2}}{5\sqrt{21}}u_{11}(\lambda)$ $+\frac{52}{5\sqrt{21}}u_{12}(\lambda)-\frac{8}{5\sqrt{21}}u_{13}(\lambda)+\frac{16\sqrt{2}}{5\sqrt{21}}u_{16}(\lambda)+\frac{8\sqrt{2}}{5\sqrt{21}}u_{17}(\lambda)-\frac{8}{5\sqrt{21}}u_{18}(\lambda)$ $+\frac{8\sqrt{2}}{5\sqrt{21}}u_{19}(\lambda)-\frac{16\sqrt{2}}{5\sqrt{21}}u_{20}(\lambda)+\frac{16}{5\sqrt{21}}u_{21}(\lambda)$
$f_{4,1}(\lambda)$	$\frac{\sqrt{3}}{\sqrt{35}}u_1(\lambda)+\frac{1}{\sqrt{105}}u_2(\lambda)-\frac{8}{\sqrt{105}}u_4(\lambda)-\frac{2}{\sqrt{105}}u_5(\lambda)-\frac{4}{\sqrt{105}}u_6(\lambda)$ $+\frac{4}{\sqrt{105}}u_8(\lambda)+\frac{16}{\sqrt{105}}u_{10}(\lambda)-\frac{8\sqrt{2}}{\sqrt{105}}u_{11}(\lambda)-\frac{8\sqrt{2}}{\sqrt{105}}u_{14}(\lambda)-\frac{4\sqrt{2}}{\sqrt{105}}u_{15}(\lambda)$ $-\frac{8\sqrt{2}}{\sqrt{105}}u_{16}(\lambda)-\frac{8\sqrt{2}}{\sqrt{105}}u_{17}(\lambda)+\frac{4}{\sqrt{105}}u_{18}(\lambda)-\frac{8\sqrt{2}}{\sqrt{105}}u_{19}(\lambda)-\frac{8\sqrt{2}}{\sqrt{105}}u_{20}(\lambda)$
$f_{4,2}(\lambda)$	$\frac{2\sqrt{2}}{3\sqrt{35}}u_1(\lambda)-\frac{4\sqrt{2}}{3\sqrt{35}}u_2(\lambda)+\frac{4\sqrt{2}}{3\sqrt{35}}u_3(\lambda)-\frac{4\sqrt{2}}{3\sqrt{35}}u_4(\lambda)-\frac{2\sqrt{2}}{3\sqrt{35}}u_5(\lambda)$ $-\frac{16\sqrt{2}}{3\sqrt{35}}u_7(\lambda)+\frac{12\sqrt{2}}{\sqrt{35}}u_9(\lambda)+\frac{8\sqrt{2}}{3\sqrt{35}}u_{10}(\lambda)+\frac{8}{3\sqrt{35}}u_{11}(\lambda)-\frac{8\sqrt{2}}{3\sqrt{35}}u_{12}(\lambda)$ $-\frac{4\sqrt{2}}{3\sqrt{35}}u_{13}(\lambda)-\frac{8}{\sqrt{35}}u_{14}(\lambda)+\frac{8}{\sqrt{35}}u_{15}(\lambda)+\frac{16}{3\sqrt{35}}u_{16}(\lambda)-\frac{8}{3\sqrt{35}}u_{17}(\lambda)$ $+\frac{8}{3\sqrt{35}}u_{19}(\lambda)+\frac{32\sqrt{2}}{3\sqrt{35}}u_{21}(\lambda)$
$f_{4,3}(\lambda)$	$\frac{\sqrt{3}}{5\sqrt{7}}u_1(\lambda)-\frac{1}{5\sqrt{21}}u_2(\lambda)-\frac{8\sqrt{3}}{5\sqrt{7}}u_3(\lambda)+\frac{8}{5\sqrt{21}}u_4(\lambda)+\frac{2}{5\sqrt{21}}u_5(\lambda)$ $+\frac{4}{5\sqrt{21}}u_6(\lambda)-\frac{8\sqrt{3}}{5\sqrt{7}}u_7(\lambda)+\frac{8\sqrt{3}}{15\sqrt{7}}u_8(\lambda)-\frac{16\sqrt{3}}{15\sqrt{7}}u_{10}(\lambda)-\frac{16\sqrt{2}}{5\sqrt{21}}u_{11}(\lambda)$ $+\frac{4\sqrt{3}}{5\sqrt{7}}u_{13}(\lambda)-\frac{6\sqrt{3}}{5\sqrt{7}}u_{14}(\lambda)+\frac{4\sqrt{2}}{3\sqrt{7}}u_{15}(\lambda)-\frac{4\sqrt{2}}{3\sqrt{7}}u_{16}(\lambda)-\frac{4\sqrt{2}}{3\sqrt{7}}u_{17}(\lambda)$ $-\frac{16\sqrt{3}}{15\sqrt{7}}u_{18}(\lambda)-\frac{4\sqrt{2}}{5\sqrt{21}}u_{19}(\lambda)+\frac{8\sqrt{2}}{5\sqrt{21}}u_{20}(\lambda)-\frac{16\sqrt{3}}{5\sqrt{7}}u_{21}(\lambda)$
$f_{4,4}(\lambda)$	$-\frac{1}{3\sqrt{7}}u_1(\lambda)-\frac{1}{3\sqrt{7}}u_2(\lambda)+\frac{8}{3\sqrt{7}}u_3(\lambda)+\frac{8}{3\sqrt{7}}u_4(\lambda)-\frac{2}{3\sqrt{7}}u_5(\lambda)$ $+\frac{8}{3\sqrt{7}}u_7(\lambda)-\frac{16}{3\sqrt{7}}u_{10}(\lambda)-\frac{8\sqrt{2}}{3\sqrt{7}}u_{11}(\lambda)+\frac{4}{3\sqrt{7}}u_{12}(\lambda)+\frac{2}{3\sqrt{7}}u_{13}(\lambda)$ $+\frac{4\sqrt{2}}{\sqrt{7}}u_{14}(\lambda)+\frac{2\sqrt{2}}{\sqrt{7}}u_{15}(\lambda)-\frac{4\sqrt{2}}{3\sqrt{7}}u_{16}(\lambda)-\frac{4\sqrt{2}}{3\sqrt{7}}u_{17}(\lambda)+\frac{4\sqrt{2}}{30\sqrt{7}}u_{19}(\lambda)$ $-\frac{16}{3\sqrt{7}}u_{21}(\lambda)$

Table 3.9 (*Cont.*)

$f_{i,j}(\lambda)$	Expresses as
$f_{4,5}(\lambda)$	$\frac{13\sqrt{2}}{15\sqrt{77}}u_1(\lambda)-\frac{\sqrt{22}}{15\sqrt{7}}u_2(\lambda)+\frac{16\sqrt{2}}{15\sqrt{77}}u_3(\lambda)+\frac{8\sqrt{22}}{15\sqrt{7}}u_4(\lambda)+\frac{2\sqrt{2}}{15\sqrt{77}}u_5(\lambda)$
	$+\frac{3\sqrt{2}}{5\sqrt{77}}u_6(\lambda)-\frac{104\sqrt{2}}{15\sqrt{77}}u_7(\lambda)+\frac{6\sqrt{2}}{5\sqrt{77}}u_8(\lambda)-\frac{16\sqrt{2}}{15\sqrt{7}}u_{10}(\lambda)-\frac{32}{15\sqrt{77}}u_{11}(\lambda)$
	$-\frac{8\sqrt{22}}{15\sqrt{7}}u_{13}(\lambda)-\frac{8\sqrt{2}}{15\sqrt{77}}u_{13}(\lambda)+\frac{32}{15\sqrt{77}}u_{16}(\lambda)-\frac{208}{15\sqrt{77}}u_{17}(\lambda)-\frac{12\sqrt{2}}{5\sqrt{77}}u_{18}(\lambda)$
	$+\frac{32\sqrt{11}}{15\sqrt{7}}u_{19}(\lambda)-\frac{48}{5\sqrt{77}}u_{20}(\lambda)+\frac{32\sqrt{22}}{15\sqrt{7}}u_{21}(\lambda)$
$f_{6,1}(\lambda)$	$-\frac{2}{\sqrt{231}}u_1(\lambda)-\frac{16}{\sqrt{231}}u_3(\lambda)-\frac{2}{\sqrt{231}}u_5(\lambda)+\frac{2}{\sqrt{231}}u_6(\lambda)+\frac{16}{\sqrt{231}}u_7(\lambda)$
	$+\frac{4}{\sqrt{231}}u_8(\lambda)+\frac{16\sqrt{2}}{\sqrt{231}}u_{11}(\lambda)+\frac{8}{\sqrt{231}}u_{13}(\lambda)-\frac{16\sqrt{2}}{\sqrt{231}}u_{16}(\lambda)+\frac{16\sqrt{2}}{\sqrt{231}}u_{17}(\lambda)$
	$-\frac{8}{\sqrt{231}}u_{18}(\lambda)-\frac{16\sqrt{2}}{\sqrt{231}}u_{20}(\lambda)$

Substitute the obtained expression into the above display and use the Rayleigh expansion (2.62). We obtain

$$\langle \boldsymbol{e}(\boldsymbol{x}), \boldsymbol{e}(\boldsymbol{y})\rangle_{ijki'j'k'} = \sum_{n=1}^{3}\int_0^\infty \sum_{q=1}^{21} N_{nq}(\lambda,\rho)L^q_{iiki'j'k'}(\boldsymbol{y}-\boldsymbol{x})\,\mathrm{d}\Phi_n(\lambda), \tag{3.93}$$

with

$$\begin{aligned}
\mathrm{d}\Phi_1(\lambda) &= 2(u_1(\lambda)+\cdots+u_5(\lambda))\,\mathrm{d}\nu(\lambda),\\
\mathrm{d}\Phi_2(\lambda) &= (2u_6(\lambda)+2u_7(\lambda)+u_8(\lambda))\,\mathrm{d}\nu(\lambda),\\
\mathrm{d}\Phi_3(\lambda) &= (2u_9(\lambda)+u_{10}(\lambda))\,\mathrm{d}\nu(\lambda).
\end{aligned}$$

Using Table 3.9, we obtain

$$\begin{aligned}
2(u_1(0)+\cdots+u_5(0)) &= \frac{2}{\sqrt{3}}(f_{0,1}(0)+f_{0,2}(0))+\frac{2}{\sqrt{5}}f_{0,4}(0)\\
&\quad+\frac{4}{\sqrt{7}}f_{0,5}(0),\\
2u_6(0)+2u_7(0)+u_8(0) &= \frac{1}{\sqrt{3}}(f_{0,1}(0)+f_{0,2}(0))+\frac{1}{\sqrt{5}}f_{0,4}(0)\\
&\quad+\frac{2}{\sqrt{7}}f_{0,5}(0).
\end{aligned} \tag{3.94}$$

To formulate the final result, we need to introduce more notation.

$$\begin{aligned}
b^{i'j'k'\ell''m''}_{ijk\ell'm',n}(\lambda) &= \mathrm{i}^{\ell'-\ell''}\sqrt{(2\ell'+1)(2\ell''+1)}\sum_{\ell=0}^{3} g^{0[0,0]}_{2\ell[\ell',\ell'']}\sum_{m=-2\ell}^{2\ell} g^{-m[m',m'']}_{2\ell[\ell',\ell'']}\\
&\quad\times\sum_{n_\ell=1}^{q_\ell}\mathsf{T}^{\rho_{2\ell},n_\ell,m}_{ijki'j'k'}\sum_r a_{2\ell,n_\ell,r}v_r(\lambda),
\end{aligned}$$

where the coefficients $a_{2\ell,n_\ell,r}$ are defined in Equation (3.92) and where the index r runs over the values 0, 1, …, 10 for $n=1$, 0, 11, …, 17 for $n=2$ and

0, 18 for $n = 3$. Let $<$ be the lexicographic order on quintuples (ℓ, m, i, j, k). Let $L^{\ell' m' i' j' k'}_{\ell m i j k, n}(\lambda)$ be infinite lower triangular matrices from Cholesky factorisation of non-negative-definite matrices $b^{i' j' k' \ell'' m''}_{i j k \ell' m', n}(\lambda)$ constructed in Hansen (2010).

Theorem 34. *The one-point correlation tensor of a homogeneous and* $(\mathrm{O}(3), 2\rho_1 \oplus \rho_2 \oplus \rho_3)$*-isotropic random field* $\mathsf{e}(\boldsymbol{x})$ *is equal to* $\mathsf{0}$. *Its two-point correlation tensor has the spectral expansion*

$$\langle \mathsf{e}(\boldsymbol{x}), \mathsf{e}(\boldsymbol{y}) \rangle_{ijki'j'k'} = \sum_{n=1}^{3} \int_0^\infty \sum_{q=1}^{21} N_{nq}(\lambda, \rho) L^q_{iiki'j'k'}(\boldsymbol{y} - \boldsymbol{x}) \, \mathrm{d}\Phi_n(\lambda),$$

where the functions $N_{nq}(\lambda, \rho)$ *are given in Table* 3.10 *and the functions* $L^q_{iiki'j'k'}$ *are given in Equations* (2.40), (2.44) *and* (2.45) *and in Table* 2.3. *The measures* $\Phi_n(\lambda)$ *satisfy the condition*

$$\Phi_1(\{0\}) = 2\Phi_2(\{0\}). \tag{3.95}$$

The spectral expansion of the field has the form

$$\begin{aligned} \mathsf{e}_{ijk}(\rho, \theta, \varphi) = 2\sqrt{\pi} \sum_{n=1}^{3} \sum_{\ell=0}^{\infty} \sum_{m=-\ell}^{\ell} \int_0^\infty j_\ell(\lambda\rho) \sum_{(\ell', m', i', j', k') \leq (\ell, m, i, j, k)} \\ \times L^{\ell' m' i' j' k'}_{\ell m i j k, n}(\lambda) \, \mathrm{d}Z^n_{ijk\ell m}(\lambda) S^m_\ell(\theta, \varphi), \end{aligned}$$

where $S^m_\ell(\theta, \varphi)$ *are real-valued spherical harmonics, and where* $Z^{n'}_{ijk\ell m}$ *are centred uncorrelated real-valued random measures on* $[0, \infty)$ *with control measures* Φ_n.

Proof. The functions $N_{nq}(\lambda, \rho)$ have been calculated using the MATLAB Symbolic Math Toolbox. Equation (3.95) follows from (3.94). The spectral expansion of the field follows from Karhunen's theorem. □

3.8 The Case of Rank 4

Lomakin (1965) formulated a partial solution to the $(\mathrm{O}(3), \mathsf{S}^2(\mathsf{S}^2(\rho_1)))$-problem. His formula is similar to (3.63), but contains 15 terms. In what follows we will prove the correct version of his result with 29 terms.

Consider the representation $(\mathsf{S}^2(\mathsf{S}^2(\rho_1)), \mathsf{S}^2(\mathsf{S}^2(\mathbb{R}^3)))$ of the group $\mathrm{O}(3)$ as a group action. The symmetry classes were calculated by Forte & Vianello (1996). They are as follows: the triclinic class $[G_0] = [Z_2^c]$, the monoclinic class $[G_1] = [Z_2 \times Z_2^c]$, the orthotropic class $[G_2] = [D_2 \times Z_2^c]$, the trigonal class $[G_3] = [D_3 \times Z_2^c]$, the tetragonal class $[G_4] = [D_4 \times Z_2^c]$, the transverse isotropic class $[G_5] = [\mathrm{O}(2) \times Z_2^c]$, the cubic class $[G_6] = [\mathcal{O} \times Z_2^c]$ and the isotropic class $[G_7] = [\mathrm{O}(3)]$. The normalisers of their representatives are $N(G_0) = \mathrm{O}(3)$, $N(G_1) = \mathrm{O}(2) \times Z_2^c$,

Table 3.10 *The functions* $N_{nq}(\lambda, \rho)$.

n	q	$N_{nq}(\lambda,\rho)$
1	1	$[\frac{1}{35} + \frac{4}{105}v_2(\lambda) - \frac{1}{15}v_3(\lambda) - \frac{31}{105}v_4(\lambda) + \frac{4\sqrt{2}}{105}v_5(\lambda) + \frac{2}{105}v_7(\lambda)$ $- \frac{4\sqrt{2}}{15}v_8(\lambda) + \frac{2\sqrt{2}}{5}v_9(\lambda) - \frac{4\sqrt{2}}{105}v_{10}(\lambda)]j_0(\lambda\rho)$ $+[\frac{1}{14} - \frac{1}{4}v_1(\lambda) - \frac{19}{84}v_2(\lambda) - \frac{19}{42}v_3(\lambda) + \frac{41}{42}v_4(\lambda) - \frac{22\sqrt{2}}{21}v_5(\lambda)$ $+ \frac{10}{7}v_6(\lambda) + \frac{4}{21}v_7(\lambda) + \frac{10\sqrt{2}}{21}v_8(\lambda) + \frac{4\sqrt{2}}{21}v_{10}(\lambda)]j_2(\lambda\rho)$ $+[\frac{24}{385} - \frac{13}{385}v_2(\lambda) - \frac{96}{385}v_3(\lambda) - \frac{288}{385}v_4(\lambda) + \frac{292\sqrt{2}}{385}v_5(\lambda) - \frac{4}{7}v_6(\lambda)$ $+ \frac{36}{385}v_7(\lambda) - \frac{44\sqrt{2}}{35}v_8(\lambda) - \frac{2\sqrt{2}}{5}v_9(\lambda) + \frac{148\sqrt{2}}{385}v_{10}(\lambda)]j_4(\lambda\rho)$ $+[\frac{3}{154} - \frac{3}{154}v_2(\lambda) + \frac{3}{22}v_3(\lambda) - \frac{3}{154}v_4(\lambda) - \frac{12\sqrt{2}}{77}v_5(\lambda) - \frac{6}{77}v_7(\lambda)$ $+ \frac{12\sqrt{2}}{77}v_{10}(\lambda)]j_6(\lambda\rho)$
1	2	$[\frac{1}{35} - \frac{1}{30}v_1(\lambda) + \frac{1}{210}v_2(\lambda) + \frac{1}{15}v_3(\lambda) + \frac{11}{105}v_4(\lambda) - \frac{2\sqrt{2}}{21}v_5(\lambda)$ $+ \frac{2}{105}v_7(\lambda) + \frac{2\sqrt{2}}{15}v_8(\lambda) - \frac{2\sqrt{2}}{15}v_9(\lambda) - \frac{4\sqrt{2}}{105}v_{10}(\lambda)]j_0(\lambda\rho)$ $+[\frac{1}{42} + \frac{13}{168}v_1(\lambda) - \frac{17}{168}v_2(\lambda) + \frac{11}{42}v_3(\lambda) - \frac{23}{42}v_4(\lambda) + \frac{3\sqrt{2}}{7}v_5(\lambda)$ $- \frac{5}{7}v_6(\lambda) - \frac{1}{21}v_7(\lambda) - \frac{5\sqrt{2}}{21}v_8(\lambda) - \frac{\sqrt{2}}{21}v_9(\lambda) - \frac{4\sqrt{2}}{21}v_{10}(\lambda)]j_2(\lambda\rho)$ $+[-\frac{6}{385} - \frac{1}{70}v_1(\lambda) + \frac{23}{770}v_2(\lambda) + \frac{46}{385}v_3(\lambda) + \frac{138}{385}v_4(\lambda) - \frac{30\sqrt{2}}{77}v_5(\lambda)$ $+ \frac{2}{7}v_6(\lambda) - \frac{9}{385}v_7(\lambda) + \frac{22\sqrt{2}}{35}v_8(\lambda) + \frac{3\sqrt{2}}{35}v_9(\lambda) - \frac{92\sqrt{2}}{385}v_{10}(\lambda)]j_4(\lambda\rho)$ $+[-\frac{5}{462} + \frac{5}{462}v_2(\lambda) - \frac{5}{66}v_3(\lambda) + \frac{5}{462}v_4(\lambda) + \frac{20\sqrt{2}}{231}v_5(\lambda) + \frac{10}{231}v_7(\lambda)$ $- \frac{20\sqrt{2}}{231}v_{10}(\lambda)]j_6(\lambda\rho)$
1	3	$[-\frac{1}{210} + \frac{1}{30}v_1(\lambda) + \frac{1}{210}v_2(\lambda) - \frac{1}{30}v_3(\lambda) - \frac{9}{70}v_4(\lambda) - \frac{2\sqrt{2}}{35}v_5(\lambda)$ $+ \frac{5}{42}v_7(\lambda) + \frac{\sqrt{2}}{5}v_8(\lambda) - \frac{1}{3\sqrt{2}}v_9(\lambda) + \frac{\sqrt{2}}{35}v_{10}(\lambda)]j_0(\lambda\rho)$ $+[-\frac{1}{42} + \frac{1}{21}v_1(\lambda) + \frac{1}{42}v_2(\lambda) + \frac{5}{42}v_3(\lambda) + \frac{5}{42}v_4(\lambda) + \frac{11\sqrt{2}}{21}v_5(\lambda)$ $- \frac{5}{7}v_6(\lambda) + \frac{1}{42}v_7(\lambda) - \frac{2\sqrt{2}}{7}v_8(\lambda) - \frac{1}{21\sqrt{2}}v_9(\lambda) - \frac{2\sqrt{2}}{21}v_{10}(\lambda)]j_2(\lambda\rho)$ $+[-\frac{23}{770} + \frac{1}{70}v_1(\lambda) + \frac{23}{770}v_2(\lambda) + \frac{59}{770}v_3(\lambda) + \frac{199}{770}v_4(\lambda) - \frac{139\sqrt{2}}{385}v_5(\lambda)$ $+ \frac{2}{7}v_6(\lambda) - \frac{4}{77}v_7(\lambda) + \frac{18\sqrt{2}}{35}v_8(\lambda) + \frac{\sqrt{2}}{7}v_9(\lambda) - \frac{81\sqrt{2}}{385}v_{10}(\lambda)]j_4(\lambda\rho)$ $+[-\frac{5}{462} + \frac{5}{462}v_2(\lambda) - \frac{5}{66}v_3(\lambda) + \frac{5}{462}v_4(\lambda) + \frac{20\sqrt{2}}{231}v_5(\lambda) + \frac{10}{231}v_7(\lambda)$ $- \frac{20\sqrt{2}}{231}v_{10}(\lambda)]j_6(\lambda\rho)$
1	4	$[-\frac{1}{210} + \frac{1}{210}v_2(\lambda) + \frac{1}{30}v_3(\lambda) + \frac{1}{14}v_4(\lambda) + \frac{6\sqrt{2}}{35}v_5(\lambda) + \frac{2}{15}v_6(\lambda)$ $- \frac{1}{21}v_7(\lambda) - \frac{2\sqrt{2}}{15}v_8(\lambda) + \frac{\sqrt{2}}{15}v_9(\lambda) - \frac{4\sqrt{2}}{105}v_{10}(\lambda)]j_0(\lambda\rho)$ $+[-\frac{1}{21}v_3(\lambda) - \frac{1}{21}v_4(\lambda) - \frac{1}{3\sqrt{2}}v_5(\lambda) + \frac{1}{21}v_6(\lambda) - \frac{1}{42}v_7(\lambda)$ $+ \frac{1}{3\sqrt{2}}v_8(\lambda) + \frac{1}{21\sqrt{2}}v_9(\lambda)]j_2(\lambda\rho)$ $+[\frac{1}{110} - \frac{1}{110}v_2(\lambda) - \frac{39}{770}v_3(\lambda) - \frac{19}{154}v_4(\lambda) + \frac{7\sqrt{2}}{55}v_5(\lambda) - \frac{3}{35}v_6(\lambda)$ $+ \frac{1}{154}v_7(\lambda) - \frac{\sqrt{2}}{5}v_8(\lambda) - \frac{3}{35\sqrt{2}}v_9(\lambda) + \frac{4\sqrt{2}}{55}v_{10}(\lambda)]j_4(\lambda\rho)$ $+[\frac{1}{231} - \frac{1}{231}v_2(\lambda) + \frac{1}{33}v_3(\lambda) - \frac{1}{231}v_4(\lambda) - \frac{8\sqrt{2}}{231}v_5(\lambda) - \frac{4}{231}v_7(\lambda)$ $+ \frac{8\sqrt{2}}{231}v_{10}(\lambda)]j_6(\lambda\rho)$
1	5	$[-\frac{1}{210} + \frac{1}{30}v_1(\lambda) - \frac{1}{35}v_2(\lambda) + \frac{1}{30}v_3(\lambda) + \frac{1}{14}v_4(\lambda) - \frac{\sqrt{2}}{35}v_5(\lambda)$ $- \frac{2}{15}v_6(\lambda) - \frac{1}{21}v_7(\lambda) - \frac{\sqrt{2}}{15}v_8(\lambda) + \frac{\sqrt{2}}{15}v_9(\lambda) + \frac{2\sqrt{2}}{21}v_{10}(\lambda)]j_0(\lambda\rho)$ $+[-\frac{13}{168}v_1(\lambda) + \frac{13}{168}v_2(\lambda) - \frac{1}{21}v_3(\lambda) - \frac{1}{21}v_4(\lambda) - \frac{13}{21\sqrt{2}}v_5(\lambda)$

Table 3.10 (*Cont.*)

n	q	$N_{nq}(\lambda,\rho)$
		$+\frac{2}{3}v_6(\lambda) - \frac{1}{42}v_7(\lambda) + \frac{5}{21\sqrt{2}}v_8(\lambda) + \frac{1}{21\sqrt{2}}v_9(\lambda) + \frac{4\sqrt{2}}{21}v_{10}(\lambda)]j_2(\lambda\rho)$
		$+[\frac{1}{110} + \frac{1}{70}v_1(\lambda) - \frac{9}{385}v_2(\lambda) - \frac{39}{770}v_3(\lambda) - \frac{19}{154}v_4(\lambda) + \frac{71\sqrt{2}}{385}v_5(\lambda)$
		$-\frac{1}{5}v_6(\lambda) + \frac{1}{154}v_7(\lambda) - \frac{11\sqrt{2}}{35}v_8(\lambda) - \frac{3\sqrt{2}}{70}v_9(\lambda) + \frac{10\sqrt{2}}{77}v_{10}(\lambda)]j_4(\lambda\rho)$
		$+[\frac{1}{231} - \frac{1}{231}v_2(\lambda) + \frac{1}{33}v_3(\lambda) - \frac{1}{231}v_4(\lambda) - \frac{8\sqrt{2}}{231}v_5(\lambda) - \frac{4}{231}v_7(\lambda)$
		$+\frac{8\sqrt{2}}{231}v_{10}(\lambda)]j_6(\lambda\rho)$
1	6	$[-\frac{1}{42} + \frac{5}{28}v_1(\lambda) + \frac{17}{84}v_2(\lambda) + \frac{17}{42}v_3(\lambda) - \frac{5}{6}v_4(\lambda) + \frac{22\sqrt{2}}{21}v_5(\lambda)$
		$-\frac{10}{7}v_6(\lambda) - \frac{4}{21}v_7(\lambda) - \frac{2\sqrt{2}}{7}v_8(\lambda) + \frac{4\sqrt{2}}{7}v_9(\lambda) - \frac{4\sqrt{2}}{21}v_{10}(\lambda)]j_2(\lambda\rho)$
		$+[-\frac{3}{77} - \frac{1}{14}v_1(\lambda) - \frac{5}{154}v_2(\lambda) + \frac{23}{77}v_3(\lambda) + \frac{13}{11}v_4(\lambda) - \frac{64\sqrt{2}}{77}v_5(\lambda)$
		$+\frac{4}{7}v_6(\lambda) - \frac{10}{77}v_7(\lambda) + \frac{12\sqrt{2}}{7}v_8(\lambda) + \frac{4\sqrt{2}}{7}v_9(\lambda) - \frac{24\sqrt{2}}{77}v_{10}(\lambda)]j_4(\lambda\rho)$
		$+[-\frac{1}{66} + \frac{1}{66}v_2(\lambda) - \frac{7}{66}v_3(\lambda) + \frac{1}{66}v_4(\lambda) + \frac{4\sqrt{2}}{33}v_5(\lambda) + \frac{2}{33}v_7(\lambda)$
		$-\frac{4\sqrt{2}}{33}v_{10}(\lambda)]j_6(\lambda\rho)$
1	7	$[-\frac{2}{21} + \frac{2}{7}v_1(\lambda) + \frac{5}{21}v_2(\lambda) + \frac{10}{21}v_3(\lambda) - \frac{22}{21}v_4(\lambda) + \frac{22\sqrt{2}}{21}v_5(\lambda)$
		$-\frac{10}{7}v_6(\lambda) - \frac{4}{21}v_7(\lambda) - \frac{4\sqrt{2}}{7}v_8(\lambda) - \frac{2\sqrt{2}}{7}v_9(\lambda) - \frac{4\sqrt{2}}{21}v_{10}(\lambda)]j_2(\lambda\rho)$
		$+[-\frac{17}{154} + \frac{1}{28}v_1(\lambda) + \frac{1}{308}v_2(\lambda) + \frac{57}{154}v_3(\lambda) + \frac{149}{154}v_4(\lambda) - \frac{64\sqrt{2}}{77}v_5(\lambda)$
		$+\frac{4}{7}v_6(\lambda) - \frac{10}{77}v_7(\lambda) + \frac{10\sqrt{2}}{7}v_8(\lambda) + \frac{5\sqrt{2}}{7}v_9(\lambda) - \frac{24\sqrt{2}}{77}v_{10}(\lambda)]j_4(\lambda\rho)$
		$+[-\frac{1}{66} + \frac{1}{66}v_2(\lambda) - \frac{7}{66}v_3(\lambda) + \frac{1}{66}v_4(\lambda) + \frac{4\sqrt{2}}{33}v_5(\lambda) + \frac{2}{33}v_7(\lambda)$
		$-\frac{4\sqrt{2}}{33}v_{10}(\lambda)]j_6(\lambda\rho)$
1	8	$[\frac{1}{42} - \frac{1}{8}v_1(\lambda) + \frac{17}{168}v_2(\lambda) - \frac{17}{42}v_3(\lambda) + \frac{17}{42}v_4(\lambda) - \frac{\sqrt{2}}{3}v_5(\lambda)$
		$+\frac{5}{7}v_6(\lambda) + \frac{1}{21}v_7(\lambda) + \frac{\sqrt{2}}{7}v_8(\lambda) - \frac{\sqrt{2}}{7}v_9(\lambda) + \frac{4\sqrt{2}}{21}v_{10}(\lambda)]j_2(\lambda\rho)$
		$+[\frac{3}{77} - \frac{3}{77}v_2(\lambda) - \frac{23}{77}v_3(\lambda) - \frac{47}{77}v_4(\lambda) + \frac{6\sqrt{2}}{11}v_5(\lambda) - \frac{2}{7}v_6(\lambda)$
		$-\frac{1}{77}v_7(\lambda) - \frac{6\sqrt{2}}{7}v_8(\lambda) - \frac{\sqrt{2}}{7}v_9(\lambda) + \frac{24\sqrt{2}}{77}v_{10}(\lambda)]j_4(\lambda\rho)$
		$+[\frac{1}{66} - \frac{1}{66}v_2(\lambda) + \frac{7}{66}v_3(\lambda) - \frac{1}{66}v_4(\lambda) - \frac{4\sqrt{2}}{33}v_5(\lambda) - \frac{2}{33}v_7(\lambda)$
		$+\frac{4\sqrt{2}}{33}v_{10}(\lambda)]j_4(\lambda\rho)$
1	9	$[\frac{1}{42} - \frac{1}{28}v_1(\lambda) - \frac{5}{84}v_2(\lambda) - \frac{5}{42}v_3(\lambda) - \frac{1}{42}v_4(\lambda) - \frac{10\sqrt{2}}{21}v_5(\lambda)$
		$+\frac{4}{7}v_6(\lambda) + \frac{5}{42}v_7(\lambda) + \frac{4\sqrt{2}}{7}v_8(\lambda) - \frac{1}{7\sqrt{2}}v_9(\lambda) + \frac{4\sqrt{2}}{21}v_{10}(\lambda)]j_2(\lambda\rho)$
		$+[\frac{3}{77} - \frac{1}{28}v_1(\lambda) - \frac{23}{308}v_2(\lambda) - \frac{1}{77}v_3(\lambda) - \frac{3}{77}v_4(\lambda) + \frac{31\sqrt{2}}{77}v_5(\lambda)$
		$-\frac{3}{7}v_6(\lambda) + \frac{9}{154}v_7(\lambda) - \frac{3\sqrt{2}}{7}v_8(\lambda) - \frac{1}{7\sqrt{2}}v_9(\lambda) + \frac{24\sqrt{2}}{77}v_{10}(\lambda)]j_4(\lambda\rho)$
		$+[\frac{1}{66} - \frac{1}{66}v_2(\lambda) + \frac{7}{66}v_3(\lambda) - \frac{1}{66}v_4(\lambda) - \frac{4\sqrt{2}}{33}v_5(\lambda) - \frac{2}{33}v_7(\lambda)$
		$+\frac{4\sqrt{2}}{33}v_{10}(\lambda)]j_6(\lambda\rho)$
1	10	$[\frac{1}{7}v_3(\lambda) + \frac{1}{7}v_4(\lambda) + \frac{1}{\sqrt{2}}v_5(\lambda) - \frac{1}{7}v_6(\lambda) + \frac{1}{14}v_7(\lambda)$
		$-\frac{1}{\sqrt{2}}v_8(\lambda) - \frac{1}{7\sqrt{2}}v_9(\lambda)]j_2(\lambda\rho)$
		$+[\frac{1}{7}v_3(\lambda) + \frac{1}{7}v_4(\lambda) - \frac{1}{7}v_6(\lambda) + \frac{1}{14}v_7(\lambda) - \frac{1}{7\sqrt{2}}v_9(\lambda)]j_4(\lambda\rho)$
1	11	$[\frac{1}{42} - \frac{1}{14}v_1(\lambda) - \frac{1}{42}v_2(\lambda) - \frac{5}{42}v_3(\lambda) - \frac{1}{42}v_4(\lambda) - \frac{13\sqrt{2}}{21}v_5(\lambda)$
		$+\frac{6}{7}v_6(\lambda) + \frac{5}{42}v_7(\lambda) - \frac{\sqrt{2}}{7}v_8(\lambda) - \frac{1}{7\sqrt{2}}v_9(\lambda) + \frac{\sqrt{2}}{21}v_{10}(\lambda)]j_2(\lambda\rho)$
		$+[\frac{3}{77} - \frac{1}{14}v_1(\lambda) - \frac{3}{77}v_2(\lambda) - \frac{1}{77}v_3(\lambda) - \frac{3}{77}v_4(\lambda) + \frac{20\sqrt{2}}{77}v_5(\lambda)$

Table 3.10 (*Cont.*)

n	q	$N_{nq}(\lambda,\rho)$
		$-\frac{1}{7}v_6(\lambda)+\frac{9}{154}v_7(\lambda)-\frac{\sqrt{2}}{7}v_8(\lambda)-\frac{1}{7\sqrt{2}}v_9(\lambda)+\frac{13\sqrt{2}}{77}v_{10}(\lambda)]j_4(\lambda\rho)$
		$+[\frac{1}{66}-\frac{1}{66}v_2(\lambda)+\frac{7}{66}v_3(\lambda)-\frac{1}{66}v_4(\lambda)-\frac{4\sqrt{2}}{33}v_5(\lambda)-\frac{2}{33}v_7(\lambda)$
		$+\frac{4\sqrt{2}}{33}v_{10}(\lambda)]j_6(\lambda\rho)$
1	12	$[\frac{3}{56}v_1(\lambda)-\frac{3}{56}v_2(\lambda)+\frac{1}{7}v_3(\lambda)+\frac{1}{7}v_4(\lambda)+\frac{\sqrt{2}}{14}v_5(\lambda)$
		$-\frac{4}{7}v_6(\lambda)+\frac{1}{14}v_7(\lambda)+\frac{1}{7\sqrt{2}}v_8(\lambda)-\frac{1}{7\sqrt{2}}v_9(\lambda)-\frac{2\sqrt{2}}{7}v_{10}(\lambda)]j_2(\lambda\rho)$
		$+[-\frac{1}{14}v_1(\lambda)+\frac{1}{14}v_2(\lambda)+\frac{1}{7}v_3(\lambda)+\frac{1}{7}v_4(\lambda)-\frac{2\sqrt{2}}{7}v_5(\lambda)+\frac{3}{7}v_6(\lambda)$
		$+\frac{1}{14}v_7(\lambda)+\frac{4\sqrt{2}}{7}v_8(\lambda)-\frac{1}{7\sqrt{2}}v_9(\lambda)-\frac{2\sqrt{2}}{7}v_{10}(\lambda)]j_4(\lambda\rho)$
1	13	$[-\frac{1}{21}-\frac{3}{56}v_1(\lambda)+\frac{17}{168}v_2(\lambda)-\frac{4}{21}v_3(\lambda)+\frac{13}{21}v_4(\lambda)-\frac{10\sqrt{2}}{21}v_5(\lambda)$
		$+\frac{5}{7}v_6(\lambda)+\frac{1}{21}v_7(\lambda)+\frac{2\sqrt{2}}{7}v_8(\lambda)+\frac{\sqrt{2}}{7}v_9(\lambda)+\frac{4\sqrt{2}}{21}v_{10}(\lambda)]j_2(\lambda\rho)$
		$+[-\frac{5}{154}+\frac{1}{14}v_1(\lambda)-\frac{3}{77}v_2(\lambda)-\frac{13}{154}v_3(\lambda)-\frac{61}{154}v_4(\lambda)+\frac{31\sqrt{2}}{77}v_5(\lambda)$
		$-\frac{2}{7}v_6(\lambda)-\frac{1}{77}v_7(\lambda)-\frac{5\sqrt{2}}{7}v_8(\lambda)+\frac{\sqrt{2}}{7}v_9(\lambda)+\frac{24\sqrt{2}}{77}v_{10}(\lambda)]j_4(\lambda\rho)$
		$+[\frac{1}{66}-\frac{1}{66}v_2(\lambda)+\frac{7}{66}v_3(\lambda)-\frac{1}{66}v_4(\lambda)-\frac{4\sqrt{2}}{33}v_5(\lambda)-\frac{2}{33}v_7(\lambda)$
		$+\frac{4\sqrt{2}}{33}v_{10}(\lambda)]j_6(\lambda\rho)$
1	14	$[\frac{1}{42}-\frac{1}{28}v_1(\lambda)+\frac{1}{84}v_2(\lambda)-\frac{5}{42}v_3(\lambda)-\frac{13}{42}v_4(\lambda)-\frac{10\sqrt{2}}{21}v_5(\lambda)$
		$+\frac{5}{7}v_6(\lambda)-\frac{13}{42}v_7(\lambda)+\frac{3\sqrt{2}}{7}v_8(\lambda)+\frac{3}{7\sqrt{2}}v_9(\lambda)+\frac{\sqrt{2}}{21}v_{10}(\lambda)]j_2(\lambda\rho)$
		$+[\frac{3}{77}-\frac{1}{28}v_1(\lambda)-\frac{1}{308}v_2(\lambda)-\frac{1}{77}v_3(\lambda)-\frac{25}{77}v_4(\lambda)+\frac{31\sqrt{2}}{77}v_5(\lambda)$
		$-\frac{2}{7}v_6(\lambda)+\frac{10}{77}v_7(\lambda)-\frac{4\sqrt{2}}{7}v_8(\lambda)-\frac{2\sqrt{2}}{7}v_9(\lambda)+\frac{13\sqrt{2}}{77}v_{10}(\lambda)]j_4(\lambda\rho)$
		$+[\frac{1}{66}-\frac{1}{66}v_2(\lambda)+\frac{7}{66}v_3(\lambda)-\frac{1}{66}v_4(\lambda)-\frac{4\sqrt{2}}{33}v_5(\lambda)-\frac{2}{33}v_7(\lambda)$
		$+\frac{4\sqrt{2}}{33}v_{10}(\lambda)]j_6(\lambda\rho)$
1	15	$[\frac{5}{56}v_1(\lambda)-\frac{5}{56}v_2(\lambda)+\frac{5\sqrt{2}}{14}v_5(\lambda)-\frac{5}{7}v_6(\lambda)-\frac{3\sqrt{2}}{14}v_8(\lambda)$
		$-\frac{\sqrt{2}}{7}v_{10}(\lambda)]j_2(\lambda\rho)$
		$+[-\frac{1}{28}v_1(\lambda)+\frac{1}{28}v_2(\lambda)-\frac{\sqrt{2}}{7}v_5(\lambda)+\frac{2}{7}v_6(\lambda)+\frac{2}{7}v_8(\lambda)$
		$-\frac{\sqrt{2}}{7}v_{10}(\lambda)]j_4(\lambda\rho)$
1	16	$[\frac{5}{11}-\frac{1}{2}v_1(\lambda)+\frac{1}{22}v_2(\lambda)-\frac{9}{11}v_3(\lambda)-\frac{5}{11}v_4(\lambda)+\frac{4\sqrt{2}}{11}v_5(\lambda)$
		$+\frac{2}{11}v_7(\lambda)-2\sqrt{2}v_9(\lambda)-\frac{4\sqrt{2}}{11}v_{10}(\lambda)]j_4(\lambda\rho)$
		$+[-\frac{1}{22}+\frac{1}{22}v_2(\lambda)-\frac{7}{22}v_3(\lambda)+\frac{1}{22}v_4(\lambda)+\frac{4\sqrt{2}}{11}v_5(\lambda)+\frac{2}{11}v_7(\lambda)$
		$-\frac{4\sqrt{2}}{11}v_{10}(\lambda)]j_6(\lambda\rho)$
1	17	$[-\frac{1}{22}+\frac{1}{4}v_1(\lambda)+\frac{13}{44}v_2(\lambda)-\frac{7}{22}v_3(\lambda)-\frac{43}{22}v_4(\lambda)+\frac{4\sqrt{2}}{11}v_5(\lambda)$
		$+\frac{2}{11}v_7(\lambda)-2\sqrt{2}v_8(\lambda)-\sqrt{2}v_9(\lambda)-\frac{4\sqrt{2}}{11}v_{10}(\lambda)]j_4(\lambda\rho)$
		$+[-\frac{1}{22}+\frac{1}{22}v_2(\lambda)-\frac{7}{22}v_3(\lambda)+\frac{1}{22}v_4(\lambda)+\frac{4\sqrt{2}}{11}v_5(\lambda)+\frac{2}{11}v_7(\lambda)$
		$-\frac{4\sqrt{2}}{11}v_{10}(\lambda)]j_6(\lambda\rho)$
1	18	$[-\frac{1}{22}+\frac{1}{22}v_2(\lambda)+\frac{15}{22}v_3(\lambda)+\frac{23}{22}v_4(\lambda)-\frac{7\sqrt{2}}{11}v_5(\lambda)$
		$+\frac{2}{11}v_7(\lambda)+\sqrt{2}v_8(\lambda)-\frac{4\sqrt{2}}{11}v_{10}(\lambda)]j_4(\lambda\rho)$

Table 3.10 (*Cont.*)

n	q	$N_{nq}(\lambda,\rho)$
		$+[-\frac{1}{22}+\frac{1}{22}v_2(\lambda)-\frac{7}{22}v_3(\lambda)+\frac{1}{22}v_4(\lambda)+\frac{4\sqrt{2}}{11}v_5(\lambda)+\frac{2}{11}v_7(\lambda)$
		$-\frac{4\sqrt{2}}{11}v_{10}(\lambda)]j_6(\lambda\rho)$
1	19	$[-\frac{1}{22}+\frac{1}{4}v_1(\lambda)-\frac{9}{44}v_2(\lambda)-\frac{7}{22}v_3(\lambda)+\frac{1}{22}v_4(\lambda)+\frac{4\sqrt{2}}{11}v_5(\lambda)$
		$-v_6(\lambda)-\frac{7}{22}v_7(\lambda)-\sqrt{2}v_8(\lambda)+\frac{1}{\sqrt{2}}v_9(\lambda)+\frac{7\sqrt{2}}{11}v_{10}(\lambda)]j_4(\lambda\rho)$
		$+[-\frac{1}{22}+\frac{1}{22}v_2(\lambda)-\frac{7}{22}v_3(\lambda)+\frac{1}{22}v_4(\lambda)+\frac{4\sqrt{2}}{11}v_5(\lambda)+\frac{2}{11}v_7(\lambda)$
		$-\frac{4\sqrt{2}}{11}v_{10}(\lambda)]j_6(\lambda\rho)$
1	20	$[-\frac{1}{22}+\frac{1}{22}v_2(\lambda)-\frac{7}{22}v_3(\lambda)+\frac{1}{22}v_4(\lambda)-\frac{7\sqrt{2}}{11}v_5(\lambda)+v_6(\lambda)$
		$-\frac{7}{22}v_7(\lambda)+\sqrt{2}v_8+(\lambda)+\frac{1}{\sqrt{2}}v_9(\lambda)-\frac{4\sqrt{2}}{11}v_{10}(\lambda)]j_4(\lambda\rho)$
		$+[-\frac{1}{22}+\frac{1}{22}v_2(\lambda)-\frac{7}{22}v_3(\lambda)+\frac{1}{22}v_4(\lambda)+\frac{4\sqrt{2}}{11}v_5(\lambda)+\frac{2}{11}v_7(\lambda)$
		$-\frac{4\sqrt{2}}{11}v_{10}(\lambda)]j_6(\lambda\rho)$
1	21	$[\frac{1}{2}-\frac{1}{2}v_1(\lambda)+\frac{7}{2}v_1(\lambda)-\frac{1}{2}v_3(\lambda)-4\sqrt{2}v_4(\lambda)-2v_6(\lambda)$
		$+4\sqrt{2}v_{10}(\lambda)]j_6(\lambda\rho)$
2	1	$[\frac{1}{105}+\frac{2}{21}v_{11}(\lambda)+\frac{1}{35}v_{12}(\lambda)+\frac{4\sqrt{2}}{105}v_{13}(\lambda)+\frac{4}{35}v_{14}(\lambda)-\frac{4\sqrt{2}}{15}v_{15}(\lambda)$
		$-\frac{4\sqrt{2}}{105}v_{16}(\lambda)]j_0(\lambda\rho)$
		$+[-\frac{1}{21}+\frac{2}{21}v_{11}(\lambda)-\frac{1}{7}v_{12}(\lambda)+\frac{8\sqrt{2}}{21}v_{13}(\lambda)+\frac{2}{7}v_{14}(\lambda)-\frac{8\sqrt{2}}{21}v_{15}(\lambda)$
		$-\frac{8\sqrt{2}}{21}v_{16}(\lambda)+\frac{16}{7}v_{17}(\lambda)]j_2(\lambda\rho)$
		$+[-\frac{37}{385}+\frac{3}{154}v_{11}(\lambda)-\frac{111}{385}v_{12}(\lambda)+\frac{72\sqrt{2}}{385}v_{13}(\lambda)+\frac{96}{385}v_{14}(\lambda)-\frac{4\sqrt{2}}{35}v_{15}(\lambda)$
		$-\frac{72\sqrt{2}}{385}v_{16}(\lambda)+\frac{16}{7}v_{17}(\lambda)]j_4(\lambda\rho)$
		$+[-\frac{3}{77}+\frac{3}{154}v_{11}(\lambda)-\frac{9}{77}v_{12}(\lambda)-\frac{12\sqrt{2}}{77}v_{13}(\lambda)+\frac{6}{77}v_{14}(\lambda)$
		$+\frac{12\sqrt{2}}{77}v_{16}(\lambda)]j_6(\lambda\rho)$
2	2	$[\frac{1}{105}-\frac{1}{210}v_{11}(\lambda)+\frac{1}{35}v_{12}(\lambda)-\frac{2\sqrt{2}}{21}v_{13}(\lambda)-\frac{2}{105}v_{14}(\lambda)+\frac{2\sqrt{2}}{15}v_{15}(\lambda)$
		$-\frac{4\sqrt{2}}{105}v_{16}(\lambda)]j_0(\lambda\rho)$
		$+[\frac{1}{21}-\frac{1}{42}v_{11}(\lambda)+\frac{1}{7}v_{12}(\lambda)-\frac{2\sqrt{2}}{7}v_{13}(\lambda)-\frac{2}{21}v_{14}(\lambda)+\frac{4\sqrt{2}}{21}v_{15}(\lambda)$
		$+\frac{2\sqrt{2}}{21}v_{16}(\lambda)-\frac{8}{7}v_{17}(\lambda)]j_2(\lambda\rho)$
		$+[\frac{23}{385}-\frac{23}{770}v_{11}(\lambda)+\frac{69}{385}v_{12}(\lambda)-\frac{8\sqrt{2}}{77}v_{13}(\lambda)-\frac{46}{385}v_{14}(\lambda)+\frac{2\sqrt{2}}{35}v_{15}(\lambda)$
		$+\frac{18\sqrt{2}}{385}v_{16}(\lambda)-\frac{8}{7}v_{17}(\lambda)]j_4(\lambda\rho)$
		$+[\frac{5}{231}-\frac{5}{462}v_{11}(\lambda)+\frac{5}{77}v_{12}(\lambda)+\frac{20\sqrt{2}}{231}v_{13}(\lambda)-\frac{10}{231}v_{14}(\lambda)$
		$-\frac{20\sqrt{2}}{231}v_{16}(\lambda)]j_6(\lambda\rho)$
2	3	$[\frac{1}{105}-\frac{4}{105}v_{11}(\lambda)-\frac{11}{105}v_{12}(\lambda)-\frac{2\sqrt{2}}{35}v_{13}(\lambda)+\frac{1}{21}v_{14}(\lambda)+\frac{\sqrt{2}}{5}v_{15}(\lambda)$
		$+\frac{\sqrt{2}}{35}v_{16}(\lambda)]j_0(\lambda\rho)$
		$+[\frac{1}{21}-\frac{1}{14}v_{11}(\lambda)-\frac{1}{21}v_{12}(\lambda)-\frac{4\sqrt{2}}{21}v_{13}(\lambda)+\frac{\sqrt{2}}{7}v_{15}(\lambda)+\frac{4\sqrt{2}}{21}v_{16}(\lambda)$
		$-\frac{8}{7}v_{17}(\lambda)]j_2(\lambda\rho)$
		$+[\frac{23}{385}-\frac{17}{385}v_{11}(\lambda)+\frac{47}{385}v_{12}(\lambda)-\frac{29\sqrt{2}}{385}v_{13}(\lambda)-\frac{1}{11}v_{14}(\lambda)-\frac{2\sqrt{2}}{35}v_{15}(\lambda)$
		$+\frac{29\sqrt{2}}{385}v_{16}(\lambda)-\frac{8}{7}v_{17}(\lambda)]j_4(\lambda\rho)$
		$+[\frac{5}{231}-\frac{5}{462}v_{11}(\lambda)+\frac{5}{77}v_{12}(\lambda)+\frac{20\sqrt{2}}{231}v_{13}(\lambda)-\frac{10}{231}v_{14}(\lambda)$
		$-\frac{20\sqrt{2}}{231}v_{16}(\lambda)]j_6(\lambda\rho)$

Table 3.10 (*Cont.*)

n	q	$N_{nq}(\lambda,\rho)$
2	4	$[\frac{1}{105} - \frac{1}{210}v_{11}(\lambda) + \frac{1}{35}v_{12}(\lambda) - \frac{\sqrt{2}}{35}v_{13}(\lambda) - \frac{2}{105}v_{14}(\lambda) - \frac{\sqrt{2}}{15}v_{15}(\lambda)$ $+\frac{2\sqrt{2}}{21}v_{16}(\lambda) + \frac{2}{15}v_{17}(\lambda)]j_0(\lambda\rho)$ $+[\frac{\sqrt{2}}{21}v_{13}(\lambda) - \frac{2\sqrt{2}}{21}v_{15}(\lambda) + \frac{\sqrt{2}}{21}v_{16}(\lambda) + \frac{10}{21}v_{17}(\lambda)]j_2(\lambda\rho)$ $+[-\frac{1}{55} + \frac{1}{110}v_{11}(\lambda) - \frac{3}{55}v_{12}(\lambda) + \frac{16\sqrt{2}}{385}v_{13}(\lambda) + \frac{2}{55}v_{14}(\lambda) - \frac{\sqrt{2}}{35}v_{15}(\lambda)$ $-\frac{\sqrt{2}}{77}v_{16}(\lambda) + \frac{12}{35}v_{17}(\lambda)]j_4(\lambda\rho)$ $+[-\frac{2}{231} + \frac{1}{231}v_{11}(\lambda) - \frac{2}{77}v_{12}(\lambda) - \frac{8\sqrt{2}}{231}v_{13}(\lambda) + \frac{4}{231}v_{14}(\lambda)$ $+\frac{8\sqrt{2}}{231}v_{16}(\lambda)]j_6(\lambda\rho)$
2	5	$[\frac{1}{105} - \frac{1}{210}v_{11}(\lambda) + \frac{2}{21}v_{12}(\lambda) + \frac{6\sqrt{2}}{35}v_{13}(\lambda) - \frac{2}{105}v_{14}(\lambda) - \frac{2\sqrt{2}}{15}v_{15}(\lambda)$ $-\frac{4\sqrt{2}}{105}v_{16}(\lambda) - \frac{2}{17}v_{17}(\lambda)]j_0(\lambda\rho)$ $+[-\frac{1}{21}v_{12}(\lambda) + \frac{4\sqrt{2}}{21}v_{13}(\lambda) - \frac{\sqrt{2}}{21}v_{15}(\lambda) - \frac{\sqrt{2}}{7}v_{16}(\lambda) + \frac{2}{3}v_{17}(\lambda)]j_2(\lambda\rho)$ $+[-\frac{1}{55} + \frac{1}{110}v_{11}(\lambda) - \frac{13}{77}v_{12}(\lambda) - \frac{6\sqrt{2}}{385}v_{13}(\lambda) + \frac{2}{55}v_{14}(\lambda) + \frac{3\sqrt{2}}{35}v_{15}(\lambda)$ $-\frac{27\sqrt{2}}{385}v_{16}(\lambda) + \frac{4}{5}v_{17}(\lambda)]j_4(\lambda\rho)$ $+[-\frac{2}{231} + \frac{1}{231}v_{11}(\lambda) - \frac{2}{77}v_{12}(\lambda) - \frac{8\sqrt{2}}{231}v_{13}(\lambda) + \frac{4}{231}v_{14}(\lambda)$ $+\frac{8\sqrt{2}}{231}v_{16}(\lambda)]j_6(\lambda\rho)$
2	6	$[\frac{1}{21} - \frac{8}{21}v_{11}(\lambda) + \frac{1}{7}v_{12}(\lambda) - \frac{8\sqrt{2}}{21}v_{13}(\lambda) - \frac{8}{21}v_{14}(\lambda) + \frac{4\sqrt{2}}{7}v_{15}(\lambda)$ $+\frac{8\sqrt{2}}{21}v_{16}(\lambda) - \frac{16}{7}v_{17}(\lambda)]j_2(\lambda\rho)$ $+[\frac{6}{77} + \frac{8}{77}v_{11}(\lambda) + \frac{18}{77}v_{12}(\lambda) - \frac{20\sqrt{2}}{77}v_{13}(\lambda) - \frac{34}{77}v_{14}(\lambda) + \frac{4\sqrt{2}}{7}v_{15}(\lambda)$ $+\frac{20\sqrt{2}}{77}v_{16}(\lambda) - \frac{16}{7}v_{17}(\lambda)]j_4(\lambda\rho)$ $+[\frac{1}{33} - \frac{1}{66}v_{11}(\lambda) + \frac{1}{11}v_{12}(\lambda) + \frac{4\sqrt{2}}{33}v_{13}(\lambda) - \frac{2}{33}v_{14}(\lambda)$ $-\frac{4\sqrt{2}}{33}v_{16}(\lambda)]j_6(\lambda\rho)$
2	7	$[\frac{1}{21} + \frac{1}{21}v_{11}(\lambda) + \frac{1}{7}v_{12}(\lambda) - \frac{8\sqrt{2}}{21}v_{13}(\lambda) - \frac{5}{21}v_{14}(\lambda) + \frac{2\sqrt{2}}{7}v_{15}(\lambda)$ $+\frac{8\sqrt{2}}{21}v_{16}(\lambda) - \frac{16}{7}v_{17}(\lambda)]j_2(\lambda\rho)$ $+[\frac{6}{77} + \frac{5}{154}v_{11}(\lambda) + \frac{18}{77}v_{12}(\lambda) - \frac{20\sqrt{2}}{77}v_{13}(\lambda) - \frac{23}{77}v_{14}(\lambda) + \frac{2\sqrt{2}}{7}v_{15}(\lambda)$ $+\frac{20\sqrt{2}}{77}v_{16}(\lambda) - \frac{16}{7}v_{17}(\lambda)]j_4(\lambda\rho)$ $+[\frac{1}{33} - \frac{1}{66}v_{11}(\lambda) + \frac{1}{11}v_{12}(\lambda) + \frac{4\sqrt{2}}{33}v_{13}(\lambda) - \frac{2}{33}v_{14}(\lambda)$ $-\frac{4\sqrt{2}}{33}v_{16}(\lambda)]j_6(\lambda\rho)$
2	8	$[-\frac{1}{21} + \frac{1}{42}v_{11}(\lambda) - \frac{1}{7}v_{12}(\lambda) + \frac{8\sqrt{2}}{21}v_{13}(\lambda) + \frac{2}{21}v_{14}(\lambda) - \frac{2\sqrt{2}}{7}v_{15}(\lambda)$ $-\frac{2\sqrt{2}}{21}v_{16}(\lambda) + \frac{8}{7}v_{17}(\lambda)]j_2(\lambda\rho)$ $+[-\frac{6}{77} + \frac{3}{77}v_{11}(\lambda) - \frac{18}{77}v_{12}(\lambda) + \frac{20\sqrt{2}}{77}v_{13}(\lambda) + \frac{12}{77}v_{14}(\lambda) - \frac{2\sqrt{2}}{7}v_{15}(\lambda)$ $+\frac{2\sqrt{2}}{77}v_{16}(\lambda) + \frac{8}{7}v_{17}(\lambda)]j_4(\lambda\rho)$ $+[-\frac{1}{33} + \frac{1}{66}v_{11}(\lambda) - \frac{1}{11}v_{12}(\lambda) - \frac{4\sqrt{2}}{33}v_{13}(\lambda) + \frac{2}{33}v_{14}(\lambda)$ $+\frac{4\sqrt{2}}{33}v_{16}(\lambda)]j_6(\lambda\rho)$
2	9	$[-\frac{1}{21} + \frac{2}{21}v_{11}(\lambda) - \frac{1}{7}v_{12}(\lambda) + \frac{2\sqrt{2}}{21}v_{13}(\lambda) - \frac{1}{21}v_{14}(\lambda) - \frac{4\sqrt{2}}{7}v_{15}(\lambda)$ $-\frac{5\sqrt{2}}{21}v_{16}(\lambda) + \frac{12}{7}v_{17}(\lambda)]j_2(\lambda\rho)$ $+[-\frac{6}{77} + \frac{17}{154}v_{11}(\lambda) - \frac{18}{77}v_{12}(\lambda) - \frac{2\sqrt{2}}{77}v_{13}(\lambda) + \frac{1}{77}v_{14}(\lambda) + \frac{3\sqrt{2}}{7}v_{15}(\lambda)$

Table 3.10 (*Cont.*)

n	q	$N_{nq}(\lambda,\rho)$
		$-\frac{9\sqrt{2}}{77}v_{16}(\lambda)+\frac{12}{7}v_{17}(\lambda)]j_4(\lambda\rho)$
		$+[-\frac{1}{33}+\frac{1}{66}v_{11}(\lambda)-\frac{1}{11}v_{12}(\lambda)-\frac{4\sqrt{2}}{33}v_{13}(\lambda)+\frac{2}{33}v_{14}(\lambda)$
		$+\frac{4\sqrt{2}}{33}v_{16}(\lambda)]j_6(\lambda\rho)$
2	10	$[-\frac{\sqrt{2}}{7}v_{13}(\lambda)+\frac{2\sqrt{2}}{7}v_{15}(\lambda)-\frac{\sqrt{2}}{7}v_{16}(\lambda)-\frac{10}{7}v_{17}(\lambda)]j_2(\lambda\rho)$
		$+[-\frac{\sqrt{2}}{7}v_{13}(\lambda)-\frac{2\sqrt{2}}{7}v_{15}(\lambda)-\frac{\sqrt{2}}{7}v_{16}(\lambda)+\frac{4}{7}v_{17}(\lambda)]j_4(\lambda\rho)$
2	11	$[-\frac{1}{21}+\frac{2}{21}v_{11}(\lambda)+\frac{1}{7}v_{12}(\lambda)+\frac{5\sqrt{2}}{21}v_{13}(\lambda)-\frac{1}{21}v_{14}(\lambda)+\frac{\sqrt{2}}{7}v_{15}(\lambda)$
		$-\frac{2\sqrt{2}}{21}v_{16}(\lambda)+\frac{4}{7}v_{17}(\lambda)]j_2(\lambda\rho)$
		$+[-\frac{6}{77}+\frac{17}{154}v_{11}(\lambda)+\frac{4}{77}v_{12}(\lambda)+\frac{9\sqrt{2}}{77}v_{13}(\lambda)+\frac{1}{77}v_{14}(\lambda)+\frac{\sqrt{2}}{7}v_{15}(\lambda)$
		$+\frac{2\sqrt{2}}{77}v_{16}(\lambda)+\frac{4}{7}v_{17}(\lambda)]j_4(\lambda\rho)$
		$+[-\frac{1}{33}+\frac{1}{66}v_{11}(\lambda)-\frac{1}{11}v_{12}(\lambda)-\frac{4\sqrt{2}}{33}v_{13}(\lambda)+\frac{2}{33}v_{14}(\lambda)$
		$+\frac{4\sqrt{2}}{33}v_{16}(\lambda)]j_6(\lambda\rho)$
2	12	$[-\frac{3}{7}v_{12}(\lambda)+\frac{\sqrt{2}}{7}v_{13}(\lambda)-\frac{2\sqrt{2}}{7}v_{15}(\lambda)+\frac{\sqrt{2}}{7}v_{16}(\lambda)+\frac{2}{7}v_{17}(\lambda)]j_2(\lambda\rho)$
		$+[\frac{4}{7}v_{12}(\lambda)+\frac{\sqrt{2}}{7}v_{13}(\lambda)-\frac{2\sqrt{2}}{7}v_{15}(\lambda)+\frac{\sqrt{2}}{7}v_{16}(\lambda)-\frac{12}{7}v_{17}(\lambda)]j_4(\lambda\rho)$
2	13	$[-\frac{1}{21}+\frac{1}{42}v_{11}(\lambda)-\frac{1}{7}v_{12}(\lambda)+\frac{5\sqrt{2}}{21}v_{13}(\lambda)+\frac{2}{21}v_{14}(\lambda)-\frac{\sqrt{2}}{7}v_{15}(\lambda)$
		$-\frac{2\sqrt{2}}{21}v_{16}(\lambda)+\frac{8}{7}v_{17}(\lambda)]j_2(\lambda\rho)$
		$+[-\frac{6}{77}+\frac{3}{77}v_{11}(\lambda)-\frac{18}{77}v_{12}(\lambda)+\frac{9\sqrt{2}}{77}v_{13}(\lambda)+\frac{12}{77}v_{14}(\lambda)-\frac{\sqrt{2}}{7}v_{15}(\lambda)$
		$+\frac{2\sqrt{2}}{77}v_{16}(\lambda)+\frac{8}{7}v_{17}(\lambda)]j_4(\lambda\rho)$
		$+[-\frac{1}{33}+\frac{1}{66}v_{11}(\lambda)-\frac{1}{11}v_{12}(\lambda)-\frac{4\sqrt{2}}{33}v_{13}(\lambda)+\frac{2}{33}v_{14}(\lambda)$
		$+\frac{4\sqrt{2}}{33}v_{16}(\lambda)]j_6(\lambda\rho)$
2	14	$[-\frac{1}{21}+\frac{1}{42}v_{11}(\lambda)+\frac{1}{7}v_{12}(\lambda)+\frac{5\sqrt{2}}{21}v_{13}(\lambda)+\frac{2}{21}v_{14}(\lambda)-\frac{5\sqrt{2}}{21}v_{16}(\lambda)$
		$+\frac{8}{7}v_{17}(\lambda)]j_2(\lambda\rho)$
		$+[-\frac{6}{77}+\frac{3}{77}v_{11}(\lambda)+\frac{4}{77}v_{12}(\lambda)+\frac{9\sqrt{2}}{77}v_{13}(\lambda)+\frac{12}{77}v_{14}(\lambda)-\frac{9\sqrt{2}}{77}v_{16}(\lambda)$
		$+\frac{8}{7}v_{17}(\lambda)]j_4(\lambda\rho)$
		$+[-\frac{1}{33}+\frac{1}{66}v_{11}(\lambda)-\frac{1}{11}v_{12}(\lambda)-\frac{4\sqrt{2}}{33}v_{13}(\lambda)+\frac{2}{33}v_{14}(\lambda)$
		$+\frac{4\sqrt{2}}{33}v_{16}(\lambda)]j_6(\lambda\rho)$
2	15	$[\frac{2}{7}v_{12}(\lambda)-\frac{5}{7\sqrt{2}}v_{13}(\lambda)+\frac{3}{7\sqrt{2}}v_{15}(\lambda)+\frac{5\sqrt{2}}{21}v_{16}(\lambda)-\frac{8}{7}v_{17}(\lambda)]j_2(\lambda\rho)$
		$+[+\frac{2}{7}v_{12}(\lambda)+\frac{\sqrt{2}}{7}v_{13}(\lambda)-\frac{2\sqrt{2}}{7}v_{15}(\lambda)+\frac{\sqrt{2}}{7}v_{16}(\lambda)-\frac{8}{7}v_{17}(\lambda)]j_4(\lambda\rho)$
2	16	$[\frac{1}{11}-\frac{1}{22}v_{11}(\lambda)+\frac{3}{11}v_{12}(\lambda)+\frac{4\sqrt{2}}{11}v_{13}(\lambda)-\frac{2}{11}v_{14}(\lambda)$
		$-\frac{4\sqrt{2}}{11}v_{16}(\lambda)]j_4(\lambda\rho)$
		$+[\frac{1}{11}-\frac{1}{22}v_{11}(\lambda)+\frac{3}{11}v_{12}(\lambda)+\frac{4\sqrt{2}}{11}v_{13}(\lambda)-\frac{2}{11}v_{14}(\lambda)$
		$-\frac{4\sqrt{2}}{11}v_{16}(\lambda)]j_6(\lambda\rho)$
2	17	$[\frac{1}{11}-\frac{6}{11}v_{11}(\lambda)+\frac{3}{11}v_{12}(\lambda)+\frac{4\sqrt{2}}{11}v_{13}(\lambda)+\frac{9}{11}v_{14}(\lambda)-2\sqrt{2}v_{15}(\lambda)$
		$-\frac{4\sqrt{2}}{11}v_{16}(\lambda)]j_4(\lambda\rho)$
		$+[\frac{1}{11}-\frac{1}{22}v_{11}(\lambda)+\frac{3}{11}v_{12}(\lambda)+\frac{4\sqrt{2}}{11}v_{13}(\lambda)-\frac{2}{11}v_{14}(\lambda)$
		$-\frac{4\sqrt{2}}{11}v_{16}(\lambda)]j_6(\lambda\rho)$

Table 3.10 (*Cont.*)

n	q	$N_{nq}(\lambda,\rho)$
2	18	$[\frac{1}{11} - \frac{1}{22}v_{11}(\lambda) + \frac{3}{11}v_{12}(\lambda) - \frac{7\sqrt{2}}{11}v_{13}(\lambda) - \frac{2}{11}v_{14}(\lambda) + \sqrt{2}v_{15}(\lambda)$ $-\frac{4\sqrt{2}}{11}v_{16}(\lambda)]j_4(\lambda\rho)$ $+[\frac{1}{11} - \frac{1}{22}v_{11}(\lambda) + \frac{3}{11}v_{12}(\lambda) + \frac{4\sqrt{2}}{11}v_{13}(\lambda) - \frac{2}{11}v_{14}(\lambda)$ $-\frac{4\sqrt{2}}{11}v_{16}(\lambda)]j_6(\lambda\rho)$
2	19	$[\frac{1}{11} - \frac{1}{22}v_{11}(\lambda) - \frac{19}{11}v_{12}(\lambda) - \frac{7\sqrt{2}}{11}v_{13}(\lambda) - \frac{2}{11}v_{14}(\lambda) + \sqrt{2}v_{15}(\lambda)$ $-\frac{4\sqrt{2}}{11}v_{16}(\lambda) + 4v_{17}(\lambda)]j_4(\lambda\rho)$ $+[\frac{1}{11} - \frac{1}{22}v_{11}(\lambda) + \frac{3}{11}v_{12}(\lambda) + \frac{4\sqrt{2}}{11}v_{13}(\lambda) - \frac{2}{11}v_{14}(\lambda)$ $-\frac{4\sqrt{2}}{11}v_{16}(\lambda)]j_6(\lambda\rho)$
2	20	$[\frac{1}{11} - \frac{1}{22}v_{11}(\lambda) + \frac{3}{11}v_{12}(\lambda) + \frac{4\sqrt{2}}{11}v_{13}(\lambda) - \frac{2}{11}v_{14}(\lambda) - \sqrt{2}v_{15}(\lambda)$ $+\frac{7\sqrt{2}}{11}v_{16}(\lambda) - v_{17}(\lambda)]j_4(\lambda\rho)$ $+[\frac{1}{11} - \frac{1}{22}v_{11}(\lambda) + \frac{3}{11}v_{12}(\lambda) + \frac{4\sqrt{2}}{11}v_{13}(\lambda) - \frac{2}{11}v_{14}(\lambda)$ $-\frac{4\sqrt{2}}{11}v_{16}(\lambda)]j_6(\lambda\rho)$
2	21	$[-1 + \frac{1}{2}v_{11}(\lambda) - 3v_{12}(\lambda) - 4\sqrt{2}v_{13}(\lambda) + 2v_{14}(\lambda)$ $+4\sqrt{2}v_{16}(\lambda)]j_6(\lambda)$
3	1	$[-\frac{4}{5} + \frac{8}{15}v_{18}(\lambda)]j_0(\lambda\rho) + [\frac{4}{7} - \frac{8}{21}v_{18}(\lambda)]j_2(\lambda\rho)$ $+[\frac{48}{35} - \frac{32}{35}v_{18}(\lambda)]j_4(\lambda\rho)$
3	2	$[\frac{2}{5} - \frac{2}{15}v_{18}(\lambda)]j_0(\lambda\rho) + [-\frac{2}{7} + \frac{2}{21}v_{18}(\lambda)]j_2(\lambda\rho)$ $+[-\frac{24}{35} + \frac{8}{35}v_{18}(\lambda)]j_4(\lambda\rho)$
3	3	$[\frac{4}{15} - \frac{2}{15}v_{18}(\lambda)]j_0(\lambda\rho) + [-\frac{4}{21} + \frac{2}{21}v_{18}(\lambda)]j_2(\lambda\rho)$ $+[-\frac{16}{35} + \frac{8}{35}v_{18}(\lambda)]j_4(\lambda\rho)$
3	4	$[-\frac{2}{15} + \frac{1}{15}v_{18}(\lambda)]j_0(\lambda\rho) + [\frac{2}{21} - \frac{1}{21}v_{18}(\lambda)]j_2(\lambda\rho)$ $+[\frac{8}{35} - \frac{4}{35}v_{18}(\lambda)]j_4(\lambda\rho)$
3	5	$-\frac{2}{15}j_0(\lambda\rho) + \frac{2}{21}j_2(\lambda\rho) + \frac{8}{35}j_4(\lambda\rho)$
3	6	$[\frac{12}{7} - \frac{16}{7}v_{18}(\lambda)]j_2(\lambda\rho) + [-\frac{16}{7} + \frac{12}{7}v_{18}(\lambda)]j_4(\lambda\rho)$
3	7	$[-\frac{12}{7} + \frac{12}{7}v_{18}(\lambda)]j_2(\lambda\rho) + [-\frac{12}{7} + \frac{12}{7}v_{18}(\lambda)]j_4(\lambda\rho)$
3	8	$[-\frac{6}{7} + \frac{10}{7}v_{18}(\lambda)]j_2(\lambda\rho) + [\frac{8}{7} - \frac{4}{7}v_{18}(\lambda)]j_4(\lambda\rho)$
3	9	$\frac{2}{7}v_{18}(\lambda)j_2(\lambda\rho) + \frac{2}{7}v_{18}(\lambda)j_4(\lambda\rho)$
3	10	$[-\frac{2}{7} + \frac{1}{7}v_{18}(\lambda)]j_2(\lambda\rho) - [\frac{2}{7} + \frac{1}{7}v_{18}(\lambda)]j_4(\lambda\rho)$
3	11	0
3	12	$[-\frac{2}{7} + \frac{4}{7}v_{18}(\lambda)]j_2(\lambda\rho) + [-\frac{2}{7} - \frac{3}{7}v_{18}(\lambda)]j_4(\lambda\rho)$
3	13	$[\frac{6}{7} - \frac{6}{7}v_{18}(\lambda)]j_2(\lambda\rho) + [\frac{6}{7} + \frac{1}{7}v_{18}(\lambda)]j_4(\lambda\rho)$
3	14	$[\frac{4}{7} - \frac{4}{7}v_{18}(\lambda)]j_2(\lambda\rho) + [\frac{4}{7} - \frac{4}{7}v_{18}(\lambda)]j_4(\lambda\rho)$
3	15	$-\frac{2}{7}v_{18}(\lambda)j_2(\lambda\rho) - \frac{2}{7}v_{18}(\lambda)j_4(\lambda\rho)$
3	16	$-4v_{18}(\lambda)j_4(\lambda\rho)$
3	17	$[4 - 4v_{18}(\lambda)]j_4(\lambda\rho)$
3	18	$[-2 - v_{18}(\lambda)]j_4(\lambda\rho)$
3	19	$2v_{18}(\lambda)j_4(\lambda\rho)$
3	20	0
3	21	0

$N(G_2) = \mathcal{O} \times Z_2^c$, $N(G_3) = D_6 \times Z_2^c$, $N(G_4) = D_8 \times Z_2^c$, $N(G_5) = \mathcal{O} \times Z_2^c$, $N(G_6) = \mathrm{O}(2) \times Z_2^c$ and $N(G_7) = \mathrm{O}(3)$.

The possibilities for the symmetry group are as follows. In the triclinic class, there exist infinitely many groups between Z_2^c and $\mathrm{O}(3)$, we choose $G_1 = Z_2^c$ and $G_2 = \mathrm{O}(3)$. Similarly, for the monoclinic class we choose $G_3 = Z_2 \times Z_2^c$ and $G_4 = \mathrm{O}(2) \times Z_2^c$. The possibilities for the orthotropic class are $G_5 = D_2 \times Z_2^c$, $G_6 = D_4 \times Z_2^c$, $G_7 = D_6 \times Z_2^c$, $G_8 = \mathcal{T} \times Z_2^c$ and $G_9 = \mathcal{O} \times Z_2^c$. In the trigonal class, we choose $G_{10} = D_3 \times Z_2^c$ and $G_{11} = D_6 \times Z_2^c$. In the tetragonal class, the possibilities are $G_{12} = D_4 \times Z_2^c$ and $G_{13} = D_8 \times Z_2^c$. In the three remaining classes, the possibilities are $G_{14} = \mathrm{O}(2) \times Z_2^c$, $G_{15} = \mathcal{O} \times Z_2^c$ and $G_{16} = \mathrm{O}(3)$.

We have $V = \mathbb{R}^3$, $G = \mathrm{O}(3)$, $\mathsf{V}_0 = \mathsf{S}^2(\mathsf{S}^2(\mathbb{R}^3))$ and $\rho_0 = \mathsf{S}^2(\mathsf{S}^2(g))$. We proceed in steps.

For each symmetry class $[G_j]$, we have to describe the space V of the isotypic component of the representation ρ_0 that corresponds to the trivial representation of the group G_j acts. We perform this by introducing an orthonormal basis $\boldsymbol{T}^1_{ijkl}, \dots, \boldsymbol{T}^{\dim \mathsf{V}}_{ijkl}$. Any space V is a subspace of the space $\mathsf{S}^2(\mathsf{S}^2(\mathbb{R}^3))$. We express the tensors $\boldsymbol{T}^1_{ijkl}, \dots, \boldsymbol{T}^{\dim \mathsf{V}}_{ijkl}$ in terms of the basis tensors of the space $\mathsf{S}^2(\mathsf{S}^2(\mathbb{R}^3))$ given by Equations (3.65), (3.67), (3.66), (3.68) and (3.69); see also Equation (3.62). The calculations are similar to those in Section 3.6.

All groups G_j are of type II. We formulate and prove only the cases of $G = D_8 \times Z_2^c$ and $G = \mathrm{O}(3)$.

First, we determine a suitable basis in the space $\mathsf{V}^{D_4 \times Z_2^c}$. According to Altmann & Herzig (1994, Table 33.5), the restriction of the representation ρ_1 of the group $\mathrm{O}(3)$ to the subgroup $D_4 \times Z_2^c$ is equal to $A_{2u} \oplus E_u$. The representation A_{2u} acts in the z-axis, while the representation E_u acts in the xy-plane. Using Altmann & Herzig (1994, Table 33.8), we determine the structure of the symmetric tensor square of the representation $A_{2u} \oplus E_u$:

$$\mathsf{S}^2(A_{2u} \oplus E_u) = 2A_{1g} \oplus E_g \oplus B_{1g} \oplus B_{2g}.$$

The first copy of the representation A_{1g} acts in the one-dimensional space generated by the matrix $T^1 = \begin{pmatrix} 0 & 0 & 0 \\ 0 & 1 & 0 \\ 0 & 0 & 0 \end{pmatrix}$, the second copy in the space generated by the matrix $T^2 = \frac{1}{\sqrt{2}}\begin{pmatrix} 1 & 0 & 0 \\ 0 & 0 & 0 \\ 0 & 0 & 1 \end{pmatrix}$. The representation E_g acts in the two-dimensional space generated by the matrices $T^3 = \frac{1}{\sqrt{2}}\begin{pmatrix} 0 & 1 & 0 \\ 1 & 0 & 0 \\ 0 & 0 & 0 \end{pmatrix}$ and $T^4 = \frac{1}{\sqrt{2}}\begin{pmatrix} 0 & 0 & 0 \\ 0 & 0 & 1 \\ 0 & 1 & 0 \end{pmatrix}$. The representation B_{1g} acts in the one-dimensional space generated by the matrix $T^5 = \frac{1}{\sqrt{2}}\begin{pmatrix} -1 & 0 & 0 \\ 0 & 0 & 0 \\ 0 & 0 & 1 \end{pmatrix}$, and the representation B_{2g} in the space generated by $T^6 = \frac{1}{\sqrt{2}}\begin{pmatrix} 0 & 0 & 1 \\ 0 & 0 & 0 \\ 1 & 0 & 0 \end{pmatrix}$. Using Altmann & Herzig (1994, Table 33.8) once more, we find that the symmetric tensor square $\mathsf{S}^2(\mathsf{S}^2(A_{2u} \oplus E_u))$ contains six copies of the trivial representation A_{1g} of the group $D_4 \times Z_2^c$, and they act in the one-dimensional spaces generated by the rank 4 tensors

$$\begin{aligned}\mathsf{T}^i &= T^i \otimes T^i, \qquad i = 1, 2, 5, 6,\\ \mathsf{T}^3 &= \frac{1}{\sqrt{2}}(T^1 \otimes T^2 + T^2 \otimes T^1),\\ \mathsf{T}^4 &= \frac{1}{\sqrt{2}}(T^3 \otimes T^3 + T^4 \otimes T^4).\end{aligned}$$

Next, consider the group $G = D_8 \times Z_2^c$. According to Altmann & Herzig (1994, Table 37.5), the restriction of the representation ρ_1 of the group O(3) to G is equal to $A_{2u} \oplus E_{1u}$. The representation A_{2u} acts in the z-axis, while the representation E_{1u} acts in the xy-plane. Using Altmann & Herzig (1994, Table 33.8), we determine the structure of the symmetric tensor square of the representation $A_{2u} \oplus E_{1u}$:

$$\mathsf{S}^2(A_{2u} \oplus E_{1u}) = 2A_{1g} \oplus E_{1g} \oplus E_{2g}.$$

The first copy of the representation A_{1g} acts in the one-dimensional space generated by the matrix T^1, the second copy in the space generated by T^2. The representation E_{1g} acts in the two-dimensional space generated by the matrices T^3 and T^4, the representation E_{2g} in the space generated by T^5 and T^6. Using Altmann & Herzig (1994, Table 33.8) once more, we find the components of the symmetric tensor square $\mathsf{S}^2(\mathsf{S}^2(A_{2u} \oplus E_{1u}))$ that act in the space $\mathsf{V}^{D_8 \times Z_2^c}$. They are as follows: four copies of A_{1g} that act in the spaces generated by the tensors $\tilde{\mathsf{T}}^i = \mathsf{T}^i$, $1 \le i \le 4$, a copy of A_{1g} that acts in the space generated by $\tilde{\mathsf{T}}^5 = \frac{1}{\sqrt{2}}(\mathsf{T}^5 + \mathsf{T}^6)$ and a copy of B_{2g} that acts in the space generated by $\tilde{\mathsf{T}}^6 = \frac{1}{\sqrt{2}}(-\mathsf{T}^5 + \mathsf{T}^6)$.

By Lemma 1, we have $\tilde{\rho} = \mathsf{S}^2(5A_{1g} \oplus B_{2g})$. By Altmann & Herzig (1994, Table 37.8), $\tilde{\rho} = 16A_{1g} \oplus 5B_{2g}$. We have to investigate the restrictions of the non-trivial representation B_{2g} to the stationary subgroups of various strata of the orbit space $\hat{\mathbb{R}}^3/D_8 \times Z_2^c$. They are given by Equation 3.31. Using Altmann & Herzig (1994, Table 37.9), we find that the restrictions of B_{2g} to the stationary subgroups of the strata $(\hat{\mathbb{R}}^3/D_8 \times Z_2^c)_3$, $(\hat{\mathbb{R}}^3/D_8 \times Z_2^c)_4$, $(\hat{\mathbb{R}}^3/D_8 \times Z_2^c)_6$ and $(\hat{\mathbb{R}}^3/D_8 \times Z_2^c)_7$ are trivial. On these strata, the function $f_0(\boldsymbol{p})$ takes values in the set $\mathcal{C}_0$ of the symmetric non-negative-definite 6×6 matrices with unit trace. We divide the group $D_8 \times Z_2^c$ into two sets. On the set

$$G_+ = \{E, C_4^+, C_4^-, C_2, C_{21}'', C_{22}'', C_{23}'', C_{24}'', i, S_4^+, S_4^-, \sigma_h, \sigma_{d1}, \sigma_{d2}, \sigma_{d3}, \sigma_{d4}\}$$

the representation B_{2g} takes value 1. On the complement set $G_- = G \setminus G_+$, the above representation takes value -1. The spherical Bessel function has the form

$$j(\boldsymbol{p}, \boldsymbol{y} - \boldsymbol{x}) = j_+(\boldsymbol{p}, \boldsymbol{y} - \boldsymbol{x}) + j_-(\boldsymbol{p}, \boldsymbol{y} - \boldsymbol{x}),$$

where

$$j_\pm(\boldsymbol{p}, \boldsymbol{y} - \boldsymbol{x}) = \frac{1}{32} \sum_{g \in G_\pm} \mathrm{e}^{\mathrm{i}(g\boldsymbol{p}, \boldsymbol{y} - \boldsymbol{x})}.$$

We calculate the matrix entries of the representation $A_{2u} \oplus E_{1u}$, using the Euler angles from Altmann & Herzig (1994, Table 37.1). Then, we calculate the functions $j_\pm(\boldsymbol{p}, \boldsymbol{y} - \boldsymbol{x})$ and obtain

$$\begin{aligned}
j_+(\boldsymbol{p}, \boldsymbol{y} - \boldsymbol{x}) &= \frac{1}{8}\cos(p_3z_3)[\cos(p_1z_1 + p_2z_2) + \cos(p_1z_2 - p_2z_1)] \\
&\quad + \frac{1}{4}\cos\left(\frac{p_1z_1 - p_2z_2}{\sqrt{2}}\right)\cos\left(\frac{p_1z_2 + p_2z_1}{\sqrt{2}}\right)\cos(p_3z_3), \\
j_-(\boldsymbol{p}, \boldsymbol{y} - \boldsymbol{x}) &= \frac{1}{8}\cos(p_3z_3)[\cos(p_1z_1 - p_2z_2) + \cos(p_1z_2 + p_2z_1)] \\
&\quad + \frac{1}{4}\cos\left(\frac{p_1z_2 - p_2z_1}{\sqrt{2}}\right)\cos\left(\frac{p_1z_1 + p_2z_2}{\sqrt{2}}\right)\cos(p_3z_3).
\end{aligned}$$

The contribution of the above strata to the two-point correlation tensor becomes

$$\int_{(\hat{\mathbb{R}}^3/D_8\times Z_2^c)_{3,4,6,7}} (j_+(\boldsymbol{p}, \boldsymbol{y} - \boldsymbol{x})f_0^+(\boldsymbol{p}) + j_-(\boldsymbol{p}, \boldsymbol{y} - \boldsymbol{x})f_0^-(\boldsymbol{p}))\,\mathrm{d}\Phi(\boldsymbol{p}),$$

where $f_0^+(\boldsymbol{p})$ is a measurable function taking values in $\mathcal{C}_0$, and $f_0^-(\boldsymbol{p})$ is the function $f_0^+(\boldsymbol{p})$, where the matrix entries $((f_0^+)_{i6}(\boldsymbol{p})$ and $((f_0^+)_{6i}(\boldsymbol{p})$, $1 \leq i \leq 5$, are multiplied by -1.

On the rest of the strata, the function $f_1(\boldsymbol{p})$ takes values in the convex compact set $\mathcal{C}_1$ of the symmetric non-negative-definite 6×6 matrices with unit trace satisfying $(f_1)_{i6}(\boldsymbol{p}) = (f_1)_{6i}(\boldsymbol{p}) = 0$, $1 \leq i \leq 5$. The contribution of the remaining strata is

$$\int_{(\hat{\mathbb{R}}^3/D_8\times Z_2^c)_{0-2,5}} (j_+(\boldsymbol{p}, \boldsymbol{y} - \boldsymbol{x}) + j_-(\boldsymbol{p}, \boldsymbol{y} - \boldsymbol{x}))f_1(\boldsymbol{p}))\,\mathrm{d}\Phi(\boldsymbol{p}).$$

Theorem 35. *The one-point correlation tensor of the homogeneous and $(D_8 \times Z_2^c, 5A_{1g} \oplus B_{2g})$-isotropic random field $\mathsf{C}(\boldsymbol{x})$ is*

$$\langle \mathsf{C}(\boldsymbol{x}) \rangle = \sum_{m=1}^{5} C_m \tilde{\mathsf{T}}^m, \qquad C_m \in \mathbb{R}.$$

Its two-point correlation tensor is

$$\begin{aligned}
\langle \mathsf{C}(\boldsymbol{x}), \mathsf{C}(\boldsymbol{y}) \rangle = &\int_{(\hat{\mathbb{R}}^3/D_8\times Z_2^c)_{3,4,6,7}} (j_+(\boldsymbol{p}, \boldsymbol{y} - \boldsymbol{x})f_0^+(\boldsymbol{p}) \\
&+ j_-(\boldsymbol{p}, \boldsymbol{y} - \boldsymbol{x})f_0^-(\boldsymbol{p}))\,\mathrm{d}\Phi(\boldsymbol{p}) \\
&+ \int_{(\hat{\mathbb{R}}^3/D_8\times Z_2^c)_{0-2,5}} (j_+(\boldsymbol{p}, \boldsymbol{y} - \boldsymbol{x}) \\
&+ j_-(\boldsymbol{p}, \boldsymbol{y} - \boldsymbol{x}))f_1(\boldsymbol{p}))\,\mathrm{d}\Phi(\boldsymbol{p}).
\end{aligned}$$

The field has the form

$$
\begin{aligned}
\mathsf{C}(\boldsymbol{x}) &= \sum_{m=1}^{5} C_m \tilde{\mathsf{T}}^m + \sum_{m=1}^{6}\sum_{n=1}^{32} \int_{(\hat{\mathbb{R}}^3/D_8\times Z_2^c)_{3,4,6,7}} u_n(\boldsymbol{p},\boldsymbol{x})\,\mathrm{d}Z^{0n}_m(\boldsymbol{p})\tilde{\mathsf{T}}^m \\
&\quad + \sum_{m=1}^{6}\sum_{n=1}^{32} \int_{(\hat{\mathbb{R}}^3/D_8\times Z_2^c)_{0-2,5}} u_n(\boldsymbol{p},\boldsymbol{x})\,\mathrm{d}Z^{1n}_m(\boldsymbol{p})\tilde{\mathsf{T}}^m,
\end{aligned}
\tag{3.96}
$$

where $u_n(\boldsymbol{p},\boldsymbol{x})$, $1 \le n \le 4$ *(resp.* $5 \le n \le 8$, *resp.* $9 \le n \le 16$, *resp.* $17 \le n \le 20$, *resp.* $21 \le n \le 24$, *resp.* $25 \le n \le 32$*) are different products of sines and cosines of* $\frac{1}{\sqrt{8}}(p_1x_1+p_2x_2)$ *and* p_3x_3 *(resp.* $\frac{1}{\sqrt{8}}(p_1x_2-p_2x_1)$ *and* p_3x_3, *resp.* $\frac{1}{2\sqrt{2}}(p_1x_1-p_2x_2)$, $\frac{1}{\sqrt{2}}(p_1x_2+p_2x_1)$ *and* p_3x_3*), resp.* $\frac{1}{\sqrt{8}}(p_1x_1-p_2x_2)$ *and* p_3x_3, *resp.* $\frac{1}{\sqrt{8}}(p_1x_2+p_2x_1)$ *and* p_3x_3, *resp.* $\frac{1}{2\sqrt{2}}(p_1x_2-p_2x_1)$, $\frac{1}{\sqrt{2}}(p_1x_1+p_2x_2)$ *and* p_3x_3*)), and* $\boldsymbol{Z}^{0n}(\boldsymbol{p}) = (Z^{0n}_1(\boldsymbol{p}),\dots,Z^{0n}_6(\boldsymbol{p}))^\top$, $1 \le n \le 16$ *(resp.* $\boldsymbol{Z}^{0n}(\boldsymbol{p}) = (Z^{0n}_1(\boldsymbol{p}),\dots,Z^{0n}_6(\boldsymbol{p}))^\top$, $17 \le n \le 32$, *resp.* $\boldsymbol{Z}^{1n}(\boldsymbol{p}) = (Z^{1n}_1(\boldsymbol{p}),\dots,Z^{1n}_6(\boldsymbol{p}))^\top$*) are centred uncorrelated random measures on* $(\hat{\mathbb{R}}^3/D_8\times Z_2^c)_{3,4,6,7}$ *(resp. on* $(\hat{\mathbb{R}}^3/D_8\times Z_2^c)_{3,4,6,7}$, *resp. on* $(\hat{\mathbb{R}}^3/D_8\times Z_2^c)_{0-2,5}$*) with control measure* $f_0^+(\boldsymbol{p})\,\mathrm{d}\Phi(\boldsymbol{p})$ *(resp.* $f_0^-(\boldsymbol{p})\,\mathrm{d}\Phi(\boldsymbol{p})$, *resp.* $f_1(\boldsymbol{p})\,\mathrm{d}\Phi(\boldsymbol{p})$*).*

Proofs of the remaining cases, except of the case of $G = \mathrm{O}(3)$, may be left to the reader.

When $G = \mathrm{O}(3)$, we have $\mathsf{S}^2(\rho_1) = \rho_0 \oplus \rho_2$. The representation ρ is

$$\rho = \mathsf{S}^2(\mathsf{S}^2(\rho_1)) = \mathsf{S}^2(\rho_1 \oplus \rho_2) = 2\rho_0 \oplus 2\rho_2 \oplus \rho_4.$$

The basis in the space V of the representation ρ is given by Equations (3.65–3.69). We calculate the basis in the space of the representation $\mathsf{S}^2(\rho)$. The result is given in Table 3.11.

The function $f(\lambda)$ takes the form

$$f_{i\cdots\ell'}(\lambda) = \sum_{t=0}^{4}\sum_{v=1}^{m_{2t}} f_{2t,v}(\lambda)\mathsf{T}^{2t,v,0}_{i\cdots\ell'} \tag{3.97}$$

with $f_{2t,v}(0) = 0$ for $t \ge 1$, where $m_0 = 7$, $m_2 = 10$, $m_4 = 8$, $m_6 = 3$ and $m_8 = 1$. When $\lambda = 0$, we obtain

$$f_{i\cdots l'}(0) = \sum_{v=1}^{7} f_{0,v}(\lambda)\mathsf{T}^{0,v,0}_{i\cdots l'}.$$

Equation (3.28), which determines the M-functions, takes the form

$$\mathsf{M}^{m,n}(\boldsymbol{p}) = \sum_{q=-m}^{m} \mathsf{T}^{m,n,q}\rho^{mg}_{q0}(\boldsymbol{p}).$$

We express the M-functions in terms of the L-functions. The result is given in Table 3.12.

Table 3.11 *The tensors of the uncoupled basis of the space* $\mathsf{S}^2(\mathsf{V})$.

Tensor	**Value**
$T^{0,1,0}_{i\cdots l'}$	$T^{0,1}_{ijkl}T^{0,1}_{i'j'k'l'}$
$T^{0,2}_{i\cdots l'}$	$\frac{1}{\sqrt{2}}(T^{0,1}_{ijkl}T^{0,2}_{i'j'k'l'}+T^{0,1}_{i'j'k'l'}T^{0,2}_{ijkl})$
$T^{0,3,0}_{i\cdots l'}$	$\sum_{q,q'=-2}^{2} g^{0[q,q']}_{0[2,2]}T^{2,1,q}_{ijkl}T^{2,1,q'}_{i'j'k'l'}$
$T^{0,4,0}_{i\cdots l'}$	$T^{0,2}_{ijkl}T^{0,2}_{i'j'k'l'}$
$T^{0,5,0}_{i\cdots l'}$	$\frac{1}{\sqrt{2}}(\sum_{q,q'=-2}^{2} g^{0[q,q']}_{0[2,2]}T^{2,1,q}_{ijkl}T^{2,2,q'}_{i'j'k'l'}+\sum_{q,q'=-2}^{2} g^{0[q',q]}_{0[2,2]}T^{2,1,q'}_{i'j'k'l'}T^{2,2,q}_{ijkl})$
$T^{0,6,0}_{i\cdots l'}$	$\sum_{q,q'=-2}^{2} g^{0[q,q']}_{0[2,2]}T^{2,2,q}_{ijkl}T^{2,2,q'}_{i'j'k'l'}$
$T^{0,7,0}_{i\cdots l'}$	$\sum_{q,q'=-4}^{4} g^{0[q,q']}_{0[4,4]}T^{4,1,q}_{ijkl}T^{4,1,q'}_{i'j'k'l'}$
$T^{2,1,u}_{i\cdots l'}$	$\frac{1}{\sqrt{2}}(T^{0,1}_{ijkl}T^{2,1,u}_{i'j'k'l'}+T^{0,1}_{i'j'k'l'}T^{2,1,u}_{ijkl})$
$T^{2,2,u}_{i\cdots l'}$	$\frac{1}{\sqrt{2}}(T^{0,2}_{ijkl}T^{2,1,u}_{i'j'k'l'}+T^{0,2}_{i'j'k'l'}T^{2,1,u}_{ijkl})$
$T^{2,3,u}_{i\cdots l'}$	$\frac{1}{\sqrt{2}}(T^{0,1}_{ijkl}T^{2,2,u}_{i'j'k'l'}+T^{0,1}_{i'j'k'l'}T^{2,2,u}_{ijkl})$
$T^{2,4,u}_{i\cdots l'}$	$\sum_{q,q'=-2}^{2} g^{u[q,q']}_{2[2,2]}T^{2,1,q}_{ijkl}T^{2,1,q'}_{i'j'k'l'}$
$T^{2,5,u}_{i\cdots l'}$	$\frac{1}{\sqrt{2}}(T^{0,2}_{ijkl}T^{2,2,u}_{i'j'k'l'}+T^{0,2}_{i'j'k'l'}T^{2,2,u}_{ijkl})$
$T^{2,6,u}_{i\cdots l'}$	$\frac{1}{\sqrt{2}}(\sum_{q=-2}^{2}\sum_{q'=-4}^{4} g^{u[q,q']}_{2[2,4]}T^{2,1,q}_{ijkl}T^{4,1,q'}_{i'j'k'l'}+\sum_{q'=-2}^{2}\sum_{q=-4}^{4} g^{u[q',q]}_{2[2,4]}T^{2,1,q'}_{i'j'k'l'}T^{4,1,q}_{ijkl})$
$T^{2,7,u}_{i\cdots l'}$	$\frac{1}{\sqrt{2}}(\sum_{q,q'=-2}^{2} g^{u[q,q']}_{2[2,2]}T^{2,2,q}_{ijkl}T^{2,1,q'}_{i'j'k'l'}+\sum_{q',q=-2}^{2} g^{u[q',q]}_{2[2,2]}T^{2,2,q'}_{i'j'k'l'}T^{2,1,q}_{ijkl})$
$T^{2,8,u}_{i\cdots l'}$	$\sum_{q,q'=-2}^{2} g^{u[q,q']}_{2[2,2]}T^{2,2,q}_{ijkl}T^{2,2,q'}_{i'j'k'l'}$
$T^{2,9,u}_{i\cdots l'}$	$\frac{1}{\sqrt{2}}(\sum_{q=-2}^{2}\sum_{q'=-4}^{4} g^{u[q,q']}_{2[2,4]}T^{2,2,q}_{ijkl}T^{4,1,q'}_{i'j'k'l'}+\sum_{q'=-2}^{2}\sum_{q=-4}^{4} g^{u[q',q]}_{2[2,4]}T^{2,2,q'}_{i'j'k'l'}T^{4,1,q}_{ijkl})$
$T^{2,10,u}_{i\cdots l'}$	$\sum_{q,q'=-4}^{4} g^{u[q,q']}_{2[4,4]}T^{4,1,q}_{ijkl}T^{4,1,q'}_{i'j'k'l'}$
$T^{4,1,u}_{i\cdots l'}$	$\frac{1}{\sqrt{2}}(T^{0,1}_{ijkl}T^{4,1,u}_{i'j'k'l'}+T^{0,1}_{i'j'k'l'}T^{4,1,u}_{ijkl})$
$T^{4,2,u}_{i\cdots l'}$	$\sum_{q,q'=-4}^{4} g^{u[q,q']}_{4[2,2]}T^{2,1,q}_{ijkl}T^{2,1,q'}_{i'j'k'l'}$

Table 3.11 (*Cont.*)

Tensor	**Value**
$T^{4,3,u}_{i\cdots l'}$	$\frac{1}{\sqrt{2}}(T^{0,2}_{ijkl}T^{4,1,u}_{i'j'k'l'}+T^{0,2}_{i'j'k'l'}T^{4,1,u}_{ijkl})$
$T^{4,4,u}_{i\cdots l'}$	$\frac{1}{\sqrt{2}}(\sum_{q,q'--4}^{4} g^{u[q,q']}_{4[2,2]}T^{2,2,q}_{ijkl}T^{2,1,q'}_{i'j'k'l'}+\sum_{q',q=-4}^{4} g^{u[q',q]}_{4[2,2]}T^{2,2,q'}_{i'j'k'l'}T^{2,1,q}_{ijkl})$
$T^{4,5,u}_{i\cdots l'}$	$\frac{1}{\sqrt{2}}(\sum_{q=-2}^{2}\sum_{q'=-4}^{4} g^{u[q,q']}_{4[2,4]}T^{2,1,q}_{ijkl}T^{4,1,q'}_{i'j'k'l'}+\sum_{q'=-2}^{2}\sum_{q=-4}^{4} g^{u[q',q]}_{4[2,4]}T^{2,1,q'}_{i'j'k'l'}T^{4,1,q}_{ijkl})$
$T^{4,6,u}_{i\cdots l'}$	$\sum_{q,q'=-4}^{4} g^{u[q,q']}_{4[2,2]}T^{2,2,q}_{ijkl}T^{2,2,q'}_{i'j'k'l'}$
$T^{4,7,u}_{i\cdots l'}$	$\frac{1}{\sqrt{2}}(\sum_{q=-2}^{2}\sum_{q'=-4}^{4} g^{u[q,q']}_{4[2,4]}T^{2,2,q}_{ijkl}T^{4,1,q'}_{i'j'k'l'}+\sum_{q'=-2}^{2}\sum_{q=-4}^{4} g^{u[q',q]}_{4[2,4]}T^{2,2,q'}_{i'j'k'l'}T^{4,1,q}_{ijkl})$
$T^{4,8,u}_{i\cdots l'}$	$\sum_{q,q'=-4}^{4} g^{u[q,q']}_{4[4,4]}T^{4,1,q}_{ijkl}T^{4,1,q'}_{i'j'k'l'}$
$T^{6,1,u}_{i\cdots l'}$	$\frac{1}{\sqrt{2}}(\sum_{q=-2}^{2}\sum_{q'=-4}^{4} g^{u[q,q']}_{6[2,4]}T^{2,1,q}_{ijkl}T^{4,1,q'}_{i'j'k'l'}+\sum_{q'=-2}^{2}\sum_{q=-4}^{4} g^{u[q',q]}_{6[2,4]}T^{2,1,q'}_{i'j'k'l'}T^{4,1,q}_{ijkl})$
$T^{6,2,u}_{i\cdots l'}$	$\frac{1}{\sqrt{2}}(\sum_{q=-2}^{2}\sum_{q'=-4}^{4} g^{u[q,q']}_{6[2,4]}T^{2,2,q}_{ijkl}T^{4,1,q'}_{i'j'k'l'}+\sum_{q'=-2}^{2}\sum_{q=-4}^{4} g^{u[q',q]}_{6[2,4]}T^{2,2,q'}_{i'j'k'l'}T^{4,1,q}_{ijkl})$
$T^{6,3,u}_{i\cdots l'}$	$\sum_{q,q'=-4}^{4} g^{u[q,q']}_{6[4,4]}T^{4,1,q}_{ijkl}T^{4,1,q'}_{i'j'k'l'}$
$T^{8,1,u}_{i\cdots l'}$	$\sum_{q,q'=-4}^{4} g^{u[q,q']}_{8[4,4]}T^{4,1,q}_{ijkl}T^{4,1,q'}_{i'j'k'l'}$

We enumerate the 21 indexes $ijk\ell$ in the following order: $-1-1-1-1$, 0000, 1111, 0101, $-11-11$, $-10-10$, $-1-100$, $-1-111$, 0011, $-11-10$, $-1-101$, 1101, 0001, $01-11$, $11-10$, $-1-1-10$, $00-10$, $01-10$, $-1-1-11$, $11-11$, $00-11$. With this order, the matrix $f_{II'}(\lambda)$ becomes block-diagonal. We chose 29 linearly independent elements of the above matrix according to Table 3.13.

The remaining non-zero entries of the matrix $f(\lambda)$ are defined by

$$
\begin{aligned}
f_{1,1}(\lambda) &= f_{1,3}(\lambda), & f_{1,2}(\lambda) &= f_{2,3}(\lambda),\\
f_{1,6}(\lambda) &= f_{3,4}(\lambda), & f_{1,7}(\lambda) &= f_{3,9}(\lambda),\\
f_{1,8}(\lambda) &= f_{3,8}(\lambda), & f_{2,4}(\lambda) &= f_{2,6}(\lambda),\\
f_{2,7}(\lambda) &= f_{2,9}(\lambda), & f_{4,4}(\lambda) &= f_{6,6}(\lambda),\\
f_{4,5}(\lambda) &= f_{5,6}(\lambda), & f_{4,9}(\lambda) &= f_{6,7}(\lambda),\\
f_{5,7}(\lambda) &= f_{5,9}(\lambda), & f_{7,7}(\lambda) &= f_{9,9}(\lambda),\\
f_{10,10}(\lambda) &= f_{14,14}(\lambda), & f_{10,11}(\lambda) &= f_{14,15}(\lambda),
\end{aligned}
$$

Table 3.12 *The functions* $\boldsymbol{M}^{n,m}(\boldsymbol{p})$ *expressed as linear combinations of the functions* $\boldsymbol{L}^{m}_{i\cdots l'}(\boldsymbol{p})$

$\boldsymbol{M}^{n,m}_{i\cdots l'}(\boldsymbol{p})$	**Linear combination**
$\boldsymbol{M}^{0,1}_{i\cdots l'}$	$\frac{1}{9}\boldsymbol{L}^{0,1}_{i\cdots l'}$
$\boldsymbol{M}^{0,2}_{i\cdots l'}$	$-\frac{\sqrt{2}}{9\sqrt{5}}\boldsymbol{L}^{1}_{i\cdots l'}+\frac{1}{6\sqrt{10}}\boldsymbol{L}^{2}_{i\cdots l'}$
$\boldsymbol{M}^{0,3}_{i\cdots l'}$	$-\frac{2}{9\sqrt{5}}\boldsymbol{L}^{1}_{i\cdots l'}+\frac{1}{12\sqrt{5}}\boldsymbol{L}^{3}_{i\cdots l'}$
$\boldsymbol{M}^{0,4}_{i\cdots l'}$	$\frac{1}{45}\boldsymbol{L}^{1}_{i\cdots l'}-\frac{1}{30}\boldsymbol{L}^{2}_{i\cdots l'}+\frac{1}{20}\boldsymbol{L}^{4}_{i\cdots l'}$
$\boldsymbol{M}^{0,5}_{i\cdots l'}$	$\frac{8}{9\sqrt{35}}\boldsymbol{L}^{1}_{i\cdots l'}-\frac{1}{3\sqrt{35}}\boldsymbol{L}^{2}_{i\cdots l'}-\frac{1}{3\sqrt{35}}\boldsymbol{L}^{3}_{i\cdots l'}+\frac{1}{8\sqrt{35}}\boldsymbol{L}^{5}_{i\cdots l'}$
$\boldsymbol{M}^{0,6}_{i\cdots l'}$	$\frac{11}{63\sqrt{5}}\boldsymbol{L}^{1}_{i\cdots l'}-\frac{1}{42\sqrt{5}}\boldsymbol{L}^{2}_{i\cdots l'}-\frac{\sqrt{5}}{42}\boldsymbol{L}^{3}_{i\cdots l'}-\frac{1}{28\sqrt{5}}\boldsymbol{L}^{4}_{i\cdots l'}+\frac{1}{28\sqrt{5}}\boldsymbol{L}^{5}_{i\cdots l'}$
	$+\frac{3}{28\sqrt{5}}\boldsymbol{L}^{6}_{i\cdots l'}-\frac{3}{56\sqrt{5}}\boldsymbol{L}^{7}_{i\cdots l'}$
$\boldsymbol{M}^{0,7}_{i\cdots l'}$	$-\frac{1}{35}\boldsymbol{L}^{1}_{i\cdots l'}+\frac{2}{105}\boldsymbol{L}^{2}_{i\cdots l'}+\frac{1}{84}\boldsymbol{L}^{3}_{i\cdots l'}-\frac{1}{210}\boldsymbol{L}^{4}_{i\cdots l'}-\frac{1}{84}\boldsymbol{L}^{5}_{i\cdots l'}$
	$+\frac{1}{168}\boldsymbol{L}^{6}_{i\cdots l'}+\frac{1}{56}\boldsymbol{L}^{7}_{i\cdots l'}$
$\boldsymbol{M}^{2,1}_{i\cdots l'}(\boldsymbol{p})$	$-\frac{\sqrt{2}}{9}\boldsymbol{L}^{1}_{i\cdots l'}+\frac{1}{6\sqrt{2}}\boldsymbol{L}^{8}_{i\cdots l'}(\boldsymbol{p})$
$\boldsymbol{M}^{2,2}_{i\cdots l'}(\boldsymbol{p})$	$\frac{\sqrt{2}}{9\sqrt{5}}\boldsymbol{L}^{1}_{i\cdots l'}-\frac{1}{6\sqrt{10}}\boldsymbol{L}^{2}_{i\cdots l'}-\frac{1}{6\sqrt{10}}\boldsymbol{L}^{8}_{i\cdots l'}(\boldsymbol{p})+\frac{1}{4\sqrt{10}}\boldsymbol{L}^{9}_{i\cdots l'}(\boldsymbol{p})$
$\boldsymbol{M}^{2,3}_{i\cdots l'}(\boldsymbol{p})$	$\frac{4}{9\sqrt{7}}\boldsymbol{L}^{1}_{i\cdots l'}-\frac{1}{6\sqrt{7}}\boldsymbol{L}^{2}_{i\cdots l'}-\frac{1}{3\sqrt{7}}\boldsymbol{L}^{8}_{i\cdots l'}(\boldsymbol{p})+\frac{1}{4\sqrt{7}}\boldsymbol{L}^{10}_{i\cdots l'}(\boldsymbol{p})$
$\boldsymbol{M}^{2,4}_{i\cdots l'}(\boldsymbol{p})$	$\frac{4\sqrt{2}}{9\sqrt{7}}\boldsymbol{L}^{1}_{i\cdots l'}-\frac{1}{6\sqrt{14}}\boldsymbol{L}^{3}_{i\cdots l'}-\frac{\sqrt{2}}{3\sqrt{7}}\boldsymbol{L}^{8}_{i\cdots l'}(\boldsymbol{p})+\frac{1}{4\sqrt{14}}\boldsymbol{L}^{11}_{i\cdots l'}(\boldsymbol{p})$
$\boldsymbol{M}^{2,5}_{i\cdots l'}(\boldsymbol{p})$	$-\frac{4}{9\sqrt{35}}\boldsymbol{L}^{1}_{i\cdots l'}+\frac{1}{2\sqrt{35}}\boldsymbol{L}^{2}_{i\cdots l'}-\frac{1}{2\sqrt{35}}\boldsymbol{L}^{4}_{i\cdots l'}+\frac{1}{3\sqrt{35}}\boldsymbol{L}^{8}_{i\cdots l'}(\boldsymbol{p})$
	$-\frac{1}{2\sqrt{35}}\boldsymbol{L}^{9}_{i\cdots l'}(\boldsymbol{p})-\frac{1}{4\sqrt{35}}\boldsymbol{L}^{10}_{i\cdots l'}(\boldsymbol{p})+\frac{3}{8\sqrt{35}}\boldsymbol{L}^{12}_{i\cdots l'}(\boldsymbol{p})$
$\boldsymbol{M}^{2,6}_{i\cdots l'}(\boldsymbol{p})$	$-\frac{\sqrt{2}}{7\sqrt{5}}\boldsymbol{L}^{1}_{i\cdots l'}+\frac{\sqrt{2}}{21\sqrt{5}}\boldsymbol{L}^{2}_{i\cdots l'}+\frac{\sqrt{5}}{84\sqrt{2}}\boldsymbol{L}^{3}_{i\cdots l'}-\frac{\sqrt{5}}{168\sqrt{2}}\boldsymbol{L}^{5}_{i\cdots l'}$
	$+\frac{3}{14\sqrt{10}}\boldsymbol{L}^{8}_{i\cdots l'}(\boldsymbol{p})-\frac{1}{42\sqrt{10}}\boldsymbol{L}^{9}_{i\cdots l'}(\boldsymbol{p})-\frac{\sqrt{5}}{42\sqrt{2}}\boldsymbol{L}^{10}_{i\cdots l'}(\boldsymbol{p})-\frac{\sqrt{5}}{42\sqrt{2}}\boldsymbol{L}^{11}_{i\cdots l'}(\boldsymbol{p})$
	$+\frac{\sqrt{5}}{168\sqrt{2}}\boldsymbol{L}^{13}_{i\cdots l'}(\boldsymbol{p})+\frac{\sqrt{5}}{56\sqrt{2}}\boldsymbol{L}^{14}_{i\cdots l'}(\boldsymbol{p})$
$\boldsymbol{M}^{2,7}_{i\cdots l'}(\boldsymbol{p})$	$-\frac{11\sqrt{2}}{63}\boldsymbol{L}^{1}_{i\cdots l'}-\frac{1}{42\sqrt{2}}\boldsymbol{L}^{2}_{i\cdots l'}-\frac{5}{42\sqrt{2}}\boldsymbol{L}^{3}_{i\cdots l'}+\frac{1}{56\sqrt{2}}\boldsymbol{L}^{5}_{i\cdots l'}$
	$-\frac{11}{42\sqrt{2}}\boldsymbol{L}^{8}_{i\cdots l'}(\boldsymbol{p})-\frac{1}{28\sqrt{2}}\boldsymbol{L}^{9}_{i\cdots l'}(\boldsymbol{p})+\frac{1}{14\sqrt{2}}\boldsymbol{L}^{10}_{i\cdots l'}(\boldsymbol{p})+\frac{1}{14\sqrt{2}}\boldsymbol{L}^{11}_{i\cdots l'}(\boldsymbol{p})$
	$+\frac{3}{28\sqrt{2}}\boldsymbol{L}^{13}_{i\cdots l'}(\boldsymbol{p})-\frac{3}{56\sqrt{2}}\boldsymbol{L}^{14}_{i\cdots l'}(\boldsymbol{p})$
$\boldsymbol{M}^{2,8}_{i\cdots l'}(\boldsymbol{p})$	$\frac{37}{63\sqrt{14}}\boldsymbol{L}^{1}_{i\cdots l'}-\frac{5}{42\sqrt{14}}\boldsymbol{L}^{2}_{i\cdots l'}-\frac{2\sqrt{2}}{21\sqrt{7}}\boldsymbol{L}^{3}_{i\cdots l'}+\frac{1}{28\sqrt{14}}\boldsymbol{L}^{4}_{i\cdots l'}$
	$+\frac{1}{28\sqrt{14}}\boldsymbol{L}^{5}_{i\cdots l'}+\frac{3}{28\sqrt{14}}\boldsymbol{L}^{6}_{i\cdots l'}-\frac{3}{56\sqrt{14}}\boldsymbol{L}^{7}_{i\cdots l'}-\frac{5}{42\sqrt{14}}\boldsymbol{L}^{8}_{i\cdots l'}(\boldsymbol{p})$
	$-\frac{5}{28\sqrt{14}}\boldsymbol{L}^{9}_{i\cdots l'}(\boldsymbol{p})+\frac{1}{28\sqrt{14}}\boldsymbol{L}^{10}_{i\cdots l'}(\boldsymbol{p})+\frac{1}{14\sqrt{14}}\boldsymbol{L}^{11}_{i\cdots l'}(\boldsymbol{p})+\frac{3}{56\sqrt{14}}\boldsymbol{L}^{12}_{i\cdots l'}(\boldsymbol{p})$
	$-\frac{3}{28\sqrt{14}}\boldsymbol{L}^{13}_{i\cdots l'}(\boldsymbol{p})+\frac{3}{56\sqrt{14}}\boldsymbol{L}^{14}_{i\cdots l'}(\boldsymbol{p})+\frac{9}{56\sqrt{14}}\boldsymbol{L}^{15}_{i\cdots l'}(\boldsymbol{p})-\frac{9}{112\sqrt{14}}\boldsymbol{L}^{16}_{i\cdots l'}(\boldsymbol{p})$
$\boldsymbol{M}^{2,9}_{i\cdots l'}(\boldsymbol{p})$	$-\frac{11}{7\sqrt{35}}\boldsymbol{L}^{1}_{i\cdots l'}+\frac{13}{2\sqrt{35}}\boldsymbol{L}^{2}_{i\cdots l'}+\frac{5\sqrt{5}}{42\sqrt{7}}\boldsymbol{L}^{3}_{i\cdots l'}+\frac{3}{28\sqrt{35}}\boldsymbol{L}^{4}_{i\cdots l'}$
	$-\frac{11\sqrt{5}}{336\sqrt{7}}\boldsymbol{L}^{5}_{i\cdots l'}-\frac{\sqrt{5}}{28\sqrt{7}}\boldsymbol{L}^{6}_{i\cdots l'}-\frac{\sqrt{5}}{56\sqrt{7}}\boldsymbol{L}^{7}_{i\cdots l'}+\frac{9}{14\sqrt{35}}\boldsymbol{L}^{8}_{i\cdots l'}(\boldsymbol{p})$

Table 3.12 (*Cont.*)

$\boldsymbol{M}^{n,m}_{i\cdots l'}(\boldsymbol{p})$	**Linear combination**
	$+\frac{1}{21\sqrt{35}}\boldsymbol{L}^{9}_{i\cdots l'}(\boldsymbol{p})+\frac{1}{42\sqrt{35}}\boldsymbol{L}^{10}_{i\cdots l'}(\boldsymbol{p})-\frac{\sqrt{5}}{42\sqrt{7}}\boldsymbol{L}^{11}_{i\cdots l'}(\boldsymbol{p})-\frac{3}{14\sqrt{35}}\boldsymbol{L}^{12}_{i\cdots l'}(\boldsymbol{p})$
	$-\frac{\sqrt{5}}{21\sqrt{7}}\boldsymbol{L}^{13}_{i\cdots l'}(\boldsymbol{p})-\frac{\sqrt{5}}{56\sqrt{7}}\boldsymbol{L}^{14}_{i\cdots l'}(\boldsymbol{p})+\frac{\sqrt{5}}{112\sqrt{7}}\boldsymbol{L}^{15}_{i\cdots l'}(\boldsymbol{p})+\frac{3\sqrt{5}}{112\sqrt{7}}\boldsymbol{L}^{16}_{i\cdots l'}(\boldsymbol{p})$
$\boldsymbol{M}^{2,10}_{i\cdots l'}(\boldsymbol{p})$	$\frac{13}{14\sqrt{77}}\boldsymbol{L}^{1}_{i\cdots l'}-\frac{65}{84\sqrt{77}}\boldsymbol{L}^{2}_{i\cdots l'}+\frac{1}{84\sqrt{77}}\boldsymbol{L}^{3}_{i\cdots l'}+\frac{67}{168\sqrt{77}}\boldsymbol{L}^{4}_{i\cdots l'}$
	$+\frac{25}{168\sqrt{77}}\boldsymbol{L}^{5}_{i\cdots l'}-\frac{65}{168\sqrt{77}}\boldsymbol{L}^{6}_{i\cdots l'}-\frac{\sqrt{11}}{112\sqrt{7}}\boldsymbol{L}^{7}_{i\cdots l'}-\frac{3}{28\sqrt{77}}\boldsymbol{L}^{8}_{i\cdots l'}(\boldsymbol{p})$
	$+\frac{19}{56\sqrt{77}}\boldsymbol{L}^{9}_{i\cdots l'}(\boldsymbol{p})+\frac{13}{56\sqrt{77}}\boldsymbol{L}^{10}_{i\cdots l'}(\boldsymbol{p})-\frac{9}{56\sqrt{77}}\boldsymbol{L}^{11}_{i\cdots l'}(\boldsymbol{p})-\frac{17}{112\sqrt{77}}\boldsymbol{L}^{12}_{i\cdots l'}(\boldsymbol{p})$
	$-\frac{25}{56\sqrt{77}}\boldsymbol{L}^{13}_{i\cdots l'}(\boldsymbol{p})+\frac{\sqrt{11}}{112\sqrt{7}}\boldsymbol{L}^{14}_{i\cdots l'}(\boldsymbol{p})+\frac{5}{112\sqrt{77}}\boldsymbol{L}^{15}_{i\cdots l'}(\boldsymbol{p})-\frac{3\sqrt{11}}{224\sqrt{7}}\boldsymbol{L}^{16}_{i\cdots l'}(\boldsymbol{p})$
	$+\frac{\sqrt{7}}{16\sqrt{11}}\boldsymbol{L}^{17}_{i\cdots l'}(\boldsymbol{p})$
$\boldsymbol{M}^{4,1}_{i\cdots l'}(\boldsymbol{p})$	$\frac{1}{6\sqrt{35}}\boldsymbol{L}^{1}_{i\cdots l'}+\frac{1}{12\sqrt{35}}\boldsymbol{L}^{2}_{i\cdots l'}-\frac{\sqrt{5}}{12\sqrt{7}}\boldsymbol{L}^{8}_{i\cdots l'}(\boldsymbol{p})-\frac{\sqrt{5}}{12\sqrt{7}}\boldsymbol{L}^{10}_{i\cdots l'}(\boldsymbol{p})$
	$+\frac{\sqrt{35}}{12}\boldsymbol{L}^{18}_{i\cdots l'}(\boldsymbol{p})$
$\boldsymbol{M}^{4,2}_{i\cdots l'}(\boldsymbol{p})$	$\frac{1}{3\sqrt{70}}\boldsymbol{L}^{1}_{i\cdots l'}+\frac{1}{12\sqrt{70}}\boldsymbol{L}^{3}_{i\cdots l'}-\frac{\sqrt{5}}{6\sqrt{14}}\boldsymbol{L}^{8}_{i\cdots l'}(\boldsymbol{p})-\frac{\sqrt{5}}{12\sqrt{14}}\boldsymbol{L}^{11}_{i\cdots l'}(\boldsymbol{p})$
	$+\frac{\sqrt{35}}{12\sqrt{2}}\boldsymbol{L}^{19}_{i\cdots l'}(\boldsymbol{p})$
$\boldsymbol{M}^{4,3}_{i\cdots l'}(\boldsymbol{p})$	$-\frac{1}{30\sqrt{7}}\boldsymbol{L}^{1}_{i\cdots l'}+\frac{1}{120\sqrt{7}}\boldsymbol{L}^{2}_{i\cdots l'}+\frac{1}{20\sqrt{7}}\boldsymbol{L}^{4}_{i\cdots l'}+\frac{1}{12\sqrt{7}}\boldsymbol{L}^{8}_{i\cdots l'}(\boldsymbol{p})$
	$-\frac{1}{8\sqrt{7}}\boldsymbol{L}^{9}_{i\cdots l'}(\boldsymbol{p})+\frac{1}{12\sqrt{7}}\boldsymbol{L}^{10}_{i\cdots l'}(\boldsymbol{p})-\frac{1}{8\sqrt{7}}\boldsymbol{L}^{12}_{i\cdots l'}(\boldsymbol{p})-\frac{\sqrt{7}}{12}\boldsymbol{L}^{18}_{i\cdots l'}(\boldsymbol{p})$
	$+\frac{\sqrt{7}}{8}\boldsymbol{L}^{20}_{i\cdots l'}(\boldsymbol{p})$
$\boldsymbol{M}^{4,4}_{i\cdots l'}(\boldsymbol{p})$	$-\frac{17\sqrt{2}}{21\sqrt{5}}\boldsymbol{L}^{1}_{i\cdots l'}+\frac{3}{7\sqrt{10}}\boldsymbol{L}^{2}_{i\cdots l'}+\frac{13}{42\sqrt{10}}\boldsymbol{L}^{3}_{i\cdots l'}-\frac{1}{14\sqrt{10}}\boldsymbol{L}^{5}_{i\cdots l'}$
	$+\frac{13\sqrt{5}}{42\sqrt{2}}\boldsymbol{L}^{8}_{i\cdots l'}(\boldsymbol{p})-\frac{\sqrt{5}}{14\sqrt{2}}\boldsymbol{L}^{9}_{i\cdots l'}(\boldsymbol{p})-\frac{3\sqrt{5}}{28\sqrt{2}}\boldsymbol{L}^{10}_{i\cdots l'}(\boldsymbol{p})-\frac{\sqrt{5}}{42\sqrt{2}}\boldsymbol{L}^{11}_{i\cdots l'}(\boldsymbol{p})$
	$-\frac{\sqrt{5}}{28\sqrt{2}}\boldsymbol{L}^{13}_{i\cdots l'}(\boldsymbol{p})+\frac{\sqrt{5}}{56\sqrt{2}}\boldsymbol{L}^{14}_{i\cdots l'}(\boldsymbol{p})-\frac{\sqrt{5}}{3\sqrt{2}}\boldsymbol{L}^{19}_{i\cdots l'}(\boldsymbol{p})+\frac{\sqrt{5}}{8\sqrt{2}}\boldsymbol{L}^{21}_{i\cdots l'}(\boldsymbol{p})$
$\boldsymbol{M}^{4,5}_{i\cdots l'}(\boldsymbol{p})$	$-\frac{13}{21\sqrt{11}}\boldsymbol{L}^{1}_{i\cdots l'}+\frac{1}{84\sqrt{11}}\boldsymbol{L}^{2}_{i\cdots l'}+\frac{1}{7\sqrt{11}}\boldsymbol{L}^{3}_{i\cdots l'}-\frac{1}{112\sqrt{11}}\boldsymbol{L}^{5}_{i\cdots l'}$
	$+\frac{67}{84\sqrt{11}}\boldsymbol{L}^{8}_{i\cdots l'}(\boldsymbol{p})+\frac{1}{56\sqrt{11}}\boldsymbol{L}^{9}_{i\cdots l'}(\boldsymbol{p})+\frac{25}{84\sqrt{11}}\boldsymbol{L}^{10}_{i\cdots l'}(\boldsymbol{p})-\frac{9}{56\sqrt{11}}\boldsymbol{L}^{11}_{i\cdots l'}(\boldsymbol{p})$
	$-\frac{5}{28\sqrt{11}}\boldsymbol{L}^{13}_{i\cdots l'}(\boldsymbol{p})-\frac{\sqrt{11}}{112}\boldsymbol{L}^{14}_{i\cdots l'}(\boldsymbol{p})-\frac{7}{3\sqrt{11}}\boldsymbol{L}^{18}_{i\cdots l'}(\boldsymbol{p})-\frac{1}{2\sqrt{11}}\boldsymbol{L}^{19}_{i\cdots l'}(\boldsymbol{p})$
	$-\frac{1}{4\sqrt{11}}\boldsymbol{L}^{21}_{i\cdots l'}(\boldsymbol{p})+\frac{7}{16\sqrt{11}}\boldsymbol{L}^{22}_{i\cdots l'}(\boldsymbol{p})$
$\boldsymbol{M}^{4,6}_{i\cdots l'}(\boldsymbol{p})$	$\frac{\sqrt{2}}{21\sqrt{35}}\boldsymbol{L}^{1}_{i\cdots l'}-\frac{3\sqrt{2}}{7\sqrt{35}}\boldsymbol{L}^{2}_{i\cdots l'}+\frac{5\sqrt{5}}{42\sqrt{14}}\boldsymbol{L}^{3}_{i\cdots l'}+\frac{3\sqrt{2}}{7\sqrt{35}}\boldsymbol{L}^{4}_{i\cdots l'}$
	$-\frac{1}{7\sqrt{70}}\boldsymbol{L}^{5}_{i\cdots l'}-\frac{3}{7\sqrt{70}}\boldsymbol{L}^{6}_{i\cdots l'}+\frac{3}{14\sqrt{70}}\boldsymbol{L}^{7}_{i\cdots l'}-\frac{17\sqrt{5}}{42\sqrt{14}}\boldsymbol{L}^{8}_{i\cdots l'}(\boldsymbol{p})$
	$+\frac{11\sqrt{10}}{56\sqrt{7}}\boldsymbol{L}^{9}_{i\cdots l'}(\boldsymbol{p})+\frac{9\sqrt{10}}{56\sqrt{7}}\boldsymbol{L}^{10}_{i\cdots l'}(\boldsymbol{p})-\frac{\sqrt{5}}{42\sqrt{14}}\boldsymbol{L}^{11}_{i\cdots l'}(\boldsymbol{p})-\frac{15\sqrt{5}}{56\sqrt{14}}\boldsymbol{L}^{12}_{i\cdots l'}(\boldsymbol{p})$
	$+\frac{\sqrt{10}}{56\sqrt{7}}\boldsymbol{L}^{13}_{i\cdots l'}(\boldsymbol{p})-\frac{\sqrt{10}}{112\sqrt{7}}\boldsymbol{L}^{14}_{i\cdots l'}(\boldsymbol{p})-\frac{3\sqrt{10}}{112\sqrt{7}}\boldsymbol{L}^{15}_{i\cdots l'}(\boldsymbol{p})+\frac{3\sqrt{10}}{224\sqrt{7}}\boldsymbol{L}^{16}_{i\cdots l'}(\boldsymbol{p})$
	$+\frac{\sqrt{10}}{3\sqrt{7}}\boldsymbol{L}^{19}_{i\cdots l'}(\boldsymbol{p})-\frac{\sqrt{10}}{4\sqrt{7}}\boldsymbol{L}^{21}_{i\cdots l'}(\boldsymbol{p})+\frac{3\sqrt{10}}{16\sqrt{7}}\boldsymbol{L}^{23}_{i\cdots l'}(\boldsymbol{p})$
$\boldsymbol{M}^{4,7}_{i\cdots l'}(\boldsymbol{p})$	$-\frac{19\sqrt{2}}{21\sqrt{77}}\boldsymbol{L}^{1}_{i\cdots l'}+\frac{17}{84\sqrt{154}}\boldsymbol{L}^{2}_{i\cdots l'}+\frac{17}{28\sqrt{154}}\boldsymbol{L}^{3}_{i\cdots l'}-\frac{1}{28\sqrt{154}}\boldsymbol{L}^{4}_{i\cdots l'}$
	$-\frac{\sqrt{11}}{84\sqrt{14}}\boldsymbol{L}^{5}_{i\cdots l'}-\frac{3}{28\sqrt{154}}\boldsymbol{L}^{6}_{i\cdots l'}+\frac{3}{56\sqrt{154}}\boldsymbol{L}^{7}_{i\cdots l'}-\frac{52\sqrt{2}}{21\sqrt{77}}\boldsymbol{L}^{8}_{i\cdots l'}(\boldsymbol{p})$
	$-\frac{37}{28\sqrt{154}}\boldsymbol{L}^{9}_{i\cdots l'}(\boldsymbol{p})-\frac{17\sqrt{2}}{21\sqrt{77}}\boldsymbol{L}^{10}_{i\cdots l'}(\boldsymbol{p})-\frac{9}{28\sqrt{154}}\boldsymbol{L}^{11}_{i\cdots l'}(\boldsymbol{p})+\frac{53}{56\sqrt{154}}\boldsymbol{L}^{12}_{i\cdots l'}(\boldsymbol{p})$

Table 3.12 (*Cont.*)

$\boldsymbol{M}^{n,m}_{i\cdots l'}(\boldsymbol{p})$	**Linear combination**
	$-\frac{67}{28\sqrt{154}}\boldsymbol{L}^{13}_{i\cdots l'}(\boldsymbol{p}) + \frac{71}{112\sqrt{154}}\boldsymbol{L}^{14}_{i\cdots l'}(\boldsymbol{p}) - \frac{3}{112\sqrt{154}}\boldsymbol{L}^{15}_{i\cdots l'}(\boldsymbol{p}) - \frac{3\sqrt{11}}{84\sqrt{14}}\boldsymbol{L}^{16}_{i\cdots l'}(\boldsymbol{p})$ $-\frac{\sqrt{7}}{6\sqrt{22}}\boldsymbol{L}^{18}_{i\cdots l'}(\boldsymbol{p}) - \frac{17}{2\sqrt{154}}\boldsymbol{L}^{19}_{i\cdots l'}(\boldsymbol{p}) - \frac{\sqrt{7}}{4\sqrt{22}}\boldsymbol{L}^{20}_{i\cdots l'}(\boldsymbol{p}) + \frac{23\sqrt{7}}{56\sqrt{22}}\boldsymbol{L}^{21}_{i\cdots l'}(\boldsymbol{p})$ $+\frac{\sqrt{7}}{8\sqrt{22}}\boldsymbol{L}^{22}_{i\cdots l'}(\boldsymbol{p}) - \frac{3}{2\sqrt{154}}\boldsymbol{L}^{23}_{i\cdots l'}(\boldsymbol{p}) + \frac{3\sqrt{7}}{4\sqrt{22}}\boldsymbol{L}^{24}_{i\cdots l'}(\boldsymbol{p}) - \frac{3\sqrt{7}}{16\sqrt{22}}\boldsymbol{L}^{25}_{i\cdots l'}(\boldsymbol{p})$
$\boldsymbol{M}^{4,8}_{i\cdots l'}(\boldsymbol{p})$	$\frac{571}{35\sqrt{2002}}\boldsymbol{L}^{1}_{i\cdots l'} + \frac{41}{35\sqrt{2002}}\boldsymbol{L}^{2}_{i\cdots l'} - \frac{457}{56\sqrt{2002}}\boldsymbol{L}^{3}_{i\cdots l'} - \frac{313}{70\sqrt{2002}}\boldsymbol{L}^{4}_{i\cdots l'}$ $+\frac{87}{56\sqrt{2002}}\boldsymbol{L}^{5}_{i\cdots l'} + \frac{127}{28\sqrt{2002}}\boldsymbol{L}^{6}_{i\cdots l'} - \frac{3\sqrt{13}}{28\sqrt{154}}\boldsymbol{L}^{7}_{i\cdots l'} - \frac{531}{14\sqrt{2002}}\boldsymbol{L}^{8}_{i\cdots l'}(\boldsymbol{p})$ $+\frac{31\sqrt{2}}{7\sqrt{1001}}\boldsymbol{L}^{9}_{i\cdots l'}(\boldsymbol{p}) + \frac{2\sqrt{26}}{7\sqrt{77}}\boldsymbol{L}^{10}_{i\cdots l'}(\boldsymbol{p}) + \frac{185}{28\sqrt{2002}}\boldsymbol{L}^{11}_{i\cdots l'}(\boldsymbol{p}) - \frac{17}{28\sqrt{2002}}\boldsymbol{L}^{12}_{i\cdots l'}(\boldsymbol{p})$ $+\frac{1055}{56\sqrt{2002}}\boldsymbol{L}^{13}_{i\cdots l'}(\boldsymbol{p}) - \frac{335}{56\sqrt{2002}}\boldsymbol{L}^{14}_{i\cdots l'}(\boldsymbol{p}) - \frac{375}{56\sqrt{2002}}\boldsymbol{L}^{15}_{i\cdots l'}(\boldsymbol{p})$ $+\frac{135}{56\sqrt{2002}}\boldsymbol{L}^{16}_{i\cdots l'}(\boldsymbol{p}) - \frac{5\sqrt{7}}{8\sqrt{286}}\boldsymbol{L}^{17}_{i\cdots l'}(\boldsymbol{p}) + \frac{2\sqrt{14}}{\sqrt{143}}\boldsymbol{L}^{18}_{i\cdots l'}(\boldsymbol{p}) + \frac{205}{4\sqrt{2002}}\boldsymbol{L}^{19}_{i\cdots l'}(\boldsymbol{p})$ $-\frac{\sqrt{7}}{\sqrt{286}}\boldsymbol{L}^{20}_{i\cdots l'}(\boldsymbol{p}) - \frac{75}{4\sqrt{2002}}\boldsymbol{L}^{21}_{i\cdots l'}(\boldsymbol{p}) - \frac{5\sqrt{7}}{4\sqrt{286}}\boldsymbol{L}^{22}_{i\cdots l'}(\boldsymbol{p}) + \frac{15}{2\sqrt{2002}}\boldsymbol{L}^{23}_{i\cdots l'}(\boldsymbol{p})$ $-\frac{25\sqrt{7}}{8\sqrt{286}}\boldsymbol{L}^{24}_{i\cdots l'}(\boldsymbol{p}) + \frac{15\sqrt{7}}{8\sqrt{286}}\boldsymbol{L}^{25}_{i\cdots l'}(\boldsymbol{p})$
$\boldsymbol{M}^{6,1}_{i\cdots l'}(\boldsymbol{p})$	$\frac{3}{2\sqrt{77}}\boldsymbol{L}^{1}_{i\cdots l'} - \frac{5}{12\sqrt{77}}\boldsymbol{L}^{2}_{i\cdots l'} - \frac{5}{12\sqrt{77}}\boldsymbol{L}^{3}_{i\cdots l'} + \frac{1}{12\sqrt{77}}\boldsymbol{L}^{5}_{i\cdots l'}$ $-\frac{8963\sqrt{11}}{9240\sqrt{7}}\boldsymbol{L}^{8}_{i\cdots l'}(\boldsymbol{p}) + \frac{\sqrt{7}}{24\sqrt{11}}\boldsymbol{L}^{9}_{i\cdots l'}(\boldsymbol{p}) + \frac{\sqrt{7}}{12\sqrt{11}}\boldsymbol{L}^{10}_{i\cdots l'}(\boldsymbol{p}) + \frac{\sqrt{7}}{12\sqrt{11}}\boldsymbol{L}^{11}_{i\cdots l'}(\boldsymbol{p})$ $+\frac{\sqrt{7}}{24\sqrt{11}}\boldsymbol{L}^{13}_{i\cdots l'}(\boldsymbol{p}) + \frac{35}{4\sqrt{77}}\boldsymbol{L}^{18}_{i\cdots l'}(\boldsymbol{p}) - \frac{\sqrt{7}}{4\sqrt{11}}\boldsymbol{L}^{19}_{i\cdots l'}(\boldsymbol{p}) - \frac{\sqrt{11}}{8\sqrt{7}}\boldsymbol{L}^{21}_{i\cdots l'}(\boldsymbol{p})$ $-\frac{\sqrt{77}}{8}\boldsymbol{L}^{26}_{i\cdots l'}(\boldsymbol{p})$
$\boldsymbol{M}^{6,2}_{i\cdots l'}(\boldsymbol{p})$	$\frac{6\sqrt{2}}{7\sqrt{11}}\boldsymbol{L}^{1}_{i\cdots l'} - \frac{4343}{980\sqrt{22}}\boldsymbol{L}^{2}_{i\cdots l'} - \frac{23}{42\sqrt{22}}\boldsymbol{L}^{3}_{i\cdots l'} + \frac{1}{10\sqrt{22}}\boldsymbol{L}^{4}_{i\cdots l'}$ $+\frac{5\sqrt{11}}{392\sqrt{2}}\boldsymbol{L}^{5}_{i\cdots l'} + \frac{1}{7\sqrt{22}}\boldsymbol{L}^{6}_{i\cdots l'} + \frac{2}{49\sqrt{22}}\boldsymbol{L}^{7}_{i\cdots l'} + \frac{97}{294\sqrt{22}}\boldsymbol{L}^{8}_{i\cdots l'}(\boldsymbol{p})$ $+\frac{377}{420\sqrt{22}}\boldsymbol{L}^{9}_{i\cdots l'}(\boldsymbol{p}) + \frac{1}{24\sqrt{22}}\boldsymbol{L}^{10}_{i\cdots l'}(\boldsymbol{p}) + \frac{1}{6\sqrt{22}}\boldsymbol{L}^{11}_{i\cdots l'}(\boldsymbol{p}) + \frac{39}{280\sqrt{22}}\boldsymbol{L}^{12}_{i\cdots l'}(\boldsymbol{p})$ $+\frac{5}{6\sqrt{22}}\boldsymbol{L}^{13}_{i\cdots l'}(\boldsymbol{p}) - \frac{1}{8\sqrt{22}}\boldsymbol{L}^{14}_{i\cdots l'}(\boldsymbol{p}) - \frac{1}{49\sqrt{22}}\boldsymbol{L}^{15}_{i\cdots l'}(\boldsymbol{p}) - \frac{195\sqrt{11}}{392\sqrt{2}}\boldsymbol{L}^{16}_{i\cdots l'}(\boldsymbol{p})$ $+\frac{2\sqrt{2}}{11\sqrt{11}}\boldsymbol{L}^{18}_{i\cdots l'}(\boldsymbol{p}) + \frac{2\sqrt{2}}{\sqrt{11}}\boldsymbol{L}^{19}_{i\cdots l'}(\boldsymbol{p}) - \frac{9}{4\sqrt{22}}\boldsymbol{L}^{20}_{i\cdots l'}(\boldsymbol{p}) - \frac{5}{8\sqrt{22}}\boldsymbol{L}^{21}_{i\cdots l'}(\boldsymbol{p})$ $-\frac{1}{4\sqrt{22}}\boldsymbol{L}^{22}_{i\cdots l'}(\boldsymbol{p}) - \frac{3}{4\sqrt{22}}\boldsymbol{L}^{23}_{i\cdots l'}(\boldsymbol{p}) - \frac{3}{2\sqrt{22}}\boldsymbol{L}^{24}_{i\cdots l'}(\boldsymbol{p}) + \frac{3}{8\sqrt{22}}\boldsymbol{L}^{25}_{i\cdots l'}(\boldsymbol{p})$ $+\frac{3\sqrt{11}}{8\sqrt{2}}\boldsymbol{L}^{27}_{i\cdots l'}(\boldsymbol{p})$
$\boldsymbol{M}^{6,3}_{i\cdots l'}(\boldsymbol{p})$	$-\frac{9}{7\sqrt{55}}\boldsymbol{L}^{1}_{i\cdots l'} - \frac{\sqrt{5}}{84\sqrt{11}}\boldsymbol{L}^{2}_{i\cdots l'} + \frac{25\sqrt{5}}{168\sqrt{11}}\boldsymbol{L}^{3}_{i\cdots l'} + \frac{19}{41\sqrt{55}}\boldsymbol{L}^{4}_{i\cdots l'}$ $-\frac{61}{336\sqrt{55}}\boldsymbol{L}^{5}_{i\cdots l'} - \frac{37}{84\sqrt{55}}\boldsymbol{L}^{6}_{i\cdots l'} + \frac{\sqrt{5}}{28\sqrt{11}}\boldsymbol{L}^{7}_{i\cdots l'} + \frac{9}{2\sqrt{55}}\boldsymbol{L}^{8}_{i\cdots l'}(\boldsymbol{p})$ $-\frac{7}{4\sqrt{55}}\boldsymbol{L}^{9}_{i\cdots l'}(\boldsymbol{p}) - \frac{2}{\sqrt{55}}\boldsymbol{L}^{10}_{i\cdots l'}(\boldsymbol{p}) - \frac{9}{16\sqrt{55}}\boldsymbol{L}^{11}_{i\cdots l'}(\boldsymbol{p}) + \frac{3}{8\sqrt{55}}\boldsymbol{L}^{12}_{i\cdots l'}(\boldsymbol{p})$ $-\frac{19}{8\sqrt{55}}\boldsymbol{L}^{13}_{i\cdots l'}(\boldsymbol{p}) + \frac{19}{16\sqrt{55}}\boldsymbol{L}^{14}_{i\cdots l'}(\boldsymbol{p}) + \frac{17}{16\sqrt{55}}\boldsymbol{L}^{15}_{i\cdots l'}(\boldsymbol{p}) - \frac{7}{16\sqrt{55}}\boldsymbol{L}^{16}_{i\cdots l'}(\boldsymbol{p})$ $+\frac{7}{16\sqrt{55}}\boldsymbol{L}^{17}_{i\cdots l'}(\boldsymbol{p}) + \frac{1}{2\sqrt{55}}\boldsymbol{L}^{18}_{i\cdots l'}(\boldsymbol{p}) + \frac{3}{16\sqrt{55}}\boldsymbol{L}^{19}_{i\cdots l'}(\boldsymbol{p}) + \frac{\sqrt{5}}{4\sqrt{11}}\boldsymbol{L}^{20}_{i\cdots l'}(\boldsymbol{p})$ $+\frac{23}{8\sqrt{55}}\boldsymbol{L}^{21}_{i\cdots l'}(\boldsymbol{p}) - \frac{43}{16\sqrt{55}}\boldsymbol{L}^{22}_{i\cdots l'}(\boldsymbol{p}) + \frac{37}{8\sqrt{55}}\boldsymbol{L}^{23}_{i\cdots l'}(\boldsymbol{p}) - \frac{13}{2\sqrt{55}}\boldsymbol{L}^{24}_{i\cdots l'}(\boldsymbol{p})$ $-\frac{3}{4\sqrt{55}}\boldsymbol{L}^{25}_{i\cdots l'}(\boldsymbol{p}) - \frac{\sqrt{11}}{2\sqrt{5}}\boldsymbol{L}^{26}_{i\cdots l'}(\boldsymbol{p}) - \frac{\sqrt{11}}{2\sqrt{5}}\boldsymbol{L}^{27}_{i\cdots l'}(\boldsymbol{p}) + \frac{7\sqrt{11}}{16\sqrt{5}}\boldsymbol{L}^{28}_{i\cdots l'}(\boldsymbol{p})$

Table 3.12 (*Cont.*)

$\boldsymbol{M}^{n,m}_{i\cdots l'}(\boldsymbol{p})$	**Linear combination**
$\boldsymbol{M}^{8,1}_{i\cdots l'}(\boldsymbol{p})$	$-\frac{36833}{5880\sqrt{1430}}\boldsymbol{L}^{1}_{i\cdots l'}+\frac{5199\sqrt{11}}{33320\sqrt{130}}\boldsymbol{L}^{2}_{i\cdots l'}+\frac{5717}{3528\sqrt{1430}}\boldsymbol{L}^{3}_{i\cdots l'}-\frac{4219\sqrt{11}}{723240\sqrt{130}}\boldsymbol{L}^{4}_{i\cdots l'}$
	$-\frac{437}{882\sqrt{1430}}\boldsymbol{L}^{5}_{i\cdots l'}+\frac{433}{3528\sqrt{1430}}\boldsymbol{L}^{6}_{i\cdots l'}+\frac{\sqrt{143}}{1764\sqrt{10}}\boldsymbol{L}^{7}_{i\cdots l'}+\frac{27}{8\sqrt{1430}}\boldsymbol{L}^{8}_{i\cdots l'}(\boldsymbol{p})$
	$-\frac{3}{8\sqrt{1430}}\boldsymbol{L}^{9}_{i\cdots l'}(\boldsymbol{p})+\frac{\sqrt{13}}{8\sqrt{110}}\boldsymbol{L}^{10}_{i\cdots l'}(\boldsymbol{p})-\frac{9}{8\sqrt{1430}}\boldsymbol{L}^{11}_{i\cdots l'}(\boldsymbol{p})-\frac{7}{8\sqrt{1430}}\boldsymbol{L}^{12}_{i\cdots l'}(\boldsymbol{p})$
	$-\frac{3}{8\sqrt{1430}}\boldsymbol{L}^{13}_{i\cdots l'}(\boldsymbol{p})-\frac{213\sqrt{11}}{2744\sqrt{130}}\boldsymbol{L}^{14}_{i\cdots l'}(\boldsymbol{p})-\frac{3}{8\sqrt{1430}}\boldsymbol{L}^{15}_{i\cdots l'}(\boldsymbol{p})$
	$+\frac{\sqrt{11}}{16\sqrt{130}}\boldsymbol{L}^{16}_{i\cdots l'}(\boldsymbol{p})-\frac{41}{56\sqrt{1430}}\boldsymbol{L}^{17}_{i\cdots l'}(\boldsymbol{p})-\frac{7\sqrt{11}}{8\sqrt{130}}\boldsymbol{L}^{18}_{i\cdots l'}(\boldsymbol{p})-\frac{657}{28\sqrt{1430}}\boldsymbol{L}^{19}_{i\cdots l'}(\boldsymbol{p})$
	$+\frac{\sqrt{55}}{8\sqrt{26}}\boldsymbol{L}^{20}_{i\cdots l'}(\boldsymbol{p})+\frac{41}{8\sqrt{1430}}\boldsymbol{L}^{21}_{i\cdots l'}(\boldsymbol{p})+\frac{829}{56\sqrt{1430}}\boldsymbol{L}^{22}_{i\cdots l'}(\boldsymbol{p})$
	$-\frac{5129}{392\sqrt{1430}}\boldsymbol{L}^{23}_{i\cdots l'}(\boldsymbol{p})+\frac{2027}{56\sqrt{1430}}\boldsymbol{L}^{24}_{i\cdots l'}(\boldsymbol{p})-\frac{83}{8\sqrt{1430}}\boldsymbol{L}^{25}_{i\cdots l'}(\boldsymbol{p})-\frac{\sqrt{143}}{8\sqrt{10}}\boldsymbol{L}^{26}_{i\cdots l'}(\boldsymbol{p})$
	$-\frac{\sqrt{143}}{8\sqrt{10}}\boldsymbol{L}^{27}_{i\cdots l'}(\boldsymbol{p})-\frac{\sqrt{143}}{8\sqrt{10}}\boldsymbol{L}^{28}_{i\cdots l'}(\boldsymbol{p})+\frac{3\sqrt{715}}{8\sqrt{2}}\boldsymbol{L}^{29}_{i\cdots l'}(\boldsymbol{p})$

Table 3.13 *The elements of the matrix* $f_{II'}(\lambda)$.

I	I'	$f_{II'}(\lambda)$
1	2	3
1	1	$\frac{1}{9}f^{0,1}(\lambda)+\frac{4}{45}f^{0,4}(\lambda)+\frac{8\sqrt{5}}{315}f^{0,6}(\lambda)+\frac{4\sqrt{5}}{45}f^{0,3}(\lambda)+\frac{8}{105}f^{0,7}(\lambda)$
		$+\frac{2\sqrt{10}}{45}f^{0,2}(\lambda)+\frac{8\sqrt{35}}{315}f^{0,5}(\lambda)-\frac{4\sqrt{14}}{441}f^{2,8}(\lambda)-\frac{2\sqrt{14}}{63}f^{2,4}(\lambda)$
		$-\frac{8\sqrt{77}}{1617}f^{2,10}(\lambda)-\frac{4\sqrt{10}}{105}f^{2,6}(\lambda)-\frac{8\sqrt{35}}{735}f^{2,9}(\lambda)-\frac{\sqrt{2}}{9}f^{2,1}(\lambda)$
		$-\frac{2\sqrt{7}}{63}f^{2,3}(\lambda)-\frac{4\sqrt{35}}{315}f^{2,5}(\lambda)-\frac{4\sqrt{2}}{63}f^{2,7}(\lambda)-\frac{2\sqrt{10}}{45}f^{2,2}(\lambda)$
		$+\frac{\sqrt{70}}{245}f^{4,6}(\lambda)+\frac{\sqrt{70}}{70}f^{4,2}(\lambda)+\frac{27\sqrt{2002}}{35035}f^{4,8}(\lambda)+\frac{2\sqrt{11}}{77}f^{4,5}(\lambda)$
		$+\frac{2\sqrt{154}}{539}f^{4,7}(\lambda)+\frac{\sqrt{7}}{35}f^{4,3}(\lambda)+\frac{\sqrt{35}}{70}f^{4,1}(\lambda)+\frac{\sqrt{10}}{35}f^{4,4}(\lambda)$
		$-\frac{\sqrt{55}}{231}f^{6,3}(\lambda)-\frac{5\sqrt{77}}{462}f^{6,1}(\lambda)-\frac{5\sqrt{22}}{462}f^{6,2}(\lambda)+\frac{7\sqrt{1430}}{6864}f^{8,1}(\lambda)$
1	2	$\frac{1}{9}f^{0,1}(\lambda)+\frac{4}{45}f^{0,4}(\lambda)-\frac{4\sqrt{5}}{315}f^{0,6}(\lambda)-\frac{2\sqrt{5}}{45}f^{0,3}(\lambda)+\frac{1}{35}f^{0,7}(\lambda)$
		$+\frac{2\sqrt{10}}{45}f^{0,2}(\lambda)-\frac{4\sqrt{35}}{315}f^{0,5}(\lambda)-\frac{4\sqrt{14}}{441}f^{2,8}(\lambda)-\frac{2\sqrt{14}}{63}f^{2,4}(\lambda)$
		$+\frac{2\sqrt{77}}{539}f^{2,10}(\lambda)-\frac{\sqrt{10}}{210}f^{2,6}(\lambda)-\frac{\sqrt{35}}{735}f^{2,9}(\lambda)+\frac{\sqrt{2}}{18}f^{2,1}(\lambda)+\frac{\sqrt{7}}{63}f^{2,3}(\lambda)$
		$+\frac{2\sqrt{35}}{315}f^{2,5}(\lambda)-\frac{4\sqrt{2}}{63}f^{2,7}(\lambda)+\frac{\sqrt{10}}{45}f^{2,2}(\lambda)-\frac{4\sqrt{70}}{735}f^{4,6}(\lambda)$
		$-\frac{2\sqrt{70}}{105}f^{4,2}(\lambda)+\frac{27\sqrt{2002}}{35035}f^{4,8}(\lambda)-\frac{\sqrt{11}}{231}f^{4,5}(\lambda)-\frac{\sqrt{154}}{1617}f^{4,7}(\lambda)$
		$+\frac{11\sqrt{7}}{210}f^{4,3}(\lambda)+\frac{11\sqrt{35}}{420}f^{4,1}(\lambda)-\frac{4\sqrt{10}}{105}f^{4,4}(\lambda)+\frac{2\sqrt{55}}{385}f^{6,3}(\lambda)$
		$-\frac{\sqrt{77}}{462}f^{6,1}(\lambda)-\frac{\sqrt{22}}{462}f^{6,2}(\lambda)+\frac{\sqrt{1430}}{715}f^{8,1}(\lambda)$
1	6	$\frac{1}{15}f^{0,4}(\lambda)+\frac{\sqrt{5}}{105}f^{0,6}(\lambda)-\frac{4}{105}f^{0,7}(\lambda)+\frac{\sqrt{10}}{60}f^{0,2}(\lambda)+\frac{\sqrt{35}}{210}f^{0,5}(\lambda)$
		$-\frac{2\sqrt{14}}{147}f^{2,8}(\lambda)-\frac{5\sqrt{77}}{1617}f^{2,10}(\lambda)+\frac{3\sqrt{10}}{140}f^{2,6}(\lambda)+\frac{\sqrt{35}}{196}f^{2,9}(\lambda)+\frac{\sqrt{7}}{84}f^{2,3}(\lambda)$
		$-\frac{\sqrt{2}}{21}f^{2,7}(\lambda)-\frac{\sqrt{10}}{60}f^{2,2}(\lambda)-\frac{\sqrt{70}}{980}f^{4,6}(\lambda)+\frac{\sqrt{2002}}{140140}f^{4,8}(\lambda)-\frac{5\sqrt{11}}{462}f^{4,5}(\lambda)$

Table 3.13 (*Cont.*)

I	I'	$f_{II'}(\lambda)$
		$+\frac{2\sqrt{154}}{1617}f^{4,7}(\lambda)-\frac{\sqrt{7}}{120}f^{4,3}(\lambda)-\frac{\sqrt{35}}{105}f^{4,1}(\lambda)-\frac{\sqrt{10}}{280}f^{4,4}(\lambda)+\frac{\sqrt{55}}{462}f^{6,3}(\lambda)$
		$+\frac{\sqrt{77}}{154}f^{6,1}(\lambda)+\frac{9\sqrt{22}}{1232}f^{6,2}(\lambda)-\frac{\sqrt{1430}}{858}f^{8,1}(\lambda)$
1	7	$\frac{1}{9}f^{0,1}(\lambda)-\frac{2}{45}f^{0,4}(\lambda)-\frac{4\sqrt{5}}{315}f^{0,6}(\lambda)+\frac{\sqrt{5}}{45}f^{0,3}(\lambda)-\frac{4}{105}f^{0,7}(\lambda)$
		$+\frac{\sqrt{10}}{90}f^{0,2}(\lambda)-\frac{\sqrt{35}}{315}f^{0,5}(\lambda)+\frac{8\sqrt{14}}{441}f^{2,8}(\lambda)-\frac{2\sqrt{14}}{63}f^{2,4}(\lambda)-\frac{5\sqrt{77}}{1617}f^{2,10}(\lambda)$
		$+\frac{2\sqrt{10}}{105}f^{2,6}(\lambda)+\frac{11\sqrt{35}}{1470}f^{2,9}(\lambda)-\frac{\sqrt{2}}{36}f^{2,1}(\lambda)-\frac{2\sqrt{7}}{63}f^{2,3}(\lambda)-\frac{\sqrt{35}}{315}f^{2,5}(\lambda)$
		$+\frac{2\sqrt{2}}{63}f^{2,7}(\lambda)+\frac{\sqrt{10}}{45}f^{2,2}(\lambda)+\frac{\sqrt{70}}{735}f^{4,6}(\lambda)-\frac{\sqrt{70}}{420}f^{4,2}(\lambda)+\frac{\sqrt{2002}}{140140}f^{4,8}(\lambda)$
		$+\frac{\sqrt{11}}{462}f^{4,5}(\lambda)-\frac{17\sqrt{154}}{3234}f^{4,7}(\lambda)-\frac{11\sqrt{7}}{420}f^{4,3}(\lambda)-\frac{\sqrt{35}}{420}f^{4,1}(\lambda)+\frac{\sqrt{10}}{420}f^{4,4}(\lambda)$
		$+\frac{\sqrt{55}}{462}f^{6,3}(\lambda)+\frac{13\sqrt{77}}{1848}f^{6,1}(\lambda)+\frac{5\sqrt{22}}{924}f^{6,2}(\lambda)-\frac{\sqrt{1430}}{858}f^{8,1}(\lambda)$
1	8	$\frac{1}{9}f^{0,1}(\lambda)-\frac{2}{45}f^{0,4}(\lambda)-\frac{4\sqrt{5}}{315}f^{0,6}(\lambda)+\frac{\sqrt{5}}{45}f^{0,3}(\lambda)-\frac{4}{105}f^{0,7}(\lambda)$
		$+\frac{\sqrt{10}}{90}f^{0,2}(\lambda)-\frac{\sqrt{35}}{315}f^{0,5}(\lambda)-\frac{4\sqrt{14}}{441}f^{2,8}(\lambda)+\frac{\sqrt{14}}{63}f^{2,4}(\lambda)+\frac{13\sqrt{77}}{1617}f^{2,10}(\lambda)$
		$-\frac{\sqrt{10}}{105}f^{2,6}(\lambda)+\frac{\sqrt{35}}{294}f^{2,9}(\lambda)-\frac{\sqrt{2}}{9}f^{2,1}(\lambda)+\frac{\sqrt{7}}{63}f^{2,3}(\lambda)+\frac{\sqrt{35}}{63}f^{2,5}(\lambda)$
		$-\frac{\sqrt{2}}{63}f^{2,7}(\lambda)-\frac{\sqrt{10}}{90}f^{2,2}(\lambda)-\frac{4\sqrt{70}}{735}f^{4,6}(\lambda)+\frac{\sqrt{70}}{105}f^{4,2}(\lambda)-\frac{109\sqrt{2002}}{140140}f^{4,8}(\lambda)$
		$-\frac{2\sqrt{11}}{231}f^{4,5}(\lambda)+\frac{5\sqrt{154}}{3234}f^{4,7}(\lambda)-\frac{\sqrt{7}}{420}f^{4,3}(\lambda)+\frac{\sqrt{35}}{105}f^{4,1}(\lambda)-\frac{\sqrt{10}}{105}f^{4,4}(\lambda)$
		$+\frac{\sqrt{55}}{462}f^{6,3}(\lambda)-\frac{\sqrt{77}}{231}f^{6,1}(\lambda)+\frac{5\sqrt{22}}{924}f^{6,2}(\lambda)+\frac{\sqrt{1430}}{6864}f^{8,1}(\lambda)$
2	2	$\frac{1}{9}f^{0,1}(\lambda)+\frac{4}{45}f^{0,4}(\lambda)+\frac{8\sqrt{5}}{315}f^{0,6}(\lambda)+\frac{4\sqrt{5}}{45}f^{0,3}(\lambda)+\frac{8}{105}f^{0,7}(\lambda)$
		$+\frac{2\sqrt{10}}{45}f^{0,2}(\lambda)+\frac{8\sqrt{35}}{315}f^{0,5}(\lambda)+\frac{8\sqrt{14}}{441}f^{2,8}(\lambda)+\frac{4\sqrt{14}}{63}f^{2,4}(\lambda)$
		$+\frac{16\sqrt{77}}{1617}f^{2,10}(\lambda)+\frac{8\sqrt{10}}{105}f^{2,6}(\lambda)+\frac{16\sqrt{35}}{735}f^{2,9}(\lambda)+\frac{2\sqrt{2}}{9}f^{2,1}(\lambda)+\frac{4\sqrt{7}}{63}f^{2,3}(\lambda)$
		$+\frac{8\sqrt{35}}{315}f^{2,5}(\lambda)+\frac{8\sqrt{2}}{63}f^{2,7}(\lambda)+\frac{4\sqrt{10}}{45}f^{2,2}(\lambda)+\frac{8\sqrt{70}}{735}f^{4,6}(\lambda)$
		$+\frac{4\sqrt{70}}{105}f^{4,2}(\lambda)+\frac{72\sqrt{2002}}{35035}f^{4,8}(\lambda)+\frac{16\sqrt{11}}{231}f^{4,5}(\lambda)+\frac{16\sqrt{154}}{1617}f^{4,7}(\lambda)$
		$+\frac{8\sqrt{7}}{105}f^{4,3}(\lambda)+\frac{4\sqrt{35}}{105}f^{4,1}(\lambda)+\frac{8\sqrt{10}}{105}f^{4,4}(\lambda)+\frac{16\sqrt{55}}{1155}f^{6,3}(\lambda)$
		$+\frac{8\sqrt{77}}{231}f^{6,1}(\lambda)+\frac{8\sqrt{22}}{231}f^{6,2}(\lambda)+\frac{8\sqrt{1430}}{2145}f^{8,1}(\lambda)$
2	4	$\frac{1}{15}f^{0,4}(\lambda)+\frac{\sqrt{5}}{105}f^{0,6}(\lambda)-\frac{4}{105}f^{0,7}(\lambda)+\frac{\sqrt{10}}{60}f^{0,2}(\lambda)+\frac{\sqrt{35}}{210}f^{0,5}(\lambda)+\frac{\sqrt{14}}{147}f^{2,8}(\lambda)$
		$-\frac{8\sqrt{77}}{1617}f^{2,10}(\lambda)-\frac{2\sqrt{10}}{105}f^{2,6}(\lambda)-\frac{\sqrt{35}}{735}f^{2,9}(\lambda)+\frac{\sqrt{7}}{84}f^{2,3}(\lambda)+\frac{\sqrt{35}}{70}f^{2,5}(\lambda)+\frac{\sqrt{2}}{42}f^{2,7}(\lambda)$
		$+\frac{\sqrt{10}}{30}f^{2,2}(\lambda)+\frac{\sqrt{70}}{245}f^{4,6}(\lambda)-\frac{36\sqrt{2002}}{35035}f^{4,8}(\lambda)-\frac{4\sqrt{11}}{231}f^{4,5}(\lambda)-\frac{\sqrt{154}}{1617}f^{4,7}(\lambda)$
		$+\frac{\sqrt{7}}{105}f^{4,3}(\lambda)-\frac{\sqrt{35}}{105}f^{4,1}(\lambda)+\frac{\sqrt{10}}{70}f^{4,4}(\lambda)-\frac{8\sqrt{55}}{1155}f^{6,3}(\lambda)-\frac{2\sqrt{77}}{231}f^{6,1}(\lambda)-\frac{\sqrt{22}}{462}f^{6,2}(\lambda)$
		$-\frac{4\sqrt{1430}}{2145}f^{8,1}(\lambda)$
2	5	$\frac{1}{15}f^{0,4}(\lambda)-\frac{2\sqrt{5}}{105}f^{0,6}(\lambda)+\frac{1}{105}f^{0,7}(\lambda)+\frac{\sqrt{10}}{60}f^{0,2}(\lambda)-\frac{\sqrt{35}}{105}f^{0,5}(\lambda)-\frac{2\sqrt{14}}{147}f^{2,8}(\lambda)$
		$+\frac{2\sqrt{77}}{1617}f^{2,10}(\lambda)+\frac{\sqrt{10}}{210}f^{2,6}(\lambda)-\frac{\sqrt{35}}{147}f^{2,9}(\lambda)-\frac{\sqrt{7}}{42}f^{2,3}(\lambda)-\frac{\sqrt{2}}{21}f^{2,7}(\lambda)+\frac{\sqrt{10}}{30}f^{2,2}(\lambda)$

Table 3.13 (*Cont.*)

I	I'	$f_{II'}(\lambda)$
		$-\frac{2\sqrt{70}}{245}f^{4,6}(\lambda)+\frac{9\sqrt{2002}}{35035}f^{4,8}(\lambda)+\frac{\sqrt{11}}{231}f^{4,5}(\lambda)-\frac{5\sqrt{154}}{1617}f^{4,7}(\lambda)+\frac{\sqrt{7}}{30}f^{4,3}(\lambda)$
		$+\frac{\sqrt{35}}{420}f^{4,1}(\lambda)-\frac{\sqrt{10}}{35}f^{4,4}(\lambda)+\frac{2\sqrt{55}}{1155}f^{6,3}(\lambda)+\frac{\sqrt{77}}{462}f^{6,1}(\lambda)-\frac{5\sqrt{22}}{462}f^{6,2}(\lambda)$
		$+\frac{\sqrt{1430}}{2145}f^{8,1}(\lambda)$
2	7	$\frac{1}{9}f^{0,1}(\lambda)-\frac{2}{45}f^{0,4}(\lambda)-\frac{4\sqrt{5}}{315}f^{0,6}(\lambda)+\frac{\sqrt{5}}{45}f^{0,3}(\lambda)-\frac{4}{105}f^{0,7}(\lambda)$
		$+\frac{\sqrt{10}}{90}f^{0,2}(\lambda)-\frac{\sqrt{35}}{315}f^{0,5}(\lambda)-\frac{4\sqrt{14}}{441}f^{2,8}(\lambda)+\frac{\sqrt{14}}{63}f^{2,4}(\lambda)-\frac{8\sqrt{77}}{1617}f^{2,10}(\lambda)$
		$-\frac{\sqrt{10}}{105}f^{2,6}(\lambda)-\frac{8\sqrt{35}}{735}f^{2,9}(\lambda)+\frac{5\sqrt{2}}{36}f^{2,1}(\lambda)+\frac{\sqrt{7}}{63}f^{2,3}(\lambda)-\frac{4\sqrt{35}}{315}f^{2,5}(\lambda)$
*		$-\frac{\sqrt{2}}{63}f^{2,7}(\lambda)-\frac{\sqrt{10}}{90}f^{2,2}(\lambda)-\frac{4\sqrt{70}}{735}f^{4,6}(\lambda)+\frac{\sqrt{70}}{105}f^{4,2}(\lambda)-\frac{36\sqrt{2002}}{35035}f^{4,8}(\lambda)$
		$-\frac{2\sqrt{11}}{231}f^{4,5}(\lambda)-\frac{8\sqrt{154}}{1617}f^{4,7}(\lambda)\quad\frac{4\sqrt{7}}{105}f^{4,3}(\lambda)+\frac{\sqrt{35}}{105}f^{4,1}(\lambda)-\frac{\sqrt{10}}{105}f^{4,4}(\lambda)$
		$-\frac{8\sqrt{55}}{1155}f^{6,3}(\lambda)-\frac{\sqrt{77}}{231}f^{6,1}(\lambda)-\frac{4\sqrt{22}}{231}f^{6,2}(\lambda)-\frac{4\sqrt{1430}}{2145}f^{8,1}(\lambda)$
4	4	$\frac{1}{20}f^{0,4}(\lambda)+\frac{\sqrt{5}}{70}f^{0,6}(\lambda)+\frac{3}{70}f^{0,5}(\lambda)-\frac{\sqrt{14}}{196}f^{2,8}(\lambda)+\frac{2\sqrt{77}}{539}f^{2,10}(\lambda)$
		$+\frac{\sqrt{35}}{980}f^{2,9}(\lambda)+\frac{\sqrt{35}}{140}f^{2,5}(\lambda)+\frac{9\sqrt{70}}{3920}f^{4,6}(\lambda)+\frac{17\sqrt{2002}}{140140}f^{4,8}(\lambda)$
		$-\frac{13\sqrt{154}}{2156}f^{4,7}(\lambda)-\frac{\sqrt{7}}{35}f^{4,3}(\lambda)-\frac{\sqrt{55}}{770}f^{6,3}(\lambda)-\frac{\sqrt{22}}{308}f^{6,2}(\lambda)+\frac{\sqrt{1430}}{715}f^{8,1}(\lambda)$
4	5	$\frac{1}{20}f^{0,4}(\lambda)-\frac{\sqrt{5}}{140}f^{0,6}(\lambda)-\frac{1}{210}f^{0,5}(\lambda)-\frac{\sqrt{14}}{196}f^{2,8}(\lambda)-\frac{\sqrt{77}}{1617}f^{2,10}(\lambda)$
		$+\frac{9\sqrt{35}}{1960}f^{2,9}(\lambda)-\frac{\sqrt{35}}{280}f^{2,5}(\lambda)-\frac{3\sqrt{70}}{980}f^{4,6}(\lambda)-\frac{9\sqrt{2002}}{70070}f^{4,8}(\lambda)$
		$+\frac{9\sqrt{154}}{4312}f^{4,7}(\lambda)-\frac{3\sqrt{7}}{280}f^{4,3}(\lambda)-\frac{\sqrt{55}}{1155}f^{6,3}(\lambda)+\frac{9\sqrt{22}}{1232}f^{6,2}(\lambda)-\frac{\sqrt{1430}}{4290}f^{8,1}(\lambda)$
4	9	$-\frac{1}{30}f^{0,4}(\lambda)-\frac{2\sqrt{5}}{105}f^{0,6}(\lambda)+\frac{3}{70}f^{0,7}(\lambda)+\frac{\sqrt{10}}{60}f^{0,2}(\lambda)+\frac{\sqrt{35}}{210}f^{0,5}(\lambda)+\frac{\sqrt{14}}{147}f^{2,8}(\lambda)$
		$+\frac{2\sqrt{77}}{539}f^{2,10}(\lambda)+\frac{\sqrt{10}}{840}f^{2,6}(\lambda)-\frac{\sqrt{35}}{5880}f^{2,9}(\lambda)+\frac{\sqrt{7}}{84}f^{2,3}(\lambda)-\frac{\sqrt{35}}{140}f^{2,5}(\lambda)-\frac{\sqrt{2}}{84}f^{2,7}(\lambda)$
		$+\frac{\sqrt{10}}{120}f^{2,2}(\lambda)-\frac{3\sqrt{70}}{980}f^{4,6}(\lambda)+\frac{17\sqrt{2002}}{140140}f^{4,8}(\lambda)-\frac{13\sqrt{11}}{924}f^{4,5}(\lambda)+\frac{13\sqrt{154}}{12936}f^{4,7}(\lambda)$
		$-\frac{\sqrt{7}}{210}f^{4,3}(\lambda)-\frac{\sqrt{35}}{105}f^{4,1}(\lambda)+\frac{3\sqrt{10}}{560}f^{4,4}(\lambda)-\frac{\sqrt{55}}{770}f^{6,3}(\lambda)-\frac{\sqrt{77}}{924}f^{6,1}(\lambda)+\frac{\sqrt{22}}{1848}f^{6,2}(\lambda)$
		$+\frac{\sqrt{1430}}{715}f^{8,1}(\lambda)$
5	5	$\frac{1}{20}f^{0,4}(\lambda)+\frac{\sqrt{5}}{70}f^{0,6}(\lambda)+\frac{3}{70}f^{0,5}(\lambda)+\frac{\sqrt{14}}{98}f^{2,8}(\lambda)-\frac{4\sqrt{77}}{539}f^{2,10}(\lambda)$
		$-\frac{\sqrt{35}}{490}f^{2,9}(\lambda)-\frac{\sqrt{35}}{70}f^{2,5}(\lambda)+\frac{3\sqrt{70}}{490}f^{4,6}(\lambda)+\frac{127\sqrt{2002}}{140140}f^{4,8}(\lambda)$
		$-\frac{\sqrt{154}}{1078}f^{4,7}(\lambda)+\frac{\sqrt{7}}{140}f^{4,3}(\lambda)-\frac{\sqrt{55}}{770}f^{6,3}(\lambda)-\frac{\sqrt{22}}{308}f^{6,2}(\lambda)+\frac{\sqrt{1430}}{11440}f^{8,1}(\lambda)$
5	7	$-\frac{1}{30}f^{0,4}(\lambda)+\frac{\sqrt{5}}{105}f^{0,6}(\lambda)-\frac{1}{210}f^{0,7}(\lambda)+\frac{\sqrt{10}}{60}f^{0,2}(\lambda)-\frac{\sqrt{35}}{420}f^{0,5}(\lambda)+\frac{\sqrt{14}}{147}f^{2,8}(\lambda)$
		$-\frac{\sqrt{77}}{1617}f^{2,10}(\lambda)+\frac{\sqrt{10}}{840}f^{2,6}(\lambda)+\frac{\sqrt{35}}{294}f^{2,9}(\lambda)-\frac{\sqrt{7}}{42}f^{2,3}(\lambda)-\frac{\sqrt{2}}{84}f^{2,7}(\lambda)+\frac{\sqrt{10}}{120}f^{2,2}(\lambda)$
		$+\frac{\sqrt{70}}{245}f^{4,6}(\lambda)-\frac{9\sqrt{2002}}{70070}f^{4,8}(\lambda)+\frac{\sqrt{11}}{924}f^{4,5}(\lambda)+\frac{5\sqrt{154}}{3234}f^{4,7}(\lambda)-\frac{\sqrt{7}}{60}f^{4,3}(\lambda)$
		$+\frac{\sqrt{35}}{420}f^{4,1}(\lambda)-\frac{\sqrt{10}}{140}f^{4,4}(\lambda)-\frac{\sqrt{55}}{1155}f^{6,3}(\lambda)+\frac{\sqrt{77}}{1848}f^{6,1}(\lambda)+\frac{5\sqrt{22}}{924}f^{6,2}(\lambda)$
		$-\frac{\sqrt{1430}}{4290}f^{8,1}(\lambda)$

Table 3.13 (*Cont.*)

I	I'	$f_{II'}(\lambda)$
7	7	$\frac{1}{9}f^{0,1}(\lambda)+\frac{1}{45}f^{0,4}(\lambda)+\frac{8\sqrt{5}}{315}f^{0,6}(\lambda)+\frac{\sqrt{5}}{45}f^{0,3}(\lambda)+\frac{3}{70}f^{0,7}(\lambda)$
		$-\frac{\sqrt{10}}{45}f^{0,2}(\lambda)-\frac{4\sqrt{35}}{315}f^{0,5}(\lambda)-\frac{4\sqrt{14}}{441}f^{2,8}(\lambda)-\frac{\sqrt{14}}{126}f^{2,4}(\lambda)+\frac{2\sqrt{77}}{539}f^{2,10}(\lambda)$
		$+\frac{\sqrt{10}}{420}f^{2,6}(\lambda)-\frac{\sqrt{35}}{735}f^{2,9}(\lambda)+\frac{\sqrt{2}}{18}f^{2,1}(\lambda)-\frac{2\sqrt{7}}{63}f^{2,3}(\lambda)+\frac{2\sqrt{35}}{315}f^{2,5}(\lambda)$
		$+\frac{2\sqrt{2}}{63}f^{2,7}(\lambda)-\frac{\sqrt{10}}{90}f^{2,2}(\lambda)+\frac{\sqrt{70}}{245}f^{4,6}(\lambda)+\frac{\sqrt{70}}{280}f^{4,2}(\lambda)+\frac{17\sqrt{2002}}{140140}f^{4,8}(\lambda)$
		$-\frac{13\sqrt{11}}{462}f^{4,5}(\lambda)+\frac{13\sqrt{154}}{1617}f^{4,7}(\lambda)+\frac{2\sqrt{7}}{105}f^{4,3}(\lambda)-\frac{2\sqrt{35}}{105}f^{4,1}(\lambda)-\frac{\sqrt{10}}{70}f^{4,4}(\lambda)$
		$-\frac{\sqrt{55}}{770}f^{6,3}(\lambda)-\frac{\sqrt{77}}{462}f^{6,1}(\lambda)+\frac{\sqrt{22}}{231}f^{6,2}(\lambda)+\frac{\sqrt{1430}}{715}f^{8,1}(\lambda)$
8	8	$\frac{1}{9}f^{0,1}(\lambda)+\frac{1}{45}f^{0,4}(\lambda)+\frac{8\sqrt{5}}{315}f^{0,6}(\lambda)+\frac{\sqrt{5}}{45}f^{0,3}(\lambda)+\frac{3}{70}f^{0,7}(\lambda)$
		$-\frac{\sqrt{10}}{45}f^{0,2}(\lambda)-\frac{4\sqrt{35}}{315}f^{0,5}(\lambda)+\frac{8\sqrt{14}}{441}f^{2,8}(\lambda)+\frac{\sqrt{14}}{63}f^{2,4}(\lambda)-\frac{4\sqrt{77}}{539}f^{2,10}(\lambda)$
		$-\frac{\sqrt{10}}{210}f^{2,6}(\lambda)+\frac{2\sqrt{35}}{735}f^{2,9}(\lambda)-\frac{\sqrt{2}}{9}f^{2,1}(\lambda)+\frac{4\sqrt{7}}{63}f^{2,3}(\lambda)-\frac{4\sqrt{35}}{315}f^{2,5}(\lambda)$
		$-\frac{4\sqrt{2}}{63}f^{2,7}(\lambda)+\frac{\sqrt{10}}{45}f^{2,2}(\lambda)+\frac{8\sqrt{70}}{735}f^{4,6}(\lambda)+\frac{\sqrt{70}}{105}f^{4,2}(\lambda)+\frac{127\sqrt{2002}}{140140}f^{4,8}(\lambda)$
		$-\frac{\sqrt{11}}{231}f^{4,5}(\lambda)+\frac{2\sqrt{154}}{1617}f^{4,7}(\lambda)-\frac{\sqrt{7}}{210}f^{4,3}(\lambda)+\frac{\sqrt{35}}{210}f^{4,1}(\lambda)-\frac{4\sqrt{10}}{105}f^{4,4}(\lambda)$
		$-\frac{\sqrt{55}}{770}f^{6,3}(\lambda)-\frac{\sqrt{77}}{462}f^{6,1}(\lambda)+\frac{\sqrt{22}}{231}f^{6,2}(\lambda)+\frac{\sqrt{1430}}{11440}f^{8,1}(\lambda)$
10	10	$\frac{3\sqrt{5}}{280}f^{0,6}(\lambda)+\frac{1}{42}f^{0,7}(\lambda)+\frac{3\sqrt{14}}{784}f^{2,8}(\lambda)-\frac{\sqrt{77}}{1617}f^{2,10}(\lambda)-\frac{\sqrt{35}}{392}f^{2,9}(\lambda)$
		$-\frac{3\sqrt{70}}{980}f^{4,6}(\lambda)-\frac{69\sqrt{2002}}{112112}f^{4,8}(\lambda)-\frac{3\sqrt{154}}{8624}f^{4,7}(\lambda)+\frac{59\sqrt{55}}{18480}f^{6,3}(\lambda)$
		$+\frac{\sqrt{22}}{308}f^{6,2}(\lambda)-\frac{\sqrt{1430}}{4290}f^{8,1}(\lambda)$
10	11	$-\frac{\sqrt{5}}{70}f^{0,6}(\lambda)+\frac{1}{42}f^{0,7}(\lambda)+\frac{\sqrt{35}}{280}f^{0,5}(\lambda)-\frac{\sqrt{14}}{196}f^{2,8}(\lambda)-\frac{\sqrt{77}}{1617}f^{2,10}(\lambda)-\frac{\sqrt{10}}{336}f^{2,6}(\lambda)$
		$+\frac{\sqrt{35}}{2352}f^{2,9}(\lambda)+\frac{\sqrt{2}}{112}f^{2,7}(\lambda)+\frac{\sqrt{70}}{245}f^{4,6}(\lambda)-\frac{69\sqrt{2002}}{112112}f^{4,8}(\lambda)-\frac{\sqrt{11}}{1232}f^{4,5}(\lambda)$
		$+\frac{\sqrt{154}}{17248}f^{4,7}(\lambda)-\frac{\sqrt{10}}{140}f^{4,4}(\lambda)+\frac{59\sqrt{55}}{18480}f^{6,3}(\lambda)+\frac{\sqrt{77}}{924}f^{6,1}(\lambda)-\frac{\sqrt{22}}{1848}f^{6,2}(\lambda)$
		$-\frac{\sqrt{1430}}{4290}f^{8,1}(\lambda)$
10	13	$\frac{\sqrt{5}}{140}f^{0,6}(\lambda)-\frac{1}{84}f^{0,7}(\lambda)+\frac{\sqrt{35}}{280}f^{0,5}(\lambda)+\frac{\sqrt{14}}{392}f^{2,8}(\lambda)-\frac{17\sqrt{77}}{12936}f^{2,10}(\lambda)$
		$-\frac{\sqrt{10}}{336}f^{2,6}(\lambda)+\frac{5\sqrt{35}}{1176}f^{2,9}(\lambda)+\frac{\sqrt{2}}{112}f^{2,7}(\lambda)$
		$-\frac{\sqrt{70}}{490}f^{4,6}(\lambda)+\frac{87\sqrt{2002}}{112112}f^{4,8}(\lambda)-\frac{\sqrt{11}}{1232}f^{4,5}(\lambda)-\frac{\sqrt{154}}{1568}f^{4,7}(\lambda)$
		$-\frac{\sqrt{10}}{140}f^{4,4}(\lambda)-\frac{61\sqrt{55}}{18480}f^{6,3}(\lambda)+\frac{\sqrt{77}}{924}f^{6,1}(\lambda)+\frac{\sqrt{22}}{168}f^{6,2}(\lambda)$
		$-\frac{\sqrt{1430}}{4290}f^{8,1}(\lambda)$
11	11	$\frac{2\sqrt{5}}{105}f^{0,6}(\lambda)+\frac{\sqrt{5}}{60}f^{0,3}(\lambda)+\frac{1}{42}f^{0,7}(\lambda)-\frac{\sqrt{35}}{105}f^{0,5}(\lambda)+\frac{\sqrt{14}}{147}f^{2,8}(\lambda)+\frac{\sqrt{14}}{168}f^{2,4}(\lambda)$
		$-\frac{\sqrt{77}}{1617}f^{2,10}(\lambda)-\frac{\sqrt{10}}{168}f^{2,6}(\lambda)+\frac{\sqrt{35}}{294}f^{2,9}(\lambda)-\frac{\sqrt{2}}{42}f^{2,7}(\lambda)-\frac{4\sqrt{70}}{735}f^{4,6}(\lambda)-\frac{\sqrt{70}}{210}f^{4,2}(\lambda)$
		$-\frac{69\sqrt{2002}}{112112}f^{4,8}(\lambda)-\frac{\sqrt{11}}{616}f^{4,5}(\lambda)+\frac{\sqrt{154}}{2156}f^{4,7}(\lambda)+\frac{2\sqrt{10}}{105}f^{4,4}(\lambda)+\frac{59\sqrt{55}}{18480}f^{6,3}(\lambda)$
		$+\frac{\sqrt{77}}{462}f^{6,1}(\lambda)-\frac{\sqrt{22}}{231}f^{6,2}(\lambda)-\frac{\sqrt{1430}}{4290}f^{8,1}(\lambda)$

Table 3.13 (*Cont.*)

I	I'	$f_{II'}(\lambda)$
16	16	$\frac{\sqrt{5}}{210}f^{0,6}(\lambda)+\frac{\sqrt{5}}{60}f^{0,3}(\lambda)+\frac{1}{21}f^{0,7}(\lambda)+\frac{\sqrt{35}}{210}f^{0,5}(\lambda)+\frac{\sqrt{14}}{588}f^{2,8}(\lambda)+\frac{\sqrt{14}}{168}f^{2,4}(\lambda)+\frac{13\sqrt{77}}{6468}f^{2,10}(\lambda)-\frac{\sqrt{10}}{56}f^{2,6}(\lambda)-\frac{\sqrt{35}}{196}f^{2,9}(\lambda)+\frac{\sqrt{2}}{84}f^{2,7}(\lambda)-\frac{\sqrt{70}}{735}f^{4,6}(\lambda)-\frac{\sqrt{70}}{210}f^{4,2}(\lambda)-\frac{3\sqrt{2002}}{10192}f^{4,8}(\lambda)-\frac{3\sqrt{11}}{616}f^{4,5}(\lambda)-\frac{3\sqrt{154}}{4312}f^{4,7}(\lambda)-\frac{\sqrt{10}}{105}f^{4,4}(\lambda)+\frac{\sqrt{55}}{336}f^{6,3}(\lambda)+\frac{\sqrt{77}}{154}f^{6,1}(\lambda)+\frac{\sqrt{22}}{154}f^{6,2}(\lambda)-\frac{\sqrt{1430}}{858}f^{8,1}(\lambda)$
12	13	$\frac{\sqrt{5}}{210}f^{0,6}(\lambda)+\frac{\sqrt{5}}{60}f^{0,3}(\lambda)-\frac{1}{28}f^{0,7}(\lambda)+\frac{\sqrt{35}}{210}f^{0,5}(\lambda)+\frac{\sqrt{14}}{588}f^{2,8}(\lambda)+\frac{\sqrt{14}}{168}f^{2,4}(\lambda)-\frac{17\sqrt{77}}{4312}f^{2,10}(\lambda)+\frac{\sqrt{10}}{336}f^{2,6}(\lambda)+\frac{\sqrt{35}}{1176}f^{2,9}(\lambda)+\frac{\sqrt{2}}{84}f^{2,7}(\lambda)-\frac{\sqrt{70}}{735}f^{4,6}(\lambda)-\frac{\sqrt{70}}{210}f^{4,2}(\lambda)-\frac{27\sqrt{2002}}{56056}f^{4,8}(\lambda)+\frac{\sqrt{11}}{1232}f^{4,5}(\lambda)+\frac{\sqrt{154}}{8624}f^{4,7}(\lambda)-\frac{\sqrt{10}}{105}f^{4,4}(\lambda)+\frac{\sqrt{55}}{3080}f^{6,3}(\lambda)-\frac{\sqrt{77}}{924}f^{6,1}(\lambda)-\frac{\sqrt{22}}{924}f^{6,2}(\lambda)+\frac{\sqrt{1430}}{715}f^{8,1}(\lambda)$
13	13	$\frac{\sqrt{5}}{210}f^{0,6}(\lambda)+\frac{\sqrt{5}}{60}f^{0,3}(\lambda)+\frac{1}{21}f^{0,7}(\lambda)+\frac{\sqrt{35}}{210}f^{0,5}(\lambda)+\frac{\sqrt{14}}{588}f^{2,8}(\lambda)+\frac{\sqrt{14}}{168}f^{2,4}(\lambda)+\frac{17\sqrt{77}}{3234}f^{2,10}(\lambda)+\frac{\sqrt{10}}{42}f^{2,6}(\lambda)+\frac{\sqrt{35}}{147}f^{2,9}(\lambda)+\frac{\sqrt{2}}{84}f^{2,7}(\lambda)-\frac{\sqrt{70}}{735}f^{4,6}(\lambda)-\frac{\sqrt{70}}{210}f^{4,2}(\lambda)+\frac{9\sqrt{2002}}{14014}f^{4,8}(\lambda)+\frac{\sqrt{11}}{154}f^{4,5}(\lambda)+\frac{\sqrt{154}}{1078}f^{4,7}(\lambda)-\frac{\sqrt{10}}{105}f^{4,4}(\lambda)-\frac{\sqrt{55}}{2310}f^{6,3}(\lambda)-\frac{2\sqrt{77}}{231}f^{6,1}(\lambda)-\frac{2\sqrt{22}}{231}f^{6,2}(\lambda)-\frac{4\sqrt{1430}}{2145}f^{8,1}(\lambda)$
18	18	$\frac{3\sqrt{5}}{280}f^{0,6}(\lambda)+\frac{1}{42}f^{0,7}(\lambda)-\frac{3\sqrt{14}}{392}f^{2,8}(\lambda)+\frac{2\sqrt{77}}{1617}f^{2,10}(\lambda)+\frac{\sqrt{35}}{196}f^{2,9}(\lambda)+\frac{3\sqrt{70}}{3920}f^{4,6}(\lambda)-\frac{\sqrt{2002}}{2548}f^{4,8}(\lambda)-\frac{9\sqrt{154}}{2156}f^{4,7}(\lambda)-\frac{\sqrt{55}}{210}f^{6,3}(\lambda)+\frac{\sqrt{22}}{308}f^{6,2}(\lambda)+\frac{\sqrt{1430}}{2145}f^{8,1}(\lambda)$
18	19	$\frac{\sqrt{5}}{140}f^{0,6}(\lambda)-\frac{1}{84}f^{0,7}(\lambda)+\frac{\sqrt{35}}{280}f^{0,5}(\lambda)-\frac{\sqrt{14}}{196}f^{2,8}(\lambda)-\frac{\sqrt{77}}{1617}f^{2,10}(\lambda)+\frac{\sqrt{10}}{168}f^{2,6}(\lambda)+\frac{\sqrt{35}}{2352}f^{2,9}(\lambda)-\frac{\sqrt{2}}{56}f^{2,7}(\lambda)+\frac{\sqrt{70}}{1960}f^{4,6}(\lambda)+\frac{\sqrt{2002}}{5096}f^{4,8}(\lambda)-\frac{3\sqrt{11}}{308}f^{4,5}(\lambda)-\frac{3\sqrt{154}}{8624}f^{4,7}(\lambda)+\frac{\sqrt{10}}{560}f^{4,4}(\lambda)+\frac{\sqrt{55}}{420}f^{6,3}(\lambda)+\frac{\sqrt{77}}{924}f^{6,1}(\lambda)+\frac{\sqrt{22}}{3696}f^{6,2}(\lambda)-\frac{\sqrt{1430}}{4290}f^{8,1}(\lambda)$
18	21	$-\frac{\sqrt{5}}{70}f^{0,6}(\lambda)+\frac{1}{42}f^{0,7}(\lambda)+\frac{\sqrt{35}}{280}f^{0,5}(\lambda)+\frac{\sqrt{14}}{98}f^{2,8}(\lambda)+\frac{2\sqrt{77}}{1617}f^{2,10}(\lambda)+\frac{\sqrt{10}}{168}f^{2,6}(\lambda)-\frac{\sqrt{35}}{1176}f^{2,9}(\lambda)-\frac{\sqrt{2}}{56}f^{2,7}(\lambda)-\frac{\sqrt{70}}{980}f^{4,6}(\lambda)-\frac{\sqrt{2002}}{2548}f^{4,8}(\lambda)-\frac{3\sqrt{11}}{308}f^{4,5}(\lambda)+\frac{3\sqrt{154}}{4312}f^{4,7}(\lambda)+\frac{\sqrt{10}}{560}f^{4,4}(\lambda)-\frac{\sqrt{55}}{210}f^{6,3}(\lambda)+\frac{\sqrt{77}}{924}f^{6,1}(\lambda)-\frac{\sqrt{22}}{1848}f^{6,2}(\lambda)+\frac{\sqrt{1430}}{2145}f^{8,1}(\lambda)$
19	19	$\frac{\sqrt{5}}{210}f^{0,6}(\lambda)+\frac{\sqrt{5}}{60}f^{0,3}(\lambda)+\frac{1}{21}f^{0,7}(\lambda)+\frac{\sqrt{35}}{210}f^{0,5}(\lambda)-\frac{\sqrt{14}}{294}f^{2,8}(\lambda)-\frac{\sqrt{14}}{84}f^{2,4}(\lambda)-\frac{47\sqrt{77}}{6468}f^{2,10}(\lambda)-\frac{\sqrt{10}}{168}f^{2,6}(\lambda)-\frac{\sqrt{35}}{588}f^{2,9}(\lambda)-\frac{\sqrt{2}}{42}f^{2,7}(\lambda)+\frac{\sqrt{70}}{2940}f^{4,6}(\lambda)+\frac{\sqrt{70}}{840}f^{4,2}(\lambda)+\frac{87\sqrt{2002}}{112112}f^{4,8}(\lambda)+\frac{3\sqrt{11}}{308}f^{4,5}(\lambda)+\frac{3\sqrt{154}}{2156}f^{4,7}(\lambda)+\frac{\sqrt{10}}{420}f^{4,4}(\lambda)-\frac{5\sqrt{55}}{1848}f^{6,3}(\lambda)-\frac{\sqrt{77}}{924}f^{6,1}(\lambda)-\frac{\sqrt{22}}{924}f^{6,2}(\lambda)+\frac{\sqrt{1430}}{6864}f^{8,1}(\lambda)$

Table 3.13 (*Cont.*)

I	I'	$f_{II'}(\lambda)$
19	21	$-\frac{\sqrt{5}}{105}f^{0,6}(\lambda)+\frac{\sqrt{5}}{60}f^{0,3}(\lambda)-\frac{1}{84}f^{0,7}(\lambda)-\frac{\sqrt{35}}{420}f^{0,5}(\lambda)+\frac{\sqrt{14}}{147}f^{2,8}(\lambda)-\frac{\sqrt{14}}{84}f^{2,4}(\lambda)$ $-\frac{\sqrt{77}}{1617}f^{2,10}(\lambda)+\frac{\sqrt{10}}{336}f^{2,6}(\lambda)+\frac{\sqrt{35}}{294}f^{2,9}(\lambda)+\frac{\sqrt{2}}{84}f^{2,7}(\lambda)-\frac{\sqrt{70}}{1470}f^{4,6}(\lambda)+\frac{\sqrt{70}}{840}f^{4,2}(\lambda)$ $+\frac{\sqrt{2002}}{5096}f^{4,8}(\lambda)-\frac{3\sqrt{11}}{616}f^{4,5}(\lambda)-\frac{3\sqrt{154}}{1078}f^{4,7}(\lambda)-\frac{\sqrt{10}}{840}f^{4,4}(\lambda)+\frac{\sqrt{55}}{420}f^{6,3}(\lambda)$ $+\frac{\sqrt{77}}{1848}f^{6,1}(\lambda)+\frac{\sqrt{22}}{462}f^{6,2}(\lambda)-\frac{\sqrt{1430}}{4290}f^{8,1}(\lambda)$
21	21	$\frac{2\sqrt{5}}{105}f^{0,6}(\lambda)+\frac{\sqrt{5}}{60}f^{0,3}(\lambda)+\frac{1}{42}f^{0,7}(\lambda)-\frac{\sqrt{35}}{105}f^{0,5}(\lambda)-\frac{2\sqrt{14}}{147}f^{2,8}(\lambda)-\frac{\sqrt{14}}{84}f^{2,4}(\lambda)$ $+\frac{2\sqrt{77}}{1617}f^{2,10}(\lambda)+\frac{\sqrt{10}}{84}f^{2,6}(\lambda)-\frac{\sqrt{35}}{147}f^{2,9}(\lambda)+\frac{\sqrt{2}}{21}f^{2,7}(\lambda)+\frac{\sqrt{70}}{735}f^{4,6}(\lambda)+\frac{\sqrt{70}}{840}f^{4,2}(\lambda)$ $-\frac{\sqrt{2002}}{2548}f^{4,8}(\lambda)-\frac{3\sqrt{11}}{154}f^{4,5}(\lambda)+\frac{3\sqrt{154}}{539}f^{4,7}(\lambda)-\frac{\sqrt{10}}{210}f^{4,4}(\lambda)-\frac{\sqrt{55}}{210}f^{6,3}(\lambda)$ $+\frac{\sqrt{77}}{462}f^{6,1}(\lambda)-\frac{\sqrt{22}}{231}f^{6,2}(\lambda)+\frac{\sqrt{1430}}{2145}f^{8,1}(\lambda)$

$$
\begin{aligned}
&f_{10,13}(\lambda)=f_{14,17}(\lambda), \quad f_{11,11}(\lambda)=f_{15,15}(\lambda),\\
&f_{12,13}(\lambda)=f_{16,17}(\lambda), \quad f_{13,13}(\lambda)=f_{17,17}(\lambda),\\
&f_{18,19}(\lambda)=f_{18,20}(\lambda), \quad f_{19,19}(\lambda)=f_{20,20}(\lambda),\\
&f_{19,21}(\lambda)=f_{20,21}(\lambda),
\end{aligned}
$$

and

$$
\begin{aligned}
f_{1,3}(\lambda)&=-f_{1,1}(\lambda)+8f_{5,5}(\lambda)-2f_{8,8}(\lambda)+4f_{1,8}(\lambda),\\
f_{1,4}(\lambda)&=f_{3,6}(\lambda)=f_{1,6}(\lambda)-4f_{18,19}(\lambda),\\
f_{1,5}(\lambda)&=f_{3,5}(\lambda)=\frac{1}{2}f_{1,1}(\lambda)-2f_{19,19}(\lambda)-\frac{1}{2}f_{1,8}(\lambda),\\
f_{1,9}(\lambda)&=f_{3,7}(\lambda)=f_{1,7}(\lambda)-4f_{19,21}(\lambda),\\
f_{2,8}(\lambda)&=f_{1,2}(\lambda)-2f_{2,5}(\lambda),\\
f_{4,6}(\lambda)&=f_{4,4}(\lambda)-2f_{18,18}(\lambda),\\
f_{4,7}(\lambda)&=f_{6,9}(\lambda)=f_{4,9}(\lambda)-2f_{18,21}(\lambda),\\
f_{4,8}(\lambda)&=f_{6,8}(\lambda)=f_{1,6}(\lambda)-2f_{4,5}(\lambda)-2f_{18,19}(\lambda),\\
f_{5,8}(\lambda)&-\frac{1}{2}f_{1,1}(\lambda)+2f_{5,5}(\lambda)-f_{8,8}(\lambda)+2f_{19,19}(\lambda)+\frac{3}{2}f_{1,8}(\lambda),\\
f_{7,8}(\lambda)&=f_{8,9}(\lambda)=f_{1,7}(\lambda)-2f_{5,7}(\lambda)-2f_{19,21}(\lambda),\\
f_{7,9}(\lambda)&=f_{7,7}(\lambda)-2f_{21,21}(\lambda),\\
f_{10,12}(\lambda)&=f_{14,16}(\lambda)=-\frac{1}{2}f_{11,11}(\lambda)+\frac{1}{2}f_{12,12}(\lambda)-f_{10,11}(\lambda),\\
f_{11,12}(\lambda)&=f_{15,16}(\lambda)=-2f_{10,10}(\lambda)+\frac{1}{2}f_{11,11}(\lambda)+\frac{1}{2}f_{12,12}(\lambda),\\
f_{11,13}(\lambda)&=f_{15,17}(\lambda)=f_{12,13}(\lambda)-2f_{10,13}(\lambda),\\
f_{19,20}(\lambda)&=\frac{1}{2}f_{1,1}(\lambda)-2f_{5,5}(\lambda)+\frac{1}{2}f_{8,8}(\lambda)-f_{19,19}(\lambda)-f_{1,8}(\lambda).
\end{aligned}
$$

Let $u_i(\lambda)$, $1 \le i \le 29$, be the following functions.

$$
\begin{aligned}
&u_1(\lambda) = 2f_{1,1}(\lambda), \quad u_2(\lambda) = f_{2,2}(\lambda), \quad u_3(\lambda) = 2f_{4,4}(\lambda),\\
&u_4(\lambda) = f_{5,5}(\lambda), \quad u_5(\lambda) = 2f_{7,7}(\lambda), \quad u_6(\lambda) = f_{8,8}(\lambda),\\
&u_7(\lambda) = f_{1,2}(\lambda), \quad u_8(\lambda) = f_{1,6}(\lambda), \quad u_9(\lambda) = f_{1,7}(\lambda),\\
&u_{10}(\lambda) = f_{1,8}(\lambda), \quad u_{11}(\lambda) = f_{2,4}(\lambda), \quad u_{12}(\lambda) = f_{2,5}(\lambda),\\
&u_{13}(\lambda) = f_{2,7}(\lambda), \quad u_{14}(\lambda) = f_{4,5}(\lambda), \quad u_{15}(\lambda) = f_{4,9}(\lambda),\\
&u_{16}(\lambda) = f_{5,9}(\lambda), \quad u_{17}(\lambda) = 2f_{10,10}(\lambda), \quad u_{18}(\lambda) = 2f_{11,11}(\lambda),\\
&u_{19}(\lambda) = 2f_{12,12}(\lambda), \quad u_{20}(\lambda) = 2f_{13,13}(\lambda), \quad u_{21}(\lambda) = f_{10,11}(\lambda),\\
&u_{22}(\lambda) = f_{10,13}(\lambda), \quad u_{23}(\lambda) = f_{12,13}(\lambda), \quad u_{24}(\lambda) = f_{18,18}(\lambda),\\
&u_{25}(\lambda) = 2f_{19,19}(\lambda), \quad u_{26}(\lambda) = f_{21,21}(\lambda), \quad u_{27}(\lambda) = f_{18,19}(\lambda),\\
&u_{28}(\lambda) = f_{18,21}(\lambda), \quad u_{29}(\lambda) = f_{19,21}(\lambda).
\end{aligned}
$$

Define the functions $v_i(\lambda)$, $1 \le i \le 26$, by

$$
v_i(\lambda) = \begin{cases}
\frac{u_i(\lambda)}{u_1(\lambda)+\cdots+u_6(\lambda)}, & \text{if } 1 \le i \le 5\\
\frac{u_{i+1}(\lambda)}{u_1(\lambda)+\cdots+u_6(\lambda)}, & \text{if } 6 \le i \le 15\\
\frac{u_{i+1}(\lambda)}{u_{17}(\lambda)+\cdots+u_{20}(\lambda)}, & \text{if } 16 \le i \le 18\\
\frac{u_{i+2}(\lambda)}{u_{17}(\lambda)+\cdots+u_{20}(\lambda)}, & \text{if } 19 \le i \le 21\\
\frac{u_{i+2}(\lambda)}{u_{24}(\lambda)+u_{25}(\lambda)+u_{26}(\lambda)}, & \text{if } 22 \le i \le 23\\
\frac{u_{i+3}(\lambda)}{u_{24}(\lambda)+u_{25}(\lambda)+u_{26}(\lambda)}, & \text{if } 24 \le i \le 26.
\end{cases}
$$

The set $\mathcal{C}_0$ of the possible values of the function $f(\lambda)$ is a convex compact. The set of extreme points of $\mathcal{C}_0$ consists of three connected components. The first one is the 14-dimensional boundary of the 15-dimensional set of all 9×9 symmetric non-negative-definite matrices with unit trace with coordinates $v_1(\lambda), \dots, v_{15}(\lambda)$. The second one is the five-dimensional boundary of the six-dimensional set of all 4×4 symmetric non-negative-definite matrices with unit trace with coordinates $v_{16}(\lambda), \dots, v_{21}(\lambda)$. Finally, the third one is the four-dimensional boundary of the five-dimensional set of all 4×4 symmetric non-negative-definite matrices with unit trace with coordinates $v_{22}(\lambda), \dots, v_{26}(\lambda)$.

The functions $f^{2t,v}(\lambda)$ are expressed in terms of $u_i(\lambda)$ according to Table 3.14.

Substitute these values in (3.97). We obtain the matrix entries $f_{i\cdots l'}(\boldsymbol{p})$ expresses in terms of $u_i(\lambda)$ and $\boldsymbol{M}^{n,m}(\boldsymbol{p})$. Using Table 3.12, we express $f_{i\cdots l'}(\boldsymbol{p})$ in terms of $u_i(\lambda)$ and $\boldsymbol{L}^q_{i\cdots l'}(\boldsymbol{p})$. Substitute the obtained expression into the familiar equation

$$
\langle \boldsymbol{C}(\boldsymbol{x}), \boldsymbol{C}(\boldsymbol{y}) \rangle = \int_{\hat{\mathbb{E}}^3} \mathrm{e}^{\mathrm{i}(\boldsymbol{p},\boldsymbol{y}-\boldsymbol{x})} f(\boldsymbol{p}) \, \mathrm{d}\Omega \, \mathrm{d}\Phi(\lambda)
$$

and use the Rayleigh expansion (2.62). To formulate the result, we need more notation.

Table 3.14 *The functions* $f^{2t,v}(\lambda)$.

Function	Value
1	2
$f^{0,1}(\lambda)$	$\frac{1}{9}u_2(\lambda) + \frac{16}{9}u_4(\lambda) + \frac{8}{9}u_5(\lambda) + \frac{8}{9}u_7(\lambda) + \frac{32}{9}u_9(\lambda) + \frac{16}{9}u_{10}(\lambda)$ $-\frac{8}{9}u_{12}(\lambda) + \frac{8}{9}u_{13}(\lambda) - \frac{32}{9}u_{16}(\lambda) - \frac{16}{9}u_{26}(\lambda) - \frac{64}{9}u_{29}(\lambda)$
$f^{0,2}(\lambda)$	$\frac{2\sqrt{2}}{9\sqrt{5}}u_2(\lambda) + \frac{56\sqrt{2}}{9\sqrt{5}}u_4(\lambda) - \frac{8\sqrt{2}}{9\sqrt{5}}u_5(\lambda) - \frac{8\sqrt{2}}{3\sqrt{5}}u_6(\lambda) + \frac{2\sqrt{10}}{9}u_7(\lambda)$ $+\frac{16\sqrt{2}}{3\sqrt{5}}u_8(\lambda) - \frac{8\sqrt{2}}{9\sqrt{5}}u_9(\lambda) + \frac{32\sqrt{2}}{9\sqrt{5}}u_{10}(\lambda) + \frac{4\sqrt{2}}{3\sqrt{5}}u_{11}(\lambda) + \frac{2\sqrt{2}}{9\sqrt{5}}u_{12}(\lambda)$ $+\frac{4\sqrt{2}}{9\sqrt{5}}u_{13}(\lambda) - \frac{16\sqrt{2}}{3\sqrt{5}}u_{14}(\lambda) + \frac{16\sqrt{2}}{3\sqrt{5}}u_{15}(\lambda) + \frac{56\sqrt{2}}{9\sqrt{5}}u_{16}(\lambda) + \frac{16\sqrt{2}}{9\sqrt{5}}u_{26}(\lambda)$ $-\frac{32\sqrt{2}}{3\sqrt{5}}u_{27}(\lambda) - \frac{16\sqrt{2}}{3\sqrt{5}}u_{28}(\lambda) + \frac{16\sqrt{2}}{9\sqrt{5}}u_{29}(\lambda)$
$f^{0,3}(\lambda)$	$\frac{4}{3\sqrt{5}}u_1(\lambda) + \frac{4}{9\sqrt{5}}u_2(\lambda) - \frac{16\sqrt{5}}{9}u_4(\lambda) + \frac{2}{9\sqrt{5}}u_5(\lambda) + \frac{8}{3\sqrt{5}}u_6(\lambda)$ $-\frac{16}{9\sqrt{5}}u_7(\lambda) - \frac{16}{9\sqrt{5}}u_9(\lambda) - \frac{32}{9\sqrt{5}}u_{10}(\lambda) + \frac{16}{9\sqrt{5}}u_{12}(\lambda) + \frac{8}{9\sqrt{5}}u_{13}(\lambda)$ $+\frac{16}{9\sqrt{5}}u_{16}(\lambda) - \frac{16}{3\sqrt{5}}u_{17}(\lambda) + \frac{8}{3\sqrt{5}}u_{18}(\lambda) + \frac{8}{3\sqrt{5}}u_{19}(\lambda) + \frac{4}{3\sqrt{5}}u_{20}(\lambda)$ $-\frac{32}{3\sqrt{5}}u_{22}(\lambda) + \frac{32}{3\sqrt{5}}u_{23}(\lambda) + \frac{4\sqrt{5}}{9}u_{26}(\lambda) + \frac{128}{9\sqrt{5}}u_{29}(\lambda)$
$f^{0,4}(\lambda)$	$\frac{2}{5}u_1(\lambda) + \frac{4}{45}u_2(\lambda) + \frac{8}{5}u_3(\lambda) + \frac{52}{45}u_4(\lambda) + \frac{8}{45}u_5(\lambda) + \frac{4}{15}u_6(\lambda)$ $+\frac{8}{45}u_7(\lambda) + \frac{16}{15}u_8(\lambda) - \frac{16}{45}u_9(\lambda) - \frac{44}{45}u_{10}(\lambda) + \frac{16}{15}u_{11}(\lambda) + \frac{8}{9}u_{12}(\lambda)$ $-\frac{16}{45}u_{13}(\lambda) + \frac{16}{3}u_{14}(\lambda) - \frac{32}{15}u_{15}(\lambda) - \frac{16}{9}u_{16}(\lambda) - \frac{16}{5}u_{24}(\lambda) - \frac{8}{5}u_{25}(\lambda)$ $-\frac{16}{45}u_{26}(\lambda) - \frac{32}{15}u_{27}(\lambda) + \frac{32}{15}u_{28}(\lambda) + \frac{32}{45}u_{29}(\lambda)$
$f^{0,5}(\lambda)$	$\frac{8}{3\sqrt{35}}u_1(\lambda) + \frac{8}{9\sqrt{35}}u_2(\lambda) - \frac{16\sqrt{7}}{9\sqrt{5}}u_4(\lambda) - \frac{8}{9\sqrt{35}}u_5(\lambda) - \frac{8}{9\sqrt{35}}u_7(\lambda)$ $-\frac{16}{3\sqrt{35}}u_8(\lambda) + \frac{8\sqrt{5}}{9\sqrt{7}}u_9(\lambda) - \frac{64}{9\sqrt{35}}u_{10}(\lambda) + \frac{8}{3\sqrt{35}}u_{11}(\lambda) - \frac{8\sqrt{5}}{9\sqrt{7}}u_{12}(\lambda)$ $-\frac{8}{9\sqrt{35}}u_{13}(\lambda) + \frac{16}{3\sqrt{35}}u_{14}(\lambda) + \frac{8}{3\sqrt{35}}u_{15}(\lambda) - \frac{88}{9\sqrt{35}}u_{16}(\lambda)$ $+\frac{16}{3\sqrt{35}}u_{17}(\lambda) - \frac{32}{3\sqrt{35}}u_{18}(\lambda) + \frac{16}{3\sqrt{35}}u_{19}(\lambda) + \frac{8}{3\sqrt{35}}u_{20}(\lambda)$ $+\frac{16\sqrt{5}}{21\sqrt{7}}u_{22}(\lambda) + \frac{16}{3\sqrt{35}}u_{23}(\lambda) - \frac{16\sqrt{5}}{9\sqrt{7}}u_{26}(\lambda) + \frac{128}{3\sqrt{35}}u_{27}(\lambda) + \frac{8\sqrt{5}}{3\sqrt{7}}u_{28}(\lambda)$ $-\frac{176}{9\sqrt{35}}u_{29}(\lambda)$
$f^{0,6}(\lambda)$	$\frac{4\sqrt{5}}{21}u_1(\lambda) + \frac{8}{63\sqrt{5}}u_2(\lambda) + \frac{4}{7\sqrt{5}}u_3(\lambda) - \frac{8\sqrt{5}}{9}u_4(\lambda) + \frac{16}{63\sqrt{5}}u_5(\lambda)$ $+\frac{8}{3\sqrt{5}}u_6(\lambda) + \frac{16}{63\sqrt{5}}u_7(\lambda) + \frac{16}{21\sqrt{5}}u_8(\lambda) - \frac{32}{63\sqrt{5}}u_9(\lambda) - \frac{8\sqrt{5}}{9}u_{10}(\lambda)$ $+\frac{16}{21\sqrt{5}}u_{11}(\lambda) - \frac{16}{9\sqrt{5}}u_{12}(\lambda) - \frac{32}{63\sqrt{5}}u_{13}(\lambda) - \frac{16}{3\sqrt{5}}u_{14}(\lambda) - \frac{32}{21\sqrt{5}}u_{15}(\lambda)$ $+\frac{32}{9\sqrt{5}}u_{16}(\lambda) + \frac{136}{21\sqrt{5}}u_{17}(\lambda) - \frac{8}{21\sqrt{5}}u_{18}(\lambda) + \frac{16}{21\sqrt{5}}u_{19}(\lambda) + \frac{8}{21\sqrt{5}}u_{20}(\lambda)$ $-\frac{96}{7\sqrt{5}}u_{21}(\lambda) + \frac{32}{3\sqrt{5}}u_{22}(\lambda) - \frac{32}{21\sqrt{5}}u_{23}(\lambda) + \frac{8\sqrt{5}}{7}u_{24}(\lambda) - \frac{16}{7\sqrt{5}}u_{25}(\lambda)$ $+\frac{32\sqrt{5}}{63}u_{26}(\lambda) + \frac{32\sqrt{5}}{21}u_{27}(\lambda) - \frac{32\sqrt{5}}{21}u_{28}(\lambda) - \frac{64\sqrt{5}}{63}u_{29}(\lambda)$
$f^{0,7}(\lambda)$	$-\frac{59}{105}u_1(\lambda) + \frac{8}{105}u_2(\lambda) + \frac{64}{105}u_3(\lambda) + \frac{24}{5}u_4(\lambda) + \frac{16}{105}u_5(\lambda) - \frac{6}{5}u_6(\lambda)$ $+\frac{16}{105}u_7(\lambda) - \frac{64}{105}u_8(\lambda) - \frac{32}{105}u_9(\lambda) + \frac{12}{5}u_{10}(\lambda) - \frac{64}{105}u_{11}(\lambda) - \frac{32}{105}u_{13}(\lambda)$ $+\frac{128}{105}u_{15}(\lambda) + \frac{16}{7}u_{17}(\lambda) + \frac{4}{7}u_{18}(\lambda) + \frac{4}{21}u_{19}(\lambda) + \frac{16}{21}u_{20}(\lambda) + \frac{32}{7}u_{21}(\lambda)$ $-\frac{64}{21}u_{23}(\lambda) + \frac{64}{35}u_{24}(\lambda) + \frac{92}{35}u_{25}(\lambda) + \frac{16}{35}u_{26}(\lambda) - \frac{64}{35}u_{27}(\lambda) + \frac{64}{35}u_{28}(\lambda)$ $-\frac{32}{35}u_{29}(\lambda)$
$f^{2,1}(\lambda)$	$\frac{2\sqrt{2}}{9}u_2(\lambda) - \frac{16\sqrt{2}}{9}u_4(\lambda) + \frac{4\sqrt{2}}{9}u_5(\lambda) + \frac{4\sqrt{2}}{9}u_7(\lambda) - \frac{8\sqrt{2}}{9}u_9(\lambda)$ $-\frac{16\sqrt{2}}{9}u_{10}(\lambda) - \frac{4\sqrt{2}}{9}u_{12}(\lambda) + \frac{10\sqrt{2}}{9}u_{13}(\lambda) + \frac{8\sqrt{2}}{9}u_{16}(\lambda) - \frac{8\sqrt{2}}{9}u_{26}(\lambda)$ $+\frac{16\sqrt{2}}{9}u_{29}(\lambda)$

Table 3.14 (*Cont.*)

Function	Value
$f^{2,2}(\lambda)$	$\frac{4\sqrt{2}}{9\sqrt{5}}u_2(\lambda) - \frac{56\sqrt{2}}{9\sqrt{5}}u_4(\lambda) - \frac{4\sqrt{2}}{9\sqrt{5}}u_5(\lambda) + \frac{8\sqrt{2}}{3\sqrt{5}}u_6(\lambda) - \frac{4\sqrt{2}}{9\sqrt{5}}u_7(\lambda)$ $- \frac{16\sqrt{2}}{3\sqrt{5}}u_8(\lambda) + \frac{4\sqrt{10}}{9}u_9(\lambda) - \frac{32\sqrt{2}}{9\sqrt{5}}u_{10}(\lambda) + \frac{8\sqrt{2}}{3\sqrt{5}}u_{11}(\lambda) + \frac{28\sqrt{2}}{9\sqrt{5}}u_{12}(\lambda)$ $- \frac{4\sqrt{2}}{9\sqrt{5}}u_{13}(\lambda) + \frac{16\sqrt{2}}{3\sqrt{5}}u_{14}(\lambda) + \frac{8\sqrt{2}}{3\sqrt{5}}u_{15}(\lambda) + \frac{4\sqrt{2}}{9\sqrt{5}}u_{16}(\lambda) + \frac{8\sqrt{2}}{9\sqrt{5}}u_{26}(\lambda)$ $+ \frac{32\sqrt{2}}{3\sqrt{5}}u_{27}(\lambda) - \frac{8\sqrt{2}}{3\sqrt{5}}u_{28}(\lambda) - \frac{8\sqrt{10}}{9}u_{29}(\lambda)$
$f^{2,3}(\lambda)$	$\frac{4}{9\sqrt{7}}u_2(\lambda) - \frac{80}{9\sqrt{7}}u_4(\lambda) - \frac{16}{9\sqrt{7}}u_5(\lambda) + \frac{16}{3\sqrt{7}}u_6(\lambda) + \frac{20}{9\sqrt{7}}u_7(\lambda)$ $+ \frac{16}{3\sqrt{7}}u_8(\lambda) - \frac{16}{9\sqrt{7}}u_9(\lambda) - \frac{32}{9\sqrt{7}}u_{10}(\lambda) + \frac{4}{3\sqrt{7}}u_{11}(\lambda) - \frac{44}{9\sqrt{7}}u_{12}(\lambda)$ $+ \frac{8}{9\sqrt{7}}u_{13}(\lambda) - \frac{16}{3\sqrt{7}}u_{14}(\lambda) + \frac{16}{3\sqrt{7}}u_{15}(\lambda) - \frac{80}{9\sqrt{7}}u_{16}(\lambda) + \frac{32}{9\sqrt{7}}u_{26}(\lambda)$ $- \frac{32}{3\sqrt{7}}u_{27}(\lambda) - \frac{16}{3\sqrt{7}}u_{28}(\lambda) + \frac{32}{9\sqrt{7}}u_{29}(\lambda)$
$f^{2,4}(\lambda)$	$- \frac{4\sqrt{2}}{3\sqrt{7}}u_1(\lambda) + \frac{4\sqrt{2}}{9\sqrt{7}}u_2(\lambda) + \frac{16\sqrt{14}}{9}u_4(\lambda) + \frac{2\sqrt{2}}{9\sqrt{7}}u_5(\lambda) - \frac{8\sqrt{2}}{3\sqrt{7}}u_6(\lambda)$ $- \frac{16\sqrt{2}}{9\sqrt{7}}u_7(\lambda) - \frac{16\sqrt{2}}{9\sqrt{7}}u_9(\lambda) + \frac{64\sqrt{2}}{9\sqrt{7}}u_{10}(\lambda) + \frac{16\sqrt{2}}{9\sqrt{7}}u_{12}(\lambda) + \frac{8\sqrt{2}}{9\sqrt{7}}u_{13}(\lambda)$ $+ \frac{16\sqrt{2}}{9\sqrt{7}}u_{16}(\lambda) - \frac{8\sqrt{2}}{3\sqrt{7}}u_{17}(\lambda) + \frac{4\sqrt{2}}{3\sqrt{7}}u_{18}(\lambda) + \frac{4\sqrt{2}}{3\sqrt{7}}u_{19}(\lambda) + \frac{2\sqrt{2}}{3\sqrt{7}}u_{20}(\lambda)$ $- \frac{16\sqrt{2}}{3\sqrt{7}}u_{22}(\lambda) + \frac{16\sqrt{2}}{3\sqrt{7}}u_{23}(\lambda) - \frac{4\sqrt{14}}{9}u_{26}(\lambda) - \frac{64\sqrt{2}}{9\sqrt{7}}u_{29}(\lambda)$
$f^{2,5}(\lambda)$	$- \frac{4}{\sqrt{35}}u_1(\lambda) + \frac{8}{9\sqrt{35}}u_2(\lambda) + \frac{8}{\sqrt{35}}u_3(\lambda) + \frac{8}{9\sqrt{35}}u_4(\lambda) + \frac{16}{9\sqrt{35}}u_5(\lambda)$ $- \frac{8}{\sqrt{35}}u_6(\lambda) + \frac{16}{9\sqrt{35}}u_7(\lambda) + \frac{8}{\sqrt{35}}u_8(\lambda) - \frac{32}{9\sqrt{35}}u_9(\lambda) + \frac{152}{9\sqrt{35}}u_{10}(\lambda)$ $+ \frac{8}{\sqrt{35}}u_{11}(\lambda) - \frac{16}{9\sqrt{35}}u_{12}(\lambda) - \frac{32}{9\sqrt{35}}u_{13}(\lambda) - \frac{24}{\sqrt{35}}u_{14}(\lambda) - \frac{16}{\sqrt{35}}u_{15}(\lambda)$ $+ \frac{32}{9\sqrt{35}}u_{16}(\lambda) - \frac{16}{\sqrt{35}}u_{24}(\lambda) + \frac{16}{\sqrt{35}}u_{25}(\lambda) - \frac{32}{9\sqrt{35}}u_{26}(\lambda) - \frac{16}{\sqrt{35}}u_{27}(\lambda)$ $+ \frac{16}{\sqrt{35}}u_{28}(\lambda) + \frac{64}{9\sqrt{35}}u_{29}(\lambda)$
$f^{2,6}(\lambda)$	$- \frac{2\sqrt{10}}{21}u_1(\lambda) + \frac{8\sqrt{2}}{21\sqrt{5}}u_2(\lambda) + \frac{16\sqrt{2}}{7\sqrt{5}}u_4(\lambda) - \frac{8\sqrt{2}}{21\sqrt{5}}u_5(\lambda) - \frac{4\sqrt{2}}{7\sqrt{5}}u_6(\lambda)$ $- \frac{8\sqrt{2}}{21\sqrt{5}}u_7(\lambda) + \frac{64\sqrt{2}}{21\sqrt{5}}u_8(\lambda) + \frac{8\sqrt{10}}{21}u_9(\lambda) + \frac{16\sqrt{2}}{21\sqrt{5}}u_{10}(\lambda) - \frac{32\sqrt{2}}{21\sqrt{5}}u_{11}(\lambda)$ $+ \frac{16\sqrt{2}}{21\sqrt{5}}u_{12}(\lambda) - \frac{8\sqrt{2}}{21\sqrt{5}}u_{13}(\lambda) - \frac{64\sqrt{2}}{21\sqrt{5}}u_{14}(\lambda) - \frac{32\sqrt{2}}{21\sqrt{5}}u_{15}(\lambda)$ $- \frac{32\sqrt{2}}{21\sqrt{5}}u_{16}(\lambda) + \frac{16\sqrt{10}}{21}u_{17}(\lambda) - \frac{4\sqrt{10}}{21}u_{18}(\lambda) - \frac{4\sqrt{10}}{7}u_{19}(\lambda)$ $+ \frac{8\sqrt{10}}{21}u_{20}(\lambda) - \frac{32\sqrt{10}}{21}u_{22}(\lambda) + \frac{16\sqrt{10}}{21}u_{23}(\lambda) + \frac{8\sqrt{2}}{3\sqrt{5}}u_{26}(\lambda)$ $+ \frac{32\sqrt{2}}{21\sqrt{5}}u_{27}(\lambda) + \frac{16\sqrt{2}}{3\sqrt{5}}u_{28}(\lambda) - \frac{8\sqrt{10}}{21}u_{29}(\lambda)$
$f^{2,7}(\lambda)$	$- \frac{8\sqrt{2}}{21}u_1(\lambda) + \frac{8\sqrt{2}}{63}u_2(\lambda) + \frac{272\sqrt{2}}{63}u_4(\lambda) - \frac{8\sqrt{2}}{63}u_5(\lambda) - \frac{32\sqrt{2}}{21}u_6(\lambda)$ $- \frac{8\sqrt{2}}{63}u_7(\lambda) - \frac{16\sqrt{2}}{21}u_8(\lambda) + \frac{40\sqrt{2}}{63}u_9(\lambda) + \frac{128\sqrt{2}}{63}u_{10}(\lambda) + \frac{8\sqrt{2}}{21}u_{11}(\lambda)$ $- \frac{40\sqrt{2}}{63}u_{12}(\lambda) - \frac{8\sqrt{2}}{63}u_{13}(\lambda) + \frac{16\sqrt{2}}{21}u_{14}(\lambda) + \frac{8\sqrt{2}}{21}u_{15}(\lambda) - \frac{88\sqrt{2}}{63}u_{16}(\lambda)$ $+ \frac{8\sqrt{2}}{21}u_{17}(\lambda) - \frac{16\sqrt{2}}{21}u_{18}(\lambda) + \frac{8\sqrt{2}}{21}u_{19}(\lambda) + \frac{4\sqrt{2}}{21}u_{20}(\lambda) + \frac{40\sqrt{2}}{21}u_{22}(\lambda)$ $+ \frac{8\sqrt{2}}{21}u_{23}(\lambda) + \frac{16\sqrt{2}}{9}u_{26}(\lambda) - \frac{64\sqrt{2}}{21}u_{27}(\lambda) - \frac{8\sqrt{2}}{3}u_{28}(\lambda) + \frac{16\sqrt{2}}{63}u_{29}(\lambda)$
$f^{2,8}(\lambda)$	$\frac{4\sqrt{2}}{21\sqrt{7}}u_1(\lambda) + \frac{8\sqrt{2}}{63\sqrt{7}}u_2(\lambda) + \frac{4\sqrt{2}}{7\sqrt{7}}u_3(\lambda) + \frac{104\sqrt{2}}{63\sqrt{7}}u_4(\lambda) + \frac{16\sqrt{2}}{63\sqrt{7}}u_5(\lambda)$ $+ \frac{8\sqrt{2}}{7\sqrt{7}}u_6(\lambda) + \frac{16\sqrt{2}}{63\sqrt{7}}u_7(\lambda) + \frac{16\sqrt{2}}{21\sqrt{7}}u_8(\lambda) - \frac{32\sqrt{2}}{63\sqrt{7}}u_9(\lambda) - \frac{88\sqrt{2}}{63\sqrt{7}}u_{10}(\lambda)$ $+ \frac{16\sqrt{2}}{21\sqrt{7}}u_{11}(\lambda) - \frac{16\sqrt{2}}{9\sqrt{7}}u_{12}(\lambda) - \frac{32\sqrt{2}}{63\sqrt{7}}u_{13}(\lambda) - \frac{16\sqrt{2}}{3\sqrt{7}}u_{14}(\lambda)$ $- \frac{32\sqrt{2}}{21\sqrt{7}}u_{15}(\lambda) + \frac{32\sqrt{2}}{9\sqrt{7}}u_{16}(\lambda) + \frac{68\sqrt{2}}{21\sqrt{7}}u_{17}(\lambda) - \frac{4\sqrt{2}}{21\sqrt{7}}u_{18}(\lambda)$ $+ \frac{8\sqrt{2}}{21\sqrt{7}}u_{19}(\lambda) + \frac{4\sqrt{2}}{21\sqrt{7}}u_{20}(\lambda) - \frac{48\sqrt{2}}{7\sqrt{7}}u_{21}(\lambda) + \frac{16\sqrt{2}}{3\sqrt{7}}u_{22}(\lambda)$ $- \frac{16\sqrt{2}}{21\sqrt{7}}u_{23}(\lambda) - \frac{8\sqrt{2}}{\sqrt{7}}u_{24}(\lambda) - \frac{16\sqrt{2}}{7\sqrt{7}}u_{25}(\lambda) - \frac{32\sqrt{2}}{9\sqrt{7}}u_{26}(\lambda) - \frac{32\sqrt{2}}{3\sqrt{7}}u_{27}(\lambda)$ $+ \frac{32\sqrt{2}}{3\sqrt{7}}u_{28}(\lambda) + \frac{64\sqrt{2}}{9\sqrt{7}}u_{29}(\lambda)$

Table 3.14 (*Cont.*)

Function	Value
$f^{2,9}(\lambda)$	$-\frac{32}{21\sqrt{35}}u_1(\lambda)+\frac{16}{21\sqrt{35}}u_2(\lambda)-\frac{32}{7\sqrt{35}}u_3(\lambda)+\frac{32}{7\sqrt{35}}u_4(\lambda)+\frac{32}{21\sqrt{35}}u_5(\lambda)$
	$-\frac{8}{7\sqrt{35}}u_6(\lambda)+\frac{32}{21\sqrt{35}}u_7(\lambda)-\frac{16}{21\sqrt{35}}u_8(\lambda)-\frac{64}{21\sqrt{35}}u_9(\lambda)$
	$+\frac{104}{21\sqrt{35}}u_{10}(\lambda)-\frac{16}{21\sqrt{35}}u_{11}(\lambda)-\frac{16}{3\sqrt{35}}u_{12}(\lambda)-\frac{64}{21\sqrt{35}}u_{13}(\lambda)$
	$+\frac{64}{3\sqrt{35}}u_{14}(\lambda)+\frac{32}{21\sqrt{35}}u_{15}(\lambda)+\frac{32}{3\sqrt{35}}u_{16}(\lambda)-\frac{64\sqrt{5}}{21\sqrt{7}}u_{17}(\lambda)$
	$+\frac{40\sqrt{5}}{21\sqrt{7}}u_{18}(\lambda)-\frac{8\sqrt{5}}{7\sqrt{7}}u_{19}(\lambda)+\frac{16\sqrt{5}}{21\sqrt{7}}u_{20}(\lambda)+\frac{32\sqrt{5}}{7\sqrt{7}}u_{21}(\lambda)+\frac{32\sqrt{5}}{3\sqrt{7}}u_{22}(\lambda)$
	$-\frac{64\sqrt{5}}{21\sqrt{7}}u_{23}(\lambda)+\frac{32}{\sqrt{35}}u_{24}(\lambda)+\frac{16}{7\sqrt{35}}u_{25}(\lambda)-\frac{32}{3\sqrt{35}}u_{26}(\lambda)+\frac{16}{3\sqrt{35}}u_{27}(\lambda)$
	$-\frac{16}{3\sqrt{35}}u_{28}(\lambda)+\frac{64}{3\sqrt{35}}u_{29}(\lambda)$
$f^{2,10}(\lambda)$	$\frac{206}{21\sqrt{77}}u_1(\lambda)+\frac{16}{21\sqrt{77}}u_2(\lambda)+\frac{128}{21\sqrt{77}}u_3(\lambda)-\frac{48\sqrt{11}}{7\sqrt{7}}u_4(\lambda)+\frac{32}{21\sqrt{77}}u_5(\lambda)$
	$+\frac{12\sqrt{11}}{7\sqrt{7}}u_6(\lambda)+\frac{32}{21\sqrt{77}}u_7(\lambda)-\frac{128}{21\sqrt{77}}u_8(\lambda)-\frac{64}{21\sqrt{77}}u_9(\lambda)$
	$-\frac{24\sqrt{11}}{7\sqrt{7}}u_{10}(\lambda)-\frac{128}{21\sqrt{77}}u_{11}(\lambda)-\frac{64}{21\sqrt{77}}u_{13}(\lambda)+\frac{256}{21\sqrt{77}}u_{15}(\lambda)$
	$-\frac{8\sqrt{11}}{7\sqrt{7}}u_{17}(\lambda)-\frac{2\sqrt{11}}{7\sqrt{7}}u_{18}(\lambda)+\frac{202}{21\sqrt{77}}u_{19}(\lambda)+\frac{136}{21\sqrt{77}}u_{20}(\lambda)$
	$-\frac{16\sqrt{11}}{7\sqrt{7}}u_{21}(\lambda)-\frac{544}{21\sqrt{77}}u_{23}(\lambda)-\frac{24\sqrt{11}}{7\sqrt{7}}u_{25}(\lambda)$
$f^{4,1}(\lambda)$	$\frac{4}{3\sqrt{35}}u_2(\lambda)+\frac{32}{3\sqrt{35}}u_4(\lambda)-\frac{16}{3\sqrt{35}}u_5(\lambda)-\frac{8}{3\sqrt{35}}u_6(\lambda)+\frac{4\sqrt{5}}{3\sqrt{7}}u_7(\lambda)$
	$-\frac{64}{3\sqrt{35}}u_8(\lambda)-\frac{16}{3\sqrt{35}}u_9(\lambda)+\frac{8}{\sqrt{35}}u_{10}(\lambda)-\frac{16}{3\sqrt{35}}u_{11}(\lambda)-\frac{16}{3\sqrt{35}}u_{12}(\lambda)$
	$+\frac{8}{3\sqrt{35}}u_{13}(\lambda)+\frac{64}{3\sqrt{35}}u_{14}(\lambda)-\frac{64}{3\sqrt{35}}u_{15}(\lambda)+\frac{32}{3\sqrt{35}}u_{16}(\lambda)$
	$+\frac{32}{3\sqrt{35}}u_{26}(\lambda)+\frac{128}{3\sqrt{35}}u_{27}(\lambda)+\frac{64}{3\sqrt{35}}u_{28}(\lambda)+\frac{32}{3\sqrt{35}}u_{29}(\lambda)$
$f^{4,2}(\lambda)$	$\frac{2\sqrt{2}}{3\sqrt{35}}u_1(\lambda)+\frac{4\sqrt{2}}{3\sqrt{35}}u_2(\lambda)+\frac{2\sqrt{2}}{3\sqrt{35}}u_5(\lambda)+\frac{4\sqrt{2}}{3\sqrt{35}}u_6(\lambda)-\frac{16\sqrt{2}}{3\sqrt{35}}u_7(\lambda)$
	$-\frac{16\sqrt{2}}{3\sqrt{35}}u_9(\lambda)+\frac{8\sqrt{2}}{3\sqrt{35}}u_{10}(\lambda)+\frac{16\sqrt{2}}{3\sqrt{35}}u_{12}(\lambda)+\frac{8\sqrt{2}}{3\sqrt{35}}u_{13}(\lambda)+\frac{16\sqrt{2}}{3\sqrt{35}}u_{16}(\lambda)$
	$+\frac{32\sqrt{2}}{3\sqrt{35}}u_{17}(\lambda)-\frac{16\sqrt{2}}{3\sqrt{35}}u_{18}(\lambda)-\frac{16\sqrt{2}}{3\sqrt{35}}u_{19}(\lambda)-\frac{8\sqrt{2}}{3\sqrt{35}}u_{20}(\lambda)$
	$+\frac{64\sqrt{2}}{3\sqrt{35}}u_{22}(\lambda)-\frac{64\sqrt{2}}{3\sqrt{35}}u_{23}(\lambda)+\frac{16\sqrt{2}}{\sqrt{35}}u_{29}(\lambda)$
$f^{4,3}(\lambda)$	$\frac{2}{5\sqrt{7}}u_1(\lambda)+\frac{8}{15\sqrt{7}}u_2(\lambda)-\frac{32}{5\sqrt{7}}u_3(\lambda)+\frac{64}{15\sqrt{7}}u_4(\lambda)+\frac{16}{15\sqrt{7}}u_5(\lambda)$
	$-\frac{16}{15\sqrt{7}}u_6(\lambda)+\frac{16}{15\sqrt{7}}u_7(\lambda)+\frac{16}{15\sqrt{7}}u_8(\lambda)-\frac{32}{15\sqrt{7}}u_9(\lambda)+\frac{4}{5\sqrt{7}}u_{10}(\lambda)$
	$+\frac{16}{21\sqrt{7}}u_{11}(\lambda)+\frac{8}{3\sqrt{7}}u_{12}(\lambda)-\frac{32}{15\sqrt{7}}u_{13}(\lambda)-\frac{32}{3\sqrt{7}}u_{14}(\lambda)-\frac{32}{15\sqrt{7}}u_{15}(\lambda)$
	$-\frac{16}{3\sqrt{7}}u_{16}(\lambda)+\frac{64}{5\sqrt{7}}u_{24}(\lambda)-\frac{8}{5\sqrt{7}}u_{25}(\lambda)-\frac{32}{15\sqrt{7}}u_{26}(\lambda)-\frac{32}{15\sqrt{7}}u_{27}(\lambda)$
	$+\frac{32}{15\sqrt{7}}u_{28}(\lambda)+\frac{64}{15\sqrt{7}}u_{29}(\lambda)$
$f^{4,4}(\lambda)$	$\frac{4\sqrt{2}}{21\sqrt{5}}u_1(\lambda)+\frac{8\sqrt{2}}{21\sqrt{5}}u_2(\lambda)+\frac{16\sqrt{2}}{7\sqrt{5}}u_4(\lambda)-\frac{8\sqrt{2}}{21\sqrt{5}}u_5(\lambda)-\frac{8\sqrt{10}}{21}u_6(\lambda)$
	$-\frac{8\sqrt{2}}{21\sqrt{5}}u_7(\lambda)-\frac{16\sqrt{2}}{7\sqrt{5}}u_8(\lambda)+\frac{8\sqrt{10}}{21}u_9(\lambda)+\frac{16\sqrt{2}}{21\sqrt{5}}u_{10}(\lambda)+\frac{8\sqrt{2}}{7\sqrt{5}}u_{11}(\lambda)$
	$-\frac{8\sqrt{10}}{21}u_{12}(\lambda)-\frac{8\sqrt{2}}{21\sqrt{5}}u_{13}(\lambda)+\frac{16\sqrt{2}}{7\sqrt{5}}u_{14}(\lambda)+\frac{8\sqrt{2}}{7\sqrt{5}}u_{15}(\lambda)-\frac{88\sqrt{2}}{21\sqrt{5}}u_{16}(\lambda)$
	$-\frac{32\sqrt{2}}{21\sqrt{5}}u_{17}(\lambda)+\frac{64\sqrt{2}}{21\sqrt{5}}u_{18}(\lambda)-\frac{32\sqrt{2}}{21\sqrt{5}}u_{19}(\lambda)-\frac{16\sqrt{2}}{21\sqrt{5}}u_{20}(\lambda)$
	$-\frac{32\sqrt{10}}{21}u_{22}(\lambda)-\frac{32\sqrt{2}}{21\sqrt{5}}u_{23}(\lambda)+\frac{48\sqrt{2}}{7\sqrt{5}}u_{27}(\lambda)-\frac{32\sqrt{2}}{7\sqrt{5}}u_{29}(\lambda)$
$f^{4,5}(\lambda)$	$\frac{12}{7\sqrt{11}}u_1(\lambda)+\frac{16}{21\sqrt{11}}u_2(\lambda)-\frac{32\sqrt{11}}{21}u_4(\lambda)-\frac{16}{21\sqrt{11}}u_5(\lambda)+\frac{8\sqrt{11}}{21}u_6(\lambda)$
	$-\frac{16}{21\sqrt{11}}u_7(\lambda)+\frac{128}{21\sqrt{11}}u_8(\lambda)+\frac{80}{21\sqrt{11}}u_9(\lambda)-\frac{64}{7\sqrt{11}}u_{10}(\lambda)$
	$-\frac{64}{21\sqrt{11}}u_{11}(\lambda)+\frac{32}{21\sqrt{11}}u_{12}(\lambda)-\frac{16}{21\sqrt{11}}u_{13}(\lambda)-\frac{128}{21\sqrt{11}}u_{14}(\lambda)-\frac{64}{21\sqrt{11}}u_{15}(\lambda)$
	$-\frac{64}{21\sqrt{11}}u_{16}(\lambda)+\frac{16}{7\sqrt{11}}u_{17}(\lambda)-\frac{4}{7\sqrt{11}}u_{18}(\lambda)-\frac{12}{7\sqrt{11}}u_{19}(\lambda)$
	$+\frac{8}{7\sqrt{11}}u_{20}(\lambda)-\frac{32}{7\sqrt{11}}u_{22}(\lambda)+\frac{16}{7\sqrt{11}}u_{23}(\lambda)-\frac{16}{3\sqrt{11}}u_{26}(\lambda)$
	$-\frac{832}{21\sqrt{11}}u_{27}(\lambda)-\frac{32}{3\sqrt{11}}u_{28}(\lambda)-\frac{304}{21\sqrt{11}}u_{29}(\lambda)$

Table 3.14 (*Cont.*)

Function	Value
$f^{4,6}(\lambda)$	$\frac{8\sqrt{10}}{21\sqrt{7}}u_1(\lambda)+\frac{8\sqrt{2}}{21\sqrt{35}}u_2(\lambda)+\frac{12\sqrt{2}}{7\sqrt{35}}u_3(\lambda)-\frac{8\sqrt{10}}{7\sqrt{7}}u_4(\lambda)+\frac{16\sqrt{2}}{21\sqrt{35}}u_5(\lambda)$ $+\frac{128\sqrt{2}}{21\sqrt{35}}u_6(\lambda)+\frac{16\sqrt{2}}{21\sqrt{35}}u_7(\lambda)+\frac{16\sqrt{2}}{7\sqrt{35}}u_8(\lambda)-\frac{32\sqrt{2}}{21\sqrt{35}}u_9(\lambda)$ $-\frac{40\sqrt{10}}{21\sqrt{7}}u_{10}(\lambda)+\frac{16\sqrt{2}}{7\sqrt{35}}u_{11}(\lambda)-\frac{16\sqrt{2}}{3\sqrt{35}}u_{12}(\lambda)-\frac{32\sqrt{2}}{21\sqrt{35}}u_{13}(\lambda)$ $-\frac{16\sqrt{2}}{\sqrt{35}}u_{14}(\lambda)-\frac{32\sqrt{2}}{7\sqrt{35}}u_{15}(\lambda)+\frac{32\sqrt{2}}{3\sqrt{35}}u_{16}(\lambda)-\frac{272\sqrt{2}}{21\sqrt{35}}u_{17}(\lambda)$ $+\frac{16\sqrt{2}}{21\sqrt{35}}u_{18}(\lambda)-\frac{32\sqrt{2}}{21\sqrt{35}}u_{19}(\lambda)-\frac{16\sqrt{2}}{21\sqrt{35}}u_{20}(\lambda)+\frac{192\sqrt{2}}{7\sqrt{35}}u_{21}(\lambda)-\frac{64\sqrt{2}}{3\sqrt{35}}u_{22}(\lambda)$ $+\frac{64\sqrt{2}}{21\sqrt{35}}u_{23}(\lambda)-\frac{48\sqrt{2}}{7\sqrt{35}}u_{25}(\lambda)$
$f^{4,7}(\lambda)$	$\frac{8\sqrt{2}}{7\sqrt{77}}u_1(\lambda)+\frac{16\sqrt{2}}{21\sqrt{77}}u_2(\lambda)-\frac{32\sqrt{2}}{7\sqrt{77}}u_3(\lambda)-\frac{32\sqrt{22}}{21\sqrt{7}}u_4(\lambda)+\frac{32\sqrt{2}}{21\sqrt{77}}u_5(\lambda)$ $+\frac{8\sqrt{22}}{21\sqrt{7}}u_6(\lambda)+\frac{32\sqrt{2}}{21\sqrt{77}}u_7(\lambda)-\frac{16\sqrt{2}}{21\sqrt{77}}u_8(\lambda)-\frac{64\sqrt{2}}{21\sqrt{77}}u_9(\lambda)-\frac{40\sqrt{2}}{7\sqrt{77}}u_{10}(\lambda)$ $-\frac{16\sqrt{2}}{21\sqrt{77}}u_{11}(\lambda)-\frac{16\sqrt{2}}{3\sqrt{77}}u_{12}(\lambda)-\frac{64\sqrt{2}}{21\sqrt{77}}u_{13}(\lambda)+\frac{64\sqrt{2}}{3\sqrt{77}}u_{14}(\lambda)$ $+\frac{32\sqrt{2}}{21\sqrt{77}}u_{15}(\lambda)+\frac{32\sqrt{2}}{3\sqrt{77}}u_{16}(\lambda)-\frac{32\sqrt{2}}{7\sqrt{77}}u_{17}(\lambda)+\frac{20\sqrt{2}}{7\sqrt{77}}u_{18}(\lambda)$ $-\frac{12\sqrt{2}}{7\sqrt{77}}u_{19}(\lambda)+\frac{8\sqrt{2}}{7\sqrt{77}}u_{20}(\lambda)+\frac{48\sqrt{2}}{7\sqrt{77}}u_{21}(\lambda)+\frac{16\sqrt{2}}{\sqrt{77}}u_{22}(\lambda)$ $-\frac{32\sqrt{2}}{7\sqrt{77}}u_{23}(\lambda)-\frac{32\sqrt{2}}{\sqrt{77}}u_{24}(\lambda)+\frac{16\sqrt{2}}{7\sqrt{77}}u_{25}(\lambda)+\frac{32\sqrt{2}}{3\sqrt{77}}u_{26}(\lambda)$ $-\frac{16\sqrt{2}}{3\sqrt{77}}u_{27}(\lambda)+\frac{16\sqrt{2}}{3\sqrt{77}}u_{28}(\lambda)-\frac{64\sqrt{2}}{3\sqrt{77}}u_{29}(\lambda)$
$f^{4,8}(\lambda)$	$-\frac{536\sqrt{2}}{35\sqrt{1001}}u_1(\lambda)+\frac{72\sqrt{2}}{35\sqrt{1001}}u_2(\lambda)+\frac{576\sqrt{2}}{35\sqrt{1001}}u_3(\lambda)+\frac{32\sqrt{286}}{35\sqrt{7}}u_4(\lambda)$ $+\frac{144\sqrt{2}}{35\sqrt{1001}}u_5(\lambda)-\frac{8\sqrt{286}}{35\sqrt{7}}u_6(\lambda)+\frac{144\sqrt{2}}{35\sqrt{1001}}u_7(\lambda)-\frac{576\sqrt{2}}{35\sqrt{1001}}u_8(\lambda)$ $-\frac{288\sqrt{2}}{35\sqrt{1001}}u_9(\lambda)-\frac{16\sqrt{286}}{35\sqrt{7}}u_{10}(\lambda)-\frac{576\sqrt{2}}{35\sqrt{1001}}u_{11}(\lambda)-\frac{288\sqrt{2}}{35\sqrt{1001}}u_{13}(\lambda)$ $+\frac{1152\sqrt{2}}{35\sqrt{1001}}u_{15}(\lambda)-\frac{48\sqrt{26}}{7\sqrt{77}}u_{17}(\lambda)-\frac{12\sqrt{26}}{7\sqrt{77}}u_{18}(\lambda)+\frac{228\sqrt{2}}{7\sqrt{1001}}u_{19}(\lambda)$ $+\frac{72\sqrt{2}}{7\sqrt{1001}}u_{20}(\lambda)-\frac{96\sqrt{26}}{7\sqrt{77}}u_{21}(\lambda)-\frac{288\sqrt{2}}{7\sqrt{1001}}u_{23}(\lambda)-\frac{32\sqrt{26}}{5\sqrt{77}}u_{24}(\lambda)$ $+\frac{148\sqrt{26}}{35\sqrt{77}}u_{25}(\lambda)-\frac{8\sqrt{26}}{5\sqrt{77}}u_{26}(\lambda)+\frac{32\sqrt{26}}{5\sqrt{77}}u_{27}(\lambda)-\frac{32\sqrt{26}}{5\sqrt{77}}u_{28}(\lambda)$ $+\frac{16\sqrt{26}}{5\sqrt{77}}u_{29}(\lambda)$
$f^{6,1}(\lambda)$	$-\frac{4}{3\sqrt{77}}u_1(\lambda)+\frac{8}{3\sqrt{77}}u_2(\lambda)-\frac{8}{3\sqrt{77}}u_5(\lambda)-\frac{8}{3\sqrt{77}}u_7(\lambda)+\frac{64}{3\sqrt{77}}u_8(\lambda)$ $+\frac{40}{3\sqrt{77}}u_9(\lambda)-\frac{8}{3\sqrt{77}}u_{10}(\lambda)-\frac{32}{3\sqrt{77}}u_{11}(\lambda)+\frac{16}{3\sqrt{77}}u_{12}(\lambda)-\frac{8}{3\sqrt{77}}u_{13}(\lambda)$ $-\frac{64}{3\sqrt{77}}u_{14}(\lambda)-\frac{32}{3\sqrt{77}}u_{15}(\lambda)-\frac{32}{3\sqrt{77}}u_{16}(\lambda)-\frac{64}{3\sqrt{77}}u_{17}(\lambda)$ $+\frac{16}{3\sqrt{77}}u_{18}(\lambda)+\frac{16}{\sqrt{77}}u_{19}(\lambda)-\frac{32}{3\sqrt{77}}u_{20}(\lambda)+\frac{128}{3\sqrt{77}}u_{22}(\lambda)-\frac{64}{3\sqrt{77}}u_{23}(\lambda)$ $+\frac{32}{3\sqrt{77}}u_{26}(\lambda)-\frac{64}{3\sqrt{77}}u_{27}(\lambda)+\frac{64}{3\sqrt{77}}u_{28}(\lambda)-\frac{64}{3\sqrt{77}}u_{29}(\lambda)$
$f^{6,2}(\lambda)$	$-\frac{10\sqrt{2}}{21\sqrt{11}}u_1(\lambda)+\frac{8\sqrt{2}}{21\sqrt{11}}u_2(\lambda)-\frac{16\sqrt{2}}{7\sqrt{11}}u_3(\lambda)+\frac{16\sqrt{2}}{21\sqrt{11}}u_5(\lambda)+\frac{16\sqrt{2}}{21\sqrt{11}}u_7(\lambda)$ $-\frac{8\sqrt{2}}{21\sqrt{11}}u_8(\lambda)-\frac{32\sqrt{2}}{21\sqrt{11}}u_9(\lambda)+\frac{4\sqrt{2}}{3\sqrt{11}}u_{10}(\lambda)-\frac{8\sqrt{2}}{21\sqrt{11}}u_{11}(\lambda)$ $-\frac{8\sqrt{2}}{3\sqrt{11}}u_{12}(\lambda)-\frac{32\sqrt{2}}{21\sqrt{11}}u_{13}(\lambda)+\frac{32\sqrt{2}}{3\sqrt{11}}u_{14}(\lambda)+\frac{16\sqrt{2}}{3\sqrt{11}}u_{15}(\lambda)$ $+\frac{16\sqrt{2}}{3\sqrt{11}}u_{16}(\lambda)+\frac{128\sqrt{2}}{21\sqrt{11}}u_{17}(\lambda)-\frac{80\sqrt{2}}{21\sqrt{11}}u_{18}(\lambda)+\frac{16\sqrt{2}}{7\sqrt{11}}u_{19}(\lambda)$ $-\frac{32\sqrt{2}}{21\sqrt{11}}u_{20}(\lambda)-\frac{64\sqrt{2}}{7\sqrt{11}}u_{21}(\lambda)-\frac{64\sqrt{2}}{3\sqrt{11}}u_{22}(\lambda)+\frac{128\sqrt{2}}{21\sqrt{11}}u_{23}(\lambda)$ $+\frac{64\sqrt{2}}{7\sqrt{11}}u_{24}(\lambda)+\frac{8\sqrt{2}}{7\sqrt{11}}u_{25}(\lambda)-\frac{64\sqrt{2}}{21\sqrt{11}}u_{26}(\lambda)+\frac{32\sqrt{2}}{21\sqrt{11}}u_{27}(\lambda)$ $-\frac{32\sqrt{2}}{21\sqrt{11}}u_{28}(\lambda)+\frac{128\sqrt{2}}{21\sqrt{11}}u_{29}(\lambda)$

Table 3.14 (*Cont.*)

Function	Value
$f^{6,3}(\lambda)$	$\frac{8}{21\sqrt{55}}u_1(\lambda)+\frac{16}{21\sqrt{55}}u_2(\lambda)+\frac{128}{21\sqrt{55}}u_3(\lambda)+\frac{32}{21\sqrt{55}}u_5(\lambda)+\frac{32}{21\sqrt{55}}u_7(\lambda)$ $-\frac{128}{21\sqrt{55}}u_8(\lambda)-\frac{64}{21\sqrt{55}}u_9(\lambda)-\frac{128}{21\sqrt{55}}u_{11}(\lambda)-\frac{64}{21\sqrt{55}}u_{13}(\lambda)$ $+\frac{256}{21\sqrt{55}}u_{15}(\lambda)+\frac{32\sqrt{5}}{7\sqrt{11}}u_{17}(\lambda)+\frac{8\sqrt{5}}{7\sqrt{11}}u_{18}(\lambda)-\frac{128}{21\sqrt{55}}u_{19}(\lambda)$ $-\frac{8}{21\sqrt{55}}u_{20}(\lambda)+\frac{64\sqrt{5}}{7\sqrt{11}}u_{21}(\lambda)+\frac{32}{21\sqrt{55}}u_{23}(\lambda)-\frac{64\sqrt{5}}{7\sqrt{11}}u_{24}(\lambda)$ $-\frac{8\sqrt{5}}{7\sqrt{11}}u_{25}(\lambda)-\frac{16\sqrt{5}}{7\sqrt{11}}u_{26}(\lambda)+\frac{64\sqrt{5}}{7\sqrt{11}}u_{27}(\lambda)-\frac{64\sqrt{5}}{7\sqrt{11}}u_{28}(\lambda)$ $+\frac{32\sqrt{5}}{7\sqrt{11}}u_{29}(\lambda)$
$f^{8,1}(\lambda)$	$\frac{4\sqrt{2}}{3\sqrt{715}}u_1(\lambda)+\frac{8\sqrt{2}}{3\sqrt{715}}u_2(\lambda)+\frac{64\sqrt{2}}{3\sqrt{715}}u_3(\lambda)+\frac{16\sqrt{2}}{3\sqrt{715}}u_5(\lambda)+\frac{16\sqrt{2}}{3\sqrt{715}}u_7(\lambda)$ $-\frac{64\sqrt{2}}{3\sqrt{715}}u_8(\lambda)-\frac{32\sqrt{2}}{3\sqrt{715}}u_9(\lambda)-\frac{64\sqrt{2}}{3\sqrt{715}}u_{11}(\lambda)-\frac{32\sqrt{2}}{3\sqrt{715}}u_{13}(\lambda)$ $+\frac{128\sqrt{2}}{3\sqrt{715}}u_{15}(\lambda)-\frac{64\sqrt{2}}{3\sqrt{715}}u_{19}(\lambda)-\frac{64\sqrt{2}}{3\sqrt{715}}u_{20}(\lambda)+\frac{256\sqrt{2}}{3\sqrt{715}}u_{23}(\lambda)$

Let $<$ be the lexicographic order on the sequences $tuijkl$, where $ijkl$ are indices that numerate the 21 component of the elasticity tensor, $t\geq 0$, and $-t\leq u\leq t$. Consider the infinite symmetric positive definite matrices given by

$$b^{t'u'i'j'k'l'}_{tuijkl}(m)=\mathrm{i}^{t'-t}\sqrt{(2t+1)(2t'+1)}\sum_{n=0}^{4}\frac{1}{4n+1}g^{0[0,0]}_{2n[t,t']}$$
$$\times\sum_{q=1}^{m_n}a_{nqm}\sum_{v=-2n}^{2n}\boldsymbol{T}^{2n,q,v}_{i\cdots l'}g^{v[u,u']}_{2n[t,t']}$$

with $1\leq m\leq 13$. Let $L(m)$ be the infinite lower triangular matrices of the Cholesky factorisation of the matrices $b^{t'u'i'j'k'l'}_{tuijkl}(m)$ constructed in Hansen (2010). Let $Z'_{mtuijkl}$ be the sequence of centred scattered random measures with Φ_m as their control measures. Define

$$Z_{mtuijkl}=\sum_{(t'u'i'j'k'l')\leq(tuijkl)}L^{t'u'i'j'k'l'}_{tuijkl}(m)Z'_{mtuijkl}.$$

Theorem 36. *The one-point correlation tensor of a homogeneous and* $(\mathrm{O}(3),2\rho_0\oplus 2\rho_2\oplus\rho_4)$*-isotropic random field* $\boldsymbol{C}(\boldsymbol{x})$ *has the form*

$$\langle\boldsymbol{C}(\boldsymbol{x})\rangle=C_1\boldsymbol{T}^{0,1}+C_2\boldsymbol{T}^{0,2},\qquad C_m\in\mathbb{R},$$

where the tensors $\boldsymbol{T}^{0,1}$ *and* $\boldsymbol{T}^{0,2}$ *are given by* (3.65) *and* (3.67). *Its two-point correlation tensor has the form*

$$\langle\boldsymbol{C}(\boldsymbol{x}),\boldsymbol{C}(\boldsymbol{y})\rangle=\sum_{n=1}^{3}\int_0^\infty\sum_{q=1}^{29}N_{nq}(\lambda,\rho)L^q_{iikli'j'k'l'}(\boldsymbol{y}-\boldsymbol{x})\,\mathrm{d}\Phi_n(\lambda).\tag{3.98}$$

The functions $N_{nq}(\lambda,\rho)$ *are given in Malyarenko & Ostoja-Starzewski (2016a, Table 5). The measures* $\Phi_n(\lambda)$ *satisfy the condition*

$$\Phi_2(\{0\}) = 2\Phi_3(\{0\}). \tag{3.99}$$

The spectral expansion of the field has the form

$$\begin{aligned}\mathsf{C}_{ijkl}(\rho,\theta,\varphi) &= C_1\delta_{ij}\delta_{kl} + C_2(\delta_{ik}\delta_{jl} + \delta_{il}\delta_{jk})\\ &\quad + 2\sqrt{\pi}\sum_{m=1}^{13}\sum_{t=0}^{\infty}\sum_{u=-t}^{t}\int_0^\infty j_t(\lambda\rho)\,\mathrm{d}Z_{mtuijkl}(\lambda)S_t^u(\theta,\varphi).\end{aligned}$$

Proof. Equations (3.65) and (3.71) show that the tensors $\boldsymbol{T}^{0,1}$ and $\boldsymbol{T}^{0,2}$ are linear combinations of the tensors $\delta_{ij}\delta_{kl}$ and $\delta_{ik}\delta_{jl}+\delta_{il}\delta_{jk}$. It follows that the expected value of the field may be represented as a linear combination of either the first or the second pair of tensors. In the second variant, the constants C_1 and C_2 have physical sense: they are *Lamé parameters.*

To prove (3.99), note that

$$\begin{aligned}\mathrm{d}\Phi_1(\lambda) &= (u_1(\lambda)+\cdots+u_6(\lambda))\,\mathrm{d}\Omega\,\mathrm{d}\Phi(\lambda),\\ \mathrm{d}\Phi_2(\lambda) &= (u_{17}(\lambda)+\cdots+u_{20}(\lambda))\,\mathrm{d}\Omega\,\mathrm{d}\Phi(\lambda),\\ \mathrm{d}\Phi_3(\lambda) &= (u_{24}(\lambda)+\cdots+u_{26}(\lambda))\,\mathrm{d}\Omega\,\mathrm{d}\Phi(\lambda).\end{aligned}$$

Using Table 3.13, we obtain

$$\begin{aligned}u_{17}(0)+\cdots+u_{20}(0) &= \frac{1}{2\sqrt{5}}f^{0,3}(0) + \frac{11}{28\sqrt{5}}f^{0,6}(0) + \frac{2}{7}f^{0,7}(0),\\ u_{24}(0)+\cdots+u_{26}(0) &= \frac{1}{4\sqrt{5}}f^{0,3}(0) + \frac{11}{56\sqrt{5}}f^{0,6}(0) + \frac{1}{7}f^{0,7}(0),\end{aligned}$$

which proves (3.99). The spectral expansion follows from Karhunen's theorem. □

3.9 Bibliographical Remarks

Let G be a topological group continuously acting on a topological space T. There exist precise links between the following areas:

- the theory of G-invariant positive-definite functions on T;
- the theory of G-invariant random fields on T;
- the theory of orthogonal and unitary representations of G.

These links are described in the books by Diaconis (1988) and Hannan (1965). For example, Theorem 14 was proved by Schoenberg (1938) as a result about invariant positive-definite functions.

Let $r \geq 2$ be a positive integer. The tensor power $V^{\otimes r}$ carries the orthogonal representations of two different groups: a subgroup G of the group $\mathrm{O}(V)$ and a subgroup Σ of the group Σ_r. The representation ρ_r of G is the rth tensor power of its defining representation, while the representation σ of Σ acts on V^r by (2.7). It is clear that $\rho_r(g)$ commutes with $\sigma_r(\sigma)$ for all $g \in G$ and $\sigma \in \Sigma$. Then, the mapping

$$(\sigma, g) \mapsto \rho_r(g)\sigma_r(\sigma) \tag{3.100}$$

is an orthogonal representation of the Cartesian product $\Sigma \times G$. This representation is equivalent to a direct sum of outer tensor products of the irreducible orthogonal representations of the groups Σ and G.

Consider a particular case, when $\Sigma = \Sigma_r$. Fix an irreducible representation τ of the group Σ and consider the direct sum W_τ of all spaces where the outer tensor products of the representation τ by an irreducible orthogonal representation of the group G act. If $\tau = \tau_0$ is the trivial representation, then W_{τ_0} is the linear space of all symmetric tensors. If $\tau = \varepsilon$ is the sign of the permutation σ, then W_ε is the linear space of all skew-symmetric tensors.

For example, when $V = \mathbb{R}^3$, $G = \mathrm{O}(3)$ and $r = 2$, the representation (3.100) is equivalent to the direct sum:

$$\tau_0 \hat{\otimes} \rho_{0,+} \oplus \tau_0 \hat{\otimes} \rho_{2,+} \oplus \varepsilon \hat{\otimes} \rho_{1,+}.$$

The space of the first irreducible component is the linear span of the identity matrix, that of the second component is the space of symmetric traceless matrices. Their direct sum is $W_{\tau_0} = \mathsf{S}^2(V)$. The space of the third component is the space of skew-symmetric matrices, $W_\varepsilon = \Lambda^2(V)$.

When $r \geq 3$, the structure of the representation (3.100) becomes more complicated; see Goodman & Wallach (2009). In particular, the representations τ of dimension more than 1 appear. The space W_τ still carries an orthogonal representation ρ of the group G, and we can formulate the (G, ρ)-problem: to find the general form of the one-point and two-point correlation tensors of the W-valued homogeneous and (G, ρ)-isotropic random field.

There exist several notations for point groups and general classes, including *Hermann–Mauguin symbols* introduced by Hermann (1928) and Mauguin (1931) and used in crystallography, *Schönfließ symbols* introduced by Schönfließ (1891) and used in spectroscopy, see also Brock (2014). We use *mathematical symbols* described by Olive & Auffray (2014, Section 2.3). The correspondence between the three systems of symbols may be found in Olive & Auffray (2014, Appendix B).

The complete solution to the $(\mathrm{O}(d), 1)$-problem was obtained by Yaglom (1961) and independently by M. Ĭ. Yadrenko in his unpublished PhD thesis.

Theorem 18 was proved by Robertson (1940). Theorem 19 was proved by Yaglom (1948) and Yaglom (1957). The spectral expansion of Theorems 25 and 20 appeared in Malyarenko & Ostoja-Starzewski (2016*b*).

In the case of rank 2, a partial solution (Theorem 24) was given by Lomakin (1964). The complete solution has been proposed by Malyarenko & Ostoja-Starzewski (2016*b*). The classification of symmetry classes is a classical result. Random fields related to these classes were described by the authors in a manuscript that will be published elsewhere.

4

Tensor Random Fields in Continuum Theories

As discussed in the Introduction and Chapter 1, the goal is to develop models of mesoscale TRFs of dependent quantities and constitutive responses. Two paradigms depicted in Figures 1.1 and 1.4 show the upscaling from a random two-phase microstructure, via the SVE playing the rôle of a continuum point, to a random field on macroscale. In Section 4.1 we outline a strategy for simulation of TRFs of various ranks, where arbitrary correlation structures (of Chapter 3) can be introduced. Next, in Section 4.2 two different concepts of ergodic TRFs are elaborated. In Sections 4.3–4.6, the consequences of field equations on dependent fields of various ranks (in 2D or 3D) in classical and micropolar continua are discussed. Section 4.7 focuses on the constitutive elastic-type responses, with two applications to stochastic partial differential equations then covered in Section 4.8. An application of TRFs to modelling stochastic damage phenomena is discussed in Section 4.9. Whereas nearly the entire book has focused on homogeneous and isotropic fields, as shown in Section 4.10, our tools can be extended to inhomogeneous and isotropic TRFs. The chapter ends with a very short selection of future research directions; many other possibilities exist.

4.1 Simulation of Homogeneous and Isotropic TRFs

4.1.1 Strategy

As an example, consider simulation of a homogeneous random field on $\mathbb{R}^3$ that takes values in the space $\mathsf{S}^2(\mathbb{R}^3)$ of symmetric 3×3 matrices and is isotropic with respect to the orthogonal representation $g \mapsto \mathsf{S}^2(g)$ of the group $\mathrm{O}(3)$. By Theorem 25, the expected value of the field is equal to $C\delta_{ij}$, which is easy to simulate. In what follows, we will simulate a centred random field.

According to the proof of Theorem 25, the set of all possible values of the function $f(\lambda)$, when $\lambda > 0$, is a compact convex set $\mathcal{C}_0$ whose set of extreme points has three connected components: two one-point sets and an ellipse. In order to simplify the formula for the rank 2 correlation tensor of the field, use the following idea. Inscribe a simplex into $\mathcal{C}_0$ and assume that the function $f(\lambda)$ takes values inside this simplex. The formula that describes the rank 2 correlation tensor of such a field, does not contain arbitrary functions but gives only *sufficient* condition for the field to be homogeneous and isotropic. The more the four-dimensional Lebesgue measure of the simplex approaches that of $\mathcal{C}_0$, the closer is our sufficient condition to the necessary one. In other words, we have to inscribe a simplex of maximal possible Lebesgue measure into $\mathcal{C}_0$.

Two of the extreme points of the simplex under question are known, these are the one-point components of the set of extreme points of $\mathcal{C}_0$. We must then inscribe a triangle of maximal possible area into an ellipse. The solution to this problem is well known; see Niven (1981, Theorem 7.3b). Taking any point on the ellipse $x^2/a^2 + y^2/b^2 = 1$ with coordinates $(a\cos\theta, b\sin\theta)$ as one of the triangle's vertices, we find the coordinates of the remaining vertices as $(a\cos(\theta + 2k\pi/3, b\sin(\theta + 2k\pi/3)$ with $k = 1, 2$.

Choose the following points on the ellipse $4(v_1(\lambda) - 1/2)^2 + 8v_2^2(\lambda) = 1$ (see (3.77)): $(1, 0)$, $(1/4, \sqrt{3}/(4\sqrt{2}))$ and $(1/4, -\sqrt{3}/(4\sqrt{2}))$. The matrix $f(\lambda)$, $\lambda > 0$, takes the form

$$f(\lambda) = \sum_{k=1}^{5} u_k(\lambda) D^k,$$

where the symmetric non-negative-definite 6×6 matrices D^k with unit trace have the following non-zero elements lying on and over the main diagonal:

$$\begin{aligned}
&D^1_{44} = D^1_{66} = \frac{1}{2},\\
&D^2_{11} = D^2_{33} = D^2_{55} = \frac{1}{3}, \quad D^2_{13} = -\frac{1}{3},\\
&D^3_{11} = D^3_{33} = D^3_{13} = \frac{1}{2},\\
&D^4_{11} = D^4_{33} = D^4_{13} = \frac{1}{8}, \quad D^4_{22} = \frac{3}{4}, \quad D^4_{12} = D^4_{23} = \frac{\sqrt{3}}{4\sqrt{2}},\\
&D^5_{11} = D^5_{33} = D^5_{13} = \frac{1}{8}, \quad D^5_{22} = \frac{3}{4}, \quad D^5_{12} = D^5_{23} = -\frac{\sqrt{3}}{4\sqrt{2}}.
\end{aligned}$$

The non-zero elements of the matrix $f(\lambda)$ lying on and over the main diagonal are

$$\begin{aligned}
f_{11}(\lambda) &= f_{33}(\lambda) = \frac{1}{3}u_2(\lambda) + \frac{1}{2}u_3(\lambda) + \frac{1}{8}(u_4(\lambda) + u_5(\lambda)),\\
f_{12}(\lambda) &= f_{23}(\lambda) = \frac{\sqrt{3}}{4\sqrt{2}}(u_4(\lambda) - u_5(\lambda)),\\
f_{13}(\lambda) &= -\frac{1}{3}u_2(\lambda) + \frac{1}{2}u_3(\lambda) + \frac{1}{8}(u_4(\lambda) + u_5(\lambda)),\\
f_{22}(\lambda) &= \frac{3}{4}(u_4(\lambda) + u_5(\lambda)),
\end{aligned}$$

$$
\begin{aligned}
f_{44}(\lambda) &= f_{66}(\lambda) = \frac{1}{2}u_1(\lambda),\\
f_{55}(\lambda) &= \frac{1}{3}u_2(\lambda).
\end{aligned}
$$

Inverting (3.73) and taking the last display into account, we obtain

$$
\begin{aligned}
f_1(\lambda) &= \frac{2}{3}u_3(\lambda) + \frac{5+2\sqrt{6}}{12}u_4(\lambda) + \frac{5-2\sqrt{6}}{12}u_5(\lambda),\\
f_2(\lambda) &= \frac{2}{\sqrt{5}}u_1(\lambda) + \frac{4}{3\sqrt{5}}u_2(\lambda) + \frac{1}{3\sqrt{5}}u_3(\lambda) + \frac{7-2\sqrt{6}}{12\sqrt{5}}u_4(\lambda)\\
&\quad + \frac{7+2\sqrt{6}}{12\sqrt{5}}u_5(\lambda),\\
f_3(\lambda) &= -\frac{2}{3}u_3(\lambda) + \frac{2\sqrt{2}+\sqrt{3}}{6\sqrt{2}}u_4(\lambda) + \frac{2\sqrt{2}-\sqrt{3}}{6\sqrt{2}}u_5(\lambda),\\
f_4(\lambda) &= \frac{\sqrt{2}}{\sqrt{7}}u_1(\lambda) - \frac{4\sqrt{2}}{3\sqrt{7}}u_2(\lambda) + \frac{\sqrt{2}}{3\sqrt{7}}u_3(\lambda) + \frac{7\sqrt{3}-6\sqrt{2}}{6\sqrt{42}}u_4(\lambda)\\
&\quad + \frac{7\sqrt{3}+6\sqrt{2}}{6\sqrt{42}}u_5(\lambda),\\
f_5(\lambda) &= -\frac{4\sqrt{2}}{\sqrt{35}}u_1(\lambda) + \frac{2\sqrt{2}}{3\sqrt{35}}u_2(\lambda) + \frac{\sqrt{2}}{\sqrt{35}}u_3(\lambda) + \frac{7-2\sqrt{6}}{2\sqrt{70}}u_4(\lambda)\\
&\quad + \frac{7+2\sqrt{6}}{2\sqrt{70}}u_5(\lambda).
\end{aligned}
$$

Using (3.79), we obtain

$$
\langle T(\boldsymbol{x}), T(\boldsymbol{y})\rangle_{ijkl} = \frac{1}{4\pi}\sum_{m=1}^{5}\int_{\hat{\mathbb{R}}^3} \mathrm{e}^{\mathrm{i}(\boldsymbol{p},\boldsymbol{y}-\boldsymbol{x})} F_m(\boldsymbol{p})\,\mathrm{d}\Omega\,\mathrm{d}\Phi_m(\lambda),
$$

where

$$
\begin{aligned}
F_1(\boldsymbol{p}) &= \frac{2}{\sqrt{5}}M^2_{ijkl} + \frac{\sqrt{2}}{\sqrt{7}}M^4_{ijkl}(\boldsymbol{p}) - \frac{4\sqrt{2}}{\sqrt{35}}M^5_{ijkl}(\boldsymbol{p}),\\
F_2(\boldsymbol{p}) &= \frac{4}{3\sqrt{5}}M^2_{ijkl} - \frac{4\sqrt{2}}{3\sqrt{7}}M^4_{ijkl}(\boldsymbol{p}) + \frac{2\sqrt{2}}{\sqrt{35}}M^5_{ijkl}(\boldsymbol{p}),\\
F_3(\boldsymbol{p}) &= \frac{2}{3}M^1_{ijkl} + \frac{1}{3\sqrt{5}}M^2_{ijkl} - \frac{2}{3\sqrt{7}}M^3_{ijkl}(\boldsymbol{p}) + \frac{\sqrt{2}}{3\sqrt{7}}M^4_{ijkl}(\boldsymbol{p})\\
&\quad + \frac{\sqrt{2}}{\sqrt{35}}M^5_{ijkl}(\boldsymbol{p}),\\
F_4(\boldsymbol{p}) &= \frac{5+2\sqrt{6}}{12}M^1_{ijkl} + \frac{7-2\sqrt{6}}{12\sqrt{5}}M^2_{ijkl} - \frac{2\sqrt{2}+\sqrt{3}}{6\sqrt{2}}M^3_{ijkl}(\boldsymbol{p})\\
&\quad + \frac{7\sqrt{3}-6\sqrt{2}}{6\sqrt{42}}M^4_{ijkl}(\boldsymbol{p}) + \frac{7-2\sqrt{6}}{2\sqrt{70}}M^5_{ijkl}(\boldsymbol{p}),
\end{aligned}
$$

$$
\begin{aligned}
F_5(\boldsymbol{p}) = {} & \frac{5-2\sqrt{6}}{12} M^1_{ijkl} + \frac{7+2\sqrt{6}}{12\sqrt{5}} M^2_{ijkl} - \frac{2\sqrt{2}-\sqrt{3}}{6\sqrt{2}} M^3_{ijkl}(\boldsymbol{p}) \\
& + \frac{7\sqrt{3}+6\sqrt{2}}{6\sqrt{42}} M^4_{ijkl}(\boldsymbol{p}) \frac{7+2\sqrt{6}}{2\sqrt{70}} M^5_{ijkl}(\boldsymbol{p}).
\end{aligned}
$$

We see that the random field $T(\boldsymbol{x})$ is a sum of five uncorrelated random fields corresponding to the five terms of the above sum. We show how to simulate the random field that corresponds to the first term of this sum, the remaining fields are simulated similarly. Denote this term by $U(\boldsymbol{x})$. The two-point correlation tensor of the above term is

$$
\begin{aligned}
\langle U(\boldsymbol{x}), U(\boldsymbol{y})\rangle_{ijkl} = {} & \frac{1}{4\pi} \int_{\hat{\mathbb{R}}^3} \mathrm{e}^{\mathrm{i}(\boldsymbol{p},\boldsymbol{y}-\boldsymbol{x})} \left(\frac{2}{\sqrt{5}} M^2_{ijkl} + \frac{\sqrt{2}}{\sqrt{7}} M^4_{ijkl}(\boldsymbol{p}) \right. \\
& \left. - \frac{4\sqrt{2}}{\sqrt{35}} M^5_{ijkl}(\boldsymbol{p}) \right) \mathrm{d}\Omega \, \mathrm{d}\Phi_1(\lambda) \\
= {} & \frac{1}{2\sqrt{\pi}} \int_{\hat{\mathbb{R}}^3} \mathrm{e}^{\mathrm{i}(\boldsymbol{p},\boldsymbol{y}-\boldsymbol{x})} \left(\frac{2}{\sqrt{5}} \frac{1}{\sqrt{5}} \sum_{n=-2}^{2} g^{n[i,j]}_{2[1,1]} g^{n[k,l]}_{2[1,1]} S^0_0(\theta_{\boldsymbol{p}}, \varphi_{\boldsymbol{p}}) \right. \\
& + \frac{\sqrt{2}}{\sqrt{7}} \sum_{n,p=-2}^{2} \sum_{q=-4}^{4} g^{q[n,p]}_{2[2,2]} g^{n[i,j]}_{2[1,1]} g^{p[k,l]}_{2[1,1]} \frac{1}{3} S^q_4(\theta_{\boldsymbol{p}}, \varphi_{\boldsymbol{p}}) \\
& \left. - \frac{4\sqrt{2}}{\sqrt{35}} \sum_{n,p=-2}^{2} \sum_{q=-4}^{4} g^{q[n,p]}_{4[2,2]} g^{n[i,j]}_{2[1,1]} g^{p[k,l]}_{2[1,1]} \frac{1}{3} S^q_4(\theta_{\boldsymbol{p}}, \varphi_{\boldsymbol{p}}) \right) \mathrm{d}\Omega \, \mathrm{d}\Phi_1(\lambda).
\end{aligned}
$$

As usual, we calculate the inner integral using the Rayleigh expansion (2.62) and the Gaunt integral (2.36). We obtain

$$
\begin{aligned}
\langle U(\boldsymbol{x}), U(\boldsymbol{y})\rangle_{ijkl} = {} & 4\pi \sum_{\ell,\ell'=0}^{\infty} \mathrm{i}^{\ell-\ell'} \sqrt{(2\ell+1)(2\ell'+1)} \sum_{m=-\ell}^{\ell} \sum_{m'=-\ell'}^{\ell'} \\
& \int_0^\infty \left(\frac{2}{5} g^{0[m,m']}_{0[\ell,\ell']} g^{0[0,0]}_{0[\ell,\ell']} \sum_{n=-2}^{2} g^{n[i,j]}_{2[1,1]} g^{n[k,l]}_{2[1,1]} \right. \\
& + \frac{\sqrt{2}}{9\sqrt{7}} g^{0[0,0]}_{4[\ell,\ell']} \sum_{n,p=-2}^{2} \sum_{q=-4}^{4} g^{q[n,p]}_{2[2,2]} g^{n[i,j]}_{2[1,1]} g^{p[k,l]}_{2[1,1]} g^{q[m,m']}_{4[\ell,\ell']} \\
& \left. - \frac{4\sqrt{2}}{9\sqrt{35}} g^{0[0,0]}_{4[\ell,\ell']} \sum_{n,p=-2}^{2} \sum_{q=-4}^{4} g^{q[n,p]}_{4[2,2]} g^{n[i,j]}_{2[1,1]} g^{p[k,l]}_{2[1,1]} g^{q[m,m']}_{4[\ell,\ell']} \right) \\
& \times j_\ell(\lambda\|\boldsymbol{y}\|) j_{\ell'}(\lambda\|\boldsymbol{x}\|) \, \mathrm{d}\Phi_1(\lambda) S^m_\ell(\theta_{\boldsymbol{y}}, \varphi_{\boldsymbol{y}}) S^{m'}_{\ell'}(\theta_{\boldsymbol{x}}, \varphi_{\boldsymbol{x}}).
\end{aligned}
\tag{4.1}
$$

Note that $g^{0[0,0]}_{0[\ell,\ell']} = 0$ if $\ell \neq \ell'$. Similarly, $g^{0[0,0]}_{4[\ell,\ell']} = 0$ if $|\ell - \ell'| \neq 4$. Then $\mathrm{i}^{\ell-\ell'} = 1$ in all the non-zero terms. Using Karhunen's theorem, we obtain

$$U(r,\theta,\varphi)=2\sqrt{\pi}\sum_{\ell=0}^{\infty}\sqrt{2\ell+1}\sum_{m=-\ell}^{\ell}\int_{0}^{\infty}j_{\ell}(\lambda r)\,\mathrm{d}Z^{m}_{ij\ell}(\lambda)S^{m}_{\ell}(\theta,\varphi),$$

where $Z^m_{ij\ell}$ are centred real-valued random measures on $[0,\infty)$ with control measure Φ_1 and cross-correlation

$$\mathsf{E}[Z^m_{ij\ell}(A)Z^{m'}_{kl\ell'}(B)]=b^{mm'}_{ijkl\ell\ell'}\Phi_1(A\cap B),$$

where $b^{mm'}_{ijkl\ell\ell'}$ is the expression in round brackets in (4.1).

To simulate this field, we follow Katafygiotis, Zerva & Malyarenko (1999). Assume that there exists an isotropic spectral density of the field, that is,

$$\Phi_1(A)=4\pi\int_A\lambda^2 f(\lambda)\,\mathrm{d}\lambda.$$

Choose an upper cutoff wavenumber Λ, above which the values of the spectral density are insignificant for practical purposes. Select two positive integers L and N and define $\Delta\lambda=\Lambda/N$. Choose a real number $\lambda_1\in[0,\Delta\lambda]$. Put $\lambda_n=\lambda_1+(n-1)\Delta\lambda$, $1\leq n\leq N$. The sample paths are simulated by the following formula

$$\tilde{U}_{ij}(r,\theta,\varphi)=8\pi\sqrt{\pi\Delta\lambda}\sum_{\ell=0}^{L}\sqrt{2\ell+1}\sum_{m=-\ell}^{\ell}\sum_{n=1}^{N}j_{\ell}(\lambda_n r)\sqrt{f(\lambda_n)}Z^m_{ij\ell n},$$

where $Z^m_{ij\ell n}$ are centred random variables with cross-correlation

$$\mathsf{E}[Z^m_{ij\ell n}Z^{m'}_{kl\ell' n'}]=\delta_{nn'}b^{mm'}_{ijkl\ell\ell'}.$$

In other words, we truncate the infinite sum and the region of integration, and use the rectangular rule of numerical integration. For the case of the stationary random process, this approach goes back to Rice (1944).

4.2 Ergodic TRFs

Let N and d be two positive integers. An (N,d)-field is a map $\boldsymbol{X}\colon\mathbb{R}^n\to\mathbb{R}^d$ such that for any $\boldsymbol{t}\in\mathbb{R}^n$ $\boldsymbol{X}(\boldsymbol{t})$ is a d-dimensional random vector. The field $\boldsymbol{X}$ is called strictly homogeneous if, for any positive integer n, and for any points $\boldsymbol{t}$, $\boldsymbol{t}_1$, ..., $\boldsymbol{t}_n$ the nd-dimensional random vectors

$$(\boldsymbol{X}(\boldsymbol{t}_1+\boldsymbol{t}),\dots,\boldsymbol{X}(\boldsymbol{t}_n+\boldsymbol{t}))^{\top}$$

and

$$(\boldsymbol{X}(\boldsymbol{t}_1),\dots,\boldsymbol{X}(\boldsymbol{t}_n))^{\top}$$

have the same distribution. Let Ω be the set of all $\mathbb{R}^d$-valued functions on $\mathbb{R}^N$, let $\mathfrak{F}$ be the smallest σ-field containing sets of the form

$$\{\,\boldsymbol{f}\colon\mathbb{R}^N\to\mathbb{R}^d\colon\boldsymbol{f}(\boldsymbol{t}_i)\in B_i,1\leq i\leq n\},$$

where n is a positive integer, $\boldsymbol{t}_i\in\mathbb{R}^N$, and the B_i are the intervals of the form $(a_1,b_1]\times\cdots\times(a_d,b_d]$. Let P be the probability measure uniquely defined by Kolmogorov's theorem by the finite-dimensional distributions of the field $\boldsymbol{X}$.

Let $i \in [1, N]$ be an integer, and let $\tau \in \mathbb{R}$. The shift transformation $T^i_\tau \colon \Omega \to \Omega$ is defined by

$$T^i_\tau \boldsymbol{f}(t_1, \dots, t_j, \dots, t_N) = \boldsymbol{f}(t_1, \dots, t_j + \tau, \dots, t_N).$$

For a strictly homogeneous random field, the shift transformation is measure-preserving, that is, $\mathsf{P}(T^i_\tau S) = \mathsf{P}(S)$. A set $S \in \mathfrak{F}$ is called invariant if for every i and τ the sets S and $T^i_\tau S$ differ, at most, by a set of P-measure 0. It is easy to see that the invariant sets form a σ-field and that all sets of probability 0 or 1 belong to this σ-field. The field $\boldsymbol{X}$ is ergodic if any invariant set has probability either 0 or 1.

Let $\sigma(T)$ be the centred ball in $\mathbb{R}^N$ of radius $T > 0$, and let B_N be the Lebesgue measure of $\sigma(1)$. The ergodic theorem, see Adler (2010, Theorem 6.5.1), says that under mild additional conditions we have

$$\lim_{T\to\infty} \frac{1}{B_N T^N} \int_{\sigma(T)} \boldsymbol{X}(\boldsymbol{t}) dt = \mathsf{E}[\boldsymbol{X}(\boldsymbol{0})] \tag{4.2}$$

P-almost surely. In engineering literature, Equation (4.2) is the definition of ergodicity.

By Adler (2010, Theorem 6.5.3), a Gaussian (N, d)-field with zero mean and almost surely continuous sample functions is ergodic if and only if its spectral distribution function is continuous.

On the other hand, the spectral formulation of Slutsky's theorem, see Yaglom (1987, p. 231), says that a $(1, 1)$-field is ergodic in the sense of (4.2) if and only if its spectral measure is continuous at the point $\boldsymbol{0}$. The same is also true for $N \geq 1$ and $d \geq 1$.

All random fields considered by us are homogeneous. Its spectral distribution function has the form $f(\boldsymbol{p})\, \mathrm{d}\mu(\boldsymbol{p})$, where μ is a finite measure on the wavenumber domain $\hat{\mathbb{R}^N}$, and $f(\boldsymbol{p})$ takes values in the cone of Hermitian non-negatively-definite linear operators in $\mathbb{C}^d$. If, in addition, the field is (G, U)-isotropic, then the field is completely determined by the values of the spectral distribution function on the orbit space $\hat{\mathbb{R}^N}\backslash G$. Specifically, we have

$$\hat{\mathbb{R}^N} = \cup_{m=0}^{M-1} \pi^{-1}((\hat{\mathbb{R}^N}\backslash G)_m),$$

where π maps each point $\boldsymbol{p} \in \hat{\mathbb{R}^N}$ to its orbit, and $(\hat{\mathbb{R}^N}\backslash G)_m$ is the set of all orbits of the mth type. Then $\pi^{-1}((\hat{\mathbb{R}^N}\backslash G)_m) = (\hat{\mathbb{R}^N}\backslash G)_m \times O_m$, the restriction of the measure μ to $\pi^{-1}((\hat{\mathbb{R}^N}\backslash G)_m)$ is the product of an arbitrary finite measure ν_m on $(\hat{\mathbb{R}^N}\backslash G)_m$ and the G-invariant probabilistic measure on the orbit O_m, and $f(g\boldsymbol{p}) = (U \otimes U)(g) f(\boldsymbol{p})$, $\boldsymbol{p} \in \hat{\mathbb{R}^N}\backslash G$.

It is easy to see that spectral distribution function $f(\boldsymbol{p})\, \mathrm{d}\mu(\boldsymbol{p})$ on $\hat{\mathbb{R}^N}$ is continuous (resp. continuous in $\boldsymbol{0}$) if and only if its restriction to $\hat{\mathbb{R}^N}\backslash G$ is continuous (resp. continuous in $\boldsymbol{0}$). In other words, a homogeneous and (G, U)-isotropic Gaussian random field with almost surely continuous sample trajectories is ergodic (resp. ergodic in the sense of (4.2)) if and only if its isotropic spectral measure (resp. the restriction of the spectral distribution function to the orbit space) is continuous (resp. continuous in $\mathbf{0}$).

4.3 Rank 1 TRF

4.3.1 Restriction Imposed by a Divergence-Free Property

Take $\boldsymbol{v}$ to be a homogeneous and isotropic vector random field. If $\boldsymbol{v}$ is assumed to satisfy the zero-divergence property

$$\operatorname{div} \boldsymbol{v} = 0$$

it may be interpreted as the continuum velocity field, and the classical treatment of turbulent incompressible flows (Batchelor 1982) may be applied. Thus, in index notation,

$$0 = \langle v_{i,i} (\mathbf{0}) v_j (\boldsymbol{z}) \rangle = R_{ij,i} (\boldsymbol{z}) \equiv \frac{\partial R_{ij} (\boldsymbol{z})}{\partial z_i}.$$

Now, recalling the representation (3.46) (i.e.

$$R_{ij} (\boldsymbol{z}) = A (z) z_i z_j + B (z) \delta_{ij} \quad \text{with} \quad z \equiv \|\boldsymbol{z}\|,$$

one finds

$$R_{ij,i} (\boldsymbol{z}) = 0 \Rightarrow 4A + zA' + \frac{1}{z} B' = 0, \tag{4.3}$$

where a prime denotes the derivative with respect to z. Next, introduce two particular correlation functions

$$\text{longitudinal: } f (z) = \frac{\langle v_p (\mathbf{0}) v_p (\boldsymbol{z}) \rangle}{\langle v_p^2 \rangle},$$
$$\text{lateral: } g (z) = \frac{\langle v_n (\mathbf{0}) v_n (\boldsymbol{z}) \rangle}{\langle v_n^2 \rangle},$$

whereby p (or n) denotes parallel (resp., normal) velocity components, while the summation convention does not apply to the terms in the denominator.

By ergodicity in the mean, we have that the square (v^2) of any velocity component v equals

$$v^2 = \langle v_p^2 \rangle = \langle v_n^2 \rangle = \frac{1}{3} v_i v_i,$$

so that

$$\begin{aligned} v^2 f (z) &= \langle v_p (\mathbf{0}) v_p (\boldsymbol{z}) \rangle = A (z) z^2 + B (z), \\ v^2 g (z) &= \langle v_n (\mathbf{0}) v_n (\boldsymbol{z}) \rangle = B (z). \end{aligned} \tag{4.4}$$

It follows from (4.3) that f and g are related through

$$g = f + \frac{1}{2} z f'.$$

Note that these two functions are accessible through experiments or computer simulations of turbulent flows.

For a vector RF $\boldsymbol{u}$ in 2D, the representation $R_{ij}(\boldsymbol{z}) = A(z) z_i z_j + B(z) \delta_{ij}$ holds again, but

$$R_{ij},_i(\boldsymbol{z}) = 0 \Rightarrow 3A + zA' + \frac{1}{z}B' = 0. \tag{4.5}$$

Since the relations (4.4) hold, (4.5) yields

$$g = f + zf'.$$

Next, consider the steady-state heat conduction. Since the heat flux satisfies

$$\operatorname{div} \boldsymbol{q} = 0,$$

there is a direct analogy to the velocity field and the same restrictions follow.

One more way to prove (3.46) is as follows. The velocity field v_i of an incompressible fluid satisfies $v_{i,i} = 0$. Therefore, there exists a potential φ such that $\varphi_{,ii} = 0$. The potential field $\varphi(\boldsymbol{z})$ must be homogeneous and isotropic by the same reasons as random fields in Section 3.1. Let

$$R_\varphi(\|\boldsymbol{z}\|) = \langle \varphi(\mathbf{0})\varphi(\boldsymbol{z})\rangle$$

be the two-point correlation function of the random field $\varphi(\boldsymbol{z})$. Recall that the two-point correlation tensor $R_{ij}(\boldsymbol{z}) = \langle \boldsymbol{v}(\mathbf{0})\boldsymbol{v}(\boldsymbol{z})\rangle$ of the velocity field is given by

$$R_{ij}(\boldsymbol{z}) = -\frac{\partial^2 R_\varphi(\|\boldsymbol{z}\|)}{\partial z_j \partial z_i}.$$

Now we denote $R' = \mathrm{d}R/\mathrm{d}\|\boldsymbol{z}\|$ and derive

$$\begin{aligned} R_{ij}(\boldsymbol{z}) &= -\frac{\partial}{\partial z_j}\left(R'_\varphi \frac{z_i}{\|\boldsymbol{z}\|}\right) = -\left(R''_\varphi \frac{z_i}{\|\boldsymbol{z}\|}\frac{z_j}{\|\boldsymbol{z}\|} + R'_\varphi \frac{z_{i,j}\|\boldsymbol{z}\| - z_i z_j/\|\boldsymbol{z}\|}{\|\boldsymbol{z}\|^2}\right) \\ &= \left(R'_\varphi \frac{1}{\|\boldsymbol{z}\|} - R''_\varphi\right)\frac{z_i z_j}{\|\boldsymbol{z}\|^2} - R'_\varphi \frac{\delta_{ij}}{\|\boldsymbol{z}\|} \\ &= A(\|\boldsymbol{z}\|) z_i z_j + B(\|\boldsymbol{z}\|)\delta_{ij}, \end{aligned}$$

where

$$A(\|\boldsymbol{z}\|) = \frac{1}{\|\boldsymbol{z}\|^3} R'_\varphi - \frac{1}{\|\boldsymbol{z}\|^2} R''_\varphi, \qquad B(\|\boldsymbol{z}\|) = -\frac{1}{\|\boldsymbol{z}\|} R'_\varphi.$$

The above is the paradigm for treatments of other TRFs in more complex situations, which can be summarised as:

- find the explicit form of the correlation function;
- impose a restriction dictated by the relevant physics;
- support the results by experiments and/or computational studies.

4.3.2 Restriction Imposed by a Curl-Free Property

Start with a homogeneous and isotropic vector random field $\boldsymbol{v}$ and assume it to satisfy the zero-curl property:

$$\mathbf{0} = \operatorname{curl} \boldsymbol{v} = \boldsymbol{e}_{ijk} v_{k},_j \boldsymbol{e}_i.$$

Hence

$$0 = \langle \mathbf{e}_{ijk} v_{k,j}(\mathbf{0})\, v_p(\mathbf{z})\rangle = \mathbf{e}_{ijk} R_{kp,j}(\mathbf{z}).$$

Given the representation (3.46), we identify two cases

$$\text{Case 1. } (i,p) = (1,1): \quad R_{31,2}(\mathbf{z}) = A'(z)\frac{1}{z} z_1 z_2 z_3 = R_{21,3}(\mathbf{z})$$

$$\text{Case 2. } (i,p) = (1,2): \quad \begin{aligned} R_{32,2}(\mathbf{z}) &= A'(z)\frac{1}{z} z_2^2 z_3 + A(z) z_2 \\ R_{22,3}(\mathbf{z}) &= A'(z)\frac{1}{z} z_2^2 z_3 + B'(z)\frac{1}{z} z_3 \end{aligned}$$

Case 1 is satisfied identically, while Case 2 implies the restriction

$$zA = B'(z).$$

4.3.3 Velocity and Stress Field Correlations in Fluid Mechanics

Returning back to $\boldsymbol{v}$ being the incompressible velocity field, if $\boldsymbol{v}'$ denotes the velocity fluctuation (i.e. rank 1 TRF), the Reynolds stress

$$R_{kl} := -\rho \langle v_k' v_l' \rangle$$

defines a symmetric rank 2 TRF. Next, consider the spatial average of the turbulence energy (in Cauchy media) defined from R_{kl} as

$$\psi = \frac{1}{2} R_{kk} = -\frac{1}{2}\rho \langle v_k' v_k' \rangle.$$

This defines a scalar RF. Its correlation follows from (3.63):

$$\langle \psi(\mathbf{0})\psi(\mathbf{z})\rangle = \frac{1}{4}\langle R_{ii}(\mathbf{0}) R_{kk}(\mathbf{z})\rangle = \frac{1}{4}\sum_{m=1}^{5} S_m(z) \boldsymbol{J}_{iikk}^{(m)}(\mathbf{z}),$$

implying an explicit link between the correlation function of energy and the five $S_m(r)$ functions of the Reynolds stress:

$$\langle \psi(\mathbf{0})\psi(\mathbf{z})\rangle = \frac{9}{4} S_1(z) + \frac{3}{2} S_2(z) + \frac{3}{2} S_3(z) + S_4(z) + \frac{1}{4} S_5(z).$$

4.4 TRFs in Classical Continuum Mechanics

4.4.1 Interpretations of Specific Correlations

Rank 2 TRFs play a very important rôle in continuum physics. Given that $\boldsymbol{T}$ has diagonal and off-diagonal components, there are five special cases of $\boldsymbol{B}_{ij}^{kl}$ which shed light on the physical meaning of K_αs:

1. $\langle T_{ij}(\mathbf{0}) T_{kl}(\boldsymbol{z})\rangle|_{i=j=k=l}$; i.e. auto-correlations of diagonal terms:
$$\langle T_{11}(\mathbf{0}) T_{11}(\boldsymbol{z})\rangle = K_0 + 2K_1 + 2z_1^2 K_2 + 4z_1^2 K_3 + z_1^4 K_4$$
and then $\langle T_{22}(\mathbf{0}) T_{22}(\boldsymbol{z})\rangle$ and $\langle T_{33}(\mathbf{0}) T_{33}(\boldsymbol{z})\rangle$ by cyclic permutations $1 \to 2 \to 3$.
2. $\langle T_{ij}(\mathbf{0}) T_{kl}(\boldsymbol{z})\rangle|_{i=j\neq k=l}$; i.e. cross-correlations of diagonal terms:
$$\langle T_{11}(\mathbf{0}) T_{22}(\boldsymbol{z})\rangle = K_0 + \left(z_2^2 + z_1^2\right) K_2 + z_2^2 z_1^2 K_4$$
and then $\langle T_{22}(\mathbf{0}) T_{33}(\boldsymbol{z})\rangle$ and $\langle T_{33}(\mathbf{0}) T_{11}(\boldsymbol{z})\rangle$ by cyclic permutations $1 \to 2 \to 3$.
3. $\langle T_{ij}(\mathbf{0}) T_{kl}(\boldsymbol{z})\rangle|_{i=k\neq j=l}$; i.e. auto-correlations of off-diagonal terms:
$$\langle T_{12}(\mathbf{0}) T_{12}(\boldsymbol{z})\rangle = K_1 + \left(z_1^2 + z_2^2\right) K_3 + z_1^2 z_2^2 K_4$$
and then $\langle T_{23}(\mathbf{0}) T_{23}(\boldsymbol{z})\rangle$ and $\langle T_{31}(\mathbf{0}) T_{31}(\boldsymbol{z})\rangle$ by cyclic permutations $1 \to 2 \to 3$.
4. $\langle T_{ij}(\mathbf{0}) T_{kl}(\boldsymbol{z})\rangle|_{j\neq i=k\neq l\neq j}$; that is cross-correlations of off-diagonal terms: $\langle T_{12}(\mathbf{0}) T_{13}(\boldsymbol{z})\rangle = z_2 z_3 K_3 + z_1^2 z_2 z_3 K_4$, then $\langle T_{13}(\mathbf{0}) T_{32}(\boldsymbol{z})\rangle$ and $\langle T_{32}(\mathbf{0}) T_{12}(\boldsymbol{z})\rangle$ by cyclic permutations $1 \to 2 \to 3$.
5. $\langle T_{ij}(\mathbf{0}) T_{kl}(\boldsymbol{z})\rangle|_{i=j=k\neq l\neq j}$; i.e. cross-correlations of diagonal with off-diagonal terms: such as $\langle T_{11}(\mathbf{0}) T_{12}(\boldsymbol{z})\rangle = z_1 z_2 (K_2 + 2K_3) + z_1 z_2^3 K_4$ and $\langle T_{12}(\mathbf{0}) T_{13}(\boldsymbol{z})\rangle = z_2 z_3 K_2 + z_1^2 z_2 z_3 K_4$ and the other ones by cyclic permutations $1 \to 2 \to 3$.

In principle, we can determine these five correlations for a specific physical situation. For example, when $\boldsymbol{T}$ is the anti-plane elasticity tensor for a given resolution (or mesoscale), we can use micromechanics or experiments and then determine the best fits of K_α $(\alpha = 1, \ldots, 5)$ coefficients.

One may determine $\boldsymbol{C}_{ij}^{kl}(\boldsymbol{r})$ through experimental measurements or by computational mechanics/physics on diverse material microstructures, in both cases following a strategy for fourth-rank TRF in 2D or 3D (Sena, Ostoja-Starzewski & Costa 2013).

Special case: the TRF is locally isotropic $T_{ik}(\boldsymbol{z}) = T(z)\delta_{ik}$ with (necessarily) $T(z) > 0$, so that we simply have a scalar random field. Then, the auto-correlation $\boldsymbol{B}_{11}^{11}$ of T is a single scalar function $C(z)$. With the variance $\mathsf{Var}(T) = \langle T(\mathbf{0})T(\boldsymbol{z})\rangle$, the correlation coefficient defined by
$$\rho(\boldsymbol{z}) := \frac{C(z)}{\mathsf{Var}(T)}$$
is constrained by a standard condition of scalar RFs $-1/d \leq \rho(\boldsymbol{z}) \leq 1$, if the model is set in $\mathbb{R}^d$. Basically, this is the correlation function of the conductivity (or diffusion) in conventional stochastic partial differential equations (SPDE) of elliptic type, conventionally set up on scalar random fields.

4.4.2 Anti-Plane Mechanics

This is a generalisation of the anti-plane elasticity of Subsection 1.3.2; recall $u_3(\boldsymbol{x}) \neq 0$ and $u_1(\boldsymbol{x}) = u_2(\boldsymbol{x}) = 0$ for $\forall \boldsymbol{x}$. Assuming quasi-statics without body forces, the static equilibrium condition implies a direct analogy with the velocity field in incompressible fluid flow, so the same conclusions apply.

The correlation tensor of the strain field ε_i is

$$E_i^j(\boldsymbol{z}) := \langle \varepsilon_i(\boldsymbol{z}_1)\varepsilon_j(\boldsymbol{z}_1+\boldsymbol{z})\rangle \,,$$

whereas the correlation function of the displacement $u \equiv u_3$ is

$$U(\boldsymbol{z}) := \langle u(\boldsymbol{z}_1)u(\boldsymbol{z}_1+\boldsymbol{z})\rangle \,.$$

On account of $\varepsilon_i = u_{,i}$, we find

$$\frac{\partial E_1^1(\boldsymbol{z})}{\partial z_2} = \frac{\partial E_1^2(\boldsymbol{z})}{\partial z_1}, \quad \frac{\partial E_2^2(\boldsymbol{z})}{\partial z_1} = \frac{\partial E_1^2(\boldsymbol{z})}{\partial z_2}.$$

Assuming $E_i^j(\boldsymbol{z})$ to have the representation (3.46), leads to this restriction

$$zA = B'.$$

The above results have not involved Hooke's law, so that the restrictions may also apply to other constitutive behaviours providing they involve small gradients/strains.

4.4.3 3D Continuum Theories

This section directly follows Lomakin (1964) and Shermergor (1971).

Three useful facts

(i) Decomposition of a second-rank tensor field

With reference to Kröner (1958), any second-rank TRF can be decomposed into potential (T_1) and birotational (T_2) parts:

$$T = T_1 + T_2 \qquad \operatorname{curl} T_1 = \mathbf{0} \qquad \operatorname{div} T_2 = \mathbf{0},$$

where T_1 is described by the vector potential and T_2 the tensor potential:

$$T = \operatorname{sym}(\boldsymbol{\nabla}\varphi) \qquad T_2 = \operatorname{curl}\boldsymbol{\Phi} \qquad \operatorname{div}\boldsymbol{\Phi} = \mathbf{0}.$$

$$\text{or} \qquad T_{ki} = \operatorname{sym}(\nabla\varphi) = \varphi_{(k,i)} \qquad \operatorname{curl}\operatorname{curl} T_{ij} = \mathbf{e}_{ink}\mathbf{e}_{jml}\Phi_{kl,nm}\,.$$

(ii) Strain field as a potential field

With reference to Equation (1.34), considering the well-known definition of small strains $\varepsilon_{ij} = u_{(i,j)}$, we see that ε_{ij} is a potential tensor field, with u_i being its potential.

(iii) Stress field as a birotational field

With reference to Equation (1.18), the balance of linear momentum in the absence of any body forces $\sigma_{ij,j} = 0$ indicates that σ_{ij} is birotational

4.4.4 Correlation of Stress TRF

For the 3D elasticity, the Hooke law reads:

$$\sigma_{ij} = \mathbf{C}_{ijkl}\varepsilon_{kl}, \quad i,j,k,l = 1,2,3.$$

Now, if σ_{ij} is taken as statistically homogeneous, the correlation function of its fluctuations is

$$\mathsf{S}_{ij}^{kl}(\boldsymbol{z}) := \left\langle \sigma'_{ij}(\boldsymbol{z}_1)\sigma'_{kl}(\boldsymbol{z}_1+\boldsymbol{z})\right\rangle, \quad i,j,k,l=1,2,3,$$

which, on account of the static equilibrium (a special case of Equation (1.18)), leads to:

$$\mathsf{S}_{ij}^{kl}{}_{,j}(\boldsymbol{z})=0. \tag{4.6}$$

Assuming σ_{ij} to be a statistically isotropic TRF, its representation is (3.64), so that we find

$$A_1 z_i \delta_{kl} + A_2(z_l\delta_{ik}+z_k\delta_{il}) + A_3 z_i z_k z_l = 0,$$

where

$$\begin{aligned}
A_1 &= S'_{13} + 2(S_{13}+S_{44}-S_{12}-S_{66})/z \\
A_2 &= S'_{44} + (3S_{44}+S_{13}-S_{12}-3S_{66})/z \\
A_3 &= S'_{33} - S'_{13} - 2S'_{44} + 2(2S_{11}+S_{33}-3S_{13}-6S_{44})/z
\end{aligned}$$

and

$$\begin{aligned}
\mathsf{S}_{ij}^{kl} &= S_{12}\boldsymbol{J}^1_{ijkl} + S_{66}\boldsymbol{J}^2_{ijkl} + (S_{13}-S_{12})\boldsymbol{J}^3_{ijkl} + (S_{44}-S_{66})\boldsymbol{J}^4_{ijkl} \\
&\quad + (S_{11}+S_{33}-2S_{13}-4S_{44})\boldsymbol{J}^5_{ijkl}
\end{aligned} \tag{4.7}$$

where the prime denotes d/dz.

Assuming, without loss of generality, the coordinate systems attached to locations $\boldsymbol{z}_1$ and $\boldsymbol{z}_1+\boldsymbol{z}$ to be related by a shift along a unit vector $\boldsymbol{n}=(n_1,0,0)$, we shall consider the following components:

$$\begin{aligned}
K_1 &= \mathsf{S}^{11}_{11} \quad K_2 = \mathsf{S}^{22}_{22} \quad K_3 = \mathsf{S}^{22}_{11} \\
K_4 &= \mathsf{S}^{33}_{22} \quad K_5 = \mathsf{S}^{12}_{12} \quad K_6 = \mathsf{S}^{23}_{23}.
\end{aligned}$$

From (4.6) we find

$$\begin{aligned}
K_1 &= a_1 + 2a_2 + 2x^2(2a_3+a_4) + x^4 a_5 \\
K_2 &= a_1 + 2a_2 \qquad K_3 = a_1 + a_4x^2 \qquad K_4 = a_1 \\
K_5 &= a_2 + a_3x^2 \qquad K_6 = a_1.
\end{aligned}$$

Note that this relation is satisfied:

$$K_6 = (K_2 - K_4)/2.$$

Solving the above system of equations for the a_is and substituting the results back into (4.7), we find

$$\begin{aligned}
\mathsf{S}_{ij}^{kl}(\boldsymbol{z}) &= K_4(z)\,\boldsymbol{J}^1_{ijkl} + K_6(z)\,\boldsymbol{J}^2_{ijkl} \\
&\quad + [K_5(z)-K_6(z)]\,\boldsymbol{J}^4_{ijkl} + [K_3(z)-K_4(z)]\,\boldsymbol{J}^3_{ijkl} \\
&\quad + [K_1(z)+K_2(z)-2K_3(z)-4K_5(z)]\,\boldsymbol{J}^5_{ijkl}.
\end{aligned} \tag{4.8}$$

Now, the stress field σ_{ij} is purely birotational, meaning that it has no potential component. Thus, on account of (4.6), we arrive at a system of three first-order differential equations for the K_is

$$\begin{aligned}
8K_1 &= (Z+2)(Z+4)K_2 \\
4K_5 &= (Z+2)K_2 - 2K_3 \qquad\qquad Z \equiv z\frac{d}{dz}. \\
8K_4 &= 8(Z+1)K_3 - Z(Z+2)K_2
\end{aligned}$$

This implies that, in an application/simulation of such a stress field, K_2 and K_3 should be chosen first.

4.4.5 Correlation of Strain TRF

An analogous strategy can be applied to the TRF of strain ε_{ij}. Thus, a correlation function of a second-rank TRF of strain

$$\mathbf{E}_{ij}^{kl}(\boldsymbol{z}_1, \boldsymbol{z}_2) := \langle \varepsilon_{ij}(\boldsymbol{z}_1)\varepsilon_{kl}(\boldsymbol{z}_2)\rangle.$$

Next, a correlation tensor of the displacement field u_i is

$$U_i^j(\boldsymbol{z}_1, \boldsymbol{z}_2) := \langle u_i(\boldsymbol{z}_1)u_j(\boldsymbol{z}_2)\rangle.$$

Assuming statistical homogeneity, we have $U_i^j(\boldsymbol{z})$, and on account of (1.34) while using an equivalent notation $U_{ij}(\boldsymbol{z}) \equiv U_i^j(\boldsymbol{z})$, we obtain

$$\mathbf{E}_{ij}^{kl} = -\nabla_{(i}U_{i)(k,l)}. \tag{4.9}$$

Assuming the representation (3.63) for $\mathbf{E}_{ij}^{kl}$ and (3.46) for U_i^j, the equation (4.9) leads to

$$\begin{aligned} z^2\mathbf{E}_{ij}^{kl} = &- U_2\boldsymbol{J}_{ijkl}^1 - \frac{1}{2}\boldsymbol{J}_{ijkl}^2(U_2 + zU_1') + (2U_2 - zU_2')\boldsymbol{J}_{ijkl}^3 \\ &+ \frac{1}{4}(6U_2 - 3zU_2' + zU_1'z^2U_1'')\boldsymbol{J}_{ijkl}^4 - (8U_2 - 5zU_2' + z^2U_2'')\boldsymbol{J}_{ijkl}^5. \end{aligned} \tag{4.10}$$

On a separate track, again without loss of generality, taking the vector $\boldsymbol{z}$ to be aligned with $\boldsymbol{n} = (n_1, 0, 0)$, we consider the following components:

$$\begin{aligned} &M_1 = \mathbf{E}_{11}^{11} = E_{11} \quad M_2 = \mathbf{E}_{22}^{22} = E_{33} \quad M_3 = \mathbf{E}_{11}^{22} = E_{13} \\ &M_4 = \mathbf{E}_{22}^{33} = E_{12} \quad M_5 = \mathbf{E}_{12}^{12} = E_{44} \quad M_6 = \mathbf{E}_{23}^{23} = E_{66}, \end{aligned}$$

This leads to

$$\begin{aligned} \mathbf{E}_{ij}^{kl}(z) = &M_4(z)\,\boldsymbol{J}_{ijkl}^1 + \frac{1}{4}M_6(z)\,\boldsymbol{J}_{ijkl}^2 \\ &+ \frac{1}{4}\left[M_5(z) - M_6(z)\right]\boldsymbol{J}_{ijkl}^4 + \left[M_3(z) - M_4(z)\right]\boldsymbol{J}_{ijkl}^3 \\ &+ \left[M_1(z) + M_2(z) - 2M_3(z) - M_5(z)\right]\boldsymbol{J}_{ijkl}^5. \end{aligned} \tag{4.11}$$

Note that we have $M_6 = 2(M_2 - M_4)$. Comparing (4.10) with (4.11), we eliminate the auxiliary functions U_1 and U_2 to obtain the system of equations

$$\begin{aligned} z^2M_{12} &= -U_2 \\ z^2M_6 &= -2(U_2 + zU_1') \\ z^2(M_3 - M_4) &= -2U_2 - zU_2' \\ z^2(M_5 - M_6) &= 6U_2 - 3zU_2' + zU_1' - z^2U_1'' \\ z^2(M_1 + M_2 - 2M_3 - M_5) &= -8U_2 - 5zU_2' + z^2U_2''. \end{aligned}$$

This results in a system of three differential equations

$$\begin{aligned} M_2 &= (Z+1)M_1 + Z(Z+1)M_4 \\ M_5 &= (Z+2)M_1 + (Z-2)M_4 \qquad Z \equiv z\frac{d}{dz}. \\ M_3 &= (Z+1)M_4 \end{aligned}$$

This implies that, in an application/simulation of such a stress field, K_2 and K_3 should be chosen first.

4.4.6 Correlations of Rotation and Curvature-Torsion Fields

Once the correlation tensors of stress and strain fields are known, one can also assess the TRF of rotations of grains. First, define the rotation vector

$$\boldsymbol{\omega} := \frac{1}{2}\operatorname{curl}\boldsymbol{u} \quad \text{or} \quad \omega_i := \frac{1}{2}\mathsf{e}_{ijk}u_{k,j}$$

and then introduce its correlation tensor

$$\Omega_i^j(\boldsymbol{z}_1, \boldsymbol{z}_2) := \langle \omega_i(\boldsymbol{z}_1)\omega_j(\boldsymbol{z}_2)\rangle.$$

It follows that

$$\Omega = \frac{1}{4}\operatorname{curl}\boldsymbol{U} \quad \text{and} \quad \operatorname{div}\Omega = 0.$$

The above, together with the representation of a statistically homogeneous isotropic RF of Ω ($\Omega_i^j = \Omega_{ij}$)

$$\Omega_i^j(\boldsymbol{z}) = \Omega_1^1\delta_{ij} + \left(\Omega_3^3 - \Omega_1^1\right)n_i n_j$$

implies

$$z\left(\Omega_3^3\right)'\delta_{ij} + 2\left(\Omega_3^3 - \Omega_1^1\right) = 0 \quad \text{and} \quad \Omega_1^1 = \frac{1}{2z}\frac{d}{dz}\left(z^2\Omega_3^3\right).$$

Next, define the curvature tensor

$$\kappa := \nabla\boldsymbol{\omega} \quad \text{or} \quad \kappa_{ij} := \omega_{i,j}$$

and then introduce its correlation tensor $\boldsymbol{K}$:

$$\boldsymbol{K}_{ij}^{kl}(\boldsymbol{z}_1, \boldsymbol{z}_2) = \langle \kappa_{ij}(\boldsymbol{z}_1)\kappa_{kl}(\boldsymbol{z}_2)\rangle.$$

It follows that

$$\boldsymbol{K}_{ij}^{kl} \equiv \boldsymbol{K}_{ijkl} = -\Omega_{ik,jl} \equiv -\boldsymbol{\theta}_{ikjl}.$$

In the above we have defined the tensor $\boldsymbol{\theta}_{ikjl}$; it is asymmetric with respect to interchange of the pair of indices $(ik) \leftrightarrow (jl)$, but symmetric with respect to an interchange within each pair:

$$\boldsymbol{\theta}_{ikjl} = \boldsymbol{\theta}_{kijl} = \boldsymbol{\theta}_{iklj} \neq \boldsymbol{\theta}_{jlik}.$$

The above implies

$$z\left(\Omega_3^3\right)'\delta_{ij} + 2\left(\Omega_3^3 - \Omega_1^1\right) = 0 \quad \text{and} \quad \Omega_1^1 = \frac{1}{2z}\frac{d}{dz}\left(z^2\Omega_3^3\right).$$

Relations of that type have relevance in a micropolar model of a random heterogeneous material.

As in the anti-plane mechanics, the above results have not involved Hooke's law, so that the restrictions may also apply to other constitutive behaviours providing they involve small gradients/strains.

4.5 TRFs in Plane

4.5.1 Group-Theoretical Considerations

Consider the $(\mathrm{O}(2),\rho^+\oplus\rho^2)$-problem. That is, we consider a random field $T(\boldsymbol{x})\colon\mathbb{R}^2\to\mathsf{S}^2(\mathbb{R}^2)$ that is homogeneous and $(\mathrm{O}(2),\rho^+\oplus\rho^2)$-isotropic. Exactly as in Section 3.6, we prove that $\langle T(\boldsymbol{x})\rangle=0$ and

$$\langle T(\boldsymbol{x}),T(\boldsymbol{y})\rangle=\frac{1}{2\pi}\int_0^\infty\int_0^{2\pi}\mathrm{e}^{\mathrm{i}(\boldsymbol{p},\boldsymbol{y}-\boldsymbol{x})}f(\lambda,\psi)\,\mathrm{d}\psi\,\mathrm{d}\Phi(\lambda), \tag{4.12}$$

where (λ,ψ) are the polar coordinates in $\hat{\mathbb{R}}^2$, $\boldsymbol{p}=\lambda(\cos\psi,\sin\psi)^\top$, and

$$f(\boldsymbol{p})\in\mathsf{S}^2(\mathsf{S}^2(\mathbb{R}^2)),\qquad f(g\boldsymbol{p})=\mathsf{S}^2(\rho^+\oplus\rho^2)(g)f(\boldsymbol{p}).$$

Consider the value of $f(0,0)$ first. The stationary subgroup of the point $(0,0)$ is $\mathrm{O}(2)$. It follows that the value of $f(0,0)$ belongs to the convex compact set $\mathcal{C}_1$, the intersection of the set of all non-negative-definite symmetric operators in $\mathsf{S}^2(\mathbb{R}^2)$ with unit trace and the isotypic subspace of the representation $\mathsf{S}^2(\rho^+\oplus\rho^2)$ that corresponds to the trivial representation of $\mathrm{O}(2)$. To describe the set $\mathcal{C}_1$, we introduce coordinates.

The basis of the space $\mathsf{S}^2(\mathbb{R}^2)$ that respects the representation $\rho=\rho_+\oplus\rho^2$ is given by the Clebsch–Gordan matrices (2.25). The representation $\rho\otimes\rho$ is as follows:

$$\rho\otimes\rho=(\rho_+\oplus\rho^2)\otimes(\rho_+\oplus\rho^2)=\rho_+\oplus2\rho^2\oplus(\rho_+\oplus\rho_-\oplus\rho^4),$$

and its symmetric part is

$$\mathsf{S}^2(\rho)=2\rho_+\oplus\rho^2\oplus\rho^4.$$

The first copy of ρ_+ acts in the space generated by the tensor

$$\mathsf{T}^{+,1}=c^+_{+[1,1]}\otimes c^+_{+[1,1]},$$

while the space of the second copy is generated by the tensor

$$\mathsf{T}^{+,2}=\sum_{i,j\in\{-1,1\}}c^{+[i,j]}_{+[2,2]}c^i_{2[1,1]}\otimes c^j_{2[1,1]},$$

The above two rank 4 tensors are the elements of the linear space $\mathsf{S}^2(\mathsf{S}^2(\mathbb{R}^2))$. In order to write down the entries of similar tensors on a two-dimensional paper, consider the following map acting from $\mathsf{S}^2(\mathsf{S}^2(\mathbb{R}^2))$ to $\mathsf{S}^2(\mathbb{R}^3)$:

$$\tau \boldsymbol{T}_{ijkl} = \begin{pmatrix} \boldsymbol{T}_{-1-1-1-1} & \sqrt{2}\boldsymbol{T}_{-1-1-11} & \boldsymbol{T}_{-1-111} \\ \sqrt{2}\boldsymbol{T}_{-1-1-11} & 2\boldsymbol{T}_{-11-11} & \sqrt{2}\boldsymbol{T}_{-1111} \\ \boldsymbol{T}_{-1-111} & \sqrt{2}\boldsymbol{T}_{-1111} & \boldsymbol{T}_{1111} \end{pmatrix}. \tag{4.13}$$

It is easy to check that the map τ is linear, one-to-one and orthogonal. From now on, we are working in the space $\mathsf{S}^2(\mathbb{R}^3)$ instead of $\mathsf{S}^2(\mathsf{S}^2(\mathbb{R}^2))$, using the isomorphism (4.13).

We obtain

$$\tau \boldsymbol{T}^{+,1} = \frac{1}{2}\begin{pmatrix} 1 & 0 & 1 \\ 0 & 0 & 0 \\ 1 & 0 & 1 \end{pmatrix}, \qquad \tau \boldsymbol{T}^{+,2} = \frac{1}{2\sqrt{2}}\begin{pmatrix} 1 & 0 & -1 \\ 0 & 2 & 0 \\ -1 & 0 & 1 \end{pmatrix}.$$

To simplify notation, we omit the symbol τ in all subsequent equations.

The element $f_{+,1}(0)\boldsymbol{T}^{+,1} + f_{+,2}(0)\boldsymbol{T}^{+,2}$ of the linear space generated by $\boldsymbol{T}^{+,1}$ and $\boldsymbol{T}^{+,2}$ is a non-negative-definite matrix with unit trace if and only if $f_{+,1}(0) = \frac{1}{2}$ and $f_{+,2}(0) = 0$. We have $\mathcal{C}_1 = \{\boldsymbol{T}^{+,1}\}$.

Consider the matrix $f(\lambda, 0)$ with $\lambda > 0$. The irreducible component ρ^2 of the representation $\mathsf{S}^2(\rho)$ acts in the linear space generated by the tensors

$$\boldsymbol{T}^{2,m} = -\frac{m}{\sqrt{2}}(c^{+}_{+[1,1]} \otimes c^{m}_{2[1,1]} + c^{m}_{2[1,1]} \otimes c^{+}_{+[1,1]}), \qquad m \in \{-1, 1\}.$$

The restriction of the representation ρ^2 to the subgroup $\mathrm{O}(1)$ contains the trivial irreducible component acting is the linear space generated by

$$\boldsymbol{T}^{2,1} = \frac{1}{\sqrt{2}}\begin{pmatrix} 1 & 0 & 0 \\ 0 & 0 & 0 \\ 0 & 0 & -1 \end{pmatrix}.$$

Similarly, the irreducible component ρ^4 of the representation $\mathsf{S}^2(\rho)$ acts in the linear space generated by the tensors

$$\boldsymbol{T}^{4,m} = \sum_{i,j \in \{-1,1\}} c^{m[i,j]}_{4[2,2]} c^{i}_{2[1,1]} \otimes c^{j}_{2[1,1]}, \qquad m \in \{-1, 1\},$$

The restriction of the representation ρ^4 to the subgroup $\mathrm{O}(1)$ contains the trivial irreducible component acting is the linear space generated by

$$\boldsymbol{T}^{4,1} = \frac{1}{2\sqrt{2}}\begin{pmatrix} 1 & 0 & -1 \\ 0 & -2 & 0 \\ -1 & 0 & 1 \end{pmatrix}.$$

Consider the element

$$f(\lambda, 0) = f_{+,1}(\lambda)\boldsymbol{T}^{+,1} + f_{+,2}(\lambda)\boldsymbol{T}^{+,2} + f_{2,1}(\lambda)\boldsymbol{T}^{2,1} + f_{4,1}(\lambda)\boldsymbol{T}^{4,1} \tag{4.14}$$

of the linear space generated by the four tensors $\boldsymbol{T}^{+,1}$, $\boldsymbol{T}^{+,2}$, $\boldsymbol{T}^{2,1}$ and $\boldsymbol{T}^{4,1}$. The matrix $f(\lambda, 0)$ has the form

$$f(\lambda, 0) = \begin{pmatrix} u_2(\lambda) & 0 & u_4(\lambda) \\ 0 & u_1(\lambda) & 0 \\ u_4(\lambda) & 0 & u_3(\lambda) \end{pmatrix},$$

where

$$
\begin{aligned}
u_1(\lambda) &= \frac{1}{\sqrt{2}}(f_{+,2}(\lambda) - f_{4,1}(\lambda)),\\
u_2(\lambda) &= \frac{1}{2}f_{+,1}(\lambda) + \frac{1}{2\sqrt{2}}f_{+,2}(\lambda) + \frac{1}{\sqrt{2}}f_{2,1}(\lambda) + \frac{1}{2\sqrt{2}}f_{4,1}(\lambda),\\
u_3(\lambda) &= \frac{1}{2}f_{+,1}(\lambda) + \frac{1}{2\sqrt{2}}f_{+,2}(\lambda) - \frac{1}{\sqrt{2}}f_{2,1}(\lambda) + \frac{1}{2\sqrt{2}}f_{4,1}(\lambda),\\
u_4(\lambda) &= \frac{1}{2}f_{+,1}(\lambda) - \frac{1}{2\sqrt{2}}f_{+,2}(\lambda) - \frac{1}{2\sqrt{2}}f_{4,1}(\lambda).
\end{aligned}
\tag{4.15}
$$

It is obvious that the matrix $f(\lambda, 0)$ is non-negative-definite with unit trace if and only if $u_i(\lambda) \geq 0$, $1 \leq i \leq 3$, $u_1(\lambda) + u_2(\lambda) + u_3(\lambda) = 1$ and $|u_4(\lambda)| \leq \sqrt{u_2(\lambda)u_3(\lambda)}$.

Define

$$
v_1(\lambda) = \frac{u_2(\lambda)}{u_2(\lambda) + u_3(\lambda)}, \qquad v_2(\lambda) = \frac{u_4(\lambda)}{u_2(\lambda) + u_3(\lambda)}. \tag{4.16}
$$

We see that the set $\mathcal{C}_0$ is the cone with the vertex

$$
\boldsymbol{D}_1 = \begin{pmatrix} 0 & 0 & 0 \\ 0 & 1 & 0 \\ 0 & 0 & 0 \end{pmatrix},
$$

which corresponds to the case of $u_1(\lambda) = 1$ and the base consisting of the symmetric matrices

$$
\boldsymbol{D}(\lambda) = \begin{pmatrix} v_1(\lambda) & 0 & v_2(\lambda) \\ 0 & 0 & 0 \\ v_2(\lambda) & 0 & 1 - v_1(\lambda) \end{pmatrix}
$$

lying in the closed disk

$$
\left(v_1(\lambda) - \frac{1}{2}\right)^2 + v_2^2(\lambda) \leq \frac{1}{4},
$$

which corresponds to the case of $u_1(\lambda) = 0$. The matrix $f(\lambda, 0)$ takes the form

$$
f(\lambda, 0) = u_1(\lambda)\boldsymbol{D}^1 + (u_2(\lambda) + u_3(\lambda))\boldsymbol{D}(\lambda).
$$

The coefficients of the expansion (4.14) are expressed in term of the functions $u_i(\lambda)$ as follows:

$$
\begin{aligned}
f_{+,1}(\lambda) &= \frac{1}{2}u_2(\lambda) + \frac{1}{2}u_3(\lambda) + u_4(\lambda),\\
f_{+,2}(\lambda) &= \frac{1}{\sqrt{2}}u_1(\lambda) + \frac{1}{2\sqrt{2}}u_2(\lambda) + \frac{1}{2\sqrt{2}}u_3(\lambda) - \frac{1}{\sqrt{2}}u_4(\lambda),\\
f_{2,1}(\lambda) &= \frac{1}{\sqrt{2}}(u_2(\lambda) - u_3(\lambda)),\\
f_{4,1}(\lambda) &= -\frac{1}{\sqrt{2}}u_1(\lambda) + \frac{1}{2\sqrt{2}}u_2(\lambda) + \frac{1}{2\sqrt{2}}u_3(\lambda) - \frac{1}{\sqrt{2}}u_4(\lambda).
\end{aligned}
\tag{4.17}
$$

Introduce the $\boldsymbol{M}$-functions

$$\begin{aligned}\boldsymbol{M}^{+,n}_{ijkl}(\boldsymbol{p}) &= \boldsymbol{T}^{+,n}_{ijkl}, \qquad n=1,2,\\ \boldsymbol{M}^{n}_{ijkl}(\boldsymbol{p}) &= \sum_{m\in\{-1,1\}} \boldsymbol{T}^{n,m}_{ijkl}\rho^n_{m1}(\boldsymbol{p}), \qquad n=2,4.\end{aligned}$$

Note that $\rho^n_{-11}(\boldsymbol{p}) = \sin(n\psi)$ and $\rho^n_{11}(\boldsymbol{p}) = \cos(n\psi)$. Exactly as in Malyarenko & Ostoja-Starzewski (2016*b*), we prove that

$$\begin{aligned}f_{ijkl}(\boldsymbol{p}) = \boldsymbol{M}^{+,1}_{ijkl}(\boldsymbol{p})f_{+,1}(\lambda) + \boldsymbol{M}^{+,2}_{ijkl}(\boldsymbol{p})f_{+,2}(\lambda)\\ + \boldsymbol{M}^{2}_{ijkl}(\boldsymbol{p})f_{2,1}(\lambda) + \boldsymbol{M}^{4}_{ijkl}(\boldsymbol{p})f_{4,1}(\lambda).\end{aligned} \tag{4.18}$$

Using (4.16) and (4.17), we obtain

$$\begin{aligned}f_{ijkl}(\boldsymbol{p}) &= \left[\frac{1}{\sqrt{2}}(\boldsymbol{M}^{+,2}_{ijkl}(\boldsymbol{p}) - \boldsymbol{M}^{4}_{ijkl}(\boldsymbol{p}))\right] u_1(\lambda)\\ &+ \left[\left(\frac{1}{2} + v_2(\lambda)\right)\boldsymbol{M}^{+,1}_{ijkl}(\boldsymbol{p}) + \left(\frac{1}{2\sqrt{2}} - \frac{1}{\sqrt{2}}v_2(\lambda)\right)\boldsymbol{M}^{+,2}_{ijkl}(\boldsymbol{p})\right.\\ &+ \left(-\frac{1}{\sqrt{2}} + \sqrt{2}v_1(\lambda)\right)\boldsymbol{M}^{2}_{ijkl}(\boldsymbol{p})\\ &+ \left.\left(\frac{1}{2\sqrt{2}} - \frac{1}{\sqrt{2}}v_2(\lambda)\right)\boldsymbol{M}^{4}_{ijkl}(\boldsymbol{p})\right](u_2(\lambda)+u_3(\lambda)).\end{aligned}$$

Apply the Jacobi–Anger expansion (2.61), substitute the result in Equation (4.12), and integrate with respect to $\mathrm{d}\psi$. We obtain

$$\begin{aligned}\langle T(\boldsymbol{x}), T(\boldsymbol{y})\rangle_{ijkl} &= \int_0^\infty \left[\frac{1}{\sqrt{2}}(J_0(\lambda\rho)\boldsymbol{M}^{+,2}_{ijkl}(\boldsymbol{z}) - J_4(\lambda\rho)\boldsymbol{M}^{4}_{ijkl}(\boldsymbol{z}))\right]\mathrm{d}\Phi_1(\lambda)\\ &+ \int_0^\infty \left[\left(\frac{1}{2} + v_2(\lambda)\right) J_0(\lambda\rho)\boldsymbol{M}^{+,1}_{ijkl}(\boldsymbol{z})\right.\\ &+ \left(\frac{1}{2\sqrt{2}} - \frac{1}{\sqrt{2}}v_2(\lambda)\right) J_0(\lambda\rho)\boldsymbol{M}^{+,2}_{ijkl}(\boldsymbol{z})\\ &+ \left(\frac{1}{\sqrt{2}} - \sqrt{2}v_1(\lambda)\right) J_2(\lambda\rho)\boldsymbol{M}^{2}_{ijkl}(\boldsymbol{z})\\ &+ \left.\left(\frac{1}{2\sqrt{2}} - \frac{1}{\sqrt{2}}v_2(\lambda)\right) J_4(\lambda\rho)\boldsymbol{M}^{4}_{ijkl}(\boldsymbol{z})\right]\mathrm{d}\Phi_2(\lambda),\end{aligned}$$

where $\boldsymbol{z} = \boldsymbol{y} - \boldsymbol{x}$, $\rho = \|\boldsymbol{z}\|$, $\mathrm{d}\Phi_1(\lambda) = u_1(\lambda)\,\mathrm{d}\Phi(\lambda)$ and $\mathrm{d}\Phi_2(\lambda) = (u_2(\lambda) + u_3(\lambda))\,\mathrm{d}\Phi(\lambda)$.

The $\boldsymbol{M}$-functions are expressed in terms of $\boldsymbol{L}$-functions as follows:

$$\begin{aligned}\boldsymbol{M}^{+,1}_{ijkl}(\boldsymbol{z}) &= \frac{1}{2}\boldsymbol{L}^{1}_{ijkl}(\boldsymbol{z}),\\ \boldsymbol{M}^{+,2}_{ijkl}(\boldsymbol{z}) &= \frac{1}{2\sqrt{2}}(-\boldsymbol{L}^{1}_{ijkl}(\boldsymbol{z}) + \boldsymbol{L}^{2}_{ijkl}(\boldsymbol{z})),\\ \boldsymbol{M}^{2}_{ijkl}(\boldsymbol{z}) &= \frac{1}{\sqrt{2}}(-\boldsymbol{L}^{1}_{ijkl}(\boldsymbol{z}) + \boldsymbol{L}^{4}_{ijkl}(\boldsymbol{z})),\\ \boldsymbol{M}^{4}_{ijkl}(\boldsymbol{z}) &= \frac{1}{2\sqrt{2}}(3\boldsymbol{L}^{1}_{ijkl}(\boldsymbol{z}) - \boldsymbol{L}^{2}_{ijkl}(\boldsymbol{z})) + \sqrt{2}(-\boldsymbol{L}^{4}_{ijkl}(\boldsymbol{z}) + 2\boldsymbol{L}^{5}_{ijkl}(\boldsymbol{z})).\end{aligned} \tag{4.19}$$

Table 4.1 *The functions* $N_{mn}(\lambda,\rho)$

m	n	$N_{mn}(\lambda,\rho)$
1	1	$\frac{1}{4}(-J_0(\lambda\rho)-3J_4(\lambda\rho))$
1	2	$\frac{1}{4}(J_0(\lambda\rho)-J_4(\lambda\rho))$
1	4	$-J_4(\lambda\rho)$
1	5	$2J_4(\lambda\rho)$
2	1	$\frac{1}{8}[(2v_2(\lambda)-1)J_0(\lambda\rho)+(8v_1(\lambda)-4)J_2(\lambda\rho)+(-6v_2(\lambda)+3)J_4(\lambda\rho)]$
2	2	$\frac{1}{8}[(-2v_2(\lambda)+1)J_0(\lambda\rho)+(2v_2(\lambda)-1)J_4(\lambda\rho)]$
2	4	$\frac{1}{2}[(-2v_1(\lambda)+1)J_2(\lambda\rho)+(2v_2(\lambda)-1)J_4(\lambda\rho)]$
2	5	$(2v_2(\lambda)-1)J_4(\lambda\rho)$

The two-point correlation tensor of the field takes the form

$$\langle T(\boldsymbol{x}),T(\boldsymbol{y})\rangle_{ijkl}=\sum_{m=1}^{2}\int_0^\infty\sum_{n=1,2,4,5}N_{mn}(\lambda,\rho)\mathbf{L}^n_{ijkl}(\varphi)\,\mathrm{d}\Phi_m(\lambda),\tag{4.20}$$

where (ρ,φ) are the polar coordinates of the point $\boldsymbol{z}$ and the functions $N_{mn}(\lambda,\rho)$ are given in Table 4.1.

Equation (4.20) may be written in the following form:

$$\langle T(\boldsymbol{x}),T(\boldsymbol{y})\rangle_{ijkl}=\sum_{n=1,2,4,5}h_n(\rho)\mathbf{L}^n_{ijkl}(\varphi),\tag{4.21}$$

where

$$h_n(\rho)=\sum_{m=1}^{2}\int_0^\infty N_{mn}(\lambda,\rho)\,\mathrm{d}\Phi_m(\lambda).$$

In order to obtain the spectral expansion of the field $T(\boldsymbol{x})$, we can proceed as follows. Write down the Jacobi–Anger expansion twice: the first one for $\mathrm{e}^{\mathrm{i}(\boldsymbol{p},\boldsymbol{y})}$ and the second one for $\mathrm{e}^{-\mathrm{i}(\boldsymbol{p},\boldsymbol{x})}$, substitute both expansions in (4.12) and integrate with respect to $\mathrm{d}\psi$. The resulting equations become complicated.

To simplify calculations and simulations, we propose the following idea. Inscribe a simplex $\mathcal{C}$ into the cone $\mathcal{C}_0$ in such a way that $\mathcal{C}_1\subset\mathcal{C}$ and $\mathcal{C}$ has the maximal possible volume. It is easy to see that the vertices of $\mathcal{C}$ are $\mathbf{D}_1$, $\mathbf{D}_2=\mathbf{T}^{+,1}$ and

$$\mathbf{D}_3=\frac{1}{4}\begin{pmatrix}2-\sqrt{3}&0&-1\\0&0&0\\-1&0&2+\sqrt{3}\end{pmatrix},\qquad\mathbf{D}_4=\frac{1}{4}\begin{pmatrix}2+\sqrt{3}&0&-1\\0&0&0\\-1&0&2-\sqrt{3}\end{pmatrix}.$$

Assume that the matrix $f(\lambda,0)$ may only take values in $\mathcal{C}$. Then it takes the form

$$f(\lambda,0)=\sum_{m=1}^{4}u_m(\lambda)\mathbf{D}_m,$$

where $u_m(\lambda)$ are the barycentric coordinates of the point $f(\lambda,0)$ in the simplex $\mathcal{C}$. Using this equation and (4.17), we obtain

$$
\begin{aligned}
f_{+,1}(\lambda) &= u_2(\lambda) + \frac{1}{4}u_3(\lambda), \\
f_{+,2}(\lambda) &= \frac{1}{\sqrt{2}}u_1(\lambda) + \frac{3}{4\sqrt{2}}u_3(\lambda) + \frac{3}{4\sqrt{2}}u_4(\lambda), \\
f_{2,1}(\lambda) &= \frac{\sqrt{3}}{2\sqrt{2}}(-u_3(\lambda) + u_4(\lambda)), \\
f_{4,1}(\lambda) &= -\frac{1}{\sqrt{2}}u_1(\lambda) + \frac{3}{4\sqrt{2}}u_3(\lambda) + \frac{3}{4\sqrt{2}}u_4(\lambda).
\end{aligned}
$$

It follows from (4.18) that

$$
\begin{aligned}
f_{ijkl}(\boldsymbol{p}) &= \frac{1}{\sqrt{2}}[\boldsymbol{M}^{+,2}_{ijkl}(\boldsymbol{p}) - \boldsymbol{M}^{4}_{ijkl}(\boldsymbol{p})]u_1(\lambda) + \boldsymbol{M}^{+,1}_{ijkl}(\boldsymbol{p})u_2(\lambda) \\
&\quad + \frac{1}{4\sqrt{2}}[\sqrt{2}\boldsymbol{M}^{+,1}_{ijkl}(\boldsymbol{p}) + 3\boldsymbol{M}^{+,2}_{ijkl}(\boldsymbol{p}) - 2\sqrt{3}\boldsymbol{M}^{2}_{ijkl}(\boldsymbol{p}) \\
&\quad + 3\boldsymbol{M}^{4}_{ijkl}(\boldsymbol{p})]u_3(\lambda) \\
&\quad + \frac{1}{4\sqrt{2}}[3\boldsymbol{M}^{+,2}_{ijkl}(\boldsymbol{p}) + 2\sqrt{3}\boldsymbol{M}^{2}_{ijkl}(\boldsymbol{p}) + 3\boldsymbol{M}^{4}_{ijkl}(\boldsymbol{p})]u_4(\lambda).
\end{aligned}
\tag{4.22}
$$

Using the Jacobi–Anger expansion and (4.12), we obtain

$$
\begin{aligned}
\langle T(\boldsymbol{x}), T(\boldsymbol{y})\rangle_{ijkl} &= \frac{1}{\sqrt{2}}\int_0^\infty [J_0(\lambda\rho)\boldsymbol{M}^{+,2}_{ijkl}(\boldsymbol{z}) - J_2(\lambda\rho)\boldsymbol{M}^{4}_{ijkl}(\boldsymbol{z})]\,\mathrm{d}\Phi_1(\lambda) \\
&\quad + \int_0^\infty J_0(\lambda\rho)\boldsymbol{M}^{+,1}_{ijkl}(\boldsymbol{z})\,\mathrm{d}\Phi_2(\lambda) \\
&\quad + \frac{1}{4\sqrt{2}}\int_0^\infty [J_0(\lambda\rho)[\sqrt{2}\boldsymbol{M}^{+,1}_{ijkl}(\boldsymbol{z}) + 3\boldsymbol{M}^{+,2}_{ijkl}(\boldsymbol{z})] \\
&\quad + 2\sqrt{3}J_2(\lambda\rho)\boldsymbol{M}^{2}_{ijkl}(\boldsymbol{z}) + 3J_4(\lambda\rho)\boldsymbol{M}^{4}_{ijkl}(\boldsymbol{z})]\,\mathrm{d}\Phi_3(\lambda) \\
&\quad + \frac{1}{4\sqrt{2}}\int_0^\infty [3J_0(\lambda\rho)\boldsymbol{M}^{+,2}_{ijkl}(\boldsymbol{z}) - 2\sqrt{3}J_2(\lambda\rho)\boldsymbol{M}^{2}_{ijkl}(\boldsymbol{z}) \\
&\quad + 3J_4(\lambda\rho)\boldsymbol{M}^{4}_{ijkl}(\boldsymbol{z})]\,\mathrm{d}\Phi_4(\lambda),
\end{aligned}
$$

where $\mathrm{d}\Phi_m(\lambda) = u_m(\lambda)\,\mathrm{d}\Phi(\lambda)$. Only the measure Φ_2 can have an atom in 0, because $\mathcal{C}_1 = \{D_2\}$. It follows from (4.19) that

$$
\langle T(\boldsymbol{x}), T(\boldsymbol{y})\rangle_{ijkl} = \sum_{m=1}^{4}\int_0^\infty \sum_{n=1,2,4,5} \tilde{N}_{mn}(\lambda,\rho)\boldsymbol{L}^n(\varphi)\,\mathrm{d}\Phi_m(\lambda), \tag{4.23}
$$

where the non-zero functions $\tilde{N}_{mn}(\lambda,\rho)$ are shown in Table 4.2.

To obtain the spectral expansion of the random field $T(\boldsymbol{x})$ in this particular case, we perform the following. Write down $\boldsymbol{M}$-functions in the form

$$
\begin{aligned}
\boldsymbol{M}^{+,n}_{ijkl}(\boldsymbol{p}) &= \boldsymbol{T}^{+,n}_{ijkl}\cos(0\psi), \qquad n = 1, 2, \\
\boldsymbol{M}^{n}_{ijkl}(\boldsymbol{p}) &= \boldsymbol{T}^{n,-1}_{ijkl}\sin(n\psi) + \boldsymbol{T}^{n,1}_{ijkl}\cos(n\psi) \qquad n = 2, 4.
\end{aligned}
$$

Table 4.2 *The non-zero functions* $\tilde{N}_{mn}(\lambda,\rho)$

m	n	$\tilde{N}_{mn}(\lambda,\rho)$
1	1	$\frac{1}{2\sqrt{2}}(-J_0(\lambda\rho)-3J_2(\lambda\rho))$
1	2	$\frac{1}{2\sqrt{2}}(J_0(\lambda\rho)+J_2(\lambda\rho))$
1	4	$\sqrt{2}J_2(\lambda\rho)$
1	5	$-2\sqrt{2}J_2(\lambda\rho)$
2	1	$\frac{1}{2}J_0(\lambda\rho)$
3	1	$-\frac{1}{16}J_0(\lambda\rho)-\sqrt{6}J_2(\lambda\rho)+\frac{9}{2\sqrt{2}}J_4(\lambda\rho)$
3	2	$\frac{3}{16}J_0(\lambda\rho)-\frac{3}{2\sqrt{2}}J_4(\lambda\rho)$
3	4	$\sqrt{6}J_2(\lambda\rho)-3\sqrt{2}J_4(\lambda\rho))$
3	5	$6\sqrt{2}J_4(\lambda\rho)$
4	1	$-\frac{3}{16}J_0(\lambda\rho)+\frac{\sqrt{3}}{4}J_2(\lambda\rho)+\frac{9}{16}J_4(\lambda\rho)$
4	2	$\frac{3}{16}J_0(\lambda\rho)-\frac{3}{16}J_4(\lambda\rho)$
4	4	$-\frac{\sqrt{3}}{4}J_2(\lambda\rho)-\frac{3}{4}J_4(\lambda\rho))$
4	5	$\frac{3}{2}J_4(\lambda\rho)$

Equation 4.22 takes the form

$$\begin{aligned}
f_{ijkl}(\lambda,\psi) &= \frac{1}{\sqrt{2}}[\boldsymbol{T}^{+,2}_{ijkl}\cos(0\psi)-\boldsymbol{T}^{4,-1}_{ijkl}\sin(4\psi)-\boldsymbol{T}^{4,1}_{ijkl}\cos(4\psi)]u_1(\lambda)\\
&\quad+\boldsymbol{T}^{+,1}_{ijkl}\cos(0\psi)u_2(\lambda)\\
&\quad+\frac{1}{4\sqrt{2}}[\sqrt{2}\boldsymbol{T}^{+,1}_{ijkl}\cos(0\psi)+3\boldsymbol{T}^{+,2}_{ijkl}\cos(0\psi)\\
&\quad-2\sqrt{3}\boldsymbol{T}^{2,-1}_{ijkl}\sin(2\psi)\\
&\quad-2\sqrt{3}\boldsymbol{T}^{2,1}_{ijkl}\cos(2\psi)+3\boldsymbol{T}^{4,-1}_{ijkl}\sin(4\psi)+3\boldsymbol{T}^{4,1}_{ijkl}\cos(4\psi)]u_3(\lambda)\\
&\quad+\frac{1}{4\sqrt{2}}[3\boldsymbol{T}^{+,2}_{ijkl}\cos(0\psi)+2\sqrt{3}\boldsymbol{T}^{2,-1}_{ijkl}\sin(2\psi)\\
&\quad+2\sqrt{3}\boldsymbol{T}^{2,1}_{ijkl}\cos(2\psi)+3\boldsymbol{T}^{4,-1}_{ijkl}\sin(4\psi)+3\boldsymbol{T}^{4,1}_{ijkl}\cos(4\psi)]u_4(\lambda).
\end{aligned}$$

Write the Jacobi–Anger expansion for $\mathrm{e}^{-\mathrm{i}(\boldsymbol{p},\boldsymbol{x})}$ and $\mathrm{e}^{\mathrm{i}(\boldsymbol{p},\boldsymbol{y})}$ separately and subsitute the result to the formula

$$\langle T(\boldsymbol{x}),T(\boldsymbol{y})\rangle_{ijkl}=\frac{1}{2\pi}\int_0^\infty\int_0^{2\pi}\mathrm{e}^{\mathrm{i}(\boldsymbol{p},\boldsymbol{y}-\boldsymbol{x})}f_{ijkl}(\lambda,\psi)\,\mathrm{d}\psi\,\mathrm{d}\Phi(\lambda),$$

calculate the inner integral and use Karhunen's theorem. We obtain the following result.

Denote by $\Delta^m_{\ell\ell'}$, ℓ, ℓ', $m\in\mathbb{Z}$, the matrices with the following entries:

$$\begin{aligned}
\Delta^0_{\ell\ell'} &= \delta_{\ell-\ell',0},\\
\Delta^m_{\ell\ell'} &= \frac{1}{2}(\delta_{\ell-\ell'-m,0}+\delta_{\ell-\ell'+m,0}), \qquad m>0,
\end{aligned}$$

$$\Delta^m_{\ell\ell'} = \frac{1}{4}(\delta_{\ell'-\ell+m,0} - \delta_{\ell'-\ell-m,0} + \delta_{\ell-\ell'+m,0} - \delta_{\ell-\ell'-m,0} + 2\delta_{\ell'+\ell+m,0} - 2\delta_{\ell'+\ell-m,0}), \qquad m < 0.$$

Let $\mathsf{B}^{\ell\ell' m}_{ijkl}$, ℓ, $\ell' \in \mathbb{Z}$, $1 \le m \le 4$ be the tensors

$$\begin{aligned}
\mathsf{B}^{\ell\ell'1}_{ijkl} &= \frac{1}{\sqrt{2}}[\mathsf{T}^{+,2}_{ijkl}\Delta^0_{\ell\ell'} - \mathsf{T}^{4,-1}_{ijkl}\Delta^{-4}_{\ell\ell'} - \mathsf{T}^{4,1}_{ijkl}\Delta^4_{\ell\ell'}],\\
\mathsf{B}^{\ell\ell'2}_{ijkl} &= \mathsf{T}^{+,1}_{ijkl}\Delta^0_{\ell\ell'},\\
\mathsf{B}^{\ell\ell'3}_{ijkl} &= \frac{1}{4\sqrt{2}}[\sqrt{2}\mathsf{T}^{+,1}_{ijkl}\Delta^0_{\ell\ell'} + 3\mathsf{T}^{+,2}_{ijkl}\Delta^0_{\ell\ell'} - 2\sqrt{3}\mathsf{T}^{2,-1}_{ijkl}\Delta^{-2}_{\ell\ell'}\\
&\quad - 2\sqrt{3}\mathsf{T}^{2,1}_{ijkl}\Delta^2_{\ell\ell'} + 3\mathsf{T}^{4,-1}_{ijkl}\Delta^{-4}_{\ell\ell'} + 3\mathsf{T}^{4,1}_{ijkl}\Delta^4_{\ell\ell'}]\\
\mathsf{B}^{\ell\ell'4}_{ijkl} &= [3\mathsf{T}^{+,2}_{ijkl}\Delta^0_{\ell\ell'} + 2\sqrt{3}\mathsf{T}^{2,-1}_{ijkl}\Delta^{-2}_{\ell\ell'} + 2\sqrt{3}\mathsf{T}^{2,1}_{ijkl}\Delta^2_{\ell\ell'} + 3\mathsf{T}^{4,-1}_{ijkl}\Delta^{-4}_{\ell\ell'}\\
&\quad + 3\mathsf{T}^{4,1}_{ijkl}\Delta^4_{\ell\ell'}].
\end{aligned}$$

Let $Z^{\ell m}_{ij}(\lambda)$ be the centred random measures on $[0,\infty)$ with

$$\mathsf{E}[Z^{\ell m}_{ij}(A)Z^{\ell' m'}_{kl}(B)] = \delta_{mm'}\mathsf{B}^{\ell\ell' m}_{ijkl}\Phi_m(A \cap B)$$

for all Borel sets A, $B \subseteq [0,\infty)$.

Theorem 37. *Assume that the matrix $f(\lambda,0)$ may only take values in $\mathcal{C}$. Then the one-point correlation tensor of the homogeneous and isotropic random field $T(\boldsymbol{x})$ is*

$$\langle T(\boldsymbol{x})\rangle_{ij} = C\delta_{ij}, \qquad C \in \mathbb{R}.$$

Its two-point correlation tensor has the form (4.23), *where Φ_m are four finite measures on $[0,\infty)$ and only Φ_2 may have an atom in 0. The field has the form*

$$\begin{aligned}
T_{ij}(\rho,\varphi) &= C\delta_{ij} + \sum_{\ell=0}^{\infty}\sum_{m=1}^{4}\int_0^\infty J_\ell(\lambda\rho)\,\mathrm{d}Z^{\ell m}_{ij}(\lambda)\cos(\ell\varphi)\\
&\quad + \sum_{\ell=-\infty}^{-1}\sum_{m=1}^{4}\int_0^\infty J_{-\ell}(\lambda\rho)\,\mathrm{d}Z^{\ell m}_{ij}(\lambda)\sin(\ell\varphi).
\end{aligned}$$

4.5.2 Applications

Interpretations of specific correlations

For reference, we write the representation above as

$$\begin{aligned}
\mathsf{T}_{ijkl}(\boldsymbol{z}) &= H_1\delta_{ij}\delta_{kl} + H_2\left(\delta_{ik}\delta_{jl} + \delta_{il}\delta_{jk}\right) + H_4\left(\delta_{ij}z_k z_l + \delta_{kl}z_i z_j\right)\\
&\quad + H_5 z_i z_j z_k z_l.
\end{aligned}$$

Given that T has diagonal and off-diagonal components, there are four special cases of T_{ijkl} which shed light on the physical meaning of H_ns:

1. auto-correlations of diagonal terms: $\mathcal{T}_{1111} = H_1 + 2H_2 + 2z_1^2 H_4 + z_1^4 H_5$ and $\mathcal{T}_{2222} = H_1 + 2H_2 + 2z_2^2 H_4 + z_2^4 H_5$.
2. cross-correlation of diagonal terms: $\mathcal{T}_{1122} = H_1 + \left(z_1^2 + z_2^2\right) H_4 + z_1^2 z_2^2 H_5$.
3. auto-correlation of an off-diagonal term: $\mathcal{T}_{1212} = H_2 + z_1^2 x_2^2 H_5$.
4. cross-correlation of a diagonal with an off-diagonal term:

$$\langle T_{11}(\mathbf{0}) T_{12}(\boldsymbol{z}) \rangle = \mathcal{T}_{1112} = z_1 x_2 H_4 + z_1^3 x_2 H_5.$$

Just as in TRF of rank 1, we can determine these four correlations for a specific physical situation. Without loss of generality (due to wide-sense isotropy), when $\boldsymbol{z} \equiv (z_1, z_2)$ is chosen equal to $(z, 0)$, we find

$$\begin{aligned} H_1 &= \mathcal{T}_{2222} - 2\mathcal{T}_{1212} \\ H_2 &= \mathcal{T}_{1212} \\ H_4 &= z^{-2} \left(\mathcal{T}_{1122} - \mathcal{T}_{2222} - 2\mathcal{T}_{1212}\right) \\ H_5 &= z^{-4} \left(\mathcal{T}_{1111} + \mathcal{T}_{2222} - 2\mathcal{T}_{1122}\right). \end{aligned} \tag{4.24}$$

For example, when $\mathcal{T}$ is the anti-plane elasticity tensor for a given resolution (or mesoscale) (Ostoja-Starzewski 2008), we can use micromechanics or experiments and then determine the best fits of H_α ($\alpha = 1, 2, 4, 5$) coefficients, providing the positive-definiteness of $\mathcal{T}$ is imposed. However, when $\mathcal{T}$ represents the dependent field quantity, then a restriction dictated by the field equation needs to be imposed.

TRF with a local isotropic property

Take $T_{ij} = T\delta_{ij}$, where the axial component T is the scalar random field describing the randomness of such a medium. Since $T_{11} = T_{22}$ and $T_{12} = 0$ must hold everywhere, $\mathcal{T}_{1111} = \mathcal{T}_{2222} = \mathcal{T}_{1122}$ and $\mathcal{T}_{1212} = 0$. Hence, $H_2 = H_4 = H_5 = 0$ and only $H_1 \neq 0$ is retained, and that is the function modelling the correlations in T. One example is the constitutive response (e.g. conductivity $k_{ij} = k\delta_{ij}$) in conventional models of stochastic partial differential equations; see Subsection 4.8.1 for a model with anisotropy.

When T_{ij} represents a dependent field quantity, then a restriction dictated by an appropriate field equation needs to be imposed. In the following, by analogy to what was reported in (Malyarenko & Ostoja-Starzewski 2014), we consider T being either the in-plane stress or the in-plane strain.

Restriction imposed on the correlation tensor when T is divergence-free

When T represents an in-plane stress field in the absence of any body forces, it is governed by $\operatorname{div} T = \mathbf{0}$ or $T_{ij},_j = 0$; here T is assumed to be symmetric ($T_{ij} = T_{ji}$). We now find two (equivalent) restrictions imposed on the correlation function of T:

$$\frac{\partial}{\partial z_j^1}\mathbf{T}_{ijkl}=0,\qquad \frac{\partial}{\partial z_l^2}\mathbf{T}_{ijkl}=0,$$

where the coordinate systems $\boldsymbol{z}^1$ and $\boldsymbol{z}^2$ are related by a shift $\boldsymbol{z}^1=\boldsymbol{z}^2+\boldsymbol{z}$.

Let us rewrite (4.21) as

$$\begin{aligned}\mathbf{T}_{ijkl}\equiv\langle T(\boldsymbol{x}),T(\boldsymbol{y})\rangle_{ijkl}&=\sum_{n=1,2,4,5}h_n(\rho)\mathbf{L}^n_{ijkl}(\varphi)\\&=\sum_{n=1,2,4,5}H_n(\rho)\mathbf{N}^n_{ijkl}(\varphi),\end{aligned}$$

where we have defined the modified functions for $n=4,5$:

$$\begin{array}{ll}H_1(z):=h_1(z) & \mathbf{N}^1_{ijkl}:=\mathbf{L}^1_{ijkl}(\boldsymbol{z})=\delta_{ij}\delta_{kl}\\ H_2(z):=h_2(z) & \mathbf{N}^2_{ijkl}:=\mathbf{L}^2_{ijkl}(\boldsymbol{z})=\delta_{ik}\delta_{jl}+\delta_{il}\delta_{jk}\\ H_4(z):=h_4(z)/z^2 & \mathbf{N}^4_{ijkl}:=z^2\mathbf{L}^4_{ijkl}(\boldsymbol{z})=\delta_{ij}z_kz_l+\delta_{kl}z_iz_j\\ H_5(z):=h_5(z)/z^4 & \mathbf{N}^5_{ijkl}:=z^4\mathbf{L}^5_{ijkl}(\boldsymbol{z})=z_iz_jz_kz_l.\end{array}$$

Now, writing $z_j\equiv z_j^1$, we obtain

$$\begin{aligned}0=\mathbf{T}_{ijkl},_j=\;&H_1'\tfrac{z_j}{z}\mathbf{N}^1_{ijkl}+H_1\tfrac{d}{dz_j}\mathbf{N}^1_{ijkl}+H_2'\tfrac{z_j}{z}\mathbf{N}^2_{ijkl}+H_2\tfrac{d}{dz_j}\mathbf{N}^2_{ijkl}\\&+H_4'\tfrac{z_j}{z}\mathbf{N}^4_{ijkl}+H_4\tfrac{d}{dz_j}\mathbf{N}^4_{ijkl}+H_5'\tfrac{z_j}{z}\mathbf{N}^5_{ijkl}+H_5\tfrac{d}{dz_j}\mathbf{N}^5_{ijkl}.\end{aligned}$$

where the prime denotes the ordinary derivative d/dz. Working it out more explicitly, on account of (4.22), this general equation results in (with $\delta_{ii}=2$ in 2D)

$$\begin{aligned}0=\mathbf{T}_{ijkl},_j=\;&H_1'\tfrac{z_i}{z}\delta_{kl}+H_2'\left(\tfrac{z_l}{z}\delta_{ik}+\tfrac{z_k}{z}\delta_{il}\right)\\&+H_4'\left(\tfrac{1}{z}z_iz_kz_l+\delta_{kl}z_iz\right)+H_4\left(\delta_{ik}z_l+\delta_{il}z_k+3\delta_{kl}z_i\right)\\&+H_5'zz_iz_kz_l+H_5z_iz_kz_l.\end{aligned}$$

Now, without loss of generality, choose the vector $\boldsymbol{z}\equiv(z_1,z_2)=(z,0)$, so that the eight non-trivial combinations become

$$0=\mathbf{T}_{1111},_1+\mathbf{T}_{1211},_2=H_1'+2H_2'+3z^2H_4'+5zH_4+z^4H_5'+5z^3H_5\tag{4.25a}$$

$$0=\mathbf{T}_{1112},_1+\mathbf{T}_{1212},_2=0\tag{4.25b}$$

$$0=\mathbf{T}_{1121},_1+\mathbf{T}_{1221},_2=0\tag{4.25c}$$

$$0=\mathbf{T}_{1122},_1+\mathbf{T}_{1222},_2=H_1'+z^2H_4'+3zH_4\tag{4.25d}$$

$$0=\mathbf{T}_{2111},_1+\mathbf{T}_{2211},_2=0\tag{4.25e}$$

$$0=\mathbf{T}_{2112},_1+\mathbf{T}_{2212},_2=2H_2'\tag{4.25f}$$

$$0=\mathbf{T}_{2121},_1+\mathbf{T}_{2221},_2=2H_2'\tag{4.25g}$$

$$0=\mathbf{T}_{2122},_1+\mathbf{T}_{2222},_2=0.\tag{4.25h}$$

Observe:

- conditions (4.25b), (4.25c), (4.25e) and (4.25h) are satisfied identically;
- conditions (4.25f) and (4.25g) imply $H_2=\text{const}$;
- the remaining three (scalar) functions H_1, H_4, H_5 have to satisfy the two equations (4.25a) and (4.25d).

Restriction imposed on the correlation tensor when T is a potential tensor field

When T represents an in-plane strain field, the displacement vector field plays the rôle of its potential

$$T = \left(\nabla \boldsymbol{u} + \nabla \boldsymbol{u}^\top\right)/2 \quad \text{or} \quad T_{ij} = \left(u_{i,j} + u_{j,i}\right)/2.$$

This dictates a relation between the cocorrelation tensor $\boldsymbol{T}_{ijkl}$ of T with the correlation tensor $U_{ik}(\boldsymbol{z}) := \langle u_i(\boldsymbol{0})\, u_k(\boldsymbol{z})\rangle$ of $\boldsymbol{u}$:

$$\boldsymbol{T}_{ijkl} = \left(U_{ik,jl} + U_{il,jk} + U_{jk,il} + U_{jl,ik}\right)/4. \tag{4.26}$$

Thus, with the representation of $\boldsymbol{T}_{ijkl}$ given above and the representation (3.46)

$$U_{ik}(\boldsymbol{z}) = \delta_{ik} K_0(\boldsymbol{z}) + x_i x_k K_2(\boldsymbol{z}),$$

we arrive at the general restriction

$$\begin{aligned}
&4 \textstyle\sum_{n=1,2,4,5} H_n(\rho) \boldsymbol{N}^n_{ijkl}(\varphi) \\
&= \{\delta_{ik} K_{0,jl} + (\delta_{ij}\delta_{kl} + \delta_{il}\delta_{jk}) K_2 + (\delta_{il} z_k + \delta_{kl} z_i) K_{2,j} + z_i z_k K_{2,jl}\} \\
&\quad + \{\delta_{il} K_{0,jk} + (\delta_{ij}\delta_{kl} + \delta_{ik}\delta_{jl}) K_2 + (\delta_{ik} z_l + \delta_{lk} z_i) K_{2,j} + z_i z_l K_{2,jk}\} \\
&\quad + \{\delta_{jk} K_{0,il} + (\delta_{ij}\delta_{kl} + \delta_{jl}\delta_{ik}) K_2 + (\delta_{jl} z_k + \delta_{lk} z_j) K_{2,i} + z_j z_k K_{2,il}\} \\
&\quad + \{\delta_{jl} K_{0,ik} + (\delta_{ij}\delta_{kl} + \delta_{jk}\delta_{il}) K_2 + (\delta_{jk} z_l + \delta_{lk} z_j) K_{2,i} + z_j z_l K_{2,ik}\}.
\end{aligned}$$

This is rewritten more explicitly as

$$\begin{aligned}
&4\left[H_1 \delta_{ij}\delta_{kl} + H_2(\delta_{ik}\delta_{jl} + \delta_{il}\delta_{jk}) + H_4(z_k z_l \delta_{ij} + z_l z_j \delta_{kl}) + H_i z_i z_j z_k z_l\right] \\
&= \left\{\delta_{ik}\left[K_0'' \frac{z_l z_j}{z^2} + K_0'\left(\delta_{jl} - \frac{z_j z_l}{z^2}\right)\right] + (\delta_{ij}\delta_{kl} + \delta_{il}\delta_{jk}) K_2 \right.\\
&\quad \left. + (\delta_{il} z_k + \delta_{kl} z_i) K_2' \frac{z_j}{z} + z_i z_k \left[K_2'' \frac{z_l z_j}{z^2} + K_2' \frac{1}{z}\left(\delta_{jl} - \frac{z_j z_l}{z^2}\right)\right]\right\} \\
&\quad + \left\{\delta_{il}\left[K_0'' \frac{z_k z_j}{z^2} + K_0'\left(\delta_{jk} - \frac{z_j z_k}{z^2}\right)\right] + (\delta_{ij}\delta_{kl} + \delta_{ik}\delta_{jl}) K_2 \right.\\
&\quad \left. + (\delta_{ik} z_l + \delta_{kl} z_i) K_2' \frac{z_j}{z} + z_i z_l \left[K_2'' \frac{z_k z_j}{z^2} + K_2' \frac{1}{z}\left(\delta_{jk} - \frac{z_j z_k}{z^2}\right)\right]\right\} \\
&\quad + \left\{\delta_{jk}\left[K_0'' \frac{z_l z_i}{z^2} + K_0'\left(\delta_{li} - \frac{z_i z_l}{z^2}\right)\right] + (\delta_{ij}\delta_{kl} + \delta_{jl}\delta_{ik}) K_2 \right.\\
&\quad \left. + (\delta_{jl} z_k + \delta_{kl} z_j) K_2' \frac{z_i}{z} + z_j z_k \left[K_2'' \frac{z_l z_i}{z^2} + K_2' \frac{1}{z}\left(\delta_{il} - \frac{z_i z_l}{z^2}\right)\right]\right\} \\
&\quad + \left\{\delta_{jl}\left[K_0'' \frac{z_k z_i}{z^2} + K_0'\left(\delta_{ki} - \frac{z_i z_k}{z^2}\right)\right] + (\delta_{ij}\delta_{kl} + \delta_{jk}\delta_{il}) K_2 \right.\\
&\quad \left. + (\delta_{jk} z_l + \delta_{kl} z_j) K_2' \frac{z_i}{z} + z_j z_l \left[K_2'' \frac{z_k z_i}{z^2} + K_2' \frac{1}{z}\left(\delta_{ik} - \frac{z_i z_k}{z^2}\right)\right]\right\}.
\end{aligned}$$

Now, choose the vector $\boldsymbol{z} \equiv (z_1, z_2) = (r, 0)$, so that the three non-trivial autocorrelations and two cross-correlations become:

$$
\begin{aligned}
&(i,j,k,l) = (1,1,1,1) : H_1 + 2H_2 + z^2 H_4 + z^4 H_5 = K_0'' + K_2 + 2zK_2' + z^2 K_2'' \\
&(i,j,k,l) = (2,2,2,2) : H_1 + 2H_2 = K_0'/z + 2K_2 \\
&(i,j,k,l) = (1,2,1,2) : 4H_2 = K_0'/z + zK_2' + K_2 + K_0'' \\
&(i,j,k,l) = (1,1,2,2) : H_1 + z^2 H_4 = K_2 + zK_2' \\
&(i,j,k,l) = (1,1,1,2) : 0 = 0.
\end{aligned}
\tag{4.27}
$$

It turns out that, effectively, the fifth equation is trivially satisfied. The first four equations can be reduced to a single relation

$$
3H_1 + 3z^2 H_4 + z^4 H_5 = 3K_2 + 2zK_2' + z^2 K_2'', \tag{4.28}
$$

showing that, in a specific model or application, K_2 needs to be chosen first. The above relation would then lead to a restriction on H_1, H_4 and H_5 functions. Once these are set, one may proceed to determine H_2 and K_0.

Note: The Euler–Cauchy equation (4.28) for K_2 is a linear inhomogeneous ordinary differential equation.

Special case #1: when the material has a very high shear stiffness relative to bulk stiffness, then the shear strains are practically equal to zero everywhere, so $\boldsymbol{T}_{1212} = 0$. Then, (4.26) simplifies to

$$
\begin{aligned}
&(i,j,k,l) = (1,1,1,1) : H_1 + 2H_2 + 2z^2 H_4 + z^4 H_5 = K_0'' + K_2 + 2zK_2' + z^2 K_2'' \\
&(i,j,k,l) = (2,2,2,2) : H_1 + 2H_2 = K_0'/z + 2K_2 \\
&(i,j,k,l) = (1,2,1,2) : 0 = K_0'/z + zK_2' + K_2 + K_0'' \\
&(i,j,k,l) = (1,1,2,2) : H_1 + z^2 H_4 = K_2 + zK_2'.
\end{aligned}
$$

Now, these four equations can be reduced to a single relation

$$
3H_1 + 4H_2 + 3z^2 H_4 + z^4 H_5 = 3K_2 + 2zK_2' + z^2 K_2''. \tag{4.29}
$$

Note: Comparing with (4.28), the inhomogeneous term in the Euler–Cauchy equation (4.29) now also contains $4H_2$.

Special case #2: when the material has a very high bulk stiffness relative to the shear stiffness, then the normal strains are practically equal to zero everywhere, so $T_{1111} = T_{2222} = T_{1122} = 0$. In this case, (4.27) simplifies to

$$
\begin{aligned}
(i,j,k,l) = (1,1,1,1) &: \quad 0 = K_0'' + K_2 + 2zK_2' + z^2 K_2'' \\
(i,j,k,l) = (2,2,2,2) &: \quad 0 = K_0'/z + 2K_2 \\
(i,j,k,l) = (1,2,1,2) &: \quad 4H_2 = K_0'/z + zK_2' + K_2 + K_0'' \\
(i,j,k,l) = (1,1,2,2) &: \quad 0 = K_2 + zK_2'.
\end{aligned}
$$

Now, these four equations can be reduced to a single relation

$$-4H_2 = 3K_2 + 2zK_2' + z^2K_2''. \tag{4.30}$$

Note: The Euler–Cauchy equation (4.30) for K_2 is a linear inhomogeneous ordinary differential equation.

4.6 TRFs in Micropolar Continuum Mechanics

4.6.1 Asymmetric Stress Fields

The conservation equations of linear and angular momenta in a micropolar continuum have been given in Chapter 1. Henceforth, focusing on a static problem in the absence of body forces and moments, we have

$$\sigma_{ji,j} = 0, \tag{4.31a}$$

$$\boldsymbol{\varepsilon}_{ijk}\sigma_{jk} + \mu_{ji,j} = 0. \tag{4.31b}$$

For a statistically homogeneous case, from (4.31b) it follows that

$$\langle \boldsymbol{\varepsilon}_{ijk}\sigma_{jk}(\boldsymbol{z})\boldsymbol{\varepsilon}_{prs}\sigma_{rs}(\boldsymbol{z}+\boldsymbol{z}_1)\rangle = \langle \mu_{ji,j}(\boldsymbol{z})\mu_{rp,r}(\boldsymbol{z}+\boldsymbol{z}_1)\rangle . \tag{4.32}$$

The left-hand side may be written as

$$LHS = \boldsymbol{\varepsilon}_{ijk}\boldsymbol{\varepsilon}_{prs}Q_{jk}^{rs},$$

where

$$\boldsymbol{Q}_{jk}^{rs}(\boldsymbol{z}) := \langle \sigma_{jk}(\boldsymbol{z}_1)\sigma_{rs}(\boldsymbol{z}_1+\boldsymbol{z})\rangle$$

is the correlation function of the stress field. Since σ_{jk} is generally asymmetric but, by assumption, statistically isotropic, we have

$$\boldsymbol{Q}_{jk}^{rs}(\boldsymbol{z}) = \boldsymbol{Q}_{rs}^{jk}(\boldsymbol{z}). \tag{4.33}$$

Now, given that a rank 2 tensor can be interpreted as a dyadic, we may begin with the general representation of a rank 4 correlation tensor (Batchelor 1982)

$$\begin{aligned}\boldsymbol{Q}_{jk}^{rs}(\boldsymbol{z}) &= An_jn_kn_rn_s + Bn_jn_k\delta_{rs} + Cn_kn_r\delta_{js} + Dn_rn_s\delta_{jk} + En_jn_s\delta_{kr} \\ &\quad + Fn_jn_r\delta_{ks} + Gn_kn_s\delta_{jr} + H\delta_{jk}\delta_{rs} + I\delta_{jr}\delta_{ks} + J\delta_{js}\delta_{kr},\end{aligned}$$

which involves 10 functions: $A(\boldsymbol{z})$, $B(\boldsymbol{z})$, ..., $J(\boldsymbol{z})$. But, by the homogeneity property, we also have

$$\begin{aligned}\boldsymbol{Q}_{rs}^{jk}(\boldsymbol{z}) &= An_jn_kn_rn_s + Bn_rn_s\delta_{jk} + Cn_jn_s\delta_{rk} + Dn_jn_k\delta_{rs} + En_kn_r\delta_{sj} \\ &\quad + Fn_rn_j\delta_{ks} + Gn_sn_k\delta_{jr} + H\delta_{rs}\delta_{jk} + I\delta_{ks}\delta_{jr} + J\delta_{kr}\delta_{js}.\end{aligned}$$

In view of (4.33), we obtain

$$B = D, \quad C = E,$$

so that $\boldsymbol{Q}_{jk}^{rs}(\boldsymbol{z})$ actually employs 8 unknown functions (A, B, C, F, G, H, I, J):

$$\begin{aligned}\boldsymbol{Q}_{jk}^{rs}(\boldsymbol{z}) = {} & An_jn_kn_rn_s + B\left(n_jn_k\delta_{rs} + n_rn_s\delta_{jk}\right) + C\left(n_kn_r\delta_{js} + n_jn_k\delta_{kr}\right) \\ & + Fn_jn_k\delta_{ks} + Gn_kn_s\delta_{jr} + H\delta_{jk}\delta_{rs} + I\delta_{jr}\delta_{ks} + J\delta_{js}\delta_{kr}.\end{aligned}$$

It follows separately from (4.32) that its right-hand side may be written as

$$RHS = R_i^p,$$

where $R_i^p(\boldsymbol{z}) := \langle \mu_{ji},_j(\boldsymbol{z})\mu_{rp},_r(\boldsymbol{z}+\boldsymbol{z}_1)\rangle$ is the correlation function of a homogeneous couple-stress TRF. This being a rank 2 tensor, its representation has a familiar form

$$R_i^p(\boldsymbol{z}) = f(\boldsymbol{z})n_in_p + g(\boldsymbol{z})\delta_{ip}$$

with

$$R_i^p(\boldsymbol{z}) = R_p^i(\boldsymbol{z}).$$

Equation 4.31a implies

$$\boldsymbol{Q}_{jk}^{rs},_j(\boldsymbol{z}) = 0, \qquad \boldsymbol{Q}_{rs}^{jk},_r(\boldsymbol{z}) = 0$$

which, respectively, lead to

$$\begin{aligned}\boldsymbol{Q}_{jk}^{rs},_j(\boldsymbol{z}) = {} & \left(A' + 2A/z\right)n_kn_rn_s + B'\left(n_k\delta_{rs} + n_kn_rn_s\right) \\ & + \tfrac{1}{r}B\left(2n_k\delta_{rs} + n_s\delta_{kr} + n_r\delta_{sk} - n_kn_rn_s\right) \\ & + C'\left(n_kn_rn_s + n_s\delta_{kr}\right) + C\left(n_r\delta_{ks} + n_k\delta_{rs} + 2n_s\delta_{kr} - n_kn_rn_s\right)/z \\ & + 2Fn_r\delta_{ks}/z + F'n_r\delta_{ks} + G'n_kn_rn_s + G\left(n_s\delta_{kr} + n_k\delta_{sr} - 2n_kn_rn_s\right)/z \\ & + H'n_k\delta_{rs} + I'n_r\delta_{ks} + J'n_s\delta_{kr}.\end{aligned}$$

By considering all the combinations of values $(1,2,3)$ of indices (k,r,s), we find a set of four equations

$$\left\{\begin{array}{c} A' + A/z + B' - 2B/z + G' - 2G/z + C' - 2C/z = 0, \\ B/z - G/z + J' + C' + 2C/z = 0, \\ B/z + 2F/z + F' + I' + C/z = 0, \\ B' + 2B/z + G/z + H' + C/z = 0. \end{array}\right. \tag{4.34}$$

Equation 4.31b implies

$$\boldsymbol{\varepsilon}_{ijk}\boldsymbol{\varepsilon}_{prs}\boldsymbol{Q}_{jk}^{rs}(\boldsymbol{z}) = R_i^p(\boldsymbol{z}),$$

or, more explicitly,

$$\begin{aligned}\boldsymbol{\varepsilon}_{ijk}\boldsymbol{\varepsilon}_{prs} & \left[An_kn_rn_s + B\left(n_jn_k\delta_{rs} + n_rn_s\delta_{jk}\right) + C\left(n_kn_r\delta_{js} + n_jn_s\delta_{kr}\right)\right. \\ & \left. + Fn_jn_r\delta_{ks} + Gn_kn_s\delta_{jr} + H\delta_{jk}\delta_{rs} + I\delta_{jr}\delta_{ks} + J\delta_{js}\delta_{kr}\right] \\ & = fn_in_p + g\delta_{ip}, \qquad i,j,k,p,r,s, = 1,2,3.\end{aligned}$$

By considering all these combinations

$$\begin{array}{ccc} (i,p) = (1,1), & (i,p) = (1,2), & (i,p) = (1,3) \\ (i,p) = (2,1), & (i,p) = (2,2), & (i,p) = (2,3) \\ & (i,p) = (3,3), & \end{array}$$

with values $(1\ldots3)$ of the indices $(k,\ r,\ s)$, we find a set of two equations

$$\begin{cases} f\left(n_1n_1+n_2n_2+n_3n_3\right)+g=2I-2J, \\ 2C-F-G=f. \end{cases} \tag{4.35}$$

To sum up, (4.34) and (4.35) represent $4+2=6$ equations for 10 unknown functions: $(A,\ B,\ C,\ F,\ G,\ H,\ I,\ J)$ and $(f,\ g)$.

4.6.2 Correlations of Displacement, Rotation and Strain TRFs

First, with reference to Equation 1.75, recall that in the micropolar media the deformation measure conjugate to the Cauchy stress is the strain. Its correlation function is

$$\boldsymbol{\Gamma}_{ij}^{pr}(\boldsymbol{z}) := \langle \gamma_{ij}(\boldsymbol{z}_1)\gamma_{pr}(\boldsymbol{z}_1+\boldsymbol{z})\rangle,$$

while the correlation function of the microrotation field is

$$\Phi_k^l(\boldsymbol{z}) := \langle \varphi_k(\boldsymbol{z}_1)\varphi_l(\boldsymbol{z}_1+\boldsymbol{z})\rangle.$$

In view of (1.75), given that $\boldsymbol{\Gamma}_{ij}^{pr}(\boldsymbol{z})$ is wide-sense homogeneous,

$$\begin{aligned}
\boldsymbol{\Gamma}_{ij}^{pr}(\boldsymbol{z}) &:= \left\langle \left(u_{j,i} - \boldsymbol{\varepsilon}_{kij}\varphi_k\right)\big|_{\mathbf{0}} \left(u_{r,p} - \boldsymbol{\varepsilon}_{lpr}\varphi_l\right)\big|_{\boldsymbol{z}}\right\rangle \\
&= \left\langle u_{j,i}\big|_{\mathbf{0}}\, u_{r,p}\big|_{\boldsymbol{z}}\right\rangle - \left\langle u_{j,i}\big|_{\mathbf{0}}\, \boldsymbol{\varepsilon}_{lpr}\varphi_l\big|_{\boldsymbol{z}}\right\rangle - \left\langle \boldsymbol{\varepsilon}_{kij}\varphi_k\big|_{\mathbf{0}}\, u_{r,p}\big|_{\boldsymbol{z}}\right\rangle \\
&\quad + \left\langle \boldsymbol{\varepsilon}_{kij}\varphi_k\big|_{\mathbf{0}}\, \boldsymbol{\varepsilon}_{lpr}\varphi_l\big|_{\boldsymbol{z}}\right\rangle \\
&= \left\langle u_{j,i}\big|_{\mathbf{0}}\, u_{r,p}\big|_{\boldsymbol{z}}\right\rangle + \left\langle \boldsymbol{\varepsilon}_{kij}\varphi_k\big|_{\mathbf{0}}\, \boldsymbol{\varepsilon}_{lpr}\varphi_l\big|_{\boldsymbol{z}}\right\rangle,
\end{aligned}$$

where the last line follows from the basic tenet of the micropolar theory that $\boldsymbol{u}$ and $\boldsymbol{\varphi}$ are independent kinematic degrees of freedom at every point of the continuum. Hence,

$$\begin{gathered}
\boldsymbol{\Gamma}_{ij}^{pr}(\boldsymbol{s}) = U_{j,i}^{r,p}(\boldsymbol{z}) + \boldsymbol{\varepsilon}_{kij}\boldsymbol{\varepsilon}_{lpr}\Phi_k^l(\boldsymbol{z}) \\
\text{or} \\
\boldsymbol{\Gamma}(\boldsymbol{z}) = \boldsymbol{\nabla}\boldsymbol{\nabla}U_j^r(\boldsymbol{z}) + \mathcal{E}\mathcal{E}\boldsymbol{\Phi}(\boldsymbol{z}),
\end{gathered} \tag{4.36}$$

where $\mathcal{E}$ is the Levi-Civita tensor. The above relation involves correlation functions of:

(i) the generally asymmetric second-rank tensor $\boldsymbol{\gamma}$ having the form (involving eight coefficients $A_\gamma \ldots J_\gamma$)

$$\begin{aligned}
\boldsymbol{\Gamma}_{ij}^{pr}(\boldsymbol{z}) &= A_\gamma n_in_jn_pn_r + B_\gamma\left(n_in_j\delta_{pr}+n_pn_r\delta_{ij}\right) + C_\gamma\left(n_jn_p\delta_{ir}+n_in_r\delta_{jp}\right) \\
&\quad + F_\gamma n_in_p\delta_{jr} + G_\gamma n_jn_r\delta_{ip} + H_\gamma\delta_{ij}\delta_{pr} + I_\gamma\delta_{ip}\delta_{jr} + J_\gamma\delta_{ir}\delta_{jp};
\end{aligned}$$

(ii) the generally asymmetric second-rank tensor $\boldsymbol{\nabla u}$ having the form (involving eight coefficients $A_u \ldots J_u$)

$$\begin{aligned}
\boldsymbol{U}_{j,i}^{r,p}(\boldsymbol{z}) &= A_un_in_jn_pn_r + B_u\left(n_in_j\delta_{pr}+n_pn_r\delta_{ij}\right) + C_u\left(n_jn_p\delta_{ir}+n_in_r\delta_{jp}\right) \\
&\quad + F_un_in_p\delta_{jr} + G_un_jn_r\delta_{ip} + H_u\delta_{ij}\delta_{pr} + I_u\delta_{ip}\delta_{jr} + J_u\delta_{ir}\delta_{jp};
\end{aligned}$$

(iii) the first rank tensor $\boldsymbol{\varphi}$ having the form (involving 2 coefficients f, g)

$$\Phi_k^l(\boldsymbol{z}) = f z_k z_l + g\delta_{kl}.$$

Now, $\boldsymbol{\varepsilon}_{kij}\boldsymbol{\varepsilon}_{lpr}\Phi_k^l(\boldsymbol{r})$ yields

$$\begin{aligned} \boldsymbol{\varepsilon}_{kij}\boldsymbol{\varepsilon}_{lpr}\Phi_k^l(\boldsymbol{z}) &= \boldsymbol{\varepsilon}_{kij}\boldsymbol{\varepsilon}_{lpr}\left[f z_k z_l + g\delta_{kl}\right] \\ = \boldsymbol{\varepsilon}_{kij}\boldsymbol{\varepsilon}_{lpr} f z_k z_l + \boldsymbol{\varepsilon}_{lij}\boldsymbol{\varepsilon}_{lpr} g &= \boldsymbol{\varepsilon}_{kij}\boldsymbol{\varepsilon}_{lpr} f z_k z_l + \left(\delta_{ip}\delta_{jr} - \delta_{ir}\delta_{jp}\right) g, \end{aligned}$$

which has to equal $\boldsymbol{\Gamma}_{ij}^{pr}(\boldsymbol{z}) - \boldsymbol{U}_{j,i}^{r,p}(\boldsymbol{z})$. Thus, rewriting (4.36) as

$$\boldsymbol{e}_{kij}\boldsymbol{\varepsilon}_{lpr}\Phi_k^l(\boldsymbol{z}) = \boldsymbol{\Gamma}_{ij}^{pr}(\boldsymbol{z}) - \boldsymbol{U}_{j,i}^{r,p}(\boldsymbol{z}),$$

we have to examine

$$\begin{aligned} &\boldsymbol{\varepsilon}_{kij}\boldsymbol{\varepsilon}_{lpr} f z_k z_l + \left(\delta_{ip}\delta_{jr} - \delta_{ir}\delta_{jp}\right) g \\ &= A_\gamma n_i n_j n_p n_r + B_\gamma\left(n_i n_j \delta_{pr} + n_p n_r \delta_{ij}\right) + C_\gamma\left(n_j n_p \delta_{ir} + n_i n_r \delta_{jp}\right) \\ &\quad + F_\gamma n_i n_p \delta_{jr} + G_\gamma n_j n_r \delta_{ip} + H_\gamma \delta_{ij}\delta_{pr} + I_\gamma \delta_{ip}\delta_{jr} + J_\gamma \delta_{ir}\delta_{jp} \\ &\quad - \left[A_u n_i n_j n_p n_r + B_u\left(n_i n_j \delta_{pr} + n_p n_r \delta_{ij}\right) + C_u\left(n_j n_p \delta_{ir} + n_i n_r \delta_{jp}\right)\right] \\ &\quad - \left[F_u n_i n_p \delta_{jr} + G_u n_j n_r \delta_{ip} + H_u \delta_{ij}\delta_{pr} + I_u \delta_{ip}\delta_{jr} + J_u \delta_{ir}\delta_{jp}\right]. \end{aligned}$$

The 45 cases are given in Ostoja-Starzewski et al. (2015). Collecting all these results, gives

$$\begin{gathered} A_\gamma - A_u = B_\gamma - B_u = H_\gamma - H_u = 0, \\ C_\gamma - C_u = f, \quad F_\gamma - F_u = G_\gamma - G_u = -f, \quad I_\gamma - I_u = -(J_\gamma - J_u), \\ I_\gamma - I_u = f\left(n_{11} + n_{22} + n_{33}\right) + g, \end{gathered}$$

so that, we have eight relations for 18 unknown coefficients in three correlation functions.

4.7 TRFs of Constitutive Responses

4.7.1 From a Random Microstructure to Mesoscale Response

With reference to Table 1.1, consider the response on a mesoscale, i.e. a scale finite relative to the heterogeneity size such as indicated in Figures 0.1 and 4.1. The basic question concerns the trend – either rapid, moderate, or slow – of mesoscale constitutive response, with L/d increasing, to the situation postulated by Hill (1963):

> a sample that (a) is structurally entirely typical of the whole mixture on average, and (b) contains a sufficient number of inclusions for the apparent overall moduli to be effectively independent of the surface values of traction and displacement, so long as these values are macroscopically uniform.

In essence, (a) is a statement about the material's statistics being spatially homogeneous and ergodic, while (b) is a pronouncement on the independence of effective constitutive response with respect to the boundary conditions. Both

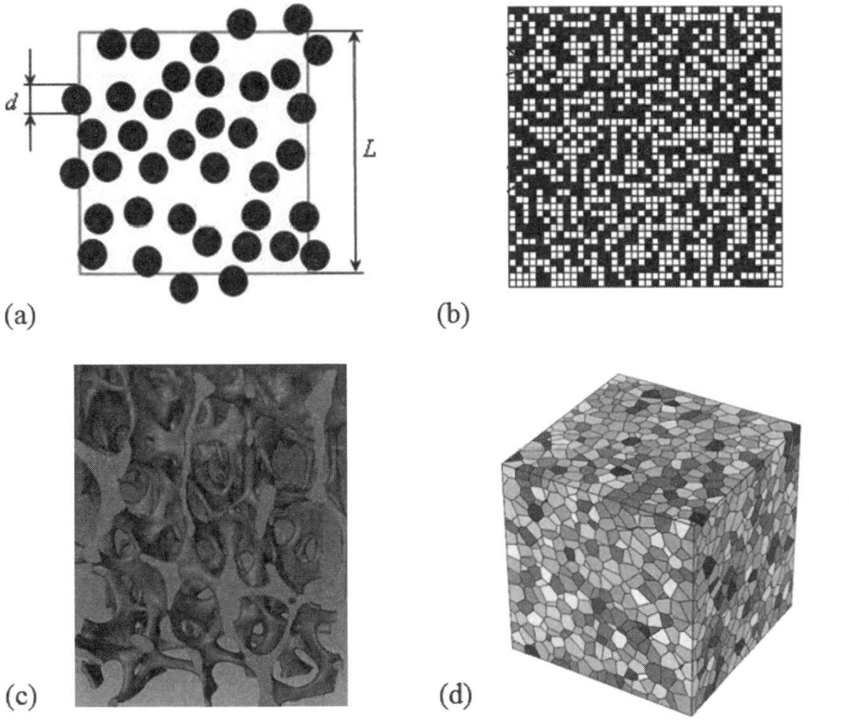

Figure 4.1 (a) Circular-inclusion composite, showing a mesoscale window. (b) Random two-phase checkerboard in 2D. (c) Deproteinised trabecular bone (Hamed et al. 2015). (d) Random polycrystal in 3D where the colour scale represents a different orientation of each grain.

of these are issues of mesoscale L of the domain of random microstructure over which smoothing (or homogenisation) is being done relative to the microscale d and macroscale L_{macro}.

Introducing a non-dimensional parameter characterising the mesoscale

$$\delta = L/d, \tag{4.37}$$

consider the random mesoscale material $\mathcal{B}_\delta$ occupying a square- (or cubic)-shaped domain $\mathcal{D}_\delta \in E^d$, $d = 2$ (3) with boundary $\partial\mathcal{D}$. The set of its possible states is

$$\mathcal{S}_\delta = \{S_\delta(\omega); \omega \in \Omega\}\,. \tag{4.38}$$

This is shown in Figure 4.1(b) with a square-shaped domain ($d = 2$). Here $S_\delta(\omega)$ is one deterministic realisation, described more specifically by a mesoscale window of the stiffness tensor field over $\mathcal{D}_\delta$. Properties on a mesoscale are also described by an adjective *apparent*, as opposed to *effective*. The latter term pertains to the limit $L/d \to \infty$ as it connotes the passage to the RVE, while, unless there is finite-scale periodicity in the medium, any finite mesoscale involves statistical scatter and, therefore, describes some *Statistical Volume Element* (SVE). Clearly, $\delta = 0$ signifies the pointwise (finest scale) description of the material, while $\delta \to \infty$ is the RVE limit.

In general, we also use $\mathcal{B}_\delta$ to denote a placement or reference configuration of the body in $\partial\mathcal{B}_\delta \subset \mathbb{R}^d$, on which the deformation map, force and stress fields are defined. The domain $\mathcal{B}_\delta$ and its boundary $\partial\mathcal{B}_\delta$ is assumed to be regular in the sense conventionally assumed in continuum mechanics ($\partial\mathcal{B}_\delta$ is piecewise smooth and Lipschitz, $\mathcal{B}_\delta$ and $\partial\mathcal{B}_\delta$ each consist of finite numbers of disjoint components).

The setting is one of quasi-static loading with negligible body forces, so that the body is governed locally by the equilibrium equation $\nabla\cdot\boldsymbol{\sigma} = \mathbf{0}$, with $\boldsymbol{\sigma}$ $(= \sigma_{ij})$ being the Cauchy stress. (The case of dynamic loading largely remains an open challenge.) Considering a body $\mathcal{B}_\delta(\omega)$ with a given microstructure, as a result of some boundary conditions (assuming the absence of body and inertia forces), there are stress and strain fields $\boldsymbol{\sigma}$ and $\boldsymbol{\varepsilon}$. If we represent them as a superposition of the means ($\overline{\boldsymbol{\sigma}}$ and $\overline{\boldsymbol{\varepsilon}}$) with the zero-mean fluctuations ($\boldsymbol{\sigma}'$ and $\boldsymbol{\varepsilon}'$)

$$\boldsymbol{\sigma}(\omega, \boldsymbol{x}) = \overline{\boldsymbol{\sigma}} + \boldsymbol{\sigma}'(\omega, \boldsymbol{x}) \qquad \boldsymbol{\varepsilon}(\omega, \boldsymbol{x}) = \overline{\boldsymbol{\varepsilon}} + \boldsymbol{\varepsilon}'(\omega, \boldsymbol{x}),$$

we find for the volume average of the energy density over $B_\delta(\omega)$

$$\overline{U} \equiv \frac{1}{2V} \int_{\mathcal{B}_\delta(\omega)} \boldsymbol{\sigma}(\omega, \boldsymbol{x}) : \boldsymbol{\varepsilon}(\omega, \boldsymbol{x}) dV = \frac{1}{2}\overline{\boldsymbol{\sigma} : \boldsymbol{\varepsilon}} = \frac{1}{2}\overline{\boldsymbol{\sigma}} : \overline{\boldsymbol{\varepsilon}} + \frac{1}{2}\overline{\boldsymbol{\sigma}' : \boldsymbol{\varepsilon}'} .$$

(While the overbar $(\bar{\cdot})$ indicates a volume average, $\langle\cdot\rangle$ stands for the ensemble (statistical) average.) Thus, we see that the volume average of a scalar product of stress and strain fields equals the product of their volume averages

$$\begin{array}{ccc} \overline{\boldsymbol{\sigma} : \boldsymbol{\varepsilon}} & = & \overline{\boldsymbol{\sigma}} : \overline{\boldsymbol{\varepsilon}} \\ \textbf{energetic interpretation} & = & \textbf{mechanical interpretation} \end{array} \tag{4.39}$$

if and only if $\overline{\boldsymbol{\sigma}' : \boldsymbol{\varepsilon}'} = 0$.

By analogy to statistics, if we replace the ensemble average by the spatial average, we may say that stress and strain fields are ‘spatially uncorrelated’. Relation (4.39) is called the *Hill–Mandel condition* in the (conventional) volume average form (Hill 1963; Dantu & Mandel 1963; Huet 1990; Sab 1991; Stolz 1986; Zohdi et al. 1996). This condition is satisfied by either of three different types of uniform boundary conditions on the mesoscale:

uniform displacement (also called kinematic, essential or Dirichlet) (d)

$$\boldsymbol{u}(\boldsymbol{x}) = \boldsymbol{\varepsilon}^0 \cdot \boldsymbol{x} \quad \forall \boldsymbol{x} \in \partial\mathcal{B}_\delta; \tag{4.40}$$

uniform traction (also called static, natural or Neumann) (t)

$$\boldsymbol{t}(\boldsymbol{x}) = \boldsymbol{\sigma}^0 \cdot \boldsymbol{n} \quad \forall \boldsymbol{x} \in \partial\mathcal{B}_\delta; \tag{4.41a}$$

uniform displacement-traction (also called *orthogonal-mixed*) (dt)

$$\left[\boldsymbol{u}(\boldsymbol{x}) - \boldsymbol{\varepsilon}^0 \cdot \boldsymbol{x}\right] \cdot \left[\boldsymbol{t}(\boldsymbol{x}) - \boldsymbol{\sigma}^0 \cdot \boldsymbol{n}\right] = 0 \quad \forall \boldsymbol{x} \in \partial\mathcal{B}_\delta. \tag{4.42}$$

Here we employ $\boldsymbol{\varepsilon}^0$ and $\boldsymbol{\sigma}^0$ to denote constant tensors, prescribed *a priori*, and note, from the average strain and stress theorems: $\boldsymbol{\varepsilon}^0 = \overline{\boldsymbol{\varepsilon}}$ and $\boldsymbol{\sigma}^0 = \overline{\boldsymbol{\sigma}}$.

Since the strain is presscribed in the boundary condition (4.40), it results in a *mesoscale* stiffness tensor $\mathsf{C}^d_\delta(\omega)$. Similarly, since the stress is prescribed in (4.41a), it yields a *mesoscale* compliance tensor $\mathsf{S}^t_\delta(\omega)$. Finally, (4.42) results in a *mesoscale* stiffness (or, depending on the interpretation, compliance) tensor $\mathsf{C}^{dt}_\delta(\omega)$. The argument ω indicates their random character, while δ their scale dependence, so that these tensors pertain to the statistical volume element (SVE) responses mentioned early in this book.

The SVE plays the role of a continuum point of TRF models of constitutive continuum theories, such as discussed in Chapter 1. The randomness vanishes as $\delta \to \infty$, and this is the *macroscale* response C_∞ of a representative volume element (RVE), where a deterministic continuum picture is obtained.

4.7.2 Hierarchy of Mesoscale Bounds

There is relation ordering the three mesoscale responses (Hazanov & Huet 1994):

$$[\boldsymbol{S}^t_\delta(\omega)]^{-1} \leq \boldsymbol{C}^{dt}_\delta(\omega) \leq \boldsymbol{C}^d_\delta(\omega), \tag{4.43}$$

The inequalities are understood in the same sense as in the definition of positive-definiteness: if $\boldsymbol{A}$ and $\boldsymbol{B}$ are the rank 4 tensors, then $\boldsymbol{A} \leq \boldsymbol{B}$ means $e : \boldsymbol{A} : e \leq e : \boldsymbol{B} : e$, for any symmetric rank 2 tensor e. From this we obtain $\langle \boldsymbol{S}^t_\delta(\omega)\rangle^{-1} \leq \boldsymbol{C}_\infty \leq \langle \boldsymbol{C}^d_\delta(\omega)\rangle$. The response $\boldsymbol{C}_\infty$ always lies between $\boldsymbol{S}^t_\delta(\omega)$ and $\boldsymbol{C}^d_\delta(\omega)$, and displays much weaker scale effects than the other two. It is identified with $\boldsymbol{C}^{eff}(\omega)$ in light of the spatial homogeneity of the material and then, by ergodicity, the ω-dependence is dropped.

Using the minimum potential energy principle for boundary value problems under displacement boundary condition (4.40), in combination with the assumption of spatial homogeneity and ergodicity of random microstructure, one arrives at the result that, the greater is the material domain, the softer is the ensemble averaged stiffness. Similarly, by the minimum complementary energy principle for the traction boundary condition (4.41a) one concludes that, simultaneously, the higher is the ensemble averaged compliance. Combining these results, we have a hierarchy of scale-dependent bounds on the RVE response

$$\langle \boldsymbol{S}^t_1\rangle^{-1} \leq \langle \boldsymbol{S}^t_{\delta'}\rangle^{-1} \leq \langle \boldsymbol{S}^t_\delta\rangle^{-1} \leq \boldsymbol{C}_\infty \leq \langle \boldsymbol{C}^d_\delta\rangle \leq \langle \boldsymbol{C}^d_{\delta'}\rangle \leq \langle \boldsymbol{C}^d_1\rangle, \quad \forall \delta' < \delta. \tag{4.44}$$

Here we recognise the Reuss bound ($\boldsymbol{C}^R = \langle \boldsymbol{S}^t_1\rangle^{-1}$) and the Voigt bound ($\boldsymbol{C}^V = \langle \boldsymbol{C}^d_1\rangle$), which clearly possess no scale dependence, and neither do the Hashin–Shtrikman bounds, which do not appear here. The hierarchy (4.44) can quantitatively be determined by computational treatment of two boundary value problems, and these have to be repeated a number of times to sample the space (4.38) in a Monte Carlo sense.

In general, for any given realisation $B_\delta(\omega)$ of (4.38), $\mathbf{C}_\delta^d(\omega)$ and $\mathbf{S}_\delta^t(\omega)$ are anisotropic. The same conclusion holds for the anti-plane tensors $C_\delta^d(\omega)$ ($C_{ij} := \mathbf{C}_{3i3j}$) and $S_\delta^t(\omega)$ ($S_{ij} := \mathbf{S}_{3i3j}$) (recall Section 1.1.2) and, by analogy, any rank 2 response tensor (recall Table 1.1). The same hierarchy as (4.44) holds for these two tensors and, on account of the statistical isotropy, there is the hierarchy of bounds on the isotropic conductivity measure

$$c^H \leq \left\langle s_{\delta'}^t \right\rangle^{-1} \leq \left\langle s_\delta^t \right\rangle^{-1} \leq c_\infty \leq \left\langle c_\delta^d \right\rangle \leq \left\langle c_{\delta'}^d \right\rangle \leq c^A, \quad \forall \delta' < \delta. \tag{4.45}$$

where each crystal's conductivity is characterised by three principal values $(c_1, c_2, c_3,)$, while

$$c^H = \left[(1/c_1 + 1/c_2 + 1/c_3)/3\right]^{-1} \quad \text{and} \quad c^A = (c_1 + c_2 + c_3)/3$$

are, respectively, the harmonic mean (Reuss type) and arithmetic mean (Voigt type) estimates.

While the hierarchy (4.45) describes the scaling trend to the first invariant of mesoscale tensors, the coefficient of variation of the second invariant for several different planar random microstructures generated by homogeneous Poisson point fields was found to be ~ 0.55 (Ostoja-Starzewski 1999).

Hierarchies generalising (4.44) and (4.45) have been obtained for a range of other materials: non-linear and or/inelastic random materials (elasto-plastic, viscoelastic, permeable and thermoelastic) of classical Cauchy type and elastic micropolar (Cosserat type); see Ostoja-Starzewski et al. (2016).

4.7.3 Scaling Function

Uncorrelated microstructures

Suppose we deal with a polycrystal: a statistically isotropic response is obtained if the crystal orientations are uniformly distributed on a unit sphere, while a deterministically isotropic response is obtained upon ensemble averaging on any mesoscale. For the latter type of response, in terms of the anti-plane tensors $C_\delta^d(\omega)$ and $S_\delta^t(\omega)$, we have

$$\left\langle C_\delta^d \right\rangle = c_\delta^d I, \qquad \left\langle S_\delta^t \right\rangle = s_\delta^t I.$$

where I is the rank 2 identity tensor. By contracting the above equation, we obtain

$$\left\langle C_\delta^d \right\rangle : \left\langle S_\delta^t \right\rangle = 3 c_\delta^d s_\delta^t \tag{4.46}$$

In the infinite volume limit $\delta \to \infty$ (RVE level) one tensor is the exact inverse of another

$$\lim_{\delta \to \infty} \left\langle C_\delta^d \right\rangle : \left\langle S_\delta^t \right\rangle = 3. \tag{4.47}$$

Now, we postulate the following relationship between the left-hand sides of (4.46) and (4.47):

$$\begin{aligned} \left\langle C_\delta^d \right\rangle : \left\langle S_\delta^t \right\rangle &= \lim\nolimits_{\delta \to \infty} \left\langle C_\delta^d \right\rangle : \left\langle S_\delta^t \right\rangle + g(c_1, c_2, c_3, \delta) \\ &\equiv \lim\nolimits_{\delta \to \infty} \left\langle C_\delta^d \right\rangle : \left\langle S_\delta^t \right\rangle + g(k_1, k_2, c_3, \delta), \end{aligned} \tag{4.48}$$

where $g(c_1, c_2, c_3, \delta)$ (or $g(k_1, k_2, c_3, \delta)$) defines the *scaling function*, with $k_1 = c_1/c_3$ and $k_2 = c_2/c_3$ being two non-dimensional parameters; see Ranganathan & Ostoja-Starzewski (2008*a*) and Ranganathan & Ostoja-Starzewski (2008*b*). The equations (4.46) and (4.47) into (4.48) jointly lead to

$$g(k_1, k_2, c_3, \delta) = g(k_1, k_2, \delta) = 3\left(c_\delta^d s_\delta^t - 1\right), \tag{4.49}$$

where c_3 has been removed since it is the only dimensional term. The scaling function of (4.49) applies to aggregates made up of biaxial single crystals (e.g. mica, gypsum, barite and rhodonite), whereas for tetragonal (e.g. urea, zircon), hexagonal (e.g. graphite) or trigonal crystals (e.g. calcite, hematite, quartz), $k_1 = k_2 = k$ and (4.49) can be rewritten as $g(k, \delta) = 3\left(c_\delta^d s_\delta^t - 1\right)$.

Theoretically, the scaling function becomes zero only when the number of grains is infinite (i.e. $\delta \to \infty$) or when the crystal is locally isotropic with $k = 1$ (such as for cubic crystals). One can further establish the following bounds on the scaling function for aggregates made up of single crystals with uniaxial thermal character ($k_1 = k_2 = k$ for trigonal, hexagonal and tetragonal single crystals, with k also being a measure of a single crystal's anisotropy)

$$0 \leq g(k, \delta) \leq \frac{2}{3}\left(\sqrt{k} - \frac{1}{\sqrt{k}}\right)^2.$$

In that case,

$$g(k, \delta) = \frac{2}{3}\left(\sqrt{k} - \frac{1}{\sqrt{k}}\right)^2 \exp\left[-0.91\sqrt{\delta - 1}\right].$$

offers a very good fit to extensive computational results. Also, on account of the hierarchy (4.45), we have $\left(c_\delta^d s_\delta^t - 1\right) \leq \left(c^A/c^H - 1\right)$; for further details see Ranganathan & Ostoja-Starzewski (2008*a*).

In the case of a composite with planar random checkerboard, at all nominal volume fractions, this stretched-exponential form provides a very good fit (Raghavan, Ranganathan & Ostoja-Starzewski 2015):

$$f(v_f, k, \delta) = 2v_f\left(1 - v_f\right)\left(\sqrt{k} - \frac{1}{\sqrt{k}}\right)^2 \exp[-0.73(\delta - 1)^{0.5}]. \tag{4.50}$$

Here $k_1 = c_1/c_2$ where c_i $(i = 1, 2)$ is the isotropic conductivity of phase i and v_f is the volume fraction.

Correlated microstructures

The foregoing concepts can be extended to spatially correlated microstructures. For example, Figure 4.2 shows a micrograph (a) and a simulation (b) of 2D interpenetrating phase composites (IPCs), where either phase is interconnected throughout the microstructure (Clarke 1992). The simulation of this statistically isotropic two-phase microstructure is based on a Gaussian-type correlation function

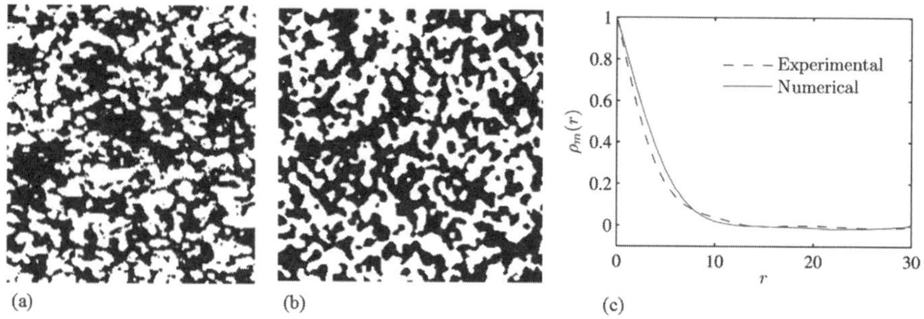

Figure 4.2 (a) A micrograph of Al_2O_3/Ni (Aldrich et al. 2000) at $v = 0.5$ and l_s calculated to be 2.92. (b) A numerically generated micrograph with $\lambda = 4.56$ that is calculated so as to have the same l_s and v as the experimental micrograph. (c) The covariance of the experimental (a) and corresponding numerically generated (b) microstructures. Reproduced from Kale et al. (2015) with the permission of AIP publishing.

$$\rho(x) = \exp\left(-\gamma r^2\right), \quad \text{with} \quad r^2 = x^2 + y^2,$$

using a Fourier filtering method based algorithm (Makse, Havlin, Schwartz & Stanley 1996; Kale et al. 2015). Here γ is the integral length scale, related to the correlation length ($\lambda = \int_0^\infty \rho(r)\, dr$) by $\gamma = \pi/4\lambda^2$. The characteristic length scale $l_s = \left[\int_0^\infty r\rho_m(r)dr\right]^{1/2}$ of the phases in a microstructure, is related to λ and v (volume fraction of one phase) by $l_s = V(v)\lambda$. Here $V(v) = cv^a(1-v)^a/B(a,a)$, where $B(a,a)$ is the beta function with a and c being the fitting constants. The two-phase field is generated by thresholding, with the threshold chosen according to the desired volume fraction v.

The scaling function $g(\delta/\lambda, k, v)$ of this microstructure, with k being the phase contrast, can be factored as

$$g(\delta/\lambda, k, v) = h(\delta/\lambda) \cdot 2v(1-v)\left(\sqrt{k} - \frac{1}{\sqrt{k}}\right)^2,$$

where $h(\delta/\lambda)$ is the normalised scaling function; $c \simeq 4.4$, $b \simeq 2.5$ and $n \simeq 1.16$.

Elastic microstructures

Hierarchies of scale-dependent bounds (4.44) carry through for rank 4 response tensors and, in the case of statistical isotropy, we obtain the hierarchies of bounds on the shear and bulk moduli:

$$\begin{array}{l} G^R \leq \langle G^t_{\delta'}\rangle^{-1} \leq \langle G^t_{\delta}\rangle^{-1} \leq G_\infty \leq \langle G^d_{\delta}\rangle \leq \langle G^d_{\delta'}\rangle \leq G^V, \\ K^R \leq \langle K^t_{\delta'}\rangle^{-1} \leq \langle K^t_{\delta}\rangle^{-1} \leq K_\infty \leq \langle K^d_{\delta}\rangle \leq \langle K^d_{\delta'}\rangle \leq K^V, \end{array} \quad \forall\delta' < \delta,$$

where (G^R, K^R) and (G^V, K^V) represent, respectively, the Reuss and Voigt estimates of shear and bulk moduli. Note that the averaged stiffness and compliance tensors can be expressed in terms of the scale- and boundry condition dependent shear and bulk moduli as follows:

$$\langle \mathbf{C}_\delta^d \rangle = 2 \langle G_\delta^d \rangle \mathbf{K} + 3 \langle K_\delta^d \rangle \mathbf{J}, \tag{4.51a}$$

$$\langle \mathbf{S}_\delta^t \rangle = \frac{1}{2 \langle G_\delta^d \rangle} \mathbf{K} + \frac{1}{3 \langle K_\delta^d \rangle} \mathbf{J}. \tag{4.51b}$$

In thc above, $\mathbf{J}$ and $\mathbf{K}$ represent the spherical and the deviatoric parts of the unit fourth-order tensor $\mathbf{I}$. By contracting (4.51a,b), we obtain

$$\langle \mathbf{C}_\delta^d \rangle : \langle \mathbf{S}_\delta^t \rangle = 5 \frac{\langle G_\delta^d \rangle}{\langle G_\delta^t \rangle} + \frac{\langle K_\delta^d \rangle}{\langle K_\delta^t \rangle}, \tag{4.52}$$

whereby we note

$$\lim_{\delta \to \infty} \langle \mathbf{C}_\delta^d \rangle : \langle \mathbf{S}_\delta^t \rangle = 6. \tag{4.53}$$

By postulating the following relationship between the left-hand sides of (4.52) and (4.53), we get

$$\langle \mathbf{C}_\delta^d \rangle : \langle \mathbf{S}_\delta^t \rangle = \lim_{\delta \to \infty} \langle \mathbf{C}_\delta^d \rangle : \langle \mathbf{S}_\delta^t \rangle + f(C_{ij}, \delta), \tag{4.54}$$

which defines the *elastic scaling function*. Clearly, $\lim_{\delta \to \infty} f(C_{ij}, \delta) = 0$. The parameter C_{ij} represents all the independent single-crystal elastic constants depending on the crystal type. For aggregates made up of cubic single crystals, $C_{ij} = (C_{11}, C_{12}, C_{44})$, while for triclinic systems C_{ij} will include all the 21 independent single-crystal constants. In light of (4.52) and (4.53), (4.54) yields

$$f(C_{ij}, \delta) = 5 \frac{G^V}{G^R} + \frac{K^V}{K^R} - 6. \tag{4.55}$$

For the special case of cubic crystals, the bulk modulus is scale-independent (Ranganathan & Ostoja-Starzewski 2008*b*; Mendelson 1981) and (4.7) can be rewritten as $f(C_{11}, C_{12}, C_{44}, \delta) = 5 \left(\langle G_\delta^d \rangle / \langle G_\delta^t \rangle - 1 \right)$. Note that the scaling function is null if the crystals are locally isotropic.

One can further establish the following bounds on the scaling function

$$f(C_{ij}, \infty) \leq f(C_{ij}, \delta) \leq f(C_{ij}, \delta') \leq f(C_{ij}, 1), \quad \forall 1 \leq \delta' < \delta \leq \infty,$$

leading to

$$0 \leq f(C_{ij}, \delta) \leq f(C_{ij}, \delta') \leq A_4^U(1), \quad \forall 1 \leq \delta' < \delta \leq \infty, \tag{4.56}$$

where

$$A_4^U(1) = 5G^V/G^R + K^V/K^R - 6 \tag{4.57}$$

is the so-called *universal anisotropy index* quantifying the single-crystal anisotropy of the fourth-rank elasticity tensor (Ranganathan & Ostoja-Starzewski 2008*c*). This index is increasingly proving to be of use to many researchers in condensed matter physics (including superconductivity), materials science and engineering and geophysics; see (Walker & Wookey 2012) for the mapping of entire Earth's surface.

Based on (4.56) one can interpret the elastic scaling function in (4.55) as the evolution of the equivalent anisotropy in the mesoscale domain, thus

$$f\left(C_{ij},\delta\right)=A_4^U\left(\delta\right). \tag{4.58}$$

The simplest form of (4.58) having a separable structure:

$$f\left(C_{ij},\delta\right)=A_4^U\left(\delta\right)=A_4^U\left(1\right)h_4\left(\delta\right)$$

is a very good approximation for single-phase aggregates made up of crystals of cubic type. The scaling function depends on the Zener anisotropy (A) and the mesoscale, and takes the following form based on numerical simulations:

$$\begin{aligned} f\left(C_{ij},\delta\right) &= A_4^U\left(\delta\right)=A_4^U\left(1\right)h_4\left(\delta\right) \\ &\simeq \frac{6}{5}\left(\sqrt{A}-\frac{1}{\sqrt{A}}\right)^2 \exp\left[-0.77\left(\delta-1\right)^{0.5}\right]. \end{aligned} \tag{4.59}$$

The stretched-exponential form also applies to viscoelastic random media (Ostoja-Starzewski et al. 2016).

4.7.4 Mesoscale TRF

Suppose we want to determine the torsional response of a square cross-section rod having the two-phase microstructure in the (x_1, x_2)-plane shown in Figure 4.3, not changing in the x_3-direction. Clearly, for significant mismatch between the black and white phases, there is no separation of scales. In light of the preceding discussion, there are three ways to define the mesoscale response: via (4.40), or (4.42), or (4.41a). Thus, three are three different TRFs of constitutive responses that one may set up, of which the first and the third have bounding characters (Ostoja-Starzewski 2008):

$$\mathcal{S}_\delta^d=\{B_\delta(\omega);\omega\in\Omega\} \quad \text{defined through } \boldsymbol{C}_\delta^d \tag{4.60a}$$

$$\mathcal{S}_\delta^{dt}=\{B_\delta(\omega);\omega\in\Omega\} \quad \text{defined through } \boldsymbol{C}_\delta^{dt} \tag{4.60b}$$

$$\mathcal{S}_\delta^t=\{B_\delta(\omega);\omega\in\Omega\} \quad \text{defined through } \boldsymbol{S}_\delta^t. \tag{4.60c}$$

In view of the variational principles involved in the derivation of $\boldsymbol{C}_\delta^d$ and $\boldsymbol{S}_\delta^t$ tensors, they bound the $\boldsymbol{C}_\delta^{dt}$ stiffness from above and below, respectively.

The TRFs defined by (4.60a) and (4.60c), separately, provide inputs to two finite-element schemes – based on minimum potential and complementary energy principles, respectively – for bounding the macroscopic response. While in the classical case of a homogeneous material, these bounds are convergent with the finite elements becoming infinitesimal, the presence of a disordered non-periodic microstructure prevents such a convergence and leads to a possibility of an optimal mesoscale. The method has been demonstrated on the torsion of the rod of Figure 4.3 (Ostoja-Starzewski 1999).

In light of Equations 1.13a,b, it is tempting to set up an average stiffness tensor random field. The simplest way to proceed is to take/estimate the average

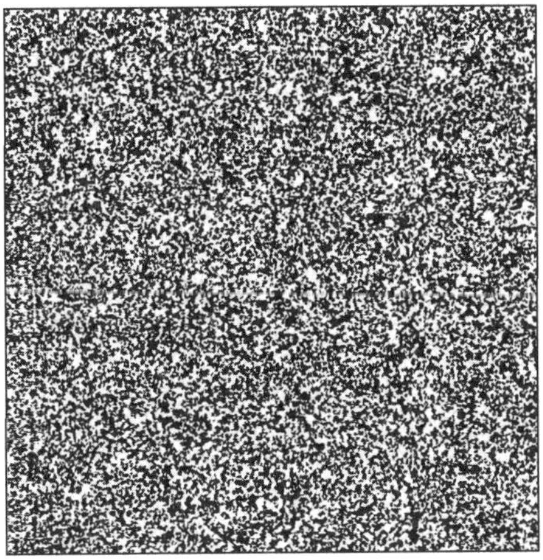

Figure 4.3 A two-phase material with a Voronoi mosaic microgeometry of a total 104,858 black and white cells, at volume fraction 50% each (Ostoja-Starzewski 1999).

stiffness as a constant tensor $\mathbf{C}_\infty = \mathbf{C}^{eff}$ of (4.44) plus a zero-mean noise corresponding to the mesoscale δ one wants to consider

$$\mathbf{C}_{ijkl}(\omega,\mathbf{x}) = \mathbf{C}^{eff}_{ijkl} + \mathbf{C}'_{ijkl}(\omega,\mathbf{x}), \qquad \left\langle \mathbf{C}'_{ijkl}(\omega,\mathbf{x})\right\rangle = 0. \tag{4.61}$$

Upon averaging of (4.61) one obtains $\mathbf{C}^{eff}_{ijkl}$, which, under the assumption of local statistical isotropy of $\mathbf{C}'_{ijkl}(\omega,\mathbf{x})$ reduces to $C_{ijkl} = \lambda\delta_{ij}\delta_{kl} + \mu\left(\delta_{ik}\delta_{jl} + \delta_{il}\delta_{jk}\right)$. However, if one simply solves a macroscopic boundary value problem lacking the separation of scales with the average stiffness field $\mathbf{C}^{eff}_{ijkl}$, the resulting macroscopic response may fall outside the range which one would get using the Dirichlet and Neumann-type bounds on mesoscales.

Consider $\mathbf{C}^d_\delta$ to be the anti-plane part of the rank 4 stiffness tensor, recall (1.45) in Chapter 1. The probability densities of half-traces of that tensor, as described by the hierarchy (2.6) are shown in terms of their histograms in Figure 4.4. Note the expected convergence of ensemble-averaged half-traces of $\mathbf{C}^d_\delta$ and $\mathbf{S}^t_\delta$ tending to a causal distribution with δ increasing. Given the strong skewness of $\mathbf{C}^d_\delta$, the non-Gaussian character of that TRF (and a similar one of $\mathbf{S}^t_\delta$) is obvious. Given the need for a finite support of the probability distribution of $\mathbf{C}_{ijkl}(\omega)$, the Beta probability distribution has been found to provide the most satisfactory and universal fits for this as well as for other types of two-phase composites over a wide range of contrasts and mesoscales.

One more avenue for defining the mesoscale constitutive response is to postulate of a periodic microstructure representative in some acceptable sense of the true (and non-periodic) random microstructure. This step then allows the imposition of periodic boundary conditions on a periodic mesoscale window:

$$u_i\left(\boldsymbol{x}+\boldsymbol{L}\right) = u_i\left(\boldsymbol{x}\right) + \varepsilon^0_{ij}L_j \quad t_i\left(\boldsymbol{x}\right) = -t_i\left(\boldsymbol{x}+\boldsymbol{L}\right) \quad \forall \boldsymbol{x}\in\partial B, \tag{4.62}$$

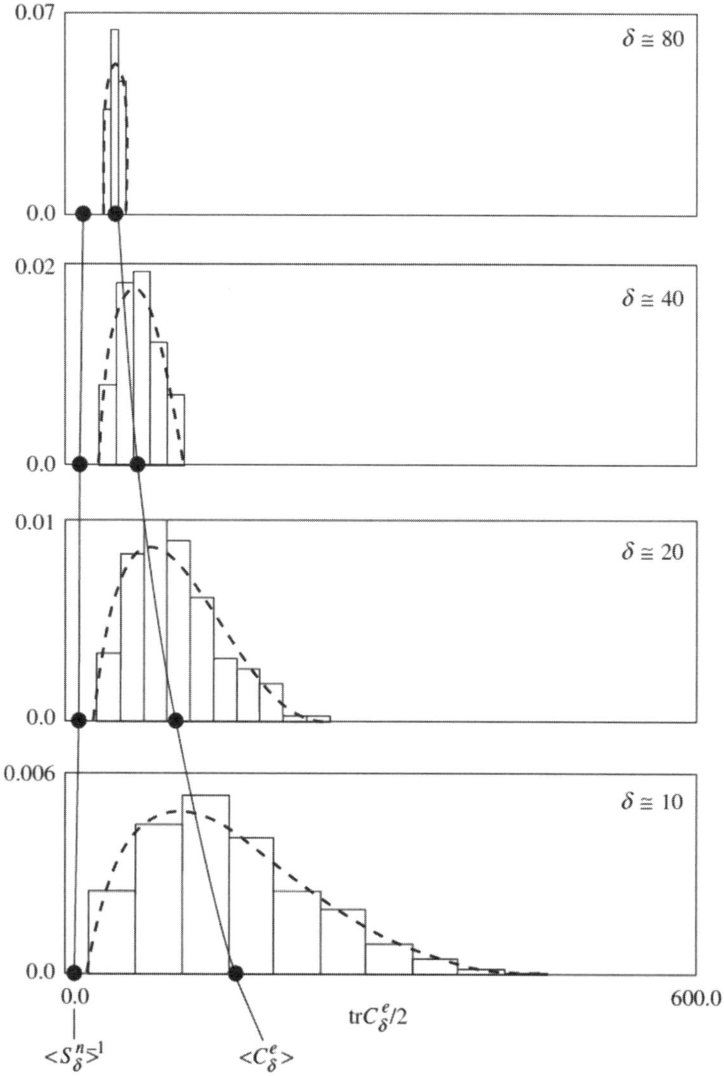

Figure 4.4 Probability densities of half-traces ($\frac{1}{2}(C_{11} + C_{22})$) of $\boldsymbol{C}_\delta^d$ are shown at four mesoscales ($\delta = 10$, 20, 40 and 80) of the two-phase microstructure of Figure 4.3 at phase mismatch 2,000. Also indicated by bold dots are the averages $\langle \boldsymbol{C}_\delta^d \rangle$ and $\langle \boldsymbol{S}_\delta^t \rangle^{-1}$; the dots are curve fitted to display two-dependent hierarchies of bounds (4.44). Figure reproduced from Ostoja-Starzewski (1999).

where ε_{ij}^0 is the applied strain, t_i is the traction on the boundary ∂B of B, and $\boldsymbol{L} = L\boldsymbol{e}_i$, with $\boldsymbol{e}_i$ being the unit base vector. Depending on the size of periodicity L, the resulting stiffness tensor will display some scale dependence, which typically abates as L is taken much larger than the microscale (say, an average grain size). With (4.62) resulting in $\boldsymbol{C}_\delta^{per}$, this leads to a fourth possible TRF of constitutive response

$$\mathcal{S}^{per} = \{B(\omega); \omega \in \Omega\} \quad \text{defined through } \boldsymbol{C}_\delta^{per} \tag{4.63}$$

While the $\boldsymbol{C}_\delta^d$ and $\boldsymbol{S}_\delta^t$ tensors lead to rather large estimates of RVE size, especially with the contrast in material properties increasing, (4.62) leads to estimates of $\boldsymbol{C}^{eff}$ with a much weaker scale dependence (Sab & Nedjar 2005); see also Bostanabad et al. (2018); Wu et al. (2018). For homogenisation of random media by micropolar continua see Trovalusci et al. (2015); Trovalusci et al. (2014).

There is also a possibility of defining the mesoscale response from a Robin boundary condition (i.e. a linear combination of Dirichlet and Neumann conditions), although it is not known what that combination should be. In general, the mesoscale constitutive response of a continuum point of a macroscopic field model is a function of the actual strain (and stress) at that point, both of which are unknown *a priori*. Ideally, if one were to use a finite element method for solving a macroscopic boundary value problem (MBVP), the constitutive response of each and every finite element should be computed by a mesoscale upscaling according to the actual stress/strain state, which would then be sent as input into the MBVP to recompute all the fields and responses of all the finite elements, and so on. Clearly, this type of an iterative procedure, even for a single realisation of random medium is very costly.

4.8 Stochastic Partial Differential Equations

4.8.1 Elliptic Equations: Example Application

With reference to Chapter 1, we start with the Dirichlet conduction problem for the T field on domain $\mathcal{D} \subset \mathbb{R}^2$, under the conductivity TRF and body force RF,

$$\begin{aligned} -\left(k_{ij}\left(\boldsymbol{x},\omega\right)T_{,j}\right)_{,i} &= f\left(\boldsymbol{x},\omega\right) & \forall \boldsymbol{x} \in \mathcal{D}, \\ T\left(\boldsymbol{x},\omega\right) &= g\left(\boldsymbol{x}\right) & \forall \boldsymbol{x} \in \partial\mathcal{D}. \end{aligned} \tag{4.64}$$

The variational formulation of (4.64) is based on a straightforward generalisation of Lord et al. (2014, Assumption 2.34); see also Demengel & Demengel (2012). Fix $\omega \in \Omega$ and consider the variational problem of determining a function $T(\boldsymbol{x},\omega) \in H^1(\mathcal{D})$ for which the functional

$$\frac{1}{2}\int_{\mathcal{D}} k_{ij}(\boldsymbol{x},\omega)T_{,i}(\boldsymbol{x},\omega)T_{,j}(\boldsymbol{x},\omega)\,\mathrm{d}\boldsymbol{x} - \int_{\mathcal{D}} f(\boldsymbol{x},\omega)T(\boldsymbol{x},\omega)\,\mathrm{d}\boldsymbol{x} \tag{4.65}$$

attains infimum on the set

$$\{\,T(\boldsymbol{x},\omega) \in H^1(\mathcal{D})\colon T = g \text{ on } \partial\mathcal{D}\,\}. \tag{4.66}$$

Here, $H^1(\mathcal{D})$ denotes the Sobolev space.

Theorem 38. *Assume that the boundary value problem* (4.64) *satisfies the following conditions:*

1. *The set $\mathcal{D}$ is bounded, open and uniformly of class C^1; see Demengel & Demengel (2012 Definition 2.66).*
2. $\mathsf{P}\{\,\omega\colon f(\boldsymbol{x},\omega)\in L^2(\mathcal{D})\,\}=1$.
3. $\mathsf{P}\{\,\omega\colon \|k(\boldsymbol{x},\omega)\|\in L^\infty(\mathcal{D})\,\}=1$.
4. *The random field $k(\boldsymbol{x})$ takes values in $\mathsf{S}^2(\mathbb{R}^2)$.*
5. *The random field $k(\boldsymbol{x})$ is uniformly elliptic P-a.s., that is, there exists an $\alpha>0$ such that for all $\boldsymbol{x}\in\mathbb{R}^2$ we have*

$$\sum_{i,j=1}^{2} k_{ij}(\boldsymbol{x},\omega)x_i x_j \geq \alpha\|\boldsymbol{x}\|^2.$$

6. $\mathsf{P}\{\,\omega\colon g(\boldsymbol{x},\omega)\in H^{1/2}(\partial\mathcal{D})\,\}=1$.

Then, there exists a unique $T(\boldsymbol{x},\omega)\in H^1(\mathcal{D})$ such that the functional (4.65) *attains infimum on the set* (4.66) P*-a.s.*

Proof. Fix $\omega\in\Omega$ outside the P-null set where conditions 2, 3, 5 and 6 fail, and follow the proof in Demengel & Demengel (2012, Subsection 5.6.1). □

Remark 4. The function T for which the functional (4.65) attains infimum on the set (4.66), solves the Dirichlet problem (4.64) in the sense of distributions. If we would like to obtain a *classical solution* to the above Dirichlet problem, we have to strengthen conditions of Theorem 38. Using Sobolev embeddings (Demengel & Demengel 2012), we conclude that T is twice continuously differentiable P-a.s. if $T\in H^4(\mathcal{D})$ P-a.s. By Demengel & Demengel (2012, Subsection 5.6.1), this can be achieved if Conditions 1, 2, 3 and 6 of Theorem 38 are replaced by the following conditions:

1′ $\mathcal{D}$ is a bounded open domain of uniform class C^4;
2′ $\mathsf{P}\{\,\omega\colon f(\boldsymbol{x},\omega)\in H^2(\mathcal{D})\,\}=1$;
3′ $\mathsf{P}\{\,\omega\colon \|k(\boldsymbol{x},\omega)\|\in C^3(\overline{\mathcal{D}})\,\}=1$;
6′ $\mathsf{P}\{\,\omega\colon g(\boldsymbol{x},\omega)\in H^{7/2}(\partial\mathcal{D})\,\}=1$.

4.8.2 Hyperbolic Equations: Example Application

The localisation of mechanical fields at the ends of structural members is treated through the Saint-Venant's Principle. Its dynamic counterpart, called Dynamic Saint-Venant's Principle, dates back to Ignaczak (1974); see also Hetnarski & Ignaczak (2011); Ostoja-Starzewski (2018). Consider a semi-infinite, inhomogeneous, anisotropic elastic cylinder, denoted by B, having a constant cross-section, aligned with the axis $\boldsymbol{k}$, having the end face C_0, and any intersection of B by a plane perpendicular to $\boldsymbol{k}$ at any distance l from C_0 denoted by C_l. Also, a semi-infinite cylinder with the end face C_0 is denoted by $B\,(l)$. A random cylinder is $\mathcal{S}=\{B(\omega);\omega\in\Omega\}$ and, analogously, we have $\mathcal{S}(l)=\{B(l,\omega);\omega\in\Omega\}$.

To state the principle, we consider one realisation $B(\omega)$, assume the body forces to be absent and the initial, boundary and asymptotic conditions of the form:

$$\sigma_{ij}(\boldsymbol{x},0)=0, \quad \dot{\sigma}_{ij}(\boldsymbol{x},0)=0 \text{ for } \boldsymbol{x}\in B \tag{4.67}$$

$$\begin{aligned} \sigma_{ij}n_j &= 0 \quad \text{on} \quad (\partial B - C_0)\times[0,\infty) \\ \sigma_{ij}n_j &= p_i \neq 0 \quad \text{on} \quad C_0\times[0,\infty) \end{aligned} \tag{4.68}$$

$$\begin{aligned} &\int_{C_l} \sigma_{ij}n_j da \to 0 \quad \text{as} \quad l\to\infty \quad \text{for} \quad t\geq 0 \\ &\int_{C_l} \boldsymbol{\varepsilon}_{ijk}x_j(\sigma_{kl}n_l)\, da \to 0 \quad \text{as} \quad l\to\infty \quad \text{for} \quad t\geq 0 \\ &\int_{C_l} \rho^{-1}\sigma_{kl,l}(\dot{\sigma}_{kj}n_j)\, da \to 0 \quad \text{as} \quad l\to\infty \quad \text{for} \quad t\geq 0. \end{aligned} \tag{4.69}$$

Here n_i stands for the outer unit normal to the surface and $p_i = p_i(\mathbf{x},t)$ is a prescribed load. The mass density RF and compliance TRF satisfy the inequalities ($\boldsymbol{e}$ is a symmetric, rank 2 tensor)

$$\begin{aligned} &0<\rho_{\min}(\omega)\leq\rho(\omega,\boldsymbol{x})\leq\rho_{\max}(\omega)<\infty, && \text{a.e. in } B, \\ &0<S_{\min}(\omega)|\boldsymbol{e}|^2\leq\boldsymbol{e}:\mathbf{S}(\omega,\boldsymbol{x}):\boldsymbol{e}\leq S_{\max}(\omega)|\boldsymbol{e}|^2<\infty, && \forall\boldsymbol{e}\neq\mathbf{0} \\ & && \text{a.e. in } B. \end{aligned} \tag{4.70}$$

Now, the Dynamic Saint-Venant's principle (Ostoja-Starzewski 2018) assumes that, given a stress field satisfying (1.43) with a zero body force the conditions (4.67), (4.68), (4.69) for the total stress energy associated with the solution and stored in the semi-infinite cylinder $B(l)$ over the time interval $[0,t]$

$$U(l,t,\omega)=\frac{1}{2}\int_0^t\int_{B(l)}\left(\frac{1}{\rho(\omega,\boldsymbol{x})}\sigma_{mn,n}\sigma_{mp,p}+\dot{\sigma}_{ij}\mathbf{S}_{ijkl}(\omega,\boldsymbol{x})\dot{\sigma}_{kl}\right)d\boldsymbol{x}d\tau$$

these estimates hold true:

$$\begin{aligned} U(l,t,\omega)&=0 \quad \text{for} \quad 0\leq t<\tfrac{l}{c}, \\ U(l,t,\omega)&=U(0,t)\exp\left(\tfrac{-l}{c(\omega)t}\right) \quad \text{for} \quad t>\tfrac{l}{c}\geq 0. \end{aligned}$$

Here

$$c(\omega)=\frac{2}{\sqrt{\rho_{\min}(\omega)\ \mathbf{S}_{\min}(\omega)}}, \tag{4.71}$$

At this point, we note that the random fields of mass density and compliance are scale-dependent: the larger is the mesoscale of Figure 0.1 in the Introduction, the weaker is the randomness. This implies that the bounds (4.70) get tighter (and tend to coincide) as the mesoscale increases (and, respectively, tends to infinity) – recall the scaling function (4.59). Thus, instead of (4.71), we should write

$$c(\omega,\delta)=\frac{2}{\sqrt{\rho_{\min}(\omega,\delta)\ S_{\min}(\omega,\delta)}},$$

As δ increases, $\rho_{\min}(\omega,\delta)$ and $S_{\min}(\omega,\delta)$ increase, so that

$$c(\omega,\delta)>c(\omega,\delta')\quad\text{for}\quad\delta<\delta';$$

implying that the lowest estimate of c is obtained in the RVE limit (recall Section 4.7)

$$\lim_{\delta\to\infty}c(\omega,\delta)=\frac{2}{\sqrt{\rho_{\mathrm{eff}}\ S_{\mathrm{eff}}}},$$

where ρ_{eff} and S_{eff} are the effective (RVE level) properties.

4.9 Damage TRF

4.9.1 Group-Theoretical Considerations

In continuum damage mechanics, damage may be described by a set $\mathcal{D}$ of scalars, vectors and tensors, see a review of damage variables in (Ganczarski et al., 2010, Table 3.1). We consider the *fabric tensors* introduced in Kanatani (1984); see also Onat & Leckie (1988), Lubarda & Krajcinovic (1993), Murakami (2012), Ganczarski et al. (2015). The idea is as follows.

Consider the spherical surface S^2 of unit radius around a material point $P(\boldsymbol{x})$, see Ganczarski et al. (2015, Figure 1.4). Let $\boldsymbol{n}$ be the direction vector drawn from $\boldsymbol{x}$ to a point on the sphere. Denote by $\xi(\boldsymbol{n})$ the directional distribution of the microvoid density. It's a function $\xi\colon S^2\to\mathbb{R}$. Assume this function is square-integrable with respect to the Lebesgue measure $\mathrm{d}\boldsymbol{n}$. Following Olive et al. (2017), denote by $\mathrm{H}^\ell(\mathbb{R}^3)$ the space of harmonic polynomials of degree $\ell\geq 0$ on the *space domain* $\mathbb{R}^3$ (homogeneous polynomials of degree ℓ with vanishing Laplacian). It is well-known that the direct sum of the above spaces over $\ell\geq 0$ is dense in the Hilbert space $L^2(S^2,\mathrm{d}\boldsymbol{n})$ of all square-integrable functions. We choose an orthonormal basis, say the basis of real-valued spherical harmonics $\{\,S^m_\ell(\boldsymbol{n})\colon -\ell\leq m\leq\ell\,\}$ in each space, and express the function $\xi(\boldsymbol{n})$ as the sum of the Fourier series

$$\xi(\boldsymbol{n})=\sum_{\ell=0}^{\infty}\sum_{m=-2\ell}^{2\ell}\xi^m_{2\ell}S^m_{2\ell}(\boldsymbol{n}).\tag{4.72}$$

Since the magnitude of $\xi(\boldsymbol{n})$ is unchanged under the inversion $\boldsymbol{n}\mapsto-\boldsymbol{n}$, we have $\xi(-\boldsymbol{n})=\xi(\boldsymbol{n})$; that's why the expansion (4.72) contains only even polynomials.

It is customary in damage mechanics to use another form of the expansion (4.72). For any polynomial $p(\boldsymbol{x}_1,\dots,\boldsymbol{x}_\ell)\in\mathrm{H}^\ell(\mathbb{R}^3)$, consider its *polarisation*, that is, the following rank ℓ tensor

$$\boldsymbol{T}(\boldsymbol{x}_1,\dots,\boldsymbol{x}_\ell)=\frac{1}{\ell!}\left.\frac{\partial^\ell}{\partial t_1\cdots\partial t_\ell}\right|_{t_1=\cdots=t_\ell=0}p(t_1\boldsymbol{x}_1+\cdots+t_\ell\boldsymbol{x}_\ell),$$

see Olive et al. (2017). It is known that the polarisation is an isomorphism between $\mathrm{H}^\ell(\mathbb{R}^3)$ and the space $\mathbb{H}^\ell(\mathbb{R}^3)$ of rank ℓ *harmonic tensors*, that are completely symmetric and completely traceless. Applying the polarisation to (4.72), we obtain

$$\xi(\boldsymbol{n}) = \mathsf{D}_0 + \sum_{\ell=1}^{\infty} \mathsf{D}_{i_1\cdots i_{2\ell}} \mathsf{f}_{i_1\cdots i_{2\ell}}(\boldsymbol{n}), \tag{4.73}$$

where the Einstein summation convention is in use. Here, $4\ell+1$ independent components of the tensor $\mathsf{f}_{i_1\cdots i_{2\ell}}(\boldsymbol{n})$ are $4\ell+1$ polarised polynomials $S^m_{2\ell}(\boldsymbol{n})$, $-2\ell \leq m \leq 2\ell$.

How to choose the harmonic tensors $\mathsf{f}_{i_1\cdots i_{2\ell}}(\boldsymbol{n})$? As a first candidate, consider the tensors

$$\{1, n_i, n_i n_j, \dots, n_{i_1}\cdots n_{i_\ell}, \dots\}.$$

They are, however, neither linearly independent nor completely traceless. For example, the contraction of $n_j n_j$ over $i=j$ gives 1. To find an *irreducible tensor basis*, that is, a set of linearly independent harmonic tensors, use Kanatani (1984, Equation 4.2):

$$\mathsf{f}_{i_1\cdots i_{2\ell}}(\boldsymbol{n}) = \sum_{m=0}^{\ell} c^\ell_m \delta_{(i_1 i_2} \cdots \delta_{i_{2m-1} i_{2m}} n_{i_{2m+1}} \cdots n_{i_{2\ell})},$$

where the brackets denote the symmetrisation of the indices, and where

$$c^\ell_m = (-1)^m \frac{\binom{2\ell}{2m}\binom{2\ell-1}{m}}{\binom{4\ell-1}{2m}}.$$

For example, $c^1_0 = 1$, $c^1_1 = -\frac{1}{3}$, and we have $\mathsf{f}_{ij}(\boldsymbol{n}) = n_i n_j - \frac{1}{3}\delta_{ij}$. The symmetrisation $\delta_{(ij} n_k n_{l)}$ is

$$\delta_{(ij} n_k n_{l)} = \frac{1}{6}(\delta_{ij} n_k n_l + \delta_{ik} n_j n_l + \delta_{il} n_j n_k + \delta_{jk} n_i n_l + \delta_{jl} n_i n_l + \delta_{kl} n_i n_j),$$

the symmetrisation $\delta_{(ij}\delta_{kl)}$ is

$$\delta_{(ij}\delta_{kl)} = \frac{1}{3}(\delta_{ij}\delta_{kl} + \delta_{ik}\delta_{jl} + \delta_{il}\delta_{jk}),$$

the coefficients are $c^2_0 = 1$, $c^2_1 = -\frac{6}{7}$, $c^2_2 = \frac{3}{35}$, and

$$\mathsf{f}_{ijkl}(\boldsymbol{n}) = n_i n_j n_k n_l - \frac{6}{7}\delta_{(ij} n_k n_{l)} + \frac{3}{35}\delta_{(ij}\delta_{kl)},$$

and so on.

The coefficients of the expansion (4.73) are called *fabric tensors*. They are calculated by

$$\mathsf{D}_{i_1\cdots i_{2\ell}} = \frac{1}{4\pi}\frac{(4\ell+1)!!}{(2\ell)!}\int_{S^2} \xi(\boldsymbol{n}) \mathsf{f}_{i_1\cdots i_{2\ell}}(\boldsymbol{n})\,\mathrm{d}\boldsymbol{n},$$

see Murakami (2012, Equation (2.29)) or Ganczarski et al. (2015, Equation (1.87)).

Recall that the directional distribution of the microvoid density depends on the material point $\boldsymbol{x}$. Then, Equation (4.73) takes the form:

$$\xi(\boldsymbol{n},\boldsymbol{x})=\boldsymbol{D}_0(\boldsymbol{x})+\sum_{\ell=1}^{\infty}\boldsymbol{D}_{i_1\cdots i_{2\ell}}(\boldsymbol{x})\boldsymbol{f}_{i_1\cdots i_{2\ell}}(\boldsymbol{n}).$$

In *stochastic* continuum damage mechanics, the fabric tensors $\boldsymbol{D}_0(\boldsymbol{x})$, $\boldsymbol{D}_{ij}(\boldsymbol{x})$, ..., $\boldsymbol{D}_{i_1\cdots i_{2\ell}}(\boldsymbol{x})$, ..., must be *random fields*.

Note that the considerations of Section 3.1 are applicable here. In particular, under an orthogonal transformation $g\in\mathrm{O}(3)$, the point $\boldsymbol{x}$ becomes the point $g\boldsymbol{x}$. The tensor $\boldsymbol{D}(\boldsymbol{x})$ transforms to the tensor $\rho_{2\ell}(g)\boldsymbol{D}(\boldsymbol{x})$. Then we have

$$\begin{aligned}\langle\boldsymbol{D}(g\boldsymbol{x})\rangle&=\rho_{2\ell}(g)\langle\boldsymbol{D}(\boldsymbol{x})\rangle,\\ \langle\boldsymbol{D}(g\boldsymbol{x}),\boldsymbol{D}(g\boldsymbol{y})\rangle&=(\rho_{2\ell}\otimes\rho_{2\ell})(g)\langle\boldsymbol{D}(\boldsymbol{x}),\boldsymbol{D}(\boldsymbol{y})\rangle,\end{aligned}\tag{4.74}$$

that is, the random field $\boldsymbol{D}(\boldsymbol{x})$ is $(\mathrm{O}(3),\rho_{2\ell})$-isotropic.

When $\ell=0$, the description of such a field is given by Theorems 14 and 15. We would like to prove a similar result for the case of $\ell\geq 1$. Note that in the case of $\ell=1$ such a result for the representation $\rho_0\oplus\rho_2$ was proved in Section 3.6 and in Malyarenko & Ostoja-Starzewski (2016*b*), and in the case of $\ell=2$ such a result for the representation $2\rho_0\oplus 2\rho_2\oplus\rho_4$ was proved in Section 3.8 and in Malyarenko & Ostoja-Starzewski (2017*b*). Our case is different.

To start with, consider the first equation in (4.74). Because $\langle\boldsymbol{D}(\boldsymbol{x})\rangle=\boldsymbol{D}$ does not depend on $\boldsymbol{x}$, we have $\boldsymbol{D}=\rho_{2\ell}(g)\boldsymbol{D}$ for all $g\in\mathrm{O}(3)$. In other words, the tensor $\boldsymbol{D}$ lies in the space, where the trivial representation of $\mathrm{O}(3)$ acts. It is well-known that the representation $\rho_{2\ell}$ is irreducible and non-trivial when $\ell\geq 1$. It follows that $\langle\boldsymbol{D}(\boldsymbol{x})\rangle=\boldsymbol{0}$.

Consider the two-point correlation tensor $\boldsymbol{B}\colon\mathbb{R}^3\to\mathbb{H}^\ell(\mathbb{R}^3)\otimes\mathbb{H}^\ell(\mathbb{R}^3)$, given by

$$\boldsymbol{B}(\boldsymbol{x})=\langle\boldsymbol{D}(\boldsymbol{0}),\boldsymbol{D}(\boldsymbol{x})\rangle.$$

Note that $\rho_1(g)=g$, then

$$\boldsymbol{B}(\rho_1(g)\boldsymbol{x})=(\rho_{2\ell}\otimes\rho_{2\ell})(g)\boldsymbol{B}(\boldsymbol{x}).$$

In other words, $\boldsymbol{B}$ is a covariant tensor of the pair of representations $(\rho_{2\ell}\otimes\rho_{2\ell},\rho_1)$. By the result of Wineman & Pipkin (1964), it has the form

$$\boldsymbol{B}(\boldsymbol{x})=\sum_{m=1}^{M}f_m(I_1,\dots,I_K)\boldsymbol{L}^m(\boldsymbol{x}),$$

where $\{I_1,\dots,I_K\}$ is an integrity basis for polynomial invariants of the representation ρ_1, and $\{\,\boldsymbol{L}^m(\boldsymbol{x})\colon 1\leq m\leq M\,\}$ is an integrity basis for polynomial covariant tensors of the above pair.

In our case, we have $K=1$ and $I_1=\|\boldsymbol{x}\|^2$. Once again, by Weyl (1997), any polynomial covariant of the group $\mathrm{O}(d)$ is a linear combination of products of Kronecker's deltas δ_{ij} and second degree homogeneous polynomials x_ix_j. The tensors $\boldsymbol{L}^m(\boldsymbol{x})$ are constructed as follows. Consider the products

$$x_{i_1i_2}\cdots x_{i_{2\ell'-1}i_{2\ell'}}\delta_{i_{2\ell'+1}i_{2\ell'+2}}\cdots\delta_{i_{4\ell-1}i_{4\ell}}, \tag{4.75}$$

where $0 \leq \ell' \leq 2\ell$. For a fixed ℓ', act on the indices of the products (4.75) by the permutation group of order $2[(2\ell)!]^2$ generated by $2\ell!$ permutations of symbols $i_1, \dots, i_{2\ell}$, $2\ell!$ permutations of symbols $i_{2\ell+1}, \dots, i_{4\ell}$, and transpositions $i_{\ell'} \to i_{\ell'+2\ell}$. The tensors $\mathsf{L}^m(\boldsymbol{x})$ are the sums of the products (4.75) over the orbits of this action. The two-point correlation tensor takes the form

$$\mathsf{B}(\boldsymbol{x}) = \sum_{m=0}^{N_\ell} f_m(\|\boldsymbol{x}\|^2)\mathsf{L}^m(\boldsymbol{x}). \tag{4.76}$$

It remains to find the general form of the functions $f_m(\|\boldsymbol{x}\|^2)$.

Exactly as in Chapter 3, we prove that the two-point correlation tensor has the form

$$\mathsf{B}(\boldsymbol{x}) = \int_0^\infty \int_{S^2} \mathrm{e}^{\mathrm{i}(\boldsymbol{p},\boldsymbol{x})}\mathsf{h}(\boldsymbol{p})\,\mathrm{d}\hat{\boldsymbol{n}}\,\mathrm{d}\Phi(\lambda), \tag{4.77}$$

where $\boldsymbol{p}$ is a point in the wavenumber domain $\hat{\mathbb{R}}^3$, $\lambda = \|\boldsymbol{p}\|$, $\hat{\boldsymbol{n}} = \frac{\boldsymbol{p}}{\lambda}$ for $\boldsymbol{p} \neq \boldsymbol{0}$, $\mathrm{d}\hat{\boldsymbol{n}}$ is the Lebesgue measure on the unit sphere S^2 in the wavenumber domain, Φ is a finite Borel measure on $[0, \infty)$, and $\mathsf{h}(\boldsymbol{p})$ is a measurable function defined on $\hat{\mathbb{R}}^3$, taking values in the convex compact set of symmetric non-negative-definite operators in $\mathbb{H}^{2\ell}(\mathbb{R}^3)$ with unit trace satisfying the following condition

$$\mathsf{h}(g\boldsymbol{p}) = \mathsf{S}^2(\rho_{2\ell}(g))\mathsf{h}(\boldsymbol{p}), \qquad g \in \mathrm{O}(3). \tag{4.78}$$

Now we choose bases in the spaces $\mathbb{H}^{2\ell}(\mathbb{R}^3)$, $\ell \geq 1$. We build them inductively.

Induction base. The basis in the space $\mathbb{H}^2(\mathbb{R}^3)$ is given by the Godunov–Gordienko matrices:

$$T^n_{i_1i_2} = g^{n[i_1,i_2]}_{2[1,1]}, \qquad -2 \leq n \leq 2.$$

Induction hypothesis. Assume that the basis $\{\,\mathsf{T}^n_{i_1\cdots i_{2\ell-2}} \colon 2-2\ell \leq n \leq 2\ell-2\,\}$ is already constructed.

Induction step. The space $\mathbb{H}^{2\ell}(\mathbb{R}^3)$ is that subspace of the tensor product $\mathbb{H}^{2\ell-2}(\mathbb{R}^3) \otimes \mathbb{H}^2(\mathbb{R}^3)$ where the irreducible representation $\rho_{2\ell}$ acts. The basis of the above space is calculated with the help of the Godunov–Gordienko coefficients as follows.

$$\mathsf{T}^n_{i_1\cdots i_{2\ell}} = \sum_{n_1=2-2\ell}^{2\ell-2}\sum_{n_2=-2}^{2} g^{n[n_1,n_2]}_{2\ell[2\ell-2,2]}\mathsf{T}^{n_1}_{i_1\cdots i_{2\ell-2}}g^{n_2[i_{2\ell-1},i_{2\ell}]}_{2[1,1]}.$$

In the chosen basis, the operator $\mathsf{h}(\boldsymbol{p})$ becomes the matrix $h_{ij}(\boldsymbol{p})$, $-2\ell \leq i, j \leq 2\ell$, so that

$$\mathsf{h}(\boldsymbol{p}) = \sum_{i,j=-2\ell}^{2\ell} h_{ij}(\boldsymbol{p})\mathsf{T}^i_{i_1\cdots i_{2\ell}} \otimes \mathsf{T}^j_{j_1\cdots j_{2\ell}}.$$

The tensor product in the right-hand side expands as follows:

$$\mathsf{T}^i_{i_1\cdots i_{2\ell}}\otimes\mathsf{T}^j_{j_1\cdots j_{2\ell}}=\sum_{\ell'=0}^{2\ell}\sum_{m=-2\ell'}^{2\ell'}g^{m[i,j]}_{2\ell'[2\ell,2\ell]}\mathsf{T}^m_{2\ell'},$$

where $\{\,\mathsf{T}^m_{2\ell'}\colon\ -2\ell'\leq m\leq 2\ell\,\}$ is the Gordienko basis in the subspace of the symmetric tensor square $\mathsf{S}^2(\mathbb{H}^{2\ell}(\mathbb{R}^3))$ where the irreducible representation $\rho_{2\ell'}$ acts, see Gordienko (2002). The expansion does not contain representations with odd indices, because they belong to the skew-symmetric part of the tensor square $\mathbb{H}^{2\ell}(\mathbb{R}^3)\otimes\mathbb{H}^{2\ell}(\mathbb{R}^3)$.

The matrix entries $h_{ij}(\boldsymbol{p})$ may be expressed as

$$h_{ij}(\boldsymbol{p})=\sum_{\ell'=0}^{2\ell}\sum_{m=-2\ell'}^{2\ell'}g^{m[i,j]}_{2\ell'[2\ell,2\ell]}h^m_{2\ell'}(\boldsymbol{p}),\tag{4.79}$$

where the vector-valued function

$$\boldsymbol{h}_{2\ell'}(\boldsymbol{p})=(h^{-2\ell'}_{2\ell'}(\boldsymbol{p}),\dots,h^{2\ell'}_{2\ell'}(\boldsymbol{p}))^\top$$

satisfies the condition

$$\boldsymbol{h}_{2\ell'}(g\boldsymbol{p})=\rho_{2\ell}(g)\boldsymbol{h}_{2\ell'}(\boldsymbol{p}),\qquad g\in\mathrm{O}(3),\tag{4.80}$$

which follows from (4.78). The advantage of (4.80) over (4.78) is that the representation $\rho_{2\ell'}$ is irreducible, while the representation $\mathsf{S}^2(\rho_{2\ell})$ is not.

Let $\boldsymbol{h}_{2\ell'}(\lambda)$ be the value of the function $\boldsymbol{h}_{2\ell'}(\boldsymbol{p})$ at the point $(0,0,\lambda)^\top\in\hat{\mathbb{R}}^3$ for $\lambda>0$. The stationary subgroup of the above point is the group $\mathrm{O}(2)$. It follows from (4.80) that for all $g\in\mathrm{O}(2)$ we have

$$\boldsymbol{h}_{2\ell'}(\lambda)=\rho_{2\ell'}(g)\boldsymbol{h}_{2\ell'}(\lambda),$$

that is, the vector $\boldsymbol{h}_{2\ell'}(\lambda)$ lies in the subspace where the trivial representation of the group $\mathrm{O}(2)$ acts. By the construction of the Gordienko basis, this subspace is generated by the tensor $\mathsf{T}^0_{2\ell'}$. Then we have

$$\boldsymbol{h}_{2\ell'}(\lambda)=(0,\dots,h^0_{2\ell'}(\lambda),0,\dots,0)^\top.$$

By (4.80), we have

$$\boldsymbol{h}_{2\ell'}(\boldsymbol{p})=\rho_{2\ell'}(g)\boldsymbol{h}_{2\ell'}(\lambda),$$

where g is an arbitrary element of the group $\mathrm{O}(3)$ satisfying $g(0,0,\lambda)^\top=\boldsymbol{p}$. In coordinates, we obtain

$$h^m_{2\ell'}(\boldsymbol{p})=\rho^{2\ell'g}_{m0}(\hat{\boldsymbol{n}})h^0_{2\ell'}(\lambda).$$

Equation 4.79 takes the form:

$$h_{ij}(\boldsymbol{p})=\sum_{\ell'=0}^{2\ell}\sum_{m=-2\ell'}^{2\ell'}g^{m[i,j]}_{2\ell'[2\ell,2\ell]}\rho^{2\ell'g}_{m0}(\hat{\boldsymbol{n}})h^0_{2\ell'}(\lambda),\qquad \boldsymbol{p}\neq\boldsymbol{0}.$$

The matrix entries $\rho^{2\ell'g}_{m0}(\hat{\boldsymbol{n}})$ are proportional to the real-valued spherical harmonics:

$$\rho^{2\ell'}_{m0}(\hat{\boldsymbol{n}}) = \frac{\sqrt{4\ell'+1}}{2\sqrt{\pi}} S^m_{2\ell'}(\hat{\boldsymbol{n}}).$$

Finally,

$$h_{ij}(\boldsymbol{p}) = \frac{1}{2\sqrt{\pi}} \sum_{\ell'=0}^{2\ell} \sqrt{4\ell'+1} \sum_{m=-2\ell'}^{2\ell'} g^{m[i,j]}_{2\ell'[2\ell,2\ell]} S^m_{2\ell'}(\hat{\boldsymbol{n}}) h^0_{2\ell'}(\lambda).$$

In particular, for a point $\boldsymbol{p} = (0,0,\lambda)^\top$ we have

$$h_{ij}(\lambda) = \sum_{\ell'=0}^{2\ell} g^{0[i,j]}_{2\ell'[2\ell,2\ell]} h^0_{2\ell'}(\lambda).$$

Note that the matrices $g^0_{2\ell'[2\ell,2\ell]} = g^{0[i,j]}_{2\ell'[2\ell,2\ell]}$ are diagonal. Moreover, we have $g^{0[-k,-k]}_{2\ell'[2\ell,2\ell]} = g^{0[k,k]}_{2\ell'[2\ell,2\ell]}$. Denote

$$\tilde{u}_k(\lambda) = \sum_{\ell'=0}^{2\ell} g^{0[k,k]}_{2\ell'[2\ell,2\ell]} h^0_{2\ell'}(\lambda), \tag{4.81}$$

then we have $\tilde{u}_{-k'}(\lambda) = \tilde{u}'_i k(\lambda)$. The matrix $h_{ij}(\lambda)$ is non-negative-definite and has unit trace if and only if

$$u_k(\lambda) \geq 0, \qquad \sum_{i=0}^{2\ell} u_k(\lambda) = 1, \tag{4.82}$$

where $u_0(\lambda) = \tilde{u}_0(\lambda)$, $u_k(\lambda) = 2\tilde{u}_k(\lambda)$ for $1 \leq k \leq 2\ell$. Geometrically, the function $h_{ij}(\lambda)$ takes values in the simplex $\mathcal{C}_0$ with $2\ell+1$ vertices. The zeroth vertex h^0 is the matrix with the only non-zero element $h^0_{00} = 1$. The kth vertex h^k is the matrix with only non-zero elements $h^k_{-k-k} = h^k_{kk} = \frac{1}{2}$, $1 \leq k \leq 2\ell$. The functions $u_k(\lambda)$ are the barycentric coordinates of the point $h_{ij}(\lambda)$ inside the simplex $\mathcal{C}_0$.

What happens when $\lambda = 0$? This time, the stationary subgroup of the point $\mathbf{0}$ is the whole of O(3). It follows from (4.80) that for all $g \in \mathrm{O}(3)$ we have

$$\boldsymbol{h}_{2\ell'}(0) = \rho_{2\ell'}(g)\boldsymbol{h}_{2\ell'}(0),$$

that is, the vector $\boldsymbol{h}_{2\ell'}(0)$ lies in the subspace where the trivial representation of the group O(3) acts. The only trivial representation among $\rho_{2\ell'}$ is ρ_0. Then $\boldsymbol{h}_{2\ell'}(0) = \mathbf{0}$ whenever $\ell' > 0$. Equation 4.79 takes the form:

$$h_{ij}(\mathbf{0}) = g^{0[i,j]}_{0[2\ell,2\ell]} h^0_0(\mathbf{0}) = \frac{1}{\sqrt{4\ell+1}} \delta_{ij} h^0_0(\mathbf{0}).$$

This matrix has unit trace if and only if $h^0_0(\mathbf{0}) = \frac{1}{\sqrt{4\ell+1}}$. Then, $h_{ij}(\mathbf{0}) = \frac{1}{4\ell+1}\delta_{ij}$. The barycentric coordinates of the point $h_{ij}(\mathbf{0})$ inside the simplex $\mathcal{C}_0$ are

$$u_0(0) = \frac{1}{4\ell+1}, \qquad u_k(0) = \frac{2}{4\ell+1}, \quad 1 \leq k \leq 2\ell. \tag{4.83}$$

Inverting Equation 4.81, we obtain

$$h^0_{2\ell'}(\lambda)=\sum_{k=0}^{2\ell}a^k_{\ell'}u_k(\lambda),$$

where $u_k(\lambda)$ are measurable functions satisfying (4.82) when $\lambda>0$ and (4.83) when $\lambda=0$. Equation 4.77 takes the form:

$$\begin{aligned}\mathsf{B}_{ij}(\boldsymbol{x})&=\frac{1}{2\sqrt{\pi}}\sum_{k=0}^{2\ell}\int_0^\infty\int_{S^2}\mathrm{e}^{\mathrm{i}(\boldsymbol{p},\boldsymbol{x})}\sum_{\ell'=0}^{2\ell}\sqrt{4\ell'+1}a^k_{\ell'}\\&\quad\times\sum_{m=-2\ell'}^{2\ell'}g^{m[i,j]}_{2\ell'[2\ell,2\ell]}S^m_{2\ell'}(\hat{\boldsymbol{n}})\,\mathrm{d}\hat{\boldsymbol{n}}\,u_k(\lambda)\,\mathrm{d}\Phi(\lambda).\end{aligned}\tag{4.84}$$

As usual, to calculate the inner integral, use the Rayleigh expansion (2.62). We obtain

$$\begin{aligned}\mathsf{B}_{ij}(\boldsymbol{x})&=2\sqrt{\pi}\sum_{k=0}^{2\ell}\int_0^\infty\sum_{\ell'=0}^{2\ell}(-1)^{\ell'}\sqrt{4\ell'+1}a^k_{\ell'}j_{2\ell'}(\lambda r)\\&\quad\times\sum_{m=-2\ell'}^{2\ell'}g^{m[i,j]}_{2\ell'[2\ell,2\ell]}S^m_{2\ell'}(\boldsymbol{n})\,\mathrm{d}\Phi_k(\lambda),\end{aligned}\tag{4.85}$$

where $\mathrm{d}\Phi_k(\lambda)=u_k(\lambda)\,\mathrm{d}\Phi(\lambda)$. Equation 4.83 means that

$$\Phi_0(\{0\})=2\Phi_k(\{0\}),\qquad 1\le k\le 2\ell,\tag{4.86}$$

otherwise Φ_k are arbitrary finite Borel measures.

Rewrite Equation (4.85) as

$$\mathsf{B}(\boldsymbol{x})=4\pi\sum_{k=0}^{2\ell}\int_0^\infty\sum_{\ell'=0}^{2\ell}(-1)^{\ell'}a^k_{\ell'}j_{2\ell'}(\lambda r)\mathsf{M}^{\ell'}(\boldsymbol{n})\,\mathrm{d}\Phi_k(\lambda),\tag{4.87}$$

where $\mathsf{M}^{\ell'}(\boldsymbol{n})$ are the M-functions:

$$\mathsf{M}^{\ell'}(\boldsymbol{n})=\sum_{i,j=-2\ell}^{2\ell}\sum_{m=-2\ell'}^{2\ell'}g^{m[i,j]}_{2\ell'[2\ell,2\ell]}U^{2\ell'}_{m0}(\boldsymbol{n})\mathsf{T}^i\otimes\mathsf{T}^j.$$

We know that the M-functions are linear combinations of the elements of the covariant tensors:

$$\mathsf{M}^{\ell'}_{i_1\cdots i_{4\ell}}(\boldsymbol{n})=\sum_{m'=0}^{N_\ell}b_{\ell'm'}\mathsf{L}^{m'}(\boldsymbol{n}).$$

The two-point correlation tensor of the random field $\mathsf{D}(\boldsymbol{x})$ indeed has the form (4.76) with

$$f_{m'}(r)=4\pi\sum_{k=0}^{2\ell}\int_0^\infty\sum_{\ell'=0}^{2\ell}(-1)^{\ell'}a^k_{\ell'}b_{\ell'm'}j_{2\ell'}(\lambda r)\,\mathrm{d}\Phi_k(\lambda).$$

To find the spectral expansion of the random field $\mathsf{D}(\boldsymbol{x})$, we write Equation (4.84) in the form

$$\begin{aligned}\mathsf{B}_{ij}(\boldsymbol{y}-\boldsymbol{x})&=\frac{1}{2\sqrt{\pi}}\sum_{k=0}^{2\ell}\int_0^\infty\int_{S^2}\mathrm{e}^{\mathrm{i}(\boldsymbol{p},\boldsymbol{y})}\overline{\mathrm{e}^{\mathrm{i}(\boldsymbol{p},\boldsymbol{x})}}\sum_{\ell'=0}^{2\ell}\sqrt{4\ell'+1}a^k_{\ell'}\\&\quad\times\sum_{m=-2\ell'}^{2\ell'}g^{m[i,j]}_{2\ell'[2\ell,2\ell]}S^m_{2\ell'}(\hat{\boldsymbol{n}})\,\mathrm{d}\hat{\boldsymbol{n}}\,\mathrm{d}\Phi_k(\lambda),\end{aligned}$$

and expand the two plane waves $\mathrm{e}^{\mathrm{i}(\boldsymbol{p},\boldsymbol{y})}$ and $\overline{\mathrm{e}^{\mathrm{i}(\boldsymbol{p},\boldsymbol{x})}}$ into spherical harmonics separately. To simplify the result, use the Gaunt integral (2.36). We obtain

$$\begin{aligned}\mathsf{B}_{ij}(\boldsymbol{y}-\boldsymbol{x})&=4\pi\sum_{k=0}^{2\ell}\int_0^\infty\sum_{\ell'=0}^{2\ell}a^k_{\ell'}\sum_{m=-2\ell'}^{2\ell'}g^{m[i,j]}_{2\ell'[2\ell,2\ell]}\sum_{\ell_1,\ell_2=0}^{\infty}\mathrm{i}^{\ell_1-\ell_2}\\&\quad\times\sqrt{(2\ell_1+1)(2\ell_2+1)}j_{\ell_1}(\lambda\|\boldsymbol{y}|)j_{\ell_2}(\lambda\|\boldsymbol{x}|)g^{0[0,0]}_{2\ell'[\ell_1,\ell_2]}\\&\quad\times\sum_{m_1=-\ell_1}^{\ell_1}\sum_{m_2=-\ell_2}^{\ell_2}S^{m_1}_{\ell_1}(\boldsymbol{n}_{\boldsymbol{y}})S^{m_2}_{\ell_2}(\boldsymbol{n}_{\boldsymbol{x}})g^{m[m_1,m_2]}_{2\ell'[\ell_1,\ell_2]}\,\mathrm{d}\Phi_k(\lambda).\end{aligned}$$

Applying Karhunen's theorem, we obtain the desired spectral expansion:

$$\begin{aligned}\mathsf{D}_{i_1\cdots i_{2\ell}}(r,\boldsymbol{n})&=2\sqrt{\pi}\sum_{i=-2\ell}^{2\ell}\sum_{k=0}^{2\ell}\sum_{\ell=0}^{\infty}\sum_{m=-\ell}^{\ell}S^m_\ell(\boldsymbol{n})\\&\quad\times\int_0^\infty j_\ell(\lambda r)\,\mathrm{d}Z^{im}_{k\ell}(\lambda)\mathsf{T}^i_{i_1\cdots i_{2\ell}},\end{aligned}\tag{4.88}$$

where $Z^{im}_{k\ell}(\lambda)$ are centred orthogonal stochastic measures on $[0,\infty)$ with

$$\begin{aligned}\mathsf{E}[Z^{im_1}_{k_1\ell_1}(A)\overline{Z^{jm_2}_{k_2\ell_2}(B)}]&=\delta_{k_1k_2}\mathrm{i}^{\ell_1-\ell_2}\sqrt{(2\ell_1+1)(2\ell_2+1)}\\&\quad\times\sum_{\ell'=0}^{2\ell}\sum_{n=-2\ell'}^{2\ell'}a^{k_1}_{\ell'}g^{0[0,0]}_{2\ell'[\ell_1,\ell_2]}g^{n[i,j]}_{2\ell'[2\ell,2\ell]}\\&\quad\times g^{n[m_1,m_2]}_{2\ell'[\ell_1,\ell_2]}\Phi_{k_1}(A\cap B)\end{aligned}\tag{4.89}$$

for all Borel subsets A and B of the set $[0,\infty)$.

Theorem 39. *Formula*

$$\mathsf{B}(\boldsymbol{x})=4\pi\sum_{k=0}^{2\ell}\int_0^\infty\sum_{\ell'=0}^{2\ell}(-1)^{\ell'}a^k_{\ell'}j_{2\ell'}(\lambda r)\sum_{m'=0}^{N_\ell}b_{\ell'm'}\mathsf{L}^{m'}(\boldsymbol{n})\,\mathrm{d}\Phi_k(\lambda)$$

establishes a one-to-one correspondence between the set of two-point correlation functions of homogeneous and isotropic $\mathbb{H}^\ell(\mathbb{R}^3)$-valued random fields on the space $\mathbb{R}^3$ and the sets of finite Borel measures Φ_k on $[0,\infty)$ satisfying (4.86). The field has the form (4.88)–(4.89).

Example 24 ($\ell = 1$)**.** The symmetric tensor square $\mathsf{S}^2(\rho_2)$ is isomorphic to the direct sum $\rho_0 \oplus \rho_2 \oplus \rho_4$. The isomorphism is given by the Godunov–Gordienko coefficients. The O(2)-invariant vector in the space of the representation ρ_0 has coefficients $g^{0[i,i]}_{0[2,2]} = \frac{1}{\sqrt{5}}$, $-2 \le i \le 2$. The O(2)-invariant vector in the space of the representation ρ_2 has coefficients $g^{0[0,0]}_{2[2,2]} = \frac{\sqrt{2}}{\sqrt{7}}$, $g^{0[\pm 1,\pm 1]}_{2[2,2]} = \frac{1}{\sqrt{14}}$ and $g^{0[\pm 2,\pm 2]}_{2[2,2]} = -\frac{\sqrt{2}}{\sqrt{7}}$. Finally, the O(2)-invariant vector in the space of the representation ρ_4 has coefficients $g^{0[0,0]}_{4[2,2]} = \frac{3\sqrt{2}}{\sqrt{35}}$, $g^{0[\pm 1,\pm 1]}_{4[2,2]} = -\frac{2\sqrt{2}}{\sqrt{35}}$ and $g^{0[\pm 2,\pm 2]}_{4[2,2]} = \frac{1}{\sqrt{70}}$. The above coefficients have been calculated using the algorithm described in Selivanova (2014).

The barycentric coordinates of a point inside the simplex take the form

$$\begin{aligned}
u_0(\lambda) &= \frac{1}{\sqrt{5}} h^0_0(\lambda) + \frac{\sqrt{2}}{\sqrt{7}} h^0_2(\lambda) + \frac{3\sqrt{2}}{\sqrt{35}} h^0_4(\lambda), \\
u_1(\lambda) &= \frac{2}{\sqrt{5}} h^0_0(\lambda) + \frac{\sqrt{2}}{\sqrt{7}} h^0_2(\lambda) - \frac{4\sqrt{2}}{\sqrt{35}} h^0_4(\lambda), \\
u_2(\lambda) &= \frac{2}{\sqrt{5}} h^0_0(\lambda) - \frac{2\sqrt{2}}{\sqrt{7}} h^0_2(\lambda) + \frac{\sqrt{2}}{\sqrt{35}} h^0_4(\lambda).
\end{aligned}$$

The functions $h^0_{2\ell'}(\lambda)$ are as follows.

$$\begin{aligned}
h^0_0(\lambda) &= \frac{1}{\sqrt{5}} u_0(\lambda) + \frac{1}{\sqrt{5}} u_1(\lambda) + \frac{1}{\sqrt{5}} u_2(\lambda), \\
h^0_2(\lambda) &= \frac{2}{\sqrt{7}} u_0(\lambda) + \frac{1}{\sqrt{14}} u_1(\lambda) - \frac{\sqrt{2}}{\sqrt{7}} u_2(\lambda), \\
h^0_4(\lambda) &= \frac{3\sqrt{2}}{\sqrt{35}} u_0(\lambda) - \frac{2\sqrt{2}}{\sqrt{35}} u_1(\lambda) + \frac{1}{\sqrt{70}} u_2(\lambda).
\end{aligned}$$

Equation (4.87) takes the form

$$\begin{aligned}
\mathsf{B}(\boldsymbol{x}) &= \frac{4\pi}{\sqrt{5}} \int_0^\infty [j_0(\lambda r)\mathsf{M}^0(\boldsymbol{n}) - j_2(\lambda r)\mathsf{M}^1(\boldsymbol{n}) + j_4(\lambda r)\mathsf{M}^2(\boldsymbol{n})]\, \mathrm{d}\Phi_0(\lambda) \\
&\quad + \frac{2\sqrt{2}\pi}{\sqrt{7}} \int_0^\infty [2j_0(\lambda r)\mathsf{M}^0(\boldsymbol{n}) - j_2(\lambda r)\mathsf{M}^1(\boldsymbol{n}) - 2j_4(\lambda r)\mathsf{M}^2(\boldsymbol{n})]\, \mathrm{d}\Phi_1(\lambda) \\
&\quad + \frac{2\sqrt{2}\pi}{\sqrt{35}} \int_0^\infty [6j_0(\lambda r)\mathsf{M}^0(\boldsymbol{n}) + 4j_2(\lambda r)\mathsf{M}^1(\boldsymbol{n}) + j_4(\lambda r)\mathsf{M}^2(\boldsymbol{n})]\, \mathrm{d}\Phi_2(\lambda),
\end{aligned} \tag{4.90}$$

where

$$\mathsf{M}^{\ell'}_{ijkl}(\boldsymbol{n}) = \sum_{n,p=-2}^{2} \sum_{m=-2\ell'}^{2\ell'} g^{m[n,p]}_{2\ell'[2,2]} g^{n[i,j]}_{2[1,1]} g^{p[k,l]}_{2[1,1]} \rho^{2\ell'}_{m0}(\boldsymbol{n}).$$

The tensors $\mathsf{L}^{m'}_{ijkl}(\boldsymbol{n})$ are given by (2.39), (2.42) and (2.43). The coefficients $b_{\ell' m'}$ were calculated in (3.81). Substituting this formulae in (4.90), we obtain

$$\mathsf{B}(\boldsymbol{x}) = \sum_{\ell=0}^{2} \int_0^\infty \sum_{m=1}^{5} \sum_{n=0}^{2} c_{\ell m n} j_{2n}(\lambda r) \mathsf{L}^m(\boldsymbol{n})\, \mathrm{d}\Phi_\ell(\lambda),$$

where the coefficients $c_{\ell mn}$ are easily calculated, for example, $c_{010} = -\frac{4\pi}{15}$, $c_{011} = -\frac{8\sqrt{2}\pi}{3\sqrt{35}}$, and so on. Equation 4.88 takes the form

$$D_{ij}(r,\boldsymbol{n}) = 2\sqrt{\pi}\sum_{n=-2}^{2}\sum_{k=0}^{2}\sum_{\ell=0}^{\infty}\sum_{m=-\ell}^{\ell} S_\ell^m(\boldsymbol{n})\int_0^\infty j_\ell(\lambda r)\,\mathrm{d}Z_{k\ell}^{nm}(\lambda) g_{2[1,1]}^{n[i,j]}.$$

Example 25 ($\ell = 2$)**.** This time we have

$$\begin{aligned}
u_0(\lambda) &= \frac{1}{3}h_0^0(\lambda) + \frac{10}{3\sqrt{77}}h_2^0(\lambda) + \frac{9\sqrt{2}}{\sqrt{1001}}h_4^0(\lambda) + \frac{2\sqrt{5}}{3\sqrt{11}}h_6^0(\lambda) + \frac{7\sqrt{10}}{3\sqrt{143}}h_8^0(\lambda),\\
u_1(\lambda) &= \frac{2}{3}h_0^0(\lambda) + \frac{17}{3\sqrt{77}}h_2^0(\lambda) + \frac{9\sqrt{2}}{\sqrt{1001}}h_4^0(\lambda) - \frac{1}{3\sqrt{55}}h_6^0(\lambda) - \frac{56\sqrt{2}}{3\sqrt{715}}h_8^0(\lambda),\\
u_2(\lambda) &= \frac{2}{3}h_0^0(\lambda) + \frac{8}{3\sqrt{77}}h_2^0(\lambda) - \frac{\sqrt{22}}{\sqrt{91}}h_4^0(\lambda) - \frac{2\sqrt{11}}{3\sqrt{5}}h_6^0(\lambda) + \frac{28\sqrt{2}}{3\sqrt{715}}h_8^0(\lambda),\\
u_3(\lambda) &= \frac{2}{3}h_0^0(\lambda) - \frac{\sqrt{7}}{3\sqrt{11}}h_2^0(\lambda) - \frac{3\sqrt{14}}{\sqrt{143}}h_4^0(\lambda) + \frac{17}{3\sqrt{55}}h_6^0(\lambda) - \frac{8\sqrt{2}}{3\sqrt{715}}h_8^0(\lambda),\\
u_4(\lambda) &= \frac{2}{3}h_0^0(\lambda) - \frac{4\sqrt{7}}{3\sqrt{11}}h_2^0(\lambda) + \frac{2\sqrt{14}}{\sqrt{143}}h_4^0(\lambda) - \frac{4}{3\sqrt{55}}h_6^0(\lambda) + \frac{\sqrt{2}}{3\sqrt{715}}h_8^0(\lambda).
\end{aligned}$$

The functions $h_{2\ell'}^0(\lambda)$ are as follows.

$$\begin{aligned}
h_0^0(\lambda) &= \frac{1}{3}[u_0(\lambda) + u_1(\lambda) + u_2(\lambda) + u_3(\lambda) + u_4(\lambda)],\\
h_2^0(\lambda) &= \frac{1}{6\sqrt{77}}[20u_0(\lambda) + 17u_1(\lambda) + 8u_2(\lambda) - 7u_3(\lambda) - 28u_4(\lambda)],\\
h_4^0(\lambda) &= \frac{1}{\sqrt{2002}}[18u_0(\lambda) + 9u_1(\lambda) - 11u_2(\lambda) - 21u_3(\lambda) + 14u_4(\lambda)],\\
h_6^0(\lambda) &= \frac{1}{6\sqrt{55}}[20u_0(\lambda) - u_1(\lambda) - 22u_2(\lambda) + 17u_3(\lambda) - 4u_4(\lambda)],\\
h_8^0(\lambda) &= \frac{1}{3\sqrt{1430}}[70u_0(\lambda) - 56u_1(\lambda) - 28u_2(\lambda) - 8u_3(\lambda) + u_4(\lambda)].
\end{aligned}$$

Equation (4.87) takes the form

$$\begin{aligned}
\boldsymbol{B}(\boldsymbol{x}) = \frac{4\pi}{3}\int_0^\infty &[j_0(\lambda r)\boldsymbol{M}^0(\boldsymbol{n}) - j_2(\lambda r)\boldsymbol{M}^1(\boldsymbol{n}) + j_4(\lambda r)\boldsymbol{M}^2(\boldsymbol{n})\\
&- j_6(\lambda r)\boldsymbol{M}^3(\boldsymbol{n}) + j_8(\lambda r)\boldsymbol{M}^4(\boldsymbol{n})]\,\mathrm{d}\Phi_0(\lambda)\\
&+ \frac{2\pi}{3\sqrt{77}}\int_0^\infty [20j_0(\lambda r)\boldsymbol{M}^0(\boldsymbol{n}) - 17j_2(\lambda)\boldsymbol{M}^1(\boldsymbol{n})\\
&+ 8j_4(\lambda r)\boldsymbol{M}^2(\boldsymbol{n}) + 7j_6(\lambda r)\boldsymbol{M}^3(\boldsymbol{n}) - 28j_8(\lambda r)\boldsymbol{M}^4(\boldsymbol{n})]\,\mathrm{d}\Phi_1(\lambda)\\
&+ \frac{2\sqrt{2}\pi}{\sqrt{1001}}\int_0^\infty [18j_0(\lambda r)\boldsymbol{M}^0(\boldsymbol{n}) - 9j_2(\lambda)\boldsymbol{M}^1(\boldsymbol{n})\\
&- 11j_4(\lambda r)\boldsymbol{M}^2(\boldsymbol{n}) + 21j_6(\lambda r)\boldsymbol{M}^3(\boldsymbol{n}) + 14j_8(\lambda r)\boldsymbol{M}^4(\boldsymbol{n})]\,\mathrm{d}\Phi_2(\lambda)
\end{aligned}$$

$$
\begin{aligned}
&+\frac{2\pi}{3\sqrt{55}}\int_0^\infty[20j_0(\lambda r)\boldsymbol{M}^0(\boldsymbol{n})+j_2(\lambda)\boldsymbol{M}^1(\boldsymbol{n})\\
&-22j_4(\lambda r)\boldsymbol{M}^2(\boldsymbol{n})-17j_6(\lambda r)\boldsymbol{M}^3(\boldsymbol{n})-4j_8(\lambda r)\boldsymbol{M}^4(\boldsymbol{n})]\,\mathrm{d}\Phi_3(\lambda)\\
&+\frac{2\sqrt{2}\pi}{3\sqrt{715}}\int_0^\infty[70j_0(\lambda r)\boldsymbol{M}^0(\boldsymbol{n})+56j_2(\lambda)\boldsymbol{M}^1(\boldsymbol{n})\\
&-28j_4(\lambda r)\boldsymbol{M}^2(\boldsymbol{n})+8j_6(\lambda r)\boldsymbol{M}^3(\boldsymbol{n})+j_8(\lambda r)\boldsymbol{M}^4(\boldsymbol{n})]\,\mathrm{d}\Phi_4(\lambda),
\end{aligned}
$$

where

$$
\boldsymbol{M}^{\ell'}_{ijkli'j'k'l'}(\boldsymbol{n})=\sum_{m,n=-4}^{4}\sum_{q=-2\ell'}^{2\ell'}g^{q[m,n]}_{2\ell'[4,4]}\boldsymbol{T}^m_{ijkl}\boldsymbol{T}^n_{i'j'k'l'}\rho^{2\ell'}_{q0}(\boldsymbol{n}),
$$

and where the tensors $\boldsymbol{T}^m_{ijkl}$ are calculated by the induction step above as

$$
\boldsymbol{T}^m_{ijkl}=\sum_{n,q=-2}^{2}g^{m[n,q]}_{4[2,2]}g^{n[i,j]}_{2[1,1]}g^{q[l,m]}_{2[1,1]}.
$$

The $N_2=29$ functions $\boldsymbol{L}_{ijkli'j'k'l'}$ can be found in Section 2.7. The coefficients $b_{\ell'm'}$ can be found in the complete version by Malyarenko & Ostoja-Starzewski (2016*a*) of the paper by Malyarenko & Ostoja-Starzewski (2017*b*).

4.9.2 Damage Tensor

With reference to Chapter 1, recall that anisotropic damage requires damage tensors of rank 2 and higher. Consider D_{ij} of (1.73) which is symmetric by definition. Its correlation function

$$
\boldsymbol{D}^{kl}_{ij}=\mathsf{E}\left[D_{ij}(\boldsymbol{z}+\boldsymbol{z}_1)D_{kl}(\boldsymbol{z}_1)\right], \tag{4.91}
$$

in the case of statistically isotropic damage has the representation (3.63).

To determine all the A_ms, without loss of generality, we may take the unit vector $\boldsymbol{n}=(n_1=1,n_2=0,n_3=0)$ co-aligned with $\boldsymbol{z}$, so that the following auto- and cross-correlations (consecutively named M_i, $i=1,\dots,7$) result

$$
\begin{aligned}
M_1&=\langle T_{11}(\mathbf{0})T_{11}(\boldsymbol{z})\rangle=S_1(z)+2S_2(z)+2S_3(z)+4S_4(z)+S_5(z)\\
M_2&=\langle T_{22}(\mathbf{0})T_{22}(\boldsymbol{z})\rangle=S_1(z)+2S_2(z)\\
M_3&=\langle T_{11}(\mathbf{0})T_{22}(\boldsymbol{z})\rangle=S_1(z)+S_3(z)\\
M_4&=\langle T_{22}(\mathbf{0})T_{33}(\boldsymbol{z})\rangle=S_1(z)\\
M_5&=\langle T_{12}(\mathbf{0})T_{12}(\boldsymbol{z})\rangle=S_2(z)+S_4(z)\\
M_6&=\langle T_{23}(\mathbf{0})T_{23}(\boldsymbol{z})\rangle=S_2(z)\\
M_7&=\langle T_{11}(\mathbf{0})T_{12}(\boldsymbol{z})\rangle=0.
\end{aligned} \tag{4.92}
$$

While the last result shows that the cross-correlation between the 11- and 12-components is always zero, we note that $M_2=M_4+2M_6$ must hold, so that only five M_is are independent, just as we have five functions S_α, $\alpha=1,\dots,5$. In principle, we can determine these five correlations for a specific physical situation. Thus, when T_{ij} is the damage tensor for a given resolution (on a given mesoscale) in a coordinate system defined by $\boldsymbol{n}$, we can use micromechanics or experiments to determine the best fits of M_is. Thus, we have a following strategy for determination of the correlation structure $\mathcal{D}_{ij}^{kl}$:

1. Measure M_i, $i = 1, \ldots, 6$.
2. Determine $S_1 = M_4$ and $S_2 = M_6$.
3. Determine $S_3 = M_3 - M_4$ and $S_4 = M_5 - M_6$.
4. Determine $S_5 = M_1 - M_3 - 4M_5 + 2M_6$.

4.10 Fractal Planetary Rings: Energy Inequalities and Random Field Model

A recent study of the photographs of Saturn's rings taken during the Cassini mission has demonstrated their fractal structure; see Li & Ostoja-Starzewski (2015). This leads us to ask these questions:

Q1: What mechanics argument in support of such a fractal structure of planetary rings is possible?

Q2: What kinematics model of such fractal rings can be formulated?

These issues are approached from the standpoint of rings' spatial structure having (i) statistical stationarity in time and (ii) statistical isotropy in space, but (iii) statistical non-stationarity in space. The reason for (i) is an extremely slow decay of rings relative to the timescale of orbiting around a planet such as Saturn. The reason for (ii) is the obviously circular, albeit disordered and fractal, pattern of rings in the radial coordinate. The reason for (iii) is the lack of invariance with respect to arbitrary shifts in Cartesian space which, on the contrary and for example, holds true for a basic model of turbulent velocity fields. Hence, the model we develop is one of rotational fields of all the particles, each travelling in its circular orbit whose radius is dictated by basic orbital mechanics.

The Q1 issue is approached in (Malyarenko & Ostoja-Starzewski, 2017*a*) by taking rings as self-generated circular structures of non-integer dimensional mass distribution. The approach is based on concepts developed in Tarasov (2005) and Tarasov (2006), while noting that, in inelastic collisions, the total momentum is conserved but the total kinetic energy is not. By comparing total energies of two rings – one of a non-fractal (called 'Euclidean') structure and another of a fractal structure, both carrying the same mass – we infer that the fractal ring is more likely to occur. We also compare their angular momenta, which leads to a consideration of a random, micropolar-type field of particles in the rings. It is important to note that mathematical modelling of fractal structures encountered in physics has evolved from those with fractal mass dimensions to those in fractal dimensional spaces. Key references in the latter area are Tarasov (2014) and Tarasov (2015). The first of these papers contains a short review of several different methods that describe fractal media and the differences between them.

The Q2 issue is approached in the following way. Assume that the angular velocity vector of a rotating particle is a single realisation of a random field. Mathematically, the above field is a random section of a special vector bundle. Using the theory of group representations, we prove that such a field is completely determined by a sequence of continuous positive-definite matrix-valued functions $\{B_k(r,s)\colon k \geq 0\}$ with

$$\sum_{k=0}^{\infty} \operatorname{tr}(B_k(r,r)) < \infty,$$

where the real-valued parameters r and s run over the radial cross-section F of a planet's rings. To reflect the observed fractal nature of Saturn's rings, Avron & Simon (1981) and independently Mandelbrot (1982) supposed that the set F is a *fat fractal subset* of the set $\mathbb{R}$ of real numbers. The set F itself is not a fractal, because its Hausdorff dimension is equal to 1. However, the topological boundary ∂F of the set F, that is, the set of points x_0 such that an arbitrarily small interval $(x_0 - \varepsilon, x_0 + \varepsilon)$ intersects with both F and its complement, $\mathbb{R} \setminus F$, is a fractal. The Hausdorff dimension of ∂F is not an integer number.

4.10.1 A Stochastic Model of Kinematics

First, we consider the particles in planet's rings at a time instant 0.

Instead of the cylindrical coordinate system, introduce the spherical coordinate system (r, φ, θ) with origin O in the centre of the planet such that the plane of planetary rings corresponds to the polar angle's value $\theta = \pi/2$. Let $\overline{\boldsymbol{\omega}}(r,\varphi) \in \mathbb{R}^3$ be the angular velocity vector of a rotating particle located at (r,φ). We assume that $\overline{\boldsymbol{\omega}}(r,\varphi)$ is a *single realisation of a random field.*

To explain the exact meaning of this construction, we proceed as follows. Let (x, y, z) be a Cartesian coordinate system with origin in the planet's centre such that the plane of planetary rings corresponds to the xy-plane, Figure 4.5. Put $G = \mathrm{O}(2) \times \mathrm{SO}(2)$, $K = \mathrm{O}(2)$. The homogeneous space $C = G/K = \mathrm{SO}(2)$ can be identified with a circle, the trajectory of a particle inside rings.

The topological space $R = \mathbb{R}^2 \setminus \{\mathbf{0}\}$ is the union of circles C_r of radii $r > 0$. Every circle determines the vector bundle $\xi_r = (E_{Ur}, \pi_r, C_r)$. Consider the vector bundle $\eta = (E, \pi, R)$, where E is the union of all E_{Ur}, and the restriction of the projection map π to E_{Ur} is equal to π_r. The random field $\overline{\boldsymbol{\omega}}(r,\varphi)$ is a *random section* of the above bundle, that is, $\overline{\boldsymbol{\omega}}(r,\varphi) \in \pi^{-1}(r,\varphi) = \mathbb{R}^3$. In what follows we assume that the random field $\overline{\boldsymbol{\omega}}(r,\varphi)$ is *second-order*, i.e. $\mathsf{E}[\|\overline{\boldsymbol{\omega}}(r,\varphi)\|^2] < \infty$ for all $(r,\varphi) \in R$.

In what follows, we will use the approach by Malyarenko (2011) to the construction of random sections of vector bundles. It is based on the following fact: the vector bundle $\eta = (E, \pi, R)$ is *homogeneous* or *equivariant.* In other words, the action of the group $\mathrm{O}(2)$ on the bundle base R induces the action of $\mathrm{O}(2)$ on the total space E by $(g_0, \boldsymbol{x}) \mapsto (gg_0, \boldsymbol{x})$. This action identifies the spaces $\pi^{-1}(r_0, \varphi)$ for all $\varphi \in [0, 2\pi)$, while the action of the multiplicative group $\mathbb{R}^+$

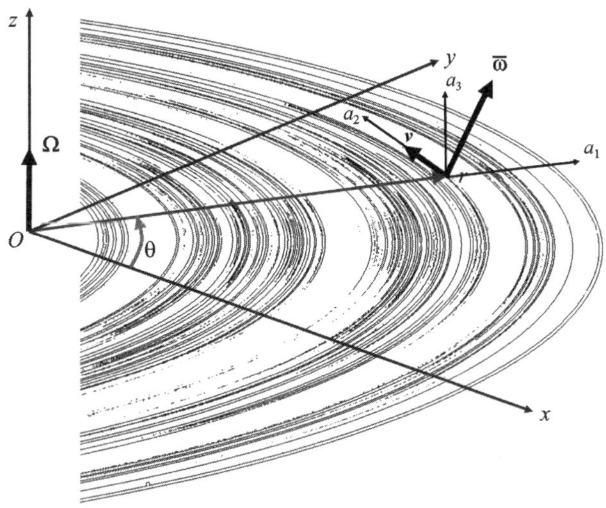

Figure 4.5 The planar ring of particles adapted from Figure 5(b) in Li & Ostoja-Starzewski (2015), showing the Saturnian (Cartesian and cylindrical) coordinate systems as well as the orbital frame of reference (a_1, a_2, a_3) and the body axes (x_1, x_2, x_3) of a typical particle.

on R, $\lambda(r, \varphi) = (\lambda r, \varphi)$, $\lambda > 0$, identifies the spaces $\pi^{-1}(r, \varphi_0)$ for all $r > 0$. We suppose that the random field $\overline{\boldsymbol{\omega}}(r, \varphi)$ is mean-square continuous, i.e.

$$\lim_{\|\boldsymbol{x}-\boldsymbol{x}_0\|\to 0} \mathsf{E}[\|\overline{\boldsymbol{\omega}}(\boldsymbol{x}) - \overline{\boldsymbol{\omega}}(\boldsymbol{x}_0))\|^2] = 0$$

for all $\boldsymbol{x}_0 \in R$.

Let $\langle\overline{\boldsymbol{\omega}}(\boldsymbol{x})\rangle = \mathsf{E}[\overline{\boldsymbol{\omega}}(\boldsymbol{x})]$ be the one-point correlation vector of the random field $\overline{\boldsymbol{\omega}}(\boldsymbol{x})$. On the one hand, under rotation and/or reflection $g \in \mathrm{O}(2)$ the point $\boldsymbol{x}$ becomes the point $g\boldsymbol{x}$. Evidently, the axial vector $\overline{\boldsymbol{\omega}}(\boldsymbol{x})$ transforms according to the representation

$$g = \begin{pmatrix} g_{11} & g_{12} \\ g_{21} & g_{22} \end{pmatrix} \mapsto \rho(g) = \begin{pmatrix} g_{11} & g_{12} & 0 \\ g_{21} & g_{22} & 0 \\ 0 & 0 & \det g \end{pmatrix}$$

and becomes $\rho(g)\overline{\boldsymbol{\omega}}(g\boldsymbol{x})$. The one-point correlation vector of the so transformed random field remains the same, i.e.

$$\langle\overline{\boldsymbol{\omega}}(g\boldsymbol{x})\rangle = \rho(g)\langle\overline{\boldsymbol{\omega}}(\boldsymbol{x})\rangle.$$

On the other hand, the one-point correlation vector of the random field $\overline{\boldsymbol{\omega}}(r, \varphi)$ should be independent upon an arbitrary choice of the x- and y-axes of the Cartesian coordinate systems, i.e. it should not depend on φ. Then we have

$$\langle\overline{\boldsymbol{\omega}}(\boldsymbol{x})\rangle = U(g)\langle\overline{\boldsymbol{\omega}}(\boldsymbol{x})\rangle$$

for all $g \in \mathrm{O}(2)$, i.e. $\langle\overline{\boldsymbol{\omega}}(\boldsymbol{x})\rangle$ belongs to the subspace of $\mathbb{R}^3$ where the trivial component of U acts. Then we obtain $\langle\overline{\boldsymbol{\omega}}(\boldsymbol{x})\rangle = \mathbf{0}$, because ρ does not contain trivial components.

Similarly, let $\langle\overline{\boldsymbol{\omega}}(\boldsymbol{x}),\overline{\boldsymbol{\omega}}(\boldsymbol{y})\rangle = \mathsf{E}[\overline{\boldsymbol{\omega}}(\boldsymbol{x})\otimes\overline{\boldsymbol{\omega}}(\boldsymbol{y})]$ be the two-point correlation tensor of the random field $\overline{\boldsymbol{\omega}}(\boldsymbol{x})$. Under the action of O(2) we should have

$$\langle\overline{\boldsymbol{\omega}}(g\boldsymbol{x}),\overline{\boldsymbol{\omega}}(g\boldsymbol{y})\rangle = (\rho\otimes\rho)(g)\langle\overline{\boldsymbol{\omega}}(\boldsymbol{x}),\overline{\boldsymbol{\omega}}(\boldsymbol{y})\rangle.$$

In other words, the random field $\overline{\boldsymbol{\omega}}(\boldsymbol{x})$ is *wide-sense isotropic* with respect to the group O(2) and its representation ρ.

Consider the restriction of the field $\overline{\boldsymbol{\omega}}(\boldsymbol{x})$ to the circle C_r, $r>0$. The spectral expansion of the field $\{\,\overline{\boldsymbol{\omega}}(r,\varphi)\colon \varphi\in C_r\,\}$ can be calculated using Malyarenko (2011, Theorem 2) or Malyarenko (2013, Theorem 2.28).

The representation ρ is the direct sum of the two irreducible representations $\lambda_-(g)=\det g$ and $\lambda_1(g)=g$. The vector bundle η is the direct sum of the vector bundles η_- and η_1, where the bundle η_- (resp. η_1) is generated by the representation λ_- (resp. λ_1). Let μ_0 be the trivial representation of the group SO(2), and let μ_k be the representation

$$\mu_k(\varphi)=\begin{pmatrix}\cos(k\varphi) & \sin(k\varphi)\\ -\sin(k\varphi) & \cos(k\varphi)\end{pmatrix}.$$

The representations $\lambda_-\otimes\mu_k$, $k\geq 0$ are all irreducible orthogonal representations of the group $G=\mathrm{O}(2)\times\mathrm{SO}(2)$ that contain λ_- after restriction to O(2). The representations $\lambda_1\otimes\mu_k$, $k\geq 0$ are all irreducible orthogonal representations of the group $G=\mathrm{O}(2)\times\mathrm{SO}(2)$ that contain λ_1 after restriction to O(2). The matrix entries of μ_0 and of the second column of μ_k form an orthogonal basis in the Hilbert space $L^2(\mathrm{SO}(2),\mathrm{d}\varphi)$. Their multiples

$$e_k(\varphi)=\begin{cases}\frac{1}{\sqrt{2\pi}}, & \text{if } k=0,\\ \frac{1}{\sqrt{\pi}}\cos(k\varphi), & \text{if } k\leq -1,\\ \frac{1}{\sqrt{\pi}}\sin(k\varphi), & \text{if } k\geq 1\end{cases}$$

form an orthonormal basis of the above space. Then we have

$$\overline{\boldsymbol{\omega}}(r,\varphi)=\sum_{k=-\infty}^{\infty}e_k(\varphi)\boldsymbol{Z}^k(r), \tag{4.93}$$

where $\{\,\boldsymbol{Z}^k(r)\colon k\in\mathbb{Z}\,\}$ is a sequence of centred stochastic processes with

$$\mathsf{E}[\boldsymbol{Z}^k(r)\otimes\boldsymbol{Z}^l(r)]=\delta_{kl}B^{(k)}(r),\qquad \sum_{k\in\mathbb{Z}}\operatorname{tr}(B^{(k)}(r))<\infty.$$

It follows that

$$\boldsymbol{Z}^k(r)=\int_0^{2\pi}\overline{\boldsymbol{\omega}}(r,\varphi)e_k(\varphi)\,\mathrm{d}\varphi.$$

Then we have

$$\mathsf{E}[\boldsymbol{Z}^k(r)\otimes\boldsymbol{Z}^l(s)]=\int_0^{2\pi}\int_0^{2\pi}\mathsf{E}[\overline{\boldsymbol{\omega}}(r,\varphi_1)\otimes\overline{\boldsymbol{\omega}}(s,\varphi_2)]e_k(\varphi_1)\,\mathrm{d}\varphi_1 e_l(\varphi_2)\,\mathrm{d}\varphi_2. \tag{4.94}$$

The field is isotropic and mean-square continuous, therefore

$$\mathsf{E}[\overline{\boldsymbol{\omega}}(r,\varphi_1)\otimes\overline{\boldsymbol{\omega}}(s,\varphi_2)]=B(r,s,\cos(\varphi_1-\varphi_2))$$

is a continuous function. Note that $e_k(\varphi)$ are spherical harmonics of degree $|k|$. Denote by $\boldsymbol{x}\cdot\boldsymbol{y}$ the standard inner product in the space $\mathbb{R}^d$, and by $\mathrm{d}\omega(\boldsymbol{y})$ the Lebesgue measure on the unit sphere $S^{d-1}=\{\,\boldsymbol{x}\in\mathbb{R}^d\colon\|\boldsymbol{x}\|=1\,\}$. Then

$$\int_{S^{d-1}}\mathrm{d}\omega(\boldsymbol{x})=\omega_d=\frac{2\pi^{d/2}}{\Gamma(d/2)},$$

where Γ is the Gamma function.

Now we use the Funk–Hecke theorem, see Andrews, Askey & Roy (1999). For any continuous function f on the interval $[-1,1]$ and for any spherical harmonic $S_k(\boldsymbol{y})$ of degree k we have

$$\int_{S^{d-1}}f(\boldsymbol{x}\cdot\boldsymbol{y})S_k(\boldsymbol{x})\,\mathrm{d}\omega(\boldsymbol{x})=\lambda_kS_k(\boldsymbol{y}),$$

where

$$\lambda_k=\omega_{d-1}\int_{-1}^{1}f(u)\frac{C_k^{(d-2)/2}(u)}{C_k^{(d-2)/2}(1)}(1-u^2)^{(d-3)/2}\,\mathrm{d}u,$$

$d\geq 3$, and $C_k^{(d-2)/2}(u)$ are Gegenbauer polynomials. To see how this theorem looks like when $d=2$, we perform a limit transition as $d\downarrow 2$. By Andrews et al. (1999, Equation (6.4.13′)),

$$\lim_{\lambda\to 0}\frac{C_k^{\lambda}(u)}{C_k^{\lambda}(1)}=T_k(u),$$

where $T_k(u)$ are Chebyshev polynomials of the first kind. We have $\omega_1=2$, $\boldsymbol{x}\cdot\boldsymbol{y}$ becomes $\cos(\varphi_1-\varphi_2)$, and $\mathrm{d}\omega(\boldsymbol{x})$ becomes $\mathrm{d}\varphi_1$. We obtain

$$\int_0^{2\pi}B(r,s,\cos(\varphi_1-\varphi_2))e_k(\varphi_1)\,\mathrm{d}\varphi_1=B^{(k)}(r,s)e_k(\varphi_2),$$

where

$$B^{(k)}(r,s)=2\int_{-1}^{1}B(r,s,u)T_{|k|}(u)(1-u^2)^{-1/2}\,\mathrm{d}u.$$

Equation (4.94) becomes

$$\mathsf{E}[\boldsymbol{Z}^k(r)\otimes\boldsymbol{Z}^l(s)]=\int_0^{2\pi}B^{(k)}(r,s)e_k(\varphi_2)e_l(\varphi_2)\,\mathrm{d}\varphi_2=\delta_{kl}B^{(k)}(r,s).$$

In particular, if $k\neq l$, then the processes $\boldsymbol{Z}^k(r)$ and $\boldsymbol{Z}^l(r)$ are uncorrelated.

Calculate the two-point correlation tensor of the random field $\overline{\boldsymbol{\omega}}(r, \varphi)$. We have

$$\begin{aligned}\mathsf{E}[\overline{\boldsymbol{\omega}}(r, \varphi_1) \otimes \overline{\boldsymbol{\omega}}(s, \varphi_2)] &= \sum_{k=-\infty}^{\infty} e_k(\varphi_1) e_k(\varphi_2) B^{(k)}(r, s) \\ &= \frac{1}{2\pi} B^{(0)}(r, s) + \frac{1}{\pi} \sum_{k=1}^{\infty} \cos(k(\varphi_1 - \varphi_2)) B^{(k)}(r, s).\end{aligned} \tag{4.95}$$

Now we add a time coordinate, t, to our considerations. A particle located at (r, φ) at time moment t, was located at $(r, \varphi - \mu t / r^{3/2})$ at time moment 0. It follows that

$$\overline{\boldsymbol{\omega}}(t, r, \varphi) = \overline{\boldsymbol{\omega}}\left(r, \varphi - \frac{\mu t}{r^{3/2}}\right).$$

Equation (4.93) gives

$$\overline{\boldsymbol{\omega}}(t, r, \varphi) = \sum_{k=-\infty}^{\infty} e_k\left(\varphi - \frac{\mu t}{r^{3/2}}\right) \boldsymbol{Z}^k(r), \tag{4.96}$$

while Equation (4.95) gives

$$\begin{aligned}\mathsf{E}[\overline{\boldsymbol{\omega}}(t_1, r, \varphi_1) \otimes \overline{\boldsymbol{\omega}}(t_2, s, \varphi_2)] &= \frac{1}{2\pi} B^{(0)}(r, s) \\ &\quad + \frac{1}{\pi} \sum_{k=1}^{\infty} \cos\left(k\left(\varphi_1 - \varphi_2 - \frac{\mu(t_1 - t_2)}{r^{3/2}}\right)\right) B^{(k)}(r, s).\end{aligned}$$

Conversely, let $\{\, B^{(k)}(r, s) \colon k \geq 0 \}$ be a sequence of continuous positive-definite matrix-valued functions with

$$\sum_{k=0}^{\infty} \operatorname{tr}(B^{(k)}(r, r)) < \infty, \qquad r \geq 0, \tag{4.97}$$

and let $\{\, \boldsymbol{Z}_k(r) \colon k \in \mathbb{Z} \}$ be a sequence of uncorrelated centred stochastic processes with

$$\mathsf{E}[\boldsymbol{Z}^k(r) \otimes \boldsymbol{Z}^l(s)] = \delta_{kl} B^{(|k|)}(r, s).$$

The random field (4.96) may describe rotating particles inside planetary rings, if all the functions $B^{(k)}(r, s)$ are equal to 0 outside the rectangle $[R_0, R_1]^2$, where R_0 (resp. R_1) is the inner (resp. outer) radius of planetary rings.

To make our model more realistic, we assume that all the functions $B^{(k)}(r, s)$ are equal to 0 outside the Cartesian square F^2, where F is a *fat fractal* subset of the interval $[R_0, R_1]$; see Umberger & Farmer (1985). Mandelbrot (1982) calls these sets *dusts of positive measure*. Such a set has a positive Lebesgue measure, its Hausdorff dimension is equal to 1, but the Hausdorff dimension of its boundary is not an integer number.

A classical example of a fat fractal is a ε-*Cantor set*; see Aliprantis & Burkinshaw (1998). Fix $\varepsilon \in (0, 1)$ and define $A_0 = [0, 1]$. In the first step, remove an open interval of length $2^{-1}(1-\varepsilon)$ from the centre of A_0 and denote the remaining set by A_1. The set A_1 consists of two disjoint closed intervals of the same length.

After n steps, we construct the set A_n: the union of 2^n disjoint close intervals all of the same length. In the $(n+1)$th step, we delete an open interval of length $2^{-2n-1}(1-\varepsilon)$ from the centre of each interval, and denote the remaining set by A_{n+1}. The ε-Cantor set is $C_\varepsilon = \cap_{n=0}^{\infty} A_n$. It is closed, nowhere dense in $[0,1]$, and its Lebesgue measure is equal to ε.

To construct an example, consider an arbitrary sequence of continuous positive-definite matrix-valued functions $\{ B^{(k)}(r,s) \colon k \geq 0 \}$ satisfying (4.97) of the following form:

$$B^{(k)}(r,s) = \sum_{i \in I_k} \boldsymbol{f}_{ik}(r) \boldsymbol{f}_{ik}^{\top}(s),$$

where $\boldsymbol{f}_{ik}(r) \colon [R_0, R_1] \to \mathbb{R}^3$ are continuous functions, satisfying the following condition: for each $r \in [R_0, R_1]$ the set $I_{kr} = \{ i \in I_k \colon f_i(r) \neq 0 \}$ is as most countable and the series

$$\sum_{i \in I_{kr}} \|\boldsymbol{f}_i(r)\|^2$$

converges. The so defined function is obviously positive-definite. Put

$$\tilde{B}^{(k)}(r,s) = \sum_{i \in I_k} \tilde{\boldsymbol{f}}_{ik}(r) \tilde{\boldsymbol{f}}_{ik}^{\top}(s), \qquad r, s \in F.$$

The functions $\tilde{B}^{(k)}(r,s)$ are the restrictions of positive-definite functions $B^{(k)}(r,s)$ to F^2 and are positive-definite themselves. Consider the centred stochastic process $\{ \tilde{\boldsymbol{Z}}^k(r) \colon r \in F \}$ with

$$\mathsf{E}[\tilde{\boldsymbol{Z}}^k(r) \otimes \tilde{\boldsymbol{Z}}^l(s)] = \delta_{kl} \tilde{B}^{(|k|)}(r,s), \qquad r, s \in F.$$

Condition (4.97) guarantees the mean-square convergence of the series

$$\overline{\boldsymbol{\omega}}(t, r, \varphi) = \sum_{k=-\infty}^{\infty} e_k \left(\varphi - \frac{\sqrt{GMt}}{r^{3/2}} \right) \tilde{\boldsymbol{Z}}^k(r)$$

for all $t \geq 0$, $r \in F$ and $\varphi \in [0, 2\pi]$.

4.11 Future Avenues

4.11.1 Quasi-Isotropic Random Fields

Consider a centred scalar-valued homogeneous random field $\tau(\boldsymbol{x})$, $\boldsymbol{x} \in \mathbb{R}^d$ with

$$\mathsf{E}[\tau(\boldsymbol{x})\tau(\boldsymbol{0})] = R(\boldsymbol{x}).$$

Let ρ_0 be the one-dimensional trivial representation of the group $\mathrm{O}(d)$. It is easy to see that the field $\tau(\boldsymbol{x})$ is $(\mathrm{O}(d), \rho_0)$-isotropic if and only if $\|\boldsymbol{x}\| = \|\boldsymbol{y}\|$ implies $R(\boldsymbol{x}) = R(\boldsymbol{y})$. Szczepankiewicz (1985) proposed the following extension of the classical definition of an isotropic random field.

Let $\boldsymbol{T} \colon \mathbb{R}^d \to \mathbb{R}^d$ be a differentiable map with non-zero Jacobian.

Definition 7. A random field $\tau(\boldsymbol{x})$ is called *quasi-isotropic* if $\|\boldsymbol{x}\| = \|\boldsymbol{T}(\boldsymbol{y})\|$ implies $R(\boldsymbol{x}) = R(\boldsymbol{y})$.

Szczepankiewicz (1985 Claim 7.3) states that the two-point correlation function of a homogeneous and quasi-isotropic random field $\tau(\boldsymbol{x})$ has the form

$$R(\boldsymbol{y}) = 2^{(d-2)/2}\Gamma(d/2)\int_0^\infty \frac{J_{(d-2)/2}(\lambda\|\boldsymbol{T}^{-1}(\boldsymbol{y})\|)}{(\lambda\|\boldsymbol{T}^{-1}(\boldsymbol{y})\|)^{(d-2)/2}}\,\mathrm{d}\Phi(\lambda).$$

A possible generalisation of the above definition can be as follows. Let $\mathsf{T}(\boldsymbol{x})$ be a centred homogeneous and (G,ρ)-isotropic random field with the two-point correlation tensor $\langle \mathsf{T}(\boldsymbol{x}), \mathsf{T}(\boldsymbol{0})\rangle$. A centred homogeneous random field $\mathsf{S}(\boldsymbol{x})$ is called *quasi-isotropic* if its two-point correlation tensor has the form

$$\langle \mathsf{S}(\boldsymbol{x}), \mathsf{S}(\boldsymbol{0})\rangle = \langle \mathsf{T}(\boldsymbol{T}^{-1}(\boldsymbol{x})), \mathsf{T}(\boldsymbol{0})\rangle.$$

It may be interesting to explore such fields.

4.11.2 Acoustic Tensor-Valued Random Fields

Let $\mathsf{C}\colon \mathbb{R}^3 \to \mathsf{S}^2(\mathsf{S}^2(\mathbb{R}^3))$ be the elasticity tensor. In a coordinate form, we have $\mathsf{C}(\boldsymbol{x}) = \mathsf{C}_{ijkl}(\boldsymbol{x})$, $\boldsymbol{x} \in \mathbb{R}^3$.

Let S^2 be the centred unit sphere in $\mathbb{R}^3$, and let $\boldsymbol{n} \in S^2$ be a vector of length 1. The *acoustic tensor* is defined as

$$A_{ik}(\boldsymbol{x}, \boldsymbol{n}) = \mathsf{C}_{ijkl}(\boldsymbol{x}) n_j n_l, \tag{4.98}$$

where we use Einstein summation convention.

Let $[G_m]$, $1 \le m \le 8$, be one of the symmetry classes of elasticity tensors, and let V_m be the corresponding fixed point set. Choose a group G lying between G_m and its normaliser $N(G_m)$. The linear space V_m is an invariant subspace of the representation $g \mapsto \mathsf{S}^2(\mathsf{S}^2(g))$ of the group G. Denote by ρ the restriction of the above representation to V_m.

Assume that $\mathsf{C}(\boldsymbol{x})$ is a second-order mean-square continuous homogeneous and (G,U)-isotropic random field taking values in V_m. Then $A_{ik}(\boldsymbol{x}, \boldsymbol{n})$ is also a random field. Our task is to find the general form of the one- and two-point correlation tensors of the random field $A_{ik}(\boldsymbol{x}, \boldsymbol{n})$ as well as its spectral expansion in terms of stochastic integrals.

Example 26. Consider the random field described in Theorem 35. That is: $G = D_8 \times Z_2^c$, $\rho(g) = 5A_{1g} \oplus A_{2g}$. Combining (3.96) and (4.98), we obtain

$$\begin{aligned} A_{ik}(\boldsymbol{x}, \boldsymbol{n}) = {} & \sum_{m=1}^{5} C_m \tilde{\mathsf{T}}^m_{ijkl} n_j n_l + \sum_{m=1}^{6}\sum_{n=1}^{32}\int_{(\hat{\mathbb{R}}^3/D_8\times Z_2^c)_{3,4,6,7}} u_n(\boldsymbol{p}, \boldsymbol{x})\,\mathrm{d}Z^{0n}_m(\boldsymbol{p}) \tilde{\mathsf{T}}^m_{ijkl} n_j n_l \\ & + \sum_{m=1}^{6}\sum_{n=1}^{32}\int_{(\hat{\mathbb{R}}^3/D_8\times Z_2^c)_{0-2,5}} u_n(\boldsymbol{p}, \boldsymbol{x})\,\mathrm{d}Z^{1n}_m(\boldsymbol{p}) \tilde{\mathsf{T}}^m_{ijkl} n_j n_l. \end{aligned}$$

It remains to calculate the matrix entries $B^m_{ik}(\boldsymbol{n}) = \tilde{\boldsymbol{T}}^m_{ijkl} n_j n_l$. Straightforward calculations give: $B^1(\boldsymbol{n}) = \begin{pmatrix} 0 & 0 & 0 \\ 0 & n_3^2 & 0 \\ 0 & 0 & 0 \end{pmatrix}$, $B^2(\boldsymbol{n}) = \frac{1}{2}\begin{pmatrix} n_1^2 & 0 & n_1 n_2 \\ 0 & 0 & 0 \\ n_2 n_1 & 0 & n_2^2 \end{pmatrix}$, $B^3(\boldsymbol{n}) = \frac{1}{2}\begin{pmatrix} 0 & n_1 n_3 & 0 \\ n_3 n_1 & 0 & n_3 n_2 \\ 0 & n_2 n_3 & 0 \end{pmatrix}$, $B^4(\boldsymbol{n}) = \frac{1}{2\sqrt{2}}\begin{pmatrix} n_3^2 & n_3 n_1 & 0 \\ n_1 n_3 & n_1^2+n_2^2 & n_2 n_3 \\ 0 & n_3 n_2 & n_3^2 \end{pmatrix}$, $B^5(\boldsymbol{n}) = \frac{1}{2\sqrt{2}}\begin{pmatrix} n_1^2+n_2^2 & 0 & 0 \\ 0 & 0 & 0 \\ 0 & 0 & n_1^2+n_1^2 \end{pmatrix}$ and $B^6(\boldsymbol{n}) = \frac{1}{2\sqrt{2}}\begin{pmatrix} n_2^2-n_1^2 & 0 & 2n_1 n_2 \\ 0 & 0 & 0 \\ 2n_2 n_1 & 0 & n_1^2-n_2^2 \end{pmatrix}$.

The one-point correlation tensor of the random field $A(\boldsymbol{x}, \boldsymbol{n})$ becomes

$$\langle A(\boldsymbol{x}, \boldsymbol{n}) \rangle = \sum_{m=1}^{5} C_m B^m(\boldsymbol{n}),$$

while its two-point correlation tensor is

$$\begin{aligned} \langle A(\boldsymbol{x}, \boldsymbol{n}_1), A(\boldsymbol{y}, \boldsymbol{n}_2) \rangle = \sum_{i,j=1}^{6} \int_{(\hat{\mathbb{R}}^3/D_8 \times Z_2^c)_{3,4,6,7}} & (j_+(\boldsymbol{p}, \boldsymbol{y}-\boldsymbol{x})(f_0^+)_{ij}(\boldsymbol{p}) \\ & + j_-(\boldsymbol{p}, \boldsymbol{y}-\boldsymbol{x})(f_0^-)_{ij}(\boldsymbol{p}))\, \mathrm{d}\Phi(\boldsymbol{p}) B^i(\boldsymbol{n}_1) \otimes B^j(\boldsymbol{n}_2) \\ + \sum_{i,j=1}^{6} \int_{(\hat{\mathbb{R}}^3/D_8 \times Z_2^c)_{0-2,5}} & (j_+(\boldsymbol{p}, \boldsymbol{y}-\boldsymbol{x}) \\ & + j_-(\boldsymbol{p}, \boldsymbol{y}-\boldsymbol{x}))(f_1)_{ij}(\boldsymbol{p}))\, \mathrm{d}\Phi(\boldsymbol{p}) B^i(\boldsymbol{n}_1) \otimes B^j(\boldsymbol{n}_2). \end{aligned}$$

Similar results may be easily obtained for the case of other symmetry classes.

4.11.3 Variogram of TRF

Define the *variogram* of a rank 1 TRF ($\boldsymbol{T} = \boldsymbol{e}_i T_i$) as the variance of the difference between field $T_i(\boldsymbol{x})$ and $T_j(\boldsymbol{y})$ values at two locations:

$$\begin{aligned} 2\gamma(\boldsymbol{x}, \boldsymbol{y}) \equiv 2\boldsymbol{e}_i \boldsymbol{e}_j \gamma_{ij}(\boldsymbol{x}, \boldsymbol{y}) &:= \mathsf{Var}(\boldsymbol{e}_i T_i(\boldsymbol{x}) - \boldsymbol{e}_j T_j(\boldsymbol{y})) \\ &= \left\langle [(\boldsymbol{e}_i T_i(\boldsymbol{x}) - \langle \boldsymbol{e}_i T_i(\boldsymbol{x}) \rangle)(\boldsymbol{e}_j T_j(\boldsymbol{y}) - \boldsymbol{e}_j \langle T_j(\boldsymbol{y}) \rangle)]^2 \right\rangle \end{aligned}$$

which is twice the *semivariogram* $\gamma_{ij}(\boldsymbol{x}, \boldsymbol{y})$. It is a rank 2 tensor.

If the TRF has the same constant mean in each component, $\langle T_i(\boldsymbol{x}) \rangle = \langle T_j(\boldsymbol{x}) \rangle = \mu$, then

$$2\gamma_{ij}(\boldsymbol{x}, \boldsymbol{y}) = \left\langle [T_i(\boldsymbol{x}) - T_j(\boldsymbol{y})]^2 \right\rangle.$$

In the case of a statistically homogeneous TRF,

$$2\gamma_{ij}(\boldsymbol{x}, \boldsymbol{y}) = 2\gamma_{ij}(\boldsymbol{x} - \boldsymbol{y}).$$

Henceforth, we can also work with $\gamma_{ij}(\boldsymbol{z})$.

If, in addition, the TRF is statistically isotropic, for any orthogonal transformation g of the vector RF $\boldsymbol{T}$,

$$\begin{aligned} \langle T_i(g\boldsymbol{x}) \rangle = \langle g T_i(\boldsymbol{x}) \rangle &= g \langle T_i(\boldsymbol{x}) \rangle \\ 2\gamma_{ij}(g\boldsymbol{z}) &= 2g\gamma_{ij}(\boldsymbol{z})\, g. \end{aligned}$$

Properties:

1. The variogram is non-negative for any i, j
$$2\gamma_{ij}(\boldsymbol{x}, \boldsymbol{y}) \geq 0;$$
2. The variogram equals zero for any i
$$2\gamma_{ii}(\boldsymbol{x}, \boldsymbol{x}) := \mathsf{Var}(T_i(\boldsymbol{x}) - T_j(\boldsymbol{x})) = 0;$$
3. For any fixed index $i = j$ (e.g. $i = j = 1$), a function $\gamma_{11}(\boldsymbol{x}_k, \boldsymbol{x}_l)$ is a semivariogram iff it is conditionally non-negative definite, i.e. for all weights $w_1, \dots, w_N$ subject to $\sum_{k=1}^{N} w_k = 0$ and any locations $\boldsymbol{x}_1, \dots, \boldsymbol{x}_N$ there holds:
$$\sum_{k=1}^{N}\sum_{l=1}^{N} w_k \gamma_{11}(\boldsymbol{x}_k, \boldsymbol{x}_l) w_l \leq 0.$$
4. If the covariance of the stationary vector RF exists, it is related to the variogram $C_{ij}(\boldsymbol{z})$ by
$$2\gamma_{ij}(\boldsymbol{z}) = C_{ij}(\boldsymbol{x}) + C_{ij}(\boldsymbol{y}) - 2C_{ij}(\boldsymbol{x}, \boldsymbol{y}).$$

4.12 Bibliographical Remarks

The random field model of fractal planetary rings has been published in Malyarenko & Ostoja-Starzewski (2017*a*).

There are at least three different (but most probably equivalent) approaches to the construction of random sections of vector bundles, the first by Geller & Marinucci (2010), the second by Malyarenko (2011) and Malyarenko (2013) and the third by Baldi & Rossi (2014).

References

Abramowitz, M. & Stegun, I. A. (1964), *Handbook of Mathematical Functions with Formulas, Graphs, and Mathematical Tables*, Vol. 55 of *National Bureau of Standards Applied Mathematics Series*, For sale by the Superintendent of Documents, US Government Printing Office, Washington, DC.

Adams, J. F. (1969), *Lectures on Lie Groups*, New York, NY and Amsterdam: W. A. Benjamin, Inc.

Adler, R. J. (2010), *The Geometry of Random Fields*, Vol. 62 of *Classics in Applied Mathematics*, Philadelphia, PA: Society for Industrial and Applied Mathematics (SIAM). Reprint of the 1981 original [MR0611857].

Aldrich, D., Fan, Z. & Mummery, P. (2000), 'Processing, microstructure, and physical properties of interpenetrating Al_2O_3/Ni composites', *Mater. Sci. Technology* **16**(7–8), 747–752.

Aliprantis, C. D. & Burkinshaw, O. (1998), *Principles of Real Analysis*, third edn, San Diego, CA: Academic Press, Inc.

Altmann, S. L. & Herzig, P. (1994), *Point-Group Theory Tables*, Oxford science publications, Oxford: Clarendon Press.

Andrews, D. L. & Ghoul, W. A. (1982), 'Irreducible fourth-rank cartesian tensors', *Phys. Rev. A* **25**, 2647–2657.

Andrews, G. E., Askey, R. & Roy, R. (1999), *Special Functions*, Vol. 71 of *Encyclopedia of Mathematics and its Applications*, Cambridge: Cambridge University Press, Cambridge.

Auffray, N., Kolev, B. & Olive, M. (2017), 'Handbook of bi-dimensional tensors: Part I: Harmonic decomposition and symmetry classes', *Math. Mech. Solids* **22**(9), 1847–1865.

Auffray, N., Kolev, B. & Petitot, M. (2014), 'On anisotropic polynomial relations for the elasticity tensor', *J. Elasticity* **115**(1), 77–103.

Avron, J. E. & Simon, B. (1981), 'Almost periodic Hill's equation and the rings of Saturn', *Phys. Rev. Lett.* **46**(17), 1166–1168.

Baldi, P. & Rossi, M. (2014), 'Representation of Gaussian isotropic spin random fields', *Stoch. Process. Appl.* **124**(5), 1910–1941.

Batchelor, G. K. (1951), 'Note on a class of solutions of the Navier–Stokes equations representing steady rotationally-symmetric flow', *Quart. J. Mech. Appl. Math.* **4**, 29–41.

Batchelor, G. K. (1982), *The Theory of Homogeneous Turbulence*, Cambridge and New York, NY: Cambridge University Press. Reprint, Cambridge Science Classics.

Bauer, G. (1859), 'Von den Coefficienten der Reihen von Kugelfunctionen einer Variablen', *J. Reine Angew. Math.* **56**, 101–121.

Bažant, Z. P. & Jirásek, M. (2002), 'Nonlocal integral formulations of plasticity and damage: Survey of progress', *J. Engng. Mech.* **128**(11), 1119–1149.

Beran, M. & McCoy, J. (1970*a*), 'Mean field variations in a statistical sample of heterogeneous linearly elastic solids', *Intern. J. Solids Struct.* **6**(8), 1035–1054.

Beran, M. & McCoy, J. (1970*b*), 'The use of strain gradient theory for analysis of random media', *Intern. J. Solids Struct.* **6**(9), 1267–1275.

Berezans'kiĭ, J. M. (1968), *Expansions in Eigenfunctions of Selfadjoint Operators*, Translated from the Russian by R. Bolstein, J. M. Danskin, J. Rovnyak and L. Shulman. Translations of Mathematical Monographs, Vol. 17, Providence, RI: American Mathematical Society.

Biedenharn, L. C. & Louck, J. D. (1981), *Angular Momentum in Quantum Physics. Theory and Application*, Vol. 8 of *Encyclopedia of Mathematics and its Applications*, With a foreword by Peter A. Carruthers. Reading, MA: Addison-Wesley Publishing Co.

Biggs, N. L. & White, A. T. (1979), *Permutation Groups and Combinatorial Structures*, Vol. 33 of *London Mathematical Society Lecture Note Series*, Cambridge and New York, NY: Cambridge University Press.

Bishop, E. & de Leeuw, K. (1959), 'The representations of linear functionals by measures on sets of extreme points', *Ann. Inst. Fourier. Grenoble* **9**, 305–331.

Boileau, M., Maillot, S. & Porti, J. (2003), *Three-Dimensional Orbifolds and their Geometric Structures*, Vol. 15 of *Panoramas et Synthèses [Panoramas and Syntheses]*, Société Mathématique de France, Paris.

Bostanabad, R., Zhang, Y., Li, X., Kearney, T., Brinson, L. C., Apley, D. W., Liu, W. K. & Chen, W. (2018), 'Computational microstructure characterization and reconstruction: Review of the state-of-the-art techniques', *Progr. Material Sci.* **95**, 1–41.

Bourbaki, N. (1998), *Algebra I. Chapters 1–3*, Elements of Mathematics. Translated from the French, Berlin: Springer-Verlag. Reprint of the 1989 English translation.

Bourbaki, N. (2004), *Integration. II. Chapters 7–9*, Elements of Mathematics. Translated from the 1963 and 1969 French originals by Sterling K. Berberian. Berlin: Springer-Verlag.

Bredon, G. E. (1972), *Introduction to Compact Transformation Groups*, Pure and Applied Mathematics, Vol. 46. New York, NY and London: Academic Press.

Brock, C., ed. (2014), *International Tables for Crystallography*, Vol. A–G, fourth edn, Wiley.

Bröcker, T. & tom Dieck, T. (1995), *Representations of Compact Lie Groups*, Vol. 98 of *Graduate Texts in Mathematics*, Translated from the German manuscript, New York, NY: Springer-Verlag. Corrected reprint of the 1985 translation.

Cameron, P. J. (1999), *Permutation Groups*, Vol. 45 of *London Mathematical Society Student Texts*, Cambridge: Cambridge University Press.

Carathéodory, C. (1907), 'Über den Variabilitätsbereich der Koeffizienten von Potenzreihen, die gegebene Werte nicht annehmen', *Math. Ann.* **64**(1), 95–115.

Cattaneo, C. (1949), 'Sulla conduzione del calore', *Atti Sem. Mat. Fis. Univ. Modena* **3**, 83–101.

Choquet, G. (1956), 'Existence des représentations intégrales au moyen des points extrémaux dans les cônes convexes', *C. R. Acad. Sci. Paris* **243**, 699–702.

Christensen, R. (2003), *Theory of Viscoelasticity*, Civil, Mechanical and Other Engineering Series, Dover Publications.

Clarke, D. R. (1992), 'Interpenetrating phase composites', *J. Amer. Ceramic Soc.* **75**(4), 739–758.

Clebsch, C. (1872), *Theorie der binären algebraischen Formen*, Leipzig: Teubner.

Condon, E. U. & Shortley, G. H. (1935), *The Theory of Atomic Spectra*, Cambridge: Cambridge University Press.

Conway, J. B. (2014), *A Course in Point Set Topology*, Undergraduate Texts in Mathematics, Cham: Springer.

Cosserat, E. & Cosserat, F. (1909), *Théorie des Corps déformables*, Paris: Hermann.

Curtis, C. W. (1999), *Pioneers of Representation Theory: Frobenius, Burnside, Schur, and Brauer*, Vol. 15 of *History of Mathematics*, Providence, RI: American Mathematical Society and London: London Mathematical Society.

Dantu, P. & Mandel, A. (1963), 'Contribution à l'étude théorique et expérimentale du coefficient d'élasticité d'un milieu hétérogène, mais statistiquement homogène', *Annales des Ponts et Chausses* **133**(2), 115–146.

Danzer, L., Grünbaum, B. & Klee, V. (1963), 'Helly's theorem and its relatives', in *Proc. Sympos. Pure Math., Vol. VII*, Providence, RI: Amer. Math. Soc., pp. 101–180.

de Groot, S. R. & Mazur, P. (1984), *Nonequilibrium Thermodynamics*, New York, NY: Dover Publications, Inc., Reprint of the 1962 original.

dell'Isola, F., Seppecher, P. & Della Corte, A. (2015), 'The postulations *à la d'Alembert* and *à la Cauchy* for higher gradient continuum theories are equivalent: a review of existing results', *Proc. R. Soc. A.* **471**(2183), 20150415, 25.

Demengel, F. & Demengel, G. (2012), *Functional Spaces for the Theory of Elliptic Partial Differential Equations*, Universitext, EDP Sciences, Les Ulis. Translated from the 2007 French original by Reinie Erné London: Springer.

Deville, M. O. & Gatski, T. B. (2012), *Mathematical Modeling for Complex Fluids and Flows*, chapter 'Tensor Analysis, Invariants, and Representations', pp. 21–46 Berlin, Heidelberg: Springer.

Diaconis, P. (1988), *Group Representations in Probability and Statistics*, Institute of Mathematical Statistics Lecture Notes – Monograph Series, 11, Hayward, CA: Institute of Mathematical Statistics.

Dixmier, J. (1984), *General Topology*, Undergraduate Texts in Mathematics, Springer-Verlag, New York. Translated from the French by Sterling K. Berberian.

Dixon, J. D. & Mortimer, B. (1996), *Permutation Groups*, Vol. 163 of *Graduate Texts in Mathematics*, New York, NY: Springer-Verlag.

Doob, J. L. (1990), *Stochastic Processes*, Wiley Classics Library, New York, NY: John Wiley & Sons, Inc. Reprint of the 1953 original, A Wiley-Interscience Publication.

Duistermaat, J. J. & Kolk, J. A. C. (2000), *Lie Groups*, Universitext, Berlin: Springer-Verlag.

Edelen, D. G. B. (1973), 'On the existence of symmetry relations and dissipation potentials', *Arch. Rational Mech. Anal.* **51**, 218–227.

Edelen, D. G. B. (1974), 'Primitive thermodynamics: a new look at the Clausius–Duhem inequality', *Internat. J. Engrg. Sci.* **12**, 121–141.

Eilenberg, S. & MacLane, S. (1945), 'General theory of natural equivalences', *Trans. Amer. Math. Soc.* **58**, 231–294.

Engelking, R. (1989), *General Topology*, Vol. 6 of *Sigma Series in Pure Mathematics*, second edn. Translated from the Polish by the author. Berlin: Heldermann Verlag.

Erdélyi, A., Magnus, W., Oberhettinger, F. & Tricomi, F. G. (1981), *Higher Transcendental Functions. Vol. II*, Based on notes left by Harry Bateman. Melbourne, FL: Robert E. Krieger Publishing Co., Inc. Reprint of the 1953 original.

Eringen, A. C. (1999), *Microcontinuum Field Theories. I. Foundations and solids*, New York, NY: Springer-Verlag.

Evans, D. J., Cohen, E. G. D. & Morriss, G. P. (1993), 'Probability of second law violations in shearing steady states', *Phys. Rev. Lett.* **71**, 2401–2404.

Evans, D. J. & Searles, D. J. (2002), 'The fluctuation theorem', *Adv. Phys.* **51**(7), 1529–1585.

Forte, S. & Vianello, M. (1996), 'Symmetry classes for elasticity tensors', *J. Elasticity* **43**(2), 81–108.

Friedmann, A. A. & Keller, L. P. (1924), 'Differentialgleichungen für Turbulente Bewegung einer Kompressiblen Flussigkeit', in *Proceedings of the First International Congress for Applied Mechanics*, Delft, pp. 395–405.

Frisch, U. (1995), *Turbulence. The legacy of A. N. Kolmogorov*, Cambridge: Cambridge University Press.

Fulton, W. & Harris, J. (1991), *Representation Theory, A First Course*, Vol. 129 of *Graduate Texts in Mathematics*, New York, NY: Springer-Verlag.

Ganczarski, A., Skrzypek, J. & Altenbach, H. (2010), *Modeling of Material Damage and Failure of Structures*, Berlin Heidelberg: Springer.

Ganczarski, A. W., Egner, H. & Skrzypek, J. J. (2015), *Mechanics of Anisotropic Materials*, Engineering Materials, chapter Introduction to Mechanics of Anisotropic Materials, Cham: Springer, pp. 1–56.

Gaunt, J. A. (1929), 'The triplets of helium', *Philos. Trans. Roy. Soc. A* **228**, 151–196.

Gegenbauer, L. (1873), 'Über die Functionen ${X_n}^m$ (13 June 1873)', *Wiener Sitzungsberichte* **LXVIII**(2), 357–367.

Gegenbauer, L. (1877*a*), 'Über die Bessel'schen Functionen (22 June 1876)', *Wiener Sitzungsberichte* **LXXIV**(2), 124–130.

Gegenbauer, L. (1877*b*), 'Zur Theorie der Bessel'schen Functionen (18 January 1877)', *Wiener Sitzungsberichte* **LXXV**(2), 218–222.

Geller, D. & Marinucci, D. (2010), 'Spin wavelets on the sphere', *J. Fourier Anal. Appl.* **16**(6), 840–884.

Geymonat, G. & Weller, T. (2002), 'Classes de symétrie des solides piézoélectriques', *C. R. Math. Acad. Sci. Paris* **335**(10), 847–852.

Gneiting, T. & Schlather, M. (2004), 'Stochastic models that separate fractal dimension and the Hurst effect', *SIAM Rev.* **46**(2), 269–282.

Goddard, J. D. (2014), 'Edelen's dissipation potentials and the visco-plasticity of particulate media', *Acta Mech.* **225**(8), 2239–2259.

Godunov, S. K. & Gordienko, V. M. (2004), 'Clebsch–Gordan coefficients in the case of various choices of bases of unitary and orthogonal representations of the groups SU(2) and SO(3)', *Sibirsk. Mat. Zh.* **45**(3), 540–557.

Golubitsky, M., Stewart, I. & Schaeffer, D. G. (1988), *Singularities and Groups in Bifurcation Theory. Vol. II*, Vol. 69 of *Applied Mathematical Sciences*, New York, NY: Springer-Verlag.

Goodman, R. & Wallach, N. R. (2009), *Symmetry, Representations, and Invariants*, Vol. 255 of *Graduate Texts in Mathematics*, Dordrecht: Springer.

Gordan, P. (1868), 'Beweis dass jede Covariante und invariante einer binären Form eine ganze Function mit numerischen Coefficienten einer endlichen Anzahl solcher Formen ist', *J. Reine U. Angew. Math.* **69**, 323–354.

Gordan, P. (1875), *Über das Formensystem binärer Formen*, Leipzig: Teubner.

Gordienko, V. M. (2002), 'Matrix elements of real representations of the groups O(3) and SO(3)', *Sibirsk. Mat. Zh.* **43**(1), 51–63.

Green, A. E. & Lindsay, K. A. (1972), 'Thermoelasticity', *J. Elasticity* **2**(1), 1–7.

Guilleminot, J., Le, T. T. & Soize, C. (2013), 'Stochastic framework for modeling the linear apparent behavior of complex materials: application to random porous materials with interphases', *Acta Mech. Sin.* **29**(6), 773–782.

Guilleminot, J., Noshadravan, A., Soize, C. & Ghanem, R. G. (2011), 'A probabilistic model for bounded elasticity tensor random fields with application to polycrystalline microstructures', *Comput. Methods Appl. Mech. Engrg.* **200**(17–20), 1637–1648.

Guilleminot, J. & Soize, C. (2011), 'Non-Gaussian positive-definite matrix-valued random fields with constrained eigenvalues: application to random elasticity tensors with uncertain material symmetries', *Internat. J. Numer. Methods Engrg.* **88**(11), 1128–1151.

Guilleminot, J. & Soize, C. (2012), 'Generalized stochastic approach for constitutive equation in linear elasticity: a random matrix model', *Internat. J. Numer. Methods Engrg.* **90**(5), 613–635.

Guilleminot, J. & Soize, C. (2013*a*), 'On the statistical dependence for the components of random elasticity tensors exhibiting material symmetry properties', *J. Elasticity* **111**(2), 109–130.

Guilleminot, J. & Soize, C. (2013*b*), 'Stochastic model and generator for random fields with symmetry properties: application to the mesoscopic modeling of elastic random media', *Multiscale Model. Simul.* **11**(3), 840–870.

Guilleminot, J. & Soize, C. (2014), 'Itô SDE-based generator for a class of non-Gaussian vector-valued random fields in uncertainty quantification', *SIAM J. Sci. Comput.* **36**(6), A2763–A2786.

Haar, A. (1933), 'Der Massbegriff in der Theorie der kontinuierlichen Gruppen', *Ann. of Math. (2)* **34**(1), 147–169.

Hamed, E., Novitskaya, E., Li, J., Jasiuk, I. & McKittrick, J. (2015), 'Experimentally-based multiscale model of the elastic moduli of bovine trabecular bone and its constituents', *Materials Science and Engineering: C* **54**, 207–216.

Hannan, E. J. (1965), *Group Representations and Applied Probability*, Methuen's Supplementary Review Series in Applied Probability, Vol. 3, London: Methuen & Co., Ltd.

Hansen, A. C. (2010), 'Infinite-dimensional numerical linear algebra: theory and applications', *Proc. R. Soc. Lond. Ser. A Math. Phys. Eng. Sci.* **466**(2124), 3539–3559.

Hazanov, S. & Huet, C. (1994), 'Order relationships for boundary conditions effect in heterogeneous bodies smaller than the representative volume', *J. Mech. Phys. Solids* **42**(12), 1995–2011.

Helland, K. N. & Atta, C. W. V. (1978), 'The "Hurst phenomenon" in grid turbulence', *J. Fluid Mech.* **85**(3), 573–589.

Helnwein, P. (2001), 'Some remarks on the compressed matrix representation of symmetric second-order and fourth-order tensors', *Comput. Methods Appl. Mech. Engrg.* **190**(22-23), 2753–2770.

Hermann, C. (1928), 'Zur systematischen Strukturtheorie I. Eine neue Raumgruppensymbolik', *Z. Kristallogr.* **68**, 257–287.

Hetnarski, R. B., ed. (2013), *Encyclopedia of Thermal Stresses*, Dordrecht; Springer.

Hetnarski, R. B. & Ignaczak, J. (2011), *The Mathematical Theory of Elasticity*, second edn, Boca Raton, FL: CRC Press.

Hilbert, D. (1890), 'Über die Theorie der algebraischen Formen', *Math. Ann.* **36**(4), 473–534.

Hilbert, D. (1893), 'Über die vollen Invariantensysteme', *Math. Ann.* **42**(3), 313–373.

Hill, R. (1963), 'Elastic properties of reinforced solids: Some theoretical principles', *J. Mech. Phys. Solids* **11**(5), 357–372.

Hofmann, K. H. & Morris, S. A. (2013), *The Structure of Compact Groups. A Primer for the Student — A Handbook for the Expert*, Vol. 25 of *De Gruyter Studies in Mathematics*, third edn, De Gruyter, Berlin.

Houlsby, G. & Puzrin, A. (2007), *Principles of Hyperplasticity: An Approach to Plasticity Theory Based on Thermodynamic Principles*, London: Springer.

Huet, C. (1990), 'Application of variational concepts to size effects in elastic heterogeneous bodies', *J. Mech. Phys. Solids* **38**(6), 813–841.

Ignaczak, J. (1963), 'A completeness problem for stess equations of motion in the linear elasticity theory', *Arch. Mech. Stos.* **15**, 225–234.

Ignaczak, J. (1974), 'A dynamic version of Saint-Venant's principle in the linear theory of elasticity', *Bull. Acad. Polon. Sci. Sér. Tech.* **22**(6), 313–319.

Ignaczak, J. & Ostoja-Starzewski, M. (2010), *Thermoelasticity with finite wave speeds*, Oxford Mathematical Monographs, Oxford University Press, Oxford.

Ivanov, A. V. & Leonenko, N. N. (1989), *Statistical Analysis of Random Fields*, Vol. 28 of *Mathematics and its Applications (Soviet Series)*, Kluwer Academic Publishers Group, Dordrecht. With a preface by A. V. Skorokhod, Translated from the Russian by A. I. Kochubinskiĭ.

Jaroszkiewicz, G. (2014), *Principles of Discrete Time Mechanics*, Cambridge Monographs on Mathematical Physics, Cambridge: Cambridge University Press.

Kale, S., Karimi, P., Sabet, F. A., Jasiuk, I. & Ostoja-Starzewski, M. (2018), 'Tunneling-percolation model of multicomponent nanocomposites', *J. Appl. Phys.* **123**(8), 085104.

Kale, S., Saharan, A., Koric, S. & Ostoja-Starzewski, M. (2015), 'Scaling and bounds in thermal conductivity of planar Gaussian correlated microstructures', *J. Appl. Phys.* **117**(10), 104301.

Kampé de Fériet, J. (1939), 'Les fonctions aléatories stationnaries et la théorie statistique de la turbulence homogène', *Ann. Soc. Sci. Bruxelles Sér. I* **59**, 145–210.

Kanatani, K.-I. (1984), 'Distribution of directional data and fabric tensors', *Internat. J. Engrg. Sci.* **22**(2), 149–164.

Karhunen, K. (1947), 'Über lineare Methoden in der Wahrscheinlichkeitsrechnung', *Ann. Acad. Sci. Fennicae. Ser. A. I. Math.-Phys.* **1947**(37), 79.

Katafygiotis, L. S., Zerva, A. & Malyarenko, A. (1999), 'Simulation of homogeneous and partially isotropic random fields', *J. Eng. Mech.* **125**, 1180–1189.

Kearsley, E. A. & Fong, J. T. (1975), 'Linearly independent sets of isotropic Cartesian tensors of ranks up to eight', *J. Res. Nat. Bur. Standards Sect. B* **79B**(1–2), 49–58.

Kelley, J. L. (1975), *General Topology*, New York, NY and Berlin: Springer-Verlag. Reprint of the 1955 edition (Van Nostrand, Toronto, Ont.), Graduate Texts in Mathematics, No. 27.

Klimyk, A. U. (1979), *Matrichnye elementy i koeffitsienty Klebsha-Gordana predstavlenii grupp*, 'Naukova Dumka', Kiev.

Krein, M. & Milman, D. (1940), 'On extreme points of regular convex sets', *Studia Math.* **9**, 133–138.

Kröner, E. (1958), *Kontinuumstheorie der Versetzungen und Eigenspannungen*, Ergebnisse der angewandten Mathematik. Bd. 5, Berlin-Göttingen-Heidelberg: Springer-Verlag.

Kuratowski, K. (1966), *Topology. Vol. I*, New edition, revised and augmented. Translated from the French by J. Jaworowski, New York, NY and London Academic Press; Warsaw: Państwowe Wydawnictwo Naukowe (Polish Scientific Publishers).

Kuratowski, K. (1968), *Topology. Vol. II*, New edition, revised and augmented. Translated from the French by A. Kirkor, New York, NY and London: Academic Press; Warsaw: Państwowe Wydawnictwo Naukowe (Polish Scientific Publishers).

Li, J. & Ostoja-Starzewski, M. (2015), 'Edges of Saturn's rings are fractal', *SpringerPlus* **4**(1), 158.

Lomakin, V. (1965), 'Deformation of microscopically nonhomogeneous elastic bodies', *J. Appl. Math. Mech.* **29**(5), 1048–1054.

Lomakin, V. A. (1964), 'Statistical description of the stressed state of a body under deformation', *Dokl. Akad. Nauk SSSR* **155**, 1274–1277.

Lord, G. J., Powell, C. E. & Shardlow, T. (2014), *An Introduction to Computational Stochastic PDEs*, Cambridge Texts in Applied Mathematics, Cambridge, NY: Cambridge University Press.

Lord, H. & Shulman, Y. (1967), 'A generalized dynamical theory of thermoelasticity', *J. Mech. Phys. Solids* **15**(5), 299–309.

Lubarda, V. & Krajcinovic, D. (1993), 'Damage tensors and the crack density distribution', *Int. J. Solids Struct.* **30**(20), 2859–2877.

Makse, H. A., Havlin, S., Schwartz, M. & Stanley, H. E. (1996), 'Method for generating long-range correlations for large systems', *Phys. Rev. E* **53**, 5445–5449.

Malyarenko, A. (2011), 'Invariant random fields in vector bundles and application to cosmology', *Ann. Inst. Henri Poincaré Probab. Stat.* **47**(4), 1068–1095.

Malyarenko, A. (2013), *Invariant Random Fields on Spaces with a Group Action*, Probability and its Applications New York, NY and Heidelberg: Springer, Heidelberg. With a foreword by Nikolai Leonenko.

Malyarenko, A. & Ostoja-Starzewski, M. (2014), 'Statistically isotropic tensor random fields: correlation structures', *Math. Mech. Complex Syst.* **2**(2), 209–231.

Malyarenko, A. & Ostoja-Starzewski, M. (2016*a*), 'A random field formulation of Hooke's law in all elasticity classes'. arXiv 1602.09066v2.

Malyarenko, A. & Ostoja-Starzewski, M. (2016*b*), 'Spectral expansions of homogeneous and isotropic tensor-valued random fields', *Z. Angew. Math. Phys.* **67**(3), Art. 59, 20.

Malyarenko, A. & Ostoja-Starzewski, M. (2017*a*), 'Fractal planetary rings: energy inequalities and random field model', *Internat. J. Modern Phys. B* **31**(30), 1750236, 14.

Malyarenko, A. & Ostoja-Starzewski, M. (2017*b*), 'A random field formulation of Hooke's law in all elasticity classes', *J. Elasticity* **127**(2), 269–302.

Mandelbrot, B. B. (1982), *The Fractal Geometry of Nature*, Schriftenreihe für den Referenten. (Series for the Referee), San Francisco, CA: W. H. Freeman and Co.

Marinucci, D. & Peccati, G. (2011), *Random fields on the Sphere. Representation, Limit Theorems and Cosmological Applications*, Vol. 389 of London Mathematical Society Lecture Note Series, Cambridge: Cambridge University Press.

Mateu, J., Porcu, E. & Nicolis, O. (2007), 'A note on decoupling of local and global behaviours for the Dagum random field', *Probab. Engng. Mech.* **22**(4), 320–329.

Maugin, G. A. (1999), *The Thermomechanics of Nonlinear Irreversible Behaviors: An Introduction*, World Scientific Series on Nonlinear Science Series A, Singapore: World Scientific.

Maugin, G. A. (2017), *Non-Classical Continuum Mechanics*, Vol. 51 of *Advanced Structured Materials*, A Dictionary, Singapore: Springer.

Mauguin, C. (1931), 'Sur le symbolisme des groupes de répetition ou de symétrie des assemblages cristallins', *Z. Kristallogr.* **76**, 542–558.

Maxwell, J. C. (1867), 'IV. On the dynamical theory of gases', *Philos. Trans. Roy. Soc. London* **157**, 49–88.

Mendelson, K. S. (1981), 'Bulk modulus of a polycrystal', *J. Phys. D* **14**(7), 1307–1309.

Minkowski, H. (1897), 'Allgemeine Lehrsätze Über konvexe Polyeder', *Nachr. Ges. Wiss. Göttingen* **1897**, 198–219.

Monchiet, V. & Bonnet, G. (2011), 'Inversion of higher order isotropic tensors with minor symmetries and solution of higher order heterogeneity problems', *Proc. R. Soc. Lond. Ser. A Math. Phys. Eng. Sci.* **467**(2126), 314–332.

Monin, A. S. & Yaglom, A. M. (2007*a*), *Statistical Fluid Mechanics: Mechanics of Turbulence. Vol. I.* Translated from the 1965 Russian original, Edited and with a preface by John L. Lumley, English edition updated, augmented and revised by the authors, Reprinted from the 1971 edition, Mineola, NY: Dover Publications, Inc.

Monin, A. S. & Yaglom, A. M. (2007*b*), *Statistical Fluid Mechanics: Mechanics of Turbulence. Vol. II.* Translated from the 1965 Russian original, Edited and with a preface by John L. Lumley, English edition updated, augmented and revised by the authors, Reprinted from the 1975 edition, Mineola, NY: Dover Publications, Inc.

Montgomery, D. & Zippin, L. (1974), *Topological Transformation Groups* Reprint of the 1955 original, Huntington, NY: Robert E. Krieger Publishing Co.

Murakami, S. (2012), *Continuum Damage Mechanics*, Vol. 185 of *Solid Mechanics and its Applications*, Dordrecht: Springer.

Murnaghan, F. D. (1963), *The Theory of Group Representations*, New York, NY: Dover Publications, Inc.

Nagata, J.-I. (1985), *Modern General Topology*, Vol. 33 of *North-Holland Mathematical Library*, second edn, Amsterdam: North-Holland Publishing Co.

Navier, C. L. M. H. (1827), 'Mémoire sur les lois de l'équilibre et du mouvement des corps solides élastiques', *Mem. Acad. Sci.* **7**, 375–394.

Niven, I. (1981), *Maxima and Minima Without Calculus*, Vol. 6 of *The Dolciani Mathematical Expositions*, Washington, DC: Mathematical Association of America.

Noshadravan, A., Ghanem, R., Guilleminot, J., Atodaria, I. & Peralta, P. (2013), 'Validation of a probabilistic model for mesoscale elasticity tensor of random polycrystals', *Int. J. Uncertain. Quantif.* **3**(1), 73–100.

Nowacki, W. (1975), *Dynamic Problems of Thermoelasticity*, Netherlands: Springer.

Nowacki, W. (1986), *Theory of Asymmetric Elasticity*, Pergamon Press, Oxford; PWN—Polish Scientific Publishers, Warsaw. Translated from the Polish by H. Zorski. Oxford: Pergamon Press and Warsaw: PWN – Polish Scientific Publishers.

Obukhov, A. (1941*a*), 'On the energy distribution in the spectrum of a turbulent flow', *C. R. (Doklady) Acad. Sci. URSS (N.S.)* **32**, 19–21.

Obukhov, A. (1941*b*), 'Über die Energieverteilung im Spektrum des Turbulenzstromes', *Bull. Acad. Sci. URSS. Sér. Géograph. Géophys. (Izvestia Akad. Nauk SSSR)* **1941**, 453–466.

Olive, M. & Auffray, N. (2013), 'Symmetry classes for even-order tensors', *Math. Mech. Complex Syst.* **1**(2), 177–210.

Olive, M. & Auffray, N. (2014), 'Symmetry classes for odd-order tensors', *ZAMM Z. Angew. Math. Mech.* **94**(5), 421–447.

Olive, M., Kolev, B., Desmorat, B. & Desmorat, R. (2017), 'Harmonic factorization and reconstruction of the elasticity tensor', *J. Elasticity*. Online.

Onat, E. T. & Leckie, F. A. (1988), 'Representation of mechanical behavior in the presence of changing internal structure', *J. Appl. Mech.* **55**(1), 1–10.

Onsager, L. (1931), 'Reciprocal relations in irreversible processes. I.', *Phys. Rev.* **37**, 405–426.

Ostoja-Starzewski, M. (1999), 'Microstructural disorder, mesoscale finite elements and macroscopic response', *Proc. R. Soc. Lond. A* **455**(1989), 3189–3199.

Ostoja-Starzewski, M. (2008), *Microstructural Randomness and Scaling in Mechanics of Materials*, CRC Series: Modern Mechanics and Mathematics, Boca Raton, FL: Chapman & Hall/CRC.

Ostoja-Starzewski, M. (2014), 'Viscothermoelasticity with finite wave speeds: thermomechanical laws', *Acta Mech.* **225**(4–5), 1277–1285.

Ostoja-Starzewski, M. (2016), 'Second law violations, continuum mechanics, and permeability', *Contin. Mech. Thermodyn.* **28**(1–2), 489–501. *Erratum* **29**, 359, 2017.

Ostoja-Starzewski, M. (2017), 'Admitting spontaneous violations of the second law in continuum thermomechanics', *Entropy* **19**(2).

Ostoja-Starzewski, M. (2018), 'Ignaczak equation of elastodynamics', *Mathematics and Mechanics of Solids* **0**(0), 1081286518757284.

Ostoja-Starzewski, M., Kale, S., Karimi, P., Malyarenko, A., Raghavan, B., Ranganathan, S. & Zhang, J. (2016), 'Scaling to RVE in random media', in S. P. Bordas & D. S. Balint, eds, *Advances in Applied Mechanics*, Vol. 49, Elsevier BV, pp. 111–211.

Ostoja-Starzewski, M. & Malyarenko, A. (2014), 'Continuum mechanics beyond the second law of thermodynamics', *Proc. R. Soc. A* **470**(2171).

Ostoja-Starzewski, M., Shen, L. & Malyarenko, A. (2015), 'Tensor random fields in conductivity and classical or microcontinuum theories', *Math. Mech. Solids* **20**(4), 418–432.

Passman, D. S. (2012), *Permutation Groups*, Revised reprint of the 1968 original, Mineola, NY: Dover Publications, Inc.

Peter, F. & Weyl, H. (1927), 'Die Vollständigkeit der primitiven Darstellungen einer geschlossenen kontinuierlichen Gruppe', *Math. Ann.* **97**, 737–755.

Plimpton, S. (1995), 'Fast parallel algorithms for short-range molecular dynamics', *J. Comp. Phys.* **117**(1), 1–19.

Pontryagin, L. S. (1966), *Topological Groups*, Translated from the second Russian edition by Arlen Brown, New York, NY, London and Paris: Gordon and Breach Science Publishers, Inc.

Porcu, E., Mateu, J., Zini, A. & Pini, R. (2007), 'Modelling spatio-temporal data: a new variogram and covariance structure proposal', *Statist. Probab. Lett.* **77**(1), 83–89.

Porcu, E. & Stein, M. L. (2012), 'On some local, global and regularity behaviour of some classes of covariance functions', in E. Porcu, J.-M. Montero & M. Schlather, eds, *Advances and Challenges in Space-Time Modelling of Natural Events*, Berlin, Heidelberg: Springer Berlin Heidelberg, pp. 221–238.

Procesi, C. (2007), *Lie Groups. An Approach through Invariants and Representations*, New York, NY: Universitext, Springer.

Raghavan, B. V., Karimi, P. & Ostoja-Starzewski, M. (2018), 'Stochastic characteristics and second law violations of atomic fluids in Couette flow', *Physica A* **496**, 90–107.

Raghavan, B. V. & Ostoja-Starzewski, M. (2017), 'Shear-thinning of molecular fluids in Couette flow', *Phys. Fluids* **29**(2), 023103.

Raghavan, B. V., Ranganathan, S. I. & Ostoja-Starzewski, M. (2015), 'Electrical properties of random checkerboards at finite scales', *AIP Advances* **5**(1), 017131.

Ranganathan, S. I. & Ostoja-Starzewski, M. (2008*a*), 'Mesoscale conductivity and scaling function in aggregates of cubic, trigonal, hexagonal, and tetragonal crystals', *Phys. Rev. B* **77**, 214308.

Ranganathan, S. I. & Ostoja-Starzewski, M. (2008*b*), 'Scaling function, anisotropy and the size of RVE in elastic random polycrystals', *J. Mech. Phys. Solids* **56**(9), 2773–2791.

Ranganathan, S. I. & Ostoja-Starzewski, M. (2008*c*), 'Universal elastic anisotropy index', *Phys. Rev. Lett.* **101**, 055504.

Rice, S. O. (1944), 'Mathematical analysis of random noise', *Bell System Tech. J.* **23**, 282–332.

Riehl, E. (2016), *Category Theory in Context*, Aurora: Dover Modern Math Originals, Dover Publications Inc.

Robertson, H. P. (1940), 'The invariant theory of isotropic turbulence', *Proc. Cambridge Philos. Soc.* **36**, 209–223.

Rossmann, W. (2002), *Lie Groups. An Introduction through Linear Groups*, Vol. 5 of Oxford Graduate Texts in Mathematics, Oxford: Oxford University Press.

Rytov, S. M., Kravtsov, Y. A. & Tatarskiĭ, V. I. (1987), *Principles of statistical Radiophysics. 1.* Elements of random process theory, Translated from the Russian by Alexander P. Repyev. Berlin: Springer-Verlag.

Sab, K. (1991), 'Principe de Hill et homogénéisation des matériaux aléatoires', *C. R. Acad. Sci. Paris Sér. II Méc. Phys. Chim. Sci. Univers Sci. Terre* **312**(1), 1–5.

Sab, K. & Nedjar, B. (2005), 'Periodization of random media and representative volume element size for linear composites', *C. R. Mécanique* **333**(2), 187–195.

Schoenberg, I. J. (1938), 'Metric spaces and completely monotone functions', *Ann. of Math. (2)* **39**(4), 811–841.

Schönfließ, A. (1891), *Krystallsysteme und Krystallstruktur*, Leipzig: Teubner.

Selivanova, S. (2014), 'Computing Clebsch–Gordan matrices with applications in elasticity theory', in *Logic, Computation, Hierarchies*, Vol. 4 of *Ontos Math. Log.*, Berlin: De Gruyter, pp. 273–295.

Sena, M. P., Ostoja-Starzewski, M. & Costa, L. (2013), 'Stiffness tensor random fields through upscaling of planar random materials', *Probab. Engng. Mech.* **34**, 131–156.

Shermergor, T. (1971), 'Relations between the components of the correlation functions of an elastic field', *J. Appl. Math. Mech.* **35**(3), 392–397.

Smith, G. F. (1994), *Constitutive Equations for Anisotropic and Isotropic Materials*, Vol. 3 of *Mechanics and Physics of Discrete Systems*, Amsterdam: North-Holland Publishing Co.

Spencer, A. (1971), 'Part III – theory of invariants', *in* A. C. Eringen, ed., *Continuum Physics I*, New York, NY: Academic Press, pp. 239–353.

Staber, B. & Guilleminot, J. (2018), 'A random field model for anisotropic strain energy functions and its application for uncertainty quantification in vascular mechanics', *Comput. Methods Appl. Mech. Engrg.* **333**, 94–113.

Staverman, A. J. & Schwarzl, F. (1956), 'Linear deformation behaviour of high polymers', in J. P. Berry, J. D. Ferry, E. Jenckel et al., eds, *Theorie und molekulare Deutung technologischer Eigenschaften von hochpolymeren Werkstoffen*, Berlin, Heidelberg: Springer Berlin Heidelberg, pp. 1–125.

Stolz, C. (1986), 'General relationships between micro and macro scales for the nonlinear behaviour of heterogeneous media', in J. Gittus & J. Zarka, eds, *Modelling Small Deformations of Polycrystals*, Dordrecht: Springer Netherlands, pp. 89–115.

Szczepankiewicz, E. (1985), *Zastosowania pól losowych*, Biblioteka Naukowa Inżyniera [Scientific Library for the Engineer], Warsaw: Państwowe Wydawnictwo Naukowe (PWN).

Tarasov, V. E. (2005), 'Dynamics of fractal solids', *Int. J. Mod. Phys. B* **19**(27), 4103–4114.

Tarasov, V. E. (2006), 'Gravitational field of fractal distribution of particles', *Celestial Mech. Dynam. Astronom.* **94**(1), 1–15.

Tarasov, V. E. (2014), 'Anisotropic fractal media by vector calculus in non-integer dimensional space', *J. Math. Phys.* **55**(8), 083510, 20.

Tarasov, V. E. (2015), 'Vector calculus in non-integer dimensional space and its applications to fractal media', *Commun. Nonlinear Sci. Numer. Simul.* **20**(2), 360–374.

Trovalusci, P., De Bellis, M. L., Ostoja-Starzewski, M. & Murrali, A. (2014), 'Particulate random composites homogenized as micropolar materials', *Mcccanica* **49**(11), 2719–2727.

Trovalusci, P., Ostoja-Starzewski, M., Bellis, M. L. D. & Murrali, A. (2015), 'Scale-dependent homogenization of random composites as micropolar continua', *Europ. J. Mech. A* **49**, 396–407.

Umberger, D. K. & Farmer, J. D. (1985), 'Fat fractals on the energy surface', *Phys. Rev. Lett.* **55**(7), 661–664.

Varshalovich, D. A., Moskalev, A. N. & Khersonskiĭ, V. K. (1988), *Quantum Theory of Angular Momentum. Irreducible Tensors, Spherical Harmonics, Vector Coupling Coefficients, 3nj Symbols*, Translated from the Russian. Teaneck, NJ: World Scientific Publishing Co., Inc.

von Kármán, T. (1937), 'On the statistical theory of turbulence', *Proc. Natl. Acad. Sci. USA* **23**, 98–105.

von Kármán, T. & Howarth, L. (1938), 'On the statistical theory of isotropic turbulence', *Proc. Roy. Soc.* **164**, 192–215.

Walker, A. M. & Wookey, J. (2012), 'MSAT — A new toolkit for the analysis of elastic and seismic anisotropy', *Computers & Geosciences* **49**, 81–90.

Weyl, H. (1997), *The Classical Groups, their Invariants and Representations*, Princeton Landmarks in Mathematics, Princeton, NJ: Princeton University Press, Fifteenth printing, Princeton Paperbacks.

Wineman, A. S. & Pipkin, A. C. (1964), 'Material symmetry restrictions on constitutive equations', *Arch. Rational Mech. Anal.* **17**, 184–214.

Wu, L., Chung, C. N., Major, Z., Adam, L. & Noels, L. (2018), 'From sem images to elastic responses: A stochastic multiscale analysis of ud fiber reinforced composites', *Composite Structures* **189**, 206–227.

Yaglom, A. M. (1948), 'Homogeneous and isotropic turbulence in a viscous compressible fluid', *Izvestiya Akad. Nauk SSSR. Ser. Geograf. Geofiz.* **12**, 501–522.

Yaglom, A. M. (1957), 'Certain types of random fields in n-dimensional space similar to stationary stochastic processes', *Teor. Veroyatnost. i Primenen.* **2**, 292–338.

Yaglom, A. M. (1961), 'Second-order homogeneous random fields', in *Proc. 4th Berkeley Sympos. Math. Statist. and Prob., Vol. II*, Berkeley, CA: University of California Press, pp. 593–622.

Yaglom, A. M. (1987), *Correlation Theory of Stationary and Related Random Functions. Vol. I. Basic Results*, Springer Series in Statistics, New York, NY: Springer-Verlag.

Ziegler, H. (1983), *An Introduction to Thermomechanics*, Vol. 21 of North-Holland Series in Applied Mathematics and Mechanics, second edn, Amsterdam: North-Holland Publishing Co.

Ziegler, H. & Wehrli, C. (1987), 'The derivation of constitutive relations from the free energy and the dissipation function', in *Advances in applied mechanics, Vol. 25*, Orlando, FL: Academic Press, pp. 183–237.

Zohdi, T. I., Oden, J. T. & Rodin, G. J. (1996), 'Hierarchical modeling of heterogeneous bodies', *Comput. Methods Appl. Mech. Engrg.* **138**(1–4), 273–298.

Index